Recurrence in Ergodic Theory and Combinatorial Number Theory

M. B. PORTER LECTURES

RICE UNIVERSITY, DEPARTMENT OF MATHEMATICS

SALOMON BOCHNER, EDITOR

Recurrence in Ergodic Theory and Combinatorial Number Theory

H. FURSTENBERG

PRINCETON UNIVERSITY PRESS

PRINCETON, NEW JERSEY

CONTENTS

PART II. RECURRENCE IN MEASURE PRESERVING SYSTEMS

PART III. DYNAMICS AND LARGE SETS OF INTEGERS

Foreword from the
Porter Lectures Committee

MILTON BROCKETT PORTER, after whom the M. B. Porter Lectures are named, was born on November 22, 1869, in Sherman, Texas, and he died, at the age of ninety, on May 27, 1960, in Austin, Texas. In 1897 he received a Ph.D. in mathematics from Harvard University where he enjoyed the attention of the renowned analysts Bôcher, Byerly, and Osgood, and in the year 1901–1902 he was assistant professor at Yale University. In 1902 he was appointed Professor of Pure Mathematics at the University of Texas at Austin, and he remained in this position until becoming emeritus in 1945.

Professor Porter was the first Texan to attain distinction in mathematics, publishing research papers, all in analysis of real variables, from 1879 until 1925, and collaborating on two textbooks. He took a lively part in academic activities of the American Mathematical Society, and at his death he was its oldest member. Due to his academic stature and fine personal qualities he also exercised considerable influence in the councils of the University of Texas.

The M. B. Porter Lectures are sustained by a fund that descendants of Professor Porter established at Rice University in memory of their prominent ancestor.

S. B.

PREFACE

This monograph is an account of some applications of the modern theory of dynamical systems to combinatorics and number theory. The underlying approach is to use various aspects and formulations of the phenomenon of *recurrence* that are natural in the framework of classical dynamical systems in the broadest possible context. For the most part this approach is used to reestablish known results, such as van der Waerden's theorem on arithmetic progressions, but some new theorems are obtained using this point of view, notably a multidimensional version of Szemerédi's theorem on arithmetic progressions in sets of integers of positive density. I have included an exposition of the elements of topological dynamics and ergodic theory, and I hope the monograph can serve as an introduction for some readers to these areas.

The material presented here is an elaboration of the Porter Lectures delivered at Rice University in 1978, and the author wishes to express his indebtedness to the Porter Foundation for making this study possible. Some of the material that follows appears in joint work with Yitzhak Katznelson and Benjamin Weiss, and their influence is implicit throughout the work. I also owe a large debt to Donald Ornstein and Jean-Paul Thouvenot who are responsible for many improvements in the exposition of the material. In addition I must mention my indebtedness to Konrad Jacobs who is responsible for kindling my interest in the combinatorial problems that are dealt with here.

Finally I would like to express my gratitude to Professor Salomon Bochner who is most directly responsible for my invitation to participate in the Porter Lecture Series. My first steps in mathematics were guided by Professor Bochner, and I have remained the beneficiary of his mathematical insight and his broad erudition. This volume is a small tribute to the many-sidedness of Professor Bochner's mathematical activity that served as an inspiration for my work and aroused in me the belief in the fundamental unity of mathematics.

Recurrence in Ergodic Theory and Combinatorial Number Theory

1. From Differentiable to Abstract Dynamics

Theoretical dynamics begins with Newton who singlehandedly set it into motion by his development of calculus, his formulation of the laws of motion, and his discovery of the universal law of gravitation. In Newton's formulation the motion of a dynamical system is governed by a system of differential equations satisfied by the parameters of the system varying in time, and for two centuries following the publication of Newton's *Principia* the subject was pursued as a chapter in the theory of differential equations. Most challenging was the problem of applying Newtonian theory to planetary motion, or, more generally, the n-body problem of determining the motion of n point masses undergoing mutual gravitational attraction. The approach most natural for the eighteenth- and nineteenth-century analysts from Euler to Jacobi was to extract as much information as one could by analytic manipulation of the specific differential equations. Hopefully one might even be able to integrate explicitly the differential equations as Newton had done in the case of the two-body problem. This hope did not materialize, however, and all the analytic machinery brought to bear on the problem including the powerful tool of infinite series did not enable one to answer the most fundamental questions regarding planetary motion.

In the last decade of the nineteenth century, Poincaré published his *Methodes nouvelles de la mécanique céleste* that brought about a decisive shift in point of view. The major ingredient of the change was bringing the geometry of the phase space—the space of "possible values" of the dynamical parameter vector—to bear on the analysis. In this global analysis one's attention is shifted from the individual solution curves to the set of all possible curves and their interrelation. While this method would not produce information regarding a specific solution curve, it sometimes yielded information about some or most solution curves. For example, using what amounts to a topological analysis of the phase space, Poincaré was able to establish in certain situations the existence of periodic solutions, thereby resolving a question of fundamental importance. A second question was whether, if we assume solution curves are not periodic they ultimately come arbitrarily close to positions already occupied (stability in the sense of Poisson). Here, using the prototype of an ergodic theoretic argument, Poincaré established that for all bounded

Hamiltonian systems, "most" solution curves are stable in the sense of Poisson.

As a result of Poincaré's qualitative approach, the focal point in dynamical theory was shifted away from the differential equations that define a dynamical system to the phase space and the group of transformations implicit in the system. Birkhoff must be credited for having made this transition most explicit, and in his 1927 treatise on dynamical systems, he discusses many dynamical phenomena in the context of transformation groups acting on general metric spaces. Among the results that Birkhoff obtains in this generality is an analogue of the aforementioned result of Poincaré. Namely, Birkhoff shows that in any dynamical system on a compact space, there necessarily exists some orbit (solution curve) that is stable in the sense of Poisson.

Ultimately two theories evolved from these developments: ergodic theory and topological dynamics. Both distilled from classical dynamical theory the notion of a one parameter group of transformations acting on an abstract space with the space endowed respectively with a notion of volume (measure space) or a topological structure. In both cases the abstract approach was initially adopted not for the sake of generality but in order to achieve greater clarity by focusing on those aspects of dynamical systems pertinent to the phenomena being studied. Nonetheless, the scope of applicability became considerably broadened, and we shall see in the sequel how this may be put to use.

2. What is a Dynamical System?

The classical definition of a dynamical system encompassing the dynamical systems of Newtonian mechanics is that of a motion whose parameters $x_1(t), x_2(t), \ldots, x_n(t)$ evolve as functions of time t in accordance with a system of differential equations

$$(0.1) \qquad \frac{dx_i}{dt} = X_i(x_1, x_2, \ldots, x_n), \qquad i = 1, 2, \ldots, n,$$

where the X_i are independent of t.

Let us suppose that in (0.1) the function X_i are locally Lipschitz functions on $\mathbf{R}^n$. Then the fundamental existence theorem for ordinary differential equations guarantees that for any point $\vec{x}^{(0)} = (x_1^{(0)}, x_2^{(0)}, \ldots, x_n^{(0)}) \in \mathbf{R}^n$, there exists a unique set of functions $x_1(t), x_2(t), \ldots, x_n(t)$ defined for $-\infty < t < \infty$ satisfying (0.1) together with the initial condition $x_1(0) = x_1^{(0)}$, $x_2(0) = x_2^{(0)}, \ldots, x_n(0) = x_n^{(0)}$. Inserting the initial position into the function, we may express the solution to (0.1) as the set of functions

$x_1(t;x_1^{(0)}, \ldots, x_n^{(0)}), \ldots, x_n(t;x_1^{(0)}, \ldots, x_n^{(0)})$, and in addition to the existence and uniqueness of these functions it is asserted that they are continuous in all $n + 1$ variables. Notice that since the functions $x_i(t + t'; \vec{x}^{(0)})$ as functions of t also satisfy (0.1) and for $t = 0$ they take on the values $x_i(t'; \vec{x}^{(0)})$, we must have by uniqueness,

(0.2) $x_i(t + t'; x_1^{(0)}, \ldots, x_n^{(0)}) = x_i(t; x_1(t'; \vec{x}^{(0)}), \ldots, x_n(t'; \vec{x}^{(0)}))$.

This can be reformulated as follows. Reverse the roles of t and $\vec{x}^{(0)}$, so that instead of regarding t as the independent variable and $\vec{x}^{(0)}$ as a parameter, let t be viewed as a parameter and $\vec{x}^{(0)}$ as the independent variable. Then the vector valued function $(x_1(t;\vec{x}^0), \ldots, x_n(t;\vec{x}^0))$ defines for each t a continuous transformation of $\mathbf{R}^n$ to $\mathbf{R}^n$. Call this transformation T_t. Equation (0.2) then asserts that

(0.3) $T_{t+t'} = T_t \circ T_{t'}$

with the circle $\circ$ representing composition of transformations. Of course, $T_0 =$ identity and again by uniqueness one can see that $T_{-t} = (T_t)^{-1}$. Thus the family of transformations $\{T_t; -\infty < t < \infty\}$ forms a group.

In the dynamical systems of classical mechanics there often existed *integrals* of the system (0.1) known as conservation laws. A function $H(x_1, \ldots, x_n)$ on $\mathbf{R}^n$ is an integral of (0.1) if

$$\sum \frac{\partial H}{\partial x_i} X_i \equiv 0.$$

This, of course, means that the function H is constant along solution curves:

$$\frac{d}{dt} H(x_1(t), \ldots, x_n(t)) = 0.$$

Alternatively, if S_c is the hypersurface $H = c$ in $\mathbf{R}^n$, then any solution curve initiating on S_c remains entirely within S_c. Thus the transformation group $\{T_t\}$ leaves each S_c invariant, and one could, if one likes, regard the underlying space of the dynamical system as the manifold S_c rather than $\mathbf{R}^n$.

It is now a natural step to redefine a dynamical system as follows. A dynamical system is a space X together with a group of transformations $\{T_t; -\infty < t < \infty\}$ of X. It is implicit that in speaking of a dynamical system one is interested in *trajectories* or *orbits* of the motion, i.e., the mapping of the reals $\mathbf{R}$ into X by $t \to T_t x_0$, for x_0 is a point of X. Also one is usually interested in long range behavior with regard to the time variable rather than local behavior.

To obtain a meaningful theory it is necessary to endow X with additional structure that reflects some aspect of the classical phase space that was usually a subset of Euclidean space. The two possibilities that have proven fruitful have been to take X to be a topological space in which case the transformations T_t are assumed to be homeomorphisms of X or to assume X to be a measure space and to suppose that the transformations T_t preserve the given measure. In the sequel we shall refer to the two kinds of systems as *dynamical systems* and *measure preserving systems* respectively.

Inasmuch as one is interested primarily in asymptotic behavior as $t \to \infty$, the dynamical aspects of behavior are often equally meaningful when the trajectories are restricted to integral values of time. Thus one is led to referring to a discrete group $\{T^n\}$ of homeomorphisms of a space—or, a measure preserving transformations of a space—as defining a dynamical system.

The ultimate step is to allow an arbitrary group, or even semigroup, G to act on a space X. There are contexts wherein it is useful to use dynamical concepts for general group actions (see Glasner [2], Margoulis [1]; and Mostow [1] for some examples). In the course of our exposition we shall never have reason to consider noncommutative groups. However, it will be quite important to allow for actions of the integer lattice $\mathbf{Z}^n$ on a space as well as the action of $\mathbf{Z}$ and $\mathbf{R}$ and also to consider some semigroup actions, in particular, the additive semigroup of positive integers.

3. Some Nonclassical Dynamical Systems

A good example of a *nonclassical dynamical system* is obtained by taking a finite set of symbols $\Lambda = \{a,b,c, \ldots\}$ and forming the space $X = \Lambda^{\mathbf{Z}}$ of all infinite sequences with entries from $\Lambda: x \in X \Leftrightarrow x = \{\ldots x(-1),x(0),x(1), \ldots\}$, $x(n) \in \Lambda$, $-\infty < n < \infty$. The space X can be regarded as a metric space, taking as metric, for example,

$$(0.4) \qquad d(x,x') = \inf\left\{\frac{1}{k+1} \,\middle|\, x(i) = x'(i) \text{ for } |i| < k\right\}.$$

Here $d(x,x') = 1$ if $x(0) \neq x'(0)$, and for $x \neq x'$, $d(x,x') = 1/(k+1)$ where $2k - 1$ is the length of the largest symmetric interval about 0 for which the two sequences coincide. With this metric X becomes a compact topological space. Now define the "shift" homeomorphism $T: X \to X$ by

$$(0.5) \qquad Tx(n) = x(n+1)$$

The study of dynamical properties of the iterates T^n of T acting on X is often referred to as *symbolic dynamics*. This subject goes back to

Hadamard [1] who first used sequences of symbols to describe qualitatively the infinite geodesic curves on a surface. These curves, incidentally, correspond to the trajectories of certain dynamical systems, the *geodesic flows*. Morse subsequently refined Hadamard's method (Morse [1]) to establish the existence of various kinds of "transitive" as well as nonperiodic but "recurrent" trajectories for particular dynamical systems. Only later did one come to think of these symbolic sequences as forming themselves the points of an abstract dynamical system, where the trajectory of a point becomes the set of translates of a given sequence. In this curious system the entire trajectory is described by the sequence constituting the given point, and dynamical properties could be deduced by a combinatorial analysis of the sequence in question.

If we assign probabilities to the symbols of $\Lambda : a_i \to p_i$ with $\sum p_i = 1$ and assume the entries of a sequence $x(n)$ are chosen independently and with probability $P(x(n) = a_i) = p_i$, then X becomes a measure space with a product measure on $\Lambda^{\mathbf{Z}}$. The transformation T is then measure preserving and what we have described is the prototype of a *Bernoulli* system, an example of which is the repeated coin toss in which $\Lambda = \{$heads, tails$\}$ and Prob(heads) = Prob(tails) = $1/2$.

A continuous analogue of symbolic dynamics is the Bebutov system (Bebutov [1]). Let $f(t)$ be a uniformly continuous bounded function on the reals. Let X be the set of all functions on $\mathbf{R}$ that arise as limits

$$g(t) = \lim_{k \to \infty} f(t + t_k)$$

for some sequence $\{t_k\}$. It may be seen that X is a compact topological space in the topology of uniform convergence on compact subsets of $\mathbf{R}$. Moreover, if $g \in X$, then the functions $T_s g$ defined by $T_s g(t) = g(t + s)$ are also in X and $g \to T_s g$ is a homeomorphism of X. Thus X together with the group $\{T_s\}$ of homeomorphism defines a dynamical system. This construction provides a wide range of examples of dynamical systems. Using functions $f(t)$ of a special kind, e.g., almost periodic functions, one could obtain examples of dynamical phenomena that were not otherwise apparent. As in the case of symbolic dynamics, the dynamics of a Bebutov system is implicit in the construction of a single function.

The idea characterizing much of what we shall do in this monograph is that the procedure described here may be reversed. Instead of constructing unorthodox dynamical systems and using function theory and combinatorial analysis in order to shed light on classical systems, we may improvise new systems and exploit general dynamical theorems in order to obtain results of a combinatorial or function-theoretic nature. In the next section we shall present the prototype of the general dynamical theorems that we shall use.

4. The Classical Recurrence Theorems

A metatheorem of dynamical theory states that whenever the underlying space X of the system is appropriately bounded, the orbits of the motion will necessarily exhibit some form of recurrence, or return close to their initial position.

The first precise result of the kind was formulated by Poincaré. Let T be a measure preserving transformation of a measure space $(X, \mathscr{B}, \mu)$ and assume that the total measure of X is finite: $\mu(X) < \infty$. If $B \in \mathscr{B}$ is an arbitrary measurable set in X with positive measure, $\mu(B) > 0$, then there is some point $x \in B$ and integer $n \geq 1$ with $T^n x \in B$. Here the boundedness of X is expressed by the finiteness of the total measure and the recurrence of orbits by the fact that in very small sets B there will be points returning to B.

This result, known as the Poincaré recurrence theorem, can be proved in a few lines. If we consider the inverse images of B under powers of $T: B, T^{-1}B, T^{-2}B, \ldots, T^{-n}B, \ldots$, we find that we have an infinite sequence of sets all having the same positive measure. If they were all disjoint, the measure of X which contains

$$\bigcup_{n=0}^{\infty} T^{-n}B$$

would be infinite. Since we assume $\mu(X) < \infty$, we must have $T^{-i}B \cap T^{-j}B \neq \varnothing$ for some $i < j$. But if $y \in T^{-i}B \cap T^{-j}B$, then $x = T^i y \in B$ and $T^{j-i}x \in B$.

Poincaré applied his result to classical Hamiltonian systems in which the phase space is Euclidean and where by a theorem of Liouville the flow of the dynamical system preserves Euclidean volume. The hypersurfaces of constant energy are also invariant and also carry an invariant measure. If the energy surface is bounded, the measure will be bounded and the recurrence theorem applies. Since every arbitrarily small set of positive measure returns to itself, one deduces that in fact almost every point $x \in X$ has some $T_t x, t \geq 1$, arbitrarily close to itself. (In fact, t may be chosen in an arbitrary cyclic subgroup of $\mathbf{R}$.) From this one finds that for some sequence $t_k \to \infty$, $T_{t_k}x \to x$. The conclusion is that almost every orbit satisfies a recurrence property known as *stability in the sense of Poisson*.

The second result of this kind is due to Birkhoff and asserts that the aforementioned phenomenon occurs under the hypothesis that the underlying space is a compact topological space. In particular if X is a compact metric space and T is a continuous map of X into itself, there exists some point $x_0 \in X$ and some sequence $n_k \to \infty$ with $T^{n_k}x_0 \to x_0$. Here the

boundedness of the underlying space is expressed in the form appropriate
to the topological context: compactness. The proof of this straightforward
statement is surprisingly recondite (it seems to require either Zorn's lemma
or the introduction of an invariant measure enabling one to use Poincaré's
theorem), and we shall present it in Chapter 1.

Notice that the topological result is weaker than the measure theoretic
result in that one is only ensured the existence of a single recurrent orbit.
Simple examples show that it may indeed happen that every orbit converges to a single fixed point in which case the fixed point provides the
unique recurrent orbit.

A major portion of our effort will be devoted to generalizing these two
basic recurrence theorems from the framework of a single transformation
to that of several commuting transformations. In the next section we
formulate such a theorem and derive one of its consequences.

5. Multiple Birkhoff Recurrence and van der Waerden's Theorem

Let $T_1, T_2, \ldots, T_l$ be l maps of a metric space X to itself. We shall say
that a point $x_0 \in X$ is *multiply recurrent* (for $T_1, T_2, \ldots, T_l$) if there exists
a sequence $n_k \to \infty$ with $T_1^{n_k} x_0 \to x_0$, $T_2^{n_k} x_0 \to x_0, \ldots, T_l^{n_k} x_0 \to x_0$ as
$k \to \infty$. In Chapter 2 we shall prove:

THEOREM (MBR). *If X is compact metric and the transformations*
$T_1, T_2, \ldots, T_l$ *of X into X commute, then there exists a multiply recurrent
point.*

Let us now show how this result implies van der Waerden's theorem on
arithmetic progressions. We might remark, incidentally, that van der
Waerden proved his theorem, conjectured by Baudet, in 1927, shortly
after Birkhoff had formulated his extension of dynamical theory and had
established his recurrence theorem. Of course, van der Waerden's proof
was entirely combinatorial (van der Waerden [1]).

Van der Waerden's theorem states (in one of several equivalent formulations) that if the integers $\mathbf{Z}$ are partitioned into finitely many sets,
$\mathbf{Z} = B_1 \cup B_2 \cup \cdots \cup B_q$, then one of these sets contains finite arithmetic
progressions of arbitrary length.

To prove van der Waerden's theorem it clearly suffices to show that for
each $l = 2, 3, \ldots$, some B_j contains an arithmetic progression of length
$l + 1$; for some, B_j will occur for infinitely many l and that will be the
desired B_j. We shall obtain this by considering the symbolic dynamical
system on the symbols $\Lambda = \{1, 2, \ldots, q\}$. Recall from Section 3 that we
had attached to Λ a dynamical system whose space X consists of all
doubly infinite sequences with entries from Λ and whose transformation

group consists of iterates of the shift: $Tx(n) = x(n + 1)$. Now each such sequence $\{x(n)\}$ determines a partition of $\mathbf{Z}$ into q sets: $B_i = \{n \,|\, x(n) = i\}$, and conversely any partition

$$\mathbf{Z} = \bigcup_{i=1}^{q} B_i$$

into sets indexed from 1 to q determines a point in X. We can formulate van der Waerden's theorem as the following assertion: For any point $x_0 \in X$ and any $l = 2,3,\ldots$, there exists $m,n \in \mathbf{Z}$ with $n > 0$ such that $x_0(m) = x_0(m + n) = x_0(m + 2n) = \cdots = x_0(m + ln)$.

If we take into account the metric (0.4) that was introduced on the sequence space X, we can reformulate this in terms of the dynamical system on X. Namely, in view of (0.4) we find that if $d(x,x') < 1$, then $x(0) = x'(0)$. Moreover, $x(n) = T^n x(0)$. We now see that van der Waerden's theorem follows from the following statement:

(*)   If X is a compact metric space, T a homeomorphism of X, $x_0 \in X$, then for any integer $l \geq 1$ and $\varepsilon > 0$, there is some point on the orbit of x_0, $y = T^m x$, which has the property that for some $n \geq 1$, the points $y, T^n y, T^{2n} y, \ldots, T^{ln} y$ are within ε of one another.

Now in fact (*) follows readily from the multiple recurrence theorem. To see this, let Y be the closure in X of the orbit of x_0:

$$Y = \overline{\{T^m x_0 \,|\, {-}\infty < m < \infty\}},$$

and let $T_1 = T, T_2 = T^2, \ldots, T_l = T^l$. Y is a T–invariant compact set, and $T_1, T_2, \ldots, T_l$ form a set of commuting transformations of Y to itself. So there exists some $y' \in Y$ with $T_1^{n_k} y' \to y'$, $T_1^{n_k} y' \to y', \ldots, T_1^{n_k} y' \to y'$, and, in particular, for some $n = n_k$, all the points $y', T_1^n y', \ldots, T_l^n y'$, or $y', T^n y', T^{2n} y', \ldots T^{ln} y'$, are within ε of one another. Having fixed n, we find this same result will be valid for any $y, T^n y, T^{2n} y, \ldots, T^{ln} y$ if y is sufficiently close to y'. But $y' \in \{T^m x_0\}_{m \in \mathbf{Z}}$, and so we obtain the desired conclusion for a point y in the orbit of x_0.

6. Variations on van der Waerden's Theme

Van der Waerden's theorem is one of a class of results in combinatorial number theory discovered in the earlier part of the century. Before discussing these, we should point out that van der Waerden's theorem has an equivalent "finitary" formulation, and this is also the form that these other results took. In the finitary formulation one considers partitions of large finite sets.

(VDW') Let $q,l = 1,2,3,\ldots$ be given. There exists $N(q,l)$ such that if the integers $\{1,2,\ldots,N(q,l)\}$ are partitioned into q sets, one of these contains arithmetic progressions of length $l + 1$.

It is clear that this version implies our previous formulation. But the opposite implication is also valid. Suppose that there are partitions of arbitrarily long blocks without any of the sets containing arithmetic progressions of length $l + 1$. We may take these blocks to be of the form $[-N,N]$ with $N \to \infty$. Regard each partition of $[-N,N]$ as a function from $[-N,N]$ to $\{1,2,\ldots,q\}$ and extend it in an arbitrary way to all of $\mathbf{Z}$. A limit point of these points in $\{1,2,\ldots,q\}^{\mathbf{Z}}$ defines a partition of $\mathbf{Z}$ into q sets, and one sees that none of the sets can contain an arithmetic progression of length $l + 1$.

Van der Waerden's theorem was preceded by the following result of I. Schur [1]:

(S') Let $q = 2,3,\ldots$ be given. There exists $N(q)$ such that if the integers $\{1,2,\ldots,N(q)\}$ are partitioned into q sets, one of them contains a solution to $x + y = z$.

From this theorem Schur deduces a negative result in connection with Fermat's problem. For any $m = 2,3,\ldots$, if p is a sufficiently large prime, there exists a nontrivial solution to the congruence $x^m + y^m \equiv z^m \pmod{p}$.

In its "infinite" formulation Schur's theorem relates to partitions of the natural numbers $\mathbf{N} = \{1,2,3,\ldots\}$. Van der Waerden's theorem can also be stated in terms of partitions of $\mathbf{N}$. It is also possible to combine Schur's result with that of van der Waerden. This was done by Schur and Brauer (Brauer [1]):

(SB) For any finite partition of the natural numbers, $\mathbf{N} = B_1 \cup B_2 \cup \cdots B_q$, one of the sets B_j contains arbitrarily long arithmetic progressions $\{a, a + d, \ldots, a + ld\}$ with the further property that the difference d is in B_j.

A few years later R. Rado published the ultimate result in this direction (Rado [2]). He noticed that in all of these results one deals with a system of linear equations

$$\sum a_{ij}x_j = 0$$

to which one hopes to find a solution set $\{x_j\}$ inside a single set of an arbitrary partition $\mathbf{N} = B_1 \cup B_2 \cup \cdots \cup B_q$. Thus in the result of Schur and Brauer we are dealing with the system

$$x_0 = x_2 - x_1 = x_3 - x_2 = \cdots = x_{l+1} - x_l.$$

Rado succeeded in finding conditions on a system of equations that were both necessary and sufficient for the phenomenon, thereby subsuming all the foregoing results. He calls such systems of equations *regular systems*, and we shall present his criterion in Chapter 8.

More recently N. Hindman obtained a generalization of Schur's theorem that is not subsumed by Rado's result (Hindman [1]). This result, conjectured by Graham and Rothschild differs fundamentally from the preceding results in that it asserts the existence of an infinite configuration of a certain type. (It is easy to describe finite partitions of $\mathbf{Z}$ where no set contains an infinite arithmetic progression!) Specifically Hindman proves:

(NH) For any finite partition of the natural numbers $\mathbf{N} = B_1 \cup B_2 \cup \cdots \cup B_q$, there exists an infinite sequence $x_1 \leq x_2 \leq x_3 \leq \cdots$. such that all elements of the sequence together with all finite sums $x_{i_1} + x_{i_2} + \cdots + x_{i_k}$ of elements with distinct indices $i_1 < i_2 < \cdots < i_k$ belong to the same set B_j of the partition.

A limited version of Hindman's result was already proved by Hilbert and deserves mention as (presumably) the first result of the van der Waerden-Schur variety (Hilbert [1]):

(DH) For any partition of the natural number $\mathbf{N} = B_1 \cup B_2 \cup \cdots \cup B_q$ and any $l = 1, 2, \ldots$, there exists a set $x_1 \leq x_2 \leq \cdots \leq x_l$ such that if $P(x_1, x_2, \ldots, x_l)$ denotes the set of 2^l sums $x_{i_1} + x_{i_2} + \cdots + x_{i_k}$, $i_1 < i_2 < \cdots < i_k$, infinitely many translates $P(x_1, x_2, \ldots, x_l) + t_n$ will belong to the same set B_j.

Hilbert used this lemma to prove his irreducibility theorem: If the polynomial $P(X, Y) \in \mathbf{Z}[X, Y]$ is irreducible, then there exists some $a \in \mathbf{N}$ with $P(a, Y) \in \mathbf{Z}[Y]$ irreducible.

There are also multidimensional versions of some of these results. Notably van der Waerden's theorem can be extended to partitions of the lattice $\mathbf{Z}^r$, as was shown by Grünwald (Rado [1]).

(G) Let $\mathbf{Z}^r = B_1 \cup B_2 \cup \cdots \cup B_q$ be an arbitrary partition of the r-dimensional lattice. One of the sets B_j has the property that if $C \subset \mathbf{Z}^r$ is an arbitrary finite configuration, B_j contains a translate of a dilation of $C: B_j \supset aC + b$, $a \in \mathbf{N}$, $b \in \mathbf{Z}^r$.

In all of these results a certain property is predicated of one of the sets of an arbitrary partition of $\mathbf{Z}, \mathbf{N}$, or $\mathbf{Z}^r$. Note that some of these properties are translation invariant, e.g., the van der Waerden property, whereas some are not, e.g. Schur's property. In the latter case the property in question is not merely a consequence of the "size" of the set, but also its position. The evens and odds have the same size but the former contains a solution to $x + y = z$ and the latter not. Where the property is translation invariant, one may conjecture that there is a measure of the size of a set that will guarantee the property. This was conjectured by Erdös and Turán in connection with the property of possessing arbitrarily long arithmetic progressions. More precisely, their conjecture asserts that any set of "positive upper density" in $\mathbf{Z}$ possesses arithmetic progressions of arbitrary finite length. This conjecture announced in the 1930s was established in stages. K. F. Roth, using analytic methods, showed in 1952 that a set of positive upper density contains arithmetic progressions of length 3. In 1969, E. Szemeredi showed that such sets contain arithmetic progressions of length 4, and finally in 1975, Szemerédi proved the full conjecture of Erdös and Turán using intricate combinatorial arguments. More precisely he showed:

(SZ) Let $B \subset \mathbf{Z}$ be a subset such that for some sequence of intervals $[a_n, b_n]$ with $b_n - a_n \to \infty$, $\dfrac{|B \cap [a_n, b_n]|}{b_n - a_n} \to d > 0$, then B contains arbitrarily long arithmetic progressions.

7. Dynamical Reformulation

From the standpoint of combinatorial theory, the foregoing list of results could be extensively enlarged, including some "Ramsey type" theorems. What the results listed in the previous section have in common is that they are all consequences of general dynamical theorems. We have already seen that van der Waerden's theorem follows from a multiple recurrence theorem for commuting homeomorphisms of a compact space. With no extra effort we shall show that the multidimensional version of van der Waerden's theorem established by Grünwald, Theorem (G), can be deduced from the same theorem.

Hilbert's lemma can be proved in a variety of ways. In particular if one regards partitions of $\mathbf{N}$ as points of a symbolic dynamical system, then Hilbert's lemma follows from Birkhoff's theorem. We shall show this in detail in Chapter 1.

The theorems of Hindman and Rado require new ideas concerning dynamical systems. One of these is the notion of *proximality*. Suppose T is a continuous map of a metric space X to itself. We say two points $x_1, x_2 \in X$ are *proximal* if for some sequence $n_k \to \infty$, $d(T^{n_k} x_1, T^{n_k} x_2) \to 0$. We shall say a point $x \in X$ is *uniformly recurrent* if for any neighborhood V of x, the sequence of values of n, $n_1 < n_2 < n_3 < \cdots$, for which $T^{n_k} x \in V$, satisfies $\{n_{k+1} - n_k\}$ is bounded. This latter property is an approximation to periodicity and is sometimes called *almost periodicity*, and in the older literature, it is simply called *recurrence*.

Now a fundamental theorem proved by J. Auslander and R. Ellis asserts that if X is compact, then every point of X is proximal to some uniformly recurrent point. In Part III we shall show how the result implies Hindman's theorem and how this result combined with a sharpened version of Theorem (MBR) gives Rado's theorem. Without going into the details of this connection, let us point out the form these results will take.

Let $\mathbf{N} = B_1 \cup B_2 \cup \cdots \cup B_q$ be a partition of $\mathbf{N}$. Attach to this partition the function $x(n)$ defined for $n \in \mathbf{N}$ by $x(n) = j \Leftrightarrow n \in B_j$. Extend this function arbitrarily to $\mathbf{Z}$ with range $\subset \{1,2,\ldots,q\}$. Regard x as a point of $\{1,2,\ldots,q\}^{\mathbf{Z}}$ on which a dynamical system is defined as in Section 3, by $Ty(n) = y(n+1)$. Apply the Auslander-Ellis theorem to this system and to the point x, and let x' be a uniformly recurrent point proximal to x. Then if $x'(0) = j$, the set B_j will satisfy the conclusion of Rado's theorem as well as Hindman's theorem. This provides a partial answer to the question which of the sets B_i possess the desired properties. In particular, if the partition $\mathbf{N} = B_1 \cup B_2 \cup \cdots \cup B_q$ is such that the function $x(n)$, $n > 0$, extends to be itself a uniformly recurrent point of $\{1,2,\ldots,q\}^{\mathbf{Z}}$, then the set $\{n:x(n) = x(0)\}$ possesses all the desired properties. In fact, a direct application of the foregoing definitions shows that for any p, the block $x(0),x(1),\ldots,x(p)$ recurs as $x(n),x(n+1),\ldots,x(n+p)$ for some n. In particular for some p, $x(p) = x(0)$ and then with n as above, $x(n+p) = x(p) = x(0)$ and $x(n) = x(0)$. This gives Schur's result in the present case.

On the other hand, if x is a uniformly recurrent point so is each $T^m x$. If follows from the foregoing that each set $\{n:x(n) = x(m)\}$ contains a solution to those Rado systems that are translation invariant. Thus each of these sets contains arbitrarily long arithmetic progressions and all the sets of the partition have the van der Waerden property. This special case gives some insight into the nature of the different properties considered.

So far we have discussed results belonging to topological dynamics and their combinatorial implications. It should not come as a surprise that the tool appropriate for handling Szemerédi's theorem in which the notion of the density of a set plays a role is *ergodic theory* or the theory of *measure preserving transformations*. Part II of this monograph will be devoted to proving a multiple recurrence analogue of Poincaré's recurrence theorem. This states

THEOREM (MPR). *Let $(X,\mathscr{B},\mu)$ be a measure space with $\mu(X) < \infty$, and let $T_1,T_2,\ldots,T_k$ be commuting measure preserving transformations of $(X,\mathscr{B},\mu)$. If $A \in \mathscr{B}$ with $\mu(A) > 0$, then for some $n \geq 1$,*

$$\mu(A \cap T_1^{-n}A \cap T_2^{-n}A \cap \cdots \cap T_k^{-n}A) > 0.$$

As in our discussion of van der Waerden's theorem we can specialize this to $T_1 = T$, $T_2 = T^2, \ldots, T_k = T^k$ and in this form the theorem is equivalent to Szemerédi's theorem, as we shall show in Part II. In addition, in its general form the theorem can be used to obtain a multidimensional analogue of Szemerédi's theorem. This last result may serve as an illustration of a combinatorial theorem first established by ergodic theoretic arguments (Furstenberg and Katznelson [1]).

Some of the general recurrence theorems to be discussed can be applied to special dynamical systems to yield other results of a diophantine nature. As we shall see later, van der Waerden's theorem can sometimes be used as a tool in diophantine approximation, so it is not surprising that certain diophantine approximation theorems fit into the framework of dynamical systems.

Let us mention two results that reflect recurrence properties.

(Hardy-Weyl)   If $p(x)$ is any real polynomial with at least one irrational coefficient, then for each $\varepsilon > 0$ the diophantine inequality $|p(n) - m| < \varepsilon$ has a solution.

(Sárközy)   Let A be set of integers of positive upper density, then there exist $a_1 < a_2$ in A with $a_2 - a_1 = b^2$ for some integer b.

8. Note on Terminology

The reader should be warned that we have taken great liberty in the choice of terminology, respecting, in many cases, neither the traditional nor the current usage of terms.

In the older literature the term *stability* is used for all forms of good behavior. Thus an orbit is "stable in the sense of Lagrange" if it has compact closure; it is "stable in the sense of Poisson" if it returns arbitrarily close to its starting point. Finally, an orbit is "stable in the sense of Liapounov" if sufficiently nearby orbits remain nearby throughout the course of (positive) time. This latter definition has become the principal connotation of stability and it is preferable to use alternative expressions for the other notions. The term *recurrence* most aptly describes the type of behavior we are going to investigate, and we shall speak of several variants of recurrence, ignoring previous conventions regarding these terms. Thus we shall speak of a *recurrent* orbit where the older literature spoke of an orbit "stable in the sense of Poisson." We shall use *uniform recurrence* where other writers (Birkhoff, Morse, Robbins) speak of "recurrence" and where current usage favors the term "almost periodicity."

PART I
RECURRENCE IN
DYNAMICAL SYSTEMS

CHAPTER 1

Recurrence and Uniform Recurrence
in Compact Spaces

1. Dynamical Systems and Recurrent Points

In this chapter we shall discuss the simplest versions of the notion of recurrence. In the course of our discussion we shall develop some of the basic concepts of topological dynamics. Throughout our discussion a *dynamical system* will consist of a compact metric space X together with a group or semigroup G acting on X by continuous transformations. To avoid problems that are not germane to our point of view, we shall assume G is discrete and abelian. Most of the time, but not exclusively, we take G to be the integers $\mathbf{Z}$ or the natural numbers $\mathbf{N}$ with their additive structure. In general we denote a dynamical system by (X,G). When G is either $\mathbf{Z}$ or $\mathbf{N}$, so that it is generated by the element 1, we denote by T the transformation on X representing the action of the element 1. We then denote the dynamical system by (X,T) and call it a *cyclic system*. When no confusion exists, we identify G with the set of transformations of X corresponding to the given action. Thus while we should denote the transformation corresponding to $g \in G$ by $T_g : X \to X$, we usually write gx instead of $T_g x$, except in the cyclic case where we denote $T_1 x$ by Tx.

The point of departure for our discussion is the following definition.

DEFINITION 1.1. *Let T be a continuous map of a topological space X into itself. A point $x \in X$ is called a recurrent point for T (or for the dynamical system (X,T)) if for any neighborhood $V \ni x$, there exists $n \geq 1$ with $T^n x \in V$.*

When X is a metric space, we can let V range through a sequence of neighborhoods of diameter $\to 0$, and we conclude that for some sequence n_k, $T^{n_k}x \to x$. This could be taken, in the metric case, to be the definition of recurrence.

Using Zorn's Lemma, we can prove that if X is a compact space, then for any continuous map $T : X \to X$, there always exist recurrent points. (In Chapter 3 we shall give another proof that doesn't use Zorn's lemma and also gives somewhat more.)

Suppose X is compact and consider the family $\mathscr{F}$ of closed subsets $\phi \neq Y \subset X$ satisfying $TY \subset Y$ and ordered by inclusion. We claim $\mathscr{F}$ has a minimal element. Namely if we have a totally ordered chain of such subsets, their intersection is nonempty and again a set in $\mathscr{F}$. Hence Zorn's lemma applies to produce a minimal element of $\mathscr{F}$. Say Y_0 is minimal. We claim each point of Y_0 is recurrent. For let $x \in Y_0$ and consider $Y = \overline{\{T^n x, n \geq 1\}}$. We call this the *forward orbit closure* of x. Now $Y \subset Y_0$ because Y_0 is closed and invariant under T. But so is Y and so, by minimality, $Y = Y_0$. That means that each neighborhood of x contains some $T^n x$ for $n \geq 1$. We have proved:

THEOREM 1.1.   *If T is a continuous map of a compact space X to itself, the set of recurrent points for T in X is nonempty.*

The simplest example of the phenomenon of recurrence and a direct generalization of periodicity is that of a Kronecker system.

DEFINITION 1.2.   *Let K be a compact group, let $a \in K$, and let $T: K \to K$ be defined by $Tx = ax$, left multiplication with the element a. We then call the dynamical system (K, T) a Kronecker system.*

The reason for the terminology is that one can reformulate a well-known theorem of Kronecker to assert that when K is a torus, the identity is a recurrent point of the corresponding Kronecker system. More generally we have

THEOREM 1.2.   *Every point in the space of a Kronecker system is recurrent.*

Proof: Some point x_0 is recurrent. Now let x be any point and write $x = x_0 u$. If V is a neighborhood of x, then Vu^{-1} is a neighborhood of x_0 and $a^n x_0 \in Vu^{-1}$ implies $a^n(x_0 u) \in V$.

2. Automorphisms and Homomorphisms of Dynamical Systems, Factors, and Extensions

Let (X, G) and (Y, G) be two dynamical systems with the same (semi-) group G of operators. A *homomorphism* from (X, G) to (Y, G) is given by a continuous map $\phi: X \to Y$ satisfying

$$(1.1) \qquad \phi(gx) = g\phi(x)$$

for $x \in X$, $g \in G$. Note that the products gx and $g\phi(x)$ refer to the action of G on two different spaces.

DEFINITION 1.3.   *A dynamical system (Y, G) is a factor of the dynamical system (X, G) if there is a homomorphism of the latter to the former given* *by a map ϕ of X onto Y. In this case we also say that (X,G) is an extension of (Y,G).*

Assume $\phi: X \to Y$ is onto. On account of the compactness of the spaces we can identify Y with the set of *fibers* $\{\phi^{-1}(y): t \in Y\}$ of the map ϕ, appropriately topologized. Equation (1.1) implies that the action of G is compatible with this identification. Thus all the information regarding a factor of a system is implicit in the system. In particular, phenomena taking place on (X,G) generally carry over to factors of (X,G). Here is a simple example.

PROPOSITION 1.3. *If ϕ determines a homomorphism of a cyclic system (X,T) to (Y,T) and $x \in X$ is recurrent for (X,T), then $\phi(x)$ is recurrent for (Y,T).*

We shall now give a class of examples where we can invert the order of the implication and assert that the *preimage* of a recurrent point is recurrent. This takes place in the case of a *group extension*, a notion that will play an important role in the sequel.

DEFINITION 1.4. *Let (Y,T) be a dynamical system, K a compact group, and $\psi: Y \to K$ a continuous mapping. Form $X = Y \times K$ and define T: $X \to X$ by $T(y,k) = (Ty, \psi(y)k)$. The resulting system (X,T) is called a group extension of (Y,T), or, sometimes, a skew product of (Y,T) with K.*

Of course, the map $(y,k) \to y$ defines a homomorphism of (X,T) to (Y,T) so that (X,T) is an extension of (Y,T).

If $(X,T) = (Y \times K, T)$ is a group extension, then the elements of K act on X by right translation:

$$(1.2) \qquad R_{k'}(y,k) = (y,kk'),$$

and the $R_{k'}$ commute with T. As a result the $R_{k'}$ constitute *automorphisms* of the system (X,T). This fact will be useful in proving the next theorem.

THEOREM 1.4. *If $y_0 \in Y$ is recurrent for (Y,T), and (X,T) is a group extension of (Y,T), $X = Y \times K$, then each of the points (y_0,k_0), $k_0 \in K$, is a recurrent point of (X,T).*

Proof: Let e denote the identity of K. We shall show that (y_0,e) is a recurrent point of (X,T), and then it follows that each $R_{k_0}(y_0,e) = (y_0,k_0)$ is recurrent, since the R_{k_0} are automorphisms. For any $x \in X$, let us denote by $Q(x)$ the forward orbit closure: $Q(x) = \overline{\{T^n x: n \geq 1\}}$. Now x is recurrent $\Leftrightarrow x \in Q(x)$. Since y_0 is recurrent for (Y,T), some $(y_0,k_1) \in Q(y_0,e)$. Apply the automorphism R_{k_1} to this inclusion and recall that $QR_{k_1}(x) = R_{k_1}Q(x)$, and we find $(y_0,k_1^2) \in Q(y_0,k_1)$. Now the relationship $x' \in Q(x)$ is transitive, and so we find $(y_0,k_1^2) \in Q(y_0,e)$. Applying R_{k_1} again and repeating, we find successively (y_0,k_1^3), (y_0,k_1^4), ... $(y_0,k_1^n) \in Q(y_0,e)$. But now by Theorem 1.2 and the fact that $Q(y_0,e)$ is closed, we conclude $(y_0,e) \in Q(y_0,e)$.[1]

Using Theorem 1.4, we can inductively obtain examples of non-Kronecker dynamical systems on a torus for which every point is recurrent. For example, consider the 2-torus

$$\mathbf{T}^2 = \{(\theta,\phi): \theta, \phi \in \mathbf{R}/\mathbf{Z}\}$$

where the components are taken as reals modulo 1. Let $T: \mathbf{T}^2 \to \mathbf{T}^2$ be given by $T(\theta,\phi) = (\theta + \alpha, \phi + 2\theta + \alpha)$. Then $(\mathbf{T}^2,T)$ is a group extension of the Kronecker system on the circle: $T\theta = \theta + \alpha$, with $\psi:\mathbf{T} \to \mathbf{T}$ given by $\psi(\theta) = 2\theta + \alpha$. By Theorems 1.2 and 1.4, every point of $\mathbf{T}^2$ is recurrent. Let us compute the orbit of $(0,0) \in \mathbf{T}^2$: $(0,0) \xrightarrow{T} (\alpha,\alpha) \xrightarrow{T} (2\alpha,4\alpha) \xrightarrow{T} \cdots \xrightarrow{T}$ $(n\alpha,n^2\alpha)$, as one verifies by induction. The fact that this comes arbitrarily close to $(0,0)$ yields the following proposition.

PROPOSITION 1.5. *For any real number α and for any $\varepsilon > 0$, we can solve the diophantine inequality $|\alpha n^2 - m| < \varepsilon$.*[2]

This also indicates that for any real number α there are "good" rational approximations using only rationals with a square in the denominator:

$$\left| \alpha - \frac{m}{n^2} \right| < \frac{\varepsilon}{n^2}.$$

Proposition 1.5 was first proved by Hardy and Littlewood [1]. Their proof was superseded by Weyl's more powerful method of trigonometric sums (Weyl [1]). We shall retrieve most of Weyl's results in the course of our discussion.

We can extend Proposition 1.5 to polynomials of higher degree. Let $p(x)$ be a polynomial of degree d with real coefficients. Write $p_d(x) = p(x)$ and form successively

$$p_{d-1}(x) = p_d(x + 1) - p_d(x), \ p_{d-2}(x) = p_{d-1}(x + 1) - p_{d-1}(x),$$

[1] An example of a group extension of (Y,T) is obtained by taking $X = Y \times \mathbf{Z}_m$ where $\mathbf{Z}_m$ is a cyclic group with m elements and $\psi(y) \equiv 1$. We shall then have $T^n(y,u) = (T^n y, n + u)$ and this can be close to (y,u) only for $m | n$. Theorem 1.4 now has the following consequence: If $x \in X$ is a recurrent point of (X,T), it is also recurrent for each (X,T^m).

[2] It is not hard to show that if α is irrational, and we have Proposition 1.5 asserting that $|\alpha n^2 - m| < \varepsilon$ can be solved for arbitrary $\varepsilon > 0$, then it follows that $|\alpha n^2 + \beta - m| < \varepsilon$ can also be solved for arbitrary β. This result first proved by Hardy and Littlewood triggered the study of the phenomenon of equidistribution. To obtain the stronger result, one considers $\alpha n^2 k^2$ for $|\alpha n^2 - m|$ small, and uses the fact that for δ small the numbers $\delta, 4\delta, 9\delta, \ldots, n^2\delta$, $n = [\delta^{-1} 2]$ form an ε-dense set in $[0,1]$ with $\varepsilon = (2n + 1)\delta \sim 2\sqrt{\delta}$. A more general result is proved in Theorem 1.26.

etc. Each $p_i(x)$ is of degree i; let $p_0(x)$ be the constant α. Now define a transformation of the d-dimensional torus $\mathbf{T}^d \to \mathbf{T}^d$ by

$$(1.3) \qquad T(\theta_1,\theta_2,\theta_3,\ldots,\theta_d) = (\theta_1 + \alpha, \theta_2 + \theta_1, \theta_3 + \theta_2, \ldots, \theta_d + \theta_{d-1}).$$

Again, this will define a group extension of a dynamical system on the $(d-1)$-torus, which in turn is a group extension of a dynamical system on the $(d-2)$-torus, etc., down to the 1-torus. We conclude that each point is recurrent. Now compute the orbit of the point $(p_1(0),p_2(0),\ldots,p_d(0))$. Since $p_i(n) + p_{i-1}(n) = p_i(n+1)$ we find that

$$T(p_1(n),p_2(n),\ldots,p_d(n)) = (p_1(n+1),p_2(n+1),\ldots,p_d(n+1))$$

and

$$T^n(p_1(0),p_2(0),\ldots,p_d(0)) = (p_1(n),p_2(n),\ldots,p_d(n)).$$

We conclude that $p_d(n) = p(n)$ comes arbitrarily close to $p(0)$ modulo 1. Thus the foregoing proposition has the following generalization.

THEOREM 1.6. *If $p(x)$ is any real polynomial with $p(0) = 0$, then for any $\varepsilon > 0$ we can solve the diophantine inequality*

$$(1.4) \qquad |p(n) - m| < \varepsilon, \qquad n > 0.$$

In the foregoing we have applied Theorem 1.4 to the special case where K is the circle group and ψ is a rather special function. These are the only cases for which we can formulate explicit results, as in Theorem 1.6, but in principle there is a much wider range of applications.

It is sometimes useful to broaden the notion of a group extension to that of an isometric extension.

DEFINITION 1.5. *Let K be a compact group of isometries of a compact metric space M, where K is topologised so that the map $K \times M \to M$ is continuous. Let (Y,T) be a dynamical system and $\psi : Y \to K$ continuous. Form $X = Y \times M$ and define $T(y,u) = (Ty,\psi(y)u)$, where $\psi(y)u$ denotes the image of u under the isometry of M determined by $\psi(y) \in K$. The system (X,T) is an isometric extension of (Y,T).*[3]

Thus M might be the sphere S^{n-1} or the ball $B^n \subset \mathbf{R}^n$ and K the orthogonal group $O(n)$.

PROPOSITION 1.7. *If (X,T) is an isometric extension of (Y,T), then $X = \bigcup X_\alpha$, where X_α is a closed T-invariant subset of X and the system (X_α,T) is a factor of a group extension of (Y,T).*

---

[3] For simplicity we have taken as the space of an isometric extension, the product of the base space Y with a metric space M. A more general definition is given in Furstenberg [2] that includes, among other things, the possibility that X be a fiber bundle over Y with fiber M.

Proof: Let $X = Y \times M$ and let $\psi: Y \to K$ as in Definition 1.5. Form the product $\tilde{X} = Y \times K$ and define $T: \tilde{X} \to \tilde{X}$ using the same function $\psi: T(y,k) = (Ty, \psi(y)k)$. Now let (y_0, u_0) be any point of X and define $\phi: \tilde{X} \to X$ by $\phi(y,k) = (y, ku_0)$. It is easy to see that ϕ defines a homomorphism of $(\tilde{X}, T)$ onto a subsystem of (X,T) and that $(y_0, u_0) = \phi(y,e)$ is in the image of this homomorphism. This proves the proposition.

As an immediate corollary we obtain

THEOREM 1.8.   *If (X,T) is an isometric extension of (Y,T), then the preimage of any recurrent point of (Y,T) is recurrent for (X,T).*

3. Recurrent Points for Bebutov Systems

Let G be a countable (semi-) group and Λ a compact metric space. Form $\Omega = \Lambda^G$, the compact metrizable space of all functions from G to Λ. We define an action of G on Ω (the *regular* action) by letting

$$g'\omega(g) = \omega(gg'), \qquad \omega \in \Omega, \; g,g' \in G.$$

The reader should convince himself that this is an action, i.e., that $g'(g''\omega) = (g'g'')\omega$.

DEFINITION 1.6.   *A Bebutov system is a subsystem of (Ω, G), i.e., it is a system (X, G) where $X \subset \Omega$ is a closed subset invariant under the regular action of G. A symbolic flow is a Bebutov system for which Λ is finite and G is either* **N** *or* **Z**.

In the case of a symbolic flow the points of Ω can be thought of as sequences of *symbols*. Sometimes we call Λ the *alphabet*.

It is sometimes useful to fix a metric on Ω. If $d(\cdot, \cdot)$ denotes the metric on Λ and $G = \{g_1, g_2, \ldots\}$ is an enumeration of the elements of G, one can define a metric on Ω by

(1.5) $D(\omega, \omega') = \sum 2^{-n} d(\omega(g_n), \omega'(g_n))$.

Equation (1.5) implies that two points $\omega, \omega' \in \Omega$ will be close if their values are close on "large" sets in G. In particular if Λ is discrete and $G = $ **N**, then the sequences are close if they agree for a large block of numbers $1, 2, \ldots, N$. If they disagree at 1 but agree at all other entries, they will still be far apart.

If $\omega_0 \in \Omega$, we can form the smallest G-invariant closed subset containing ω_0; this is the orbit closure of ω_0 in Ω and we denote it X_{ω_0}. The system (X_{ω_0}, G) will be referred to as the *dynamical system* or the *Bebutov system generated by* ω_0.

We state without proof the following theorem that ties together the theory of almost periodic functions with the theory of Kronecker systems.

THEOREM 1.9. *Let $f(n)$ be an almost periodic function on the integers the sense of Bohr* (see, e.g., Bohr [1] and Maak [1]), *and let Λ be a compact subset of the complex plane containing the range of f. Then the subsystem $(X_f, \mathbf{Z})$ of $(\Lambda^{\mathbf{Z}}, \mathbf{Z})$ is a Kronecker system. Conversely, if $(X, \mathbf{Z})$ is a Kronecker system generated by T and $\phi : X \to \mathbf{C}$ is any continuous function, $x \in X$, then $f(n) = \phi(T^n x)$ is a Bohr almost periodic function.*[4]

From this theorem and Theorem 1.2 we conclude that an almost periodic function $f(n)$ constitutes a recurrent point of the Bebutov system to which it belongs. This fact can, of course, be retrieved readily from the definition of almost periodicity.

We turn now to symbolic flows (Λ finite) and inquire what is the condition for a sequence

$$\omega_0 = \{\omega(1), \omega(2), \omega(3), \ldots\} \in \Lambda^{\mathbf{N}} = \Omega$$

to be a recurrent point of Ω. Fixing Λ, we shall speak of a *word* as any finite sequence of elements in Λ, $w = \{w_1, w_2, \ldots, w_l\}$. The number l is called the *length* of the word. A word w of length k *occurs* in another word w' if for some j, $w_1 = w'_{j+1}$, $w_2 = w'_{j+2}, \ldots, w_k = w'_{j+k}$. Similarly we speak of a word w occurring in an infinite sequence $\omega \in \Omega$. The following is then quite evident.

PROPOSITION 1.10. *In a symbolic flow, a sequence $\omega \in \Lambda^{\mathbf{N}}$ is recurrent if and only if every word occurring in ω occurs a second time.*

Of course, each word then occurs infinitely often. Note that it suffices to deal with initial words $\{\omega(1), \omega(2), \ldots, \omega(n)\}$ since these contain all words occurring in ω. Using Proposition 1.10, we can describe the most general recurrent point of $\Lambda^{\mathbf{N}}$. Namely such a point ω has the following structure:

$$(1.6) \qquad \omega = \left[(aw^{(1)}a)w^{(2)}(aw^{(1)}a)\right]w^{(3)}\left[(aw^{(1)}a)w^{(2)}(aw^{(1)}a)\right] \ldots$$

where $a \in \Lambda$ and $w^{(1)}, w^{(2)}, w^{(3)}, \ldots$ denote arbitrary words composed of elements in Λ. We may use this characterization to prove the lemma of Hilbert alluded to in the Introduction, Section 6 (Hilbert [1]).

---

[4] To obtain Theorem 1.9 one uses the notion of an *equicontinuous* dynamical system. This is a system for which the group of transformations $\{T^n\}$ is equicontinuous with respect to some metric on X. One proves that: (i) if (X, T) is equicontinuous, then X is the union of minimal sets, (ii) if (X, T) is also minimal, then it is a Kronecker system. By (i) it follows that if some orbit is dense, then (X, T) is minimal and so by (ii) it is Kronecker. It is not difficult to deduce from the definition of almost periodicity that if $f(n)$ is almost periodic, then $(X_f, \mathbf{Z})$ is equicontinuous. Since $(X_f, \mathbf{Z})$ always has a dense orbit, it follows from the foregoing that $(X_f, \mathbf{Z})$ is Kronecker. The second part of the theorem follows readily from the definitions.

LEMMA. (Hilbert). *If* $\mathbf{N} = B_1 \cup B_2 \cup \cdots \cup B_q$ *is any partition and* $l = 1,2,\ldots$, *there exists a set* $m_1 \leq m_2 \leq m_3 \leq \cdots \leq m_l$ *such that if* $P(m_1, m_2, \ldots, m_l)$ *denotes the set of* 2^l *sums* $m_{i_1} + m_{i_2} + \cdots + m_{i_k}$, $i_1 < i_2 < \cdots < i_k$, *then infinitely many translates of* $P(m_1, m_2, \ldots, m_l)$ *belong to the same* B_j.

Proof: Let $\xi(n) = i \Leftrightarrow n \in B_i$, $i = 1,2,\ldots,q$. Let $\Lambda = \{1,2,\ldots,q\}$ so that ξ is a point of the Bebutov system $\Lambda^{\mathbf{N}}$. Assume first that ξ is a recurrent point so that it has the form (1.6). Define the words $W_0 = a$, $W_1 = W_0 w^{(1)} W_0$, $W_2 = W_1 w^{(2)} W_1, \ldots, W_n = W_{n-1} w^{(n)} W_{n-1}$. The sequence ξ is is then the limit of the finite sequences W_n. Let m_{n+1} be the length of $W_n w^{(n+1)}$. Suppose some symbol occurs at position p in W_n; then it occurs at positions p and $p + m_n$ in $W_{n+1} = W_n w^{n+1} W_n$. Thus the symbol a occurs at positions $1, 1 + m_1, 1 + m_2, 1 + m_1 + m_2, \ldots$, and generally as can be verified, at positions $\in 1 + P(m_1, m_2, \ldots, m_l)$ for any l. Since every configuration occurs infinitely often, it is clear in this case that B_a contains infinitely many translates of $P(m_1, m_2, \ldots, m_l)$.

In the general case we invoke Theorem 1.1 to obtain a recurrent point in the Bebutov system generated by ξ. That means that ω has the form (1.6) and for some sequence n_k, $T^{n_k}\xi \to \omega$. Then again with a as the leading symbol in ω we find that a occurs in ω at positions $\in 1 + P(m_1, m_2, \ldots, m_l)$. But now we can refine $\{n_k\}$ so that $T^{n_k}\xi$ is sufficiently close to ω so that the two sequences agree for $n \leq 1 + m_1 + m_2 + \cdots + m_l$. Then $\xi(n_k + p) = a$ whenever $p \in 1 + P(m_1, m_2, \ldots, m_l)$. We can assume $n_k \to \infty$; otherwise a finite translate of ω would be recurrent and we can invoke the first case. We now have $1 + n_k + P(m_1, m_2, \ldots, m_l) \subset B_a$ for an infinite sequence n_k, and this proves the lemma.

We now formulate some other simple results that enable one to construct recurrent points of Bebutov systems. A general procedure is the following. Let (X,T) be any cyclic dynamical system and $x_0 \in X$ a recurrent point for (X,T). Let $f: X \to \mathbf{C}$ be any continuous function. If $\Lambda \supset f(X)$, then the map $x \to (f(x), f(Tx), f(T^2x), \ldots, f(T^nx), \ldots)$ is a homomorphism of $(X,T) \to (\Lambda^{\mathbf{N}}, T)$ and so x_0 is mapped into a recurrent point.

Another construction proceeds on the basis of the following proposition.

PROPOSITION 1.11. *Let* Λ_1, Λ_2 *be compact spaces and* $\phi: \Lambda_1 \to \Lambda_2$ *a map continuous at some but not all points of* Λ_1. *If* $\omega \in \Lambda_1^{\mathbf{N}}$ *is a recurrent point of the Bebutov system to which it belongs and* ϕ *is continuous at each* $\omega(n) \in \Lambda_1$, $n = 1,2,\ldots$, *then* ω' *defined by* $\omega'(n) = \phi(\omega(n))$ *is a recurrent point of* $\Lambda_2^{\mathbf{N}}$.

Proof: If $T^{n_k}\omega \to \omega$, then $T^{n_k}\omega' \to \omega'$.

EXAMPLE.   Let α be irrational and define $\xi(n) = 1$ for $\cos \pi n^2 \alpha > 0$, and $\xi(n) = -1$ for $\cos \pi n^2 \alpha < 0$. The sequence $\cos \pi n^2 \alpha$ is obtained by evaluating a continuous function on the orbit of a recurrent point in the torus system described in Section 2. Hence $\{\cos \pi n^2 \alpha\}$ represents a recurrent point in $[-1,1]^N$. Applying Proposition 1.11, we find that the sequence $\xi(n)$ also constitutes a recurrent point.

Finally we can use the device of group extensions to obtain recurrent points.

PROPOSITION 1.12.   *Let K be a compact group and $\xi = \{\xi(1),\xi(2), \ldots\}$ a recurrent point of K^N. Let $\eta(n) = \xi(n)\xi(n-1) \cdots \xi(1) \in K$, Then $\eta = \{\eta(1),\eta(2), \ldots\}$ is again a recurrent point of K^N.*

Proof: Define a group extension of the Bebutov system (K^N, T) by setting $\psi(\omega(1)\omega(2) \cdots) = \omega(1)$ and $T(\omega,k) = (T\omega, \psi(\omega)k)$. Here $T : K^N \to K^N$ is the shift map, and we obtain

$$T(\xi,e) = (T\xi, \xi(1))$$
$$T^2(\xi,e) = (T^2\xi, \xi(2)\xi(1))$$
$$- - -$$
$$T^n(\xi,e) = (T^n\xi, \xi(n)\xi(n-1) \cdots \xi(1)).$$

By Theorem 1.4 (ξ,e) is a recurrent point and reading the second coordinate of $T^n(\xi e)$ we obtain a recurrent point of K^N.

EXAMPLE.   Let $m \geq 1$ and choose a sequence $a_1, a_2, a_3, \ldots$ of integers modulo m. Form the sequence

(1.7) $\xi: a_1, a_2, a_1, a_3, a_1, a_2, a_1, a_4, a_1, a_2, a_1, a_3, a_1, a_2, a_1, \ldots,$

and let η be the sequence of partial sums modulo m:

$$\eta: a_1, (a_1 + a_2), (2a_1 + a_2), (2a_1 + a_2 + a_3), (3a_1 + a_2 + a_3), \ldots .$$

Here $K = \mathbf{Z}/m\mathbf{Z}$ and ξ is a recurrent point of K^N. Consequently, so is η, and therefore η also has the pattern (1.6). The reader may amuse himself proving this directly.

4. Uniform Recurrence and Minimal Systems

The notion of recurrence discussed in the previous sections constitutes a first approximation to periodicity. A closer approximation is that of uniform recurrence. While this condition is a significant strengthening of ordinary recurrence, we shall still find that every compact dynamical system possesses uniformly recurrent points. In a later chapter we shall

$$Sk^{-1} := \{g \in G : gk \in S\} \Big/ G = \bigcup_{k \in K} Sk^{-1}.$$

discuss a number of even stronger notions of recurrence that are of theoretical interest, but which are not always present.

The idea of analyzing the various degrees of recurrence possessed by a point in a dynamical system by studying the set of group elements that applied to the given point bring it close to itself (the "return times") is due to Gottschalk and Hedlund [1]. Simple recurrence means that this set does not merely reduce to the identity. Uniform recurrence will mean that the set is large in the sense of the following definition.

DEFINITION 1.7. *A subset S of an abelian topological (semi-) group G is syndetic if there exists a compact set $K \subset G$ such that for each element $g \in G$, there exists $k \in K$ with $gk \in S$.*

If G is discrete, as it will be throughout our discussion, K will be finite, and so for a discrete group, a set S is syndetic if finitely many translates of it fill G. A subset of $\mathbf{R}^n$ is syndetic if there exists $R > 0$ such that every ball of radius R meets the set. A subset $S \subset \mathbf{N}$ is syndetic if it can be arranged as an increasing sequences $s_1 < s_2 < s_3 < \cdots$ with bounded gaps $s_{n+1} - s_n$. Such sets have sometimes been called *relatively dense* sets. Syndetic sets play a fundamental role in the theory of almost periodic functions.

DEFINITION 1.8. *Let (X,G) be a dynamical system. A point $x \in X$ is uniformly recurrent for (X,G) if for any neighborhood $V \in x$, the set*

$$\{g \in G : gx \in V\}$$

is syndetic.

Note that whereas simple recurrence was defined only for cyclic dynamical systems, uniform recurrence is, in a natural way, definable for more general systems.

The phenomenon of uniform recurrence is closely related to the existence of minimal systems. We now proceed to discuss these.

DEFINITION 1.9. *A dynamical system (X,G) is called minimal if no proper closed subset of X is invariant under the action of G.*

The following useful characterization is obvious.

LEMMA 1.13. *The system (X,G) is minimal, if every orbit Gx is dense in X.* Another characterization is the following.

LEMMA 1.14. *The system (X,G) is minimal if for every open set $V \subset X$, there exist finitely many elements $g_1, g_2, \ldots, g_n \in G$ with*

$$\bigcup_{i=1}^{n} g_i^{-1} V = X.$$

Proof: If (X,G) is not minimal and Y is a closed invariant subset of X, set $V = X - Y$. Then

$$\bigcup_{g \in G} g^{-1}V \neq X.$$

Hence the condition is sufficient. Conversely suppose (X,G) is minimal. Then for any open $V \neq \varnothing$,

$$\bigcup_{g \in G} g^{-1}V = X,$$

and by compactness, some finite union covers X.

We now have the following connection between minimality and uniform recurrence.

THEOREM 1.15. *If (X,G) is minimal, every point $x \in X$ is uniformly recurrent.*

Proof: We shall prove a bit more than is necessary. Namely if (X,G) is minimal, we shall have that for any $x \in X$, V open $\subset X$, the set $N(x,V) = \{g : gx \in V\}$ is syndetic. For, by Lemma 1.14, $\bigcup g_i^{-1}V = X$ for some finite set $\{g_1, \ldots, g_n\}$, and so for any g, some $g_i(gx) \in V$, or, $g_i g \in N(x,V)$.

Now by Zorn's lemma (see proof of Theorem 1.1), every compact dynamical system contains minimal subsystems. This proves the following theorem which is the version in which Birkhoff proved his recurrence theorem.

THEOREM 1.16. *For any dynamical system (X,G) with compact X, the set of uniformly recurrent points is nonempty.*

There is a converse to Theorem 1.15.

THEOREM 1.17. *If x is a uniformly recurrent point of a dynamical system (X,G), then the orbit closure $\overline{Gx}$ is a minimal G-invariant closed subset of X.*

Proof: It suffices to prove that if $y \in \overline{Gx}$, then $x \in \overline{Gy}$. Assume otherwise, so that $x \notin \overline{Gy}$ and let V be an open neighborhood of x whose closure is disjoint from $\overline{Gy}$. Since x is uniformly recurrent, there is a finite set $\{g_1, \ldots, g_n\}$ so that for each g, some $g_i gx \in V$; in other words, each

$$gx \in \bigcup_{i=1}^{n} g_i^{-1}V.$$

Hence

$$Gx \subset \bigcup_{i=1}^{n} g_i^{-1}V$$

and

$$y \in \bigcup_{i=1}^{n} g_i^{-1} \bar{V}.$$

But then $Gy \cap \bar{V} \neq \varnothing$ contrary to our assumption.

Some of the arguments given earlier for ordinary recurrence may be applied to uniform recurrence. For example, if $X = K$ is a compact group and (X,T) is a Kronecker system ($Tx = ax$ for some $a \in K$), we can now argue that each point is uniformly recurrent. For, some point x_0 is uniformly recurrent, and if $T^n x_0 \in V x_0$ where V is a neighborhood of the identity, then $T^n x = a^n x = a^n x_0 \cdot x_0^{-1} x \in V x_0 \cdot x_0^{-1} x = Vx$, so that every point is uniformly recurrent.

THEOREM 1.18. *For a Kronecker system every point is uniformly recurrent.*

Theorem 1.18 is a special case of the following.

THEOREM 1.19. *If $(Y \times K, T)$ is a group extension of (Y,T) and $y_0 \in Y$ is a uniformly recurrent point for (Y,T), then each (y_0,k_0) is uniformly recurrent for $(Y \times K, T)$.*

Proof: Since the orbit closure of y_0 in Y is minimal by Theorem 1.17, we may assume that, to begin with, (Y,T) is minimal. Now let $Z \subset Y \times K$ be minimal for the transformation $T(y,k) = (Ty, \psi(y)k)$, and let

$$\pi : Y \times K \to Y$$

denote the projection onto the first component. The mapping π defines a homomorphism of dynamical systems and so $\pi(Z)$ is a closed T-invariant subset of Y. Since Y is minimal, $\pi(Z) = Y$. Now defining $R_{k'}(yk) = (y,kk')$, we find $R_{k'}$ is again an automorphism of $(Y \times K, T)$ and each $R_{k'}(Z)$ is a minimal subset of $Y \times K$. But

$$\bigcup R_{k'}(Z) = \pi^{-1}\{\pi(Z)\} = \pi^{-1} Y = Y \times K,$$

and so every point of $Y \times K$ is uniformly recurrent. This proves the theorem.

If we now combine the foregoing with Proposition 1.7 we can extend this result to isometric extensions:

THEOREM 1.20. *If $(Y \times M, T)$ is an isometric extension of (Y,T) and y_0 is a uniformly recurrent point of (Y,T), then each (y_0,u_0), $u_0 \in M$, is uniformly recurrent for $(Y \times M, T)$.*

We now find that the skew-product transformations defined on the d-torus in Section 2 produce dynamical systems for which each point is uniformly recurrent. More generally one obtains inductively that for any integer matrix

$$
\begin{pmatrix}
1 & 0 & 0 & \cdots & 0 \\
b_{21} & 1 & 0 & \cdots & 0 \\
 & & \cdot & \cdots & \cdot \\
b_{d1} & b_{d2} & \cdots & b_{d,d-1} & 1
\end{pmatrix}
$$

and real numbers $a_1, a_2, \ldots, a_d$, the transformation $T : \mathbf{T}^d \to \mathbf{T}^d$ with $T(\theta_1, \theta_2, \ldots, \theta_d) = (\theta_1', \theta_2', \ldots, \theta_d')$ where

$$
\theta_i' = \theta_i + b_{i1}\theta_1 + b_{i2}\theta_2 + \cdots + b_{i,\theta_{i-1}}\theta_{i-1} + a_i
$$

defines a dynamical system where each point is uniformly recurrent. In particular we deduce the following by specializing the choice of b_{ij}, a_i:

THEOREM 1.21.   *Let* $p_1(x), \ldots, p_k(x)$ *be real polynomials. For any* $\varepsilon > 0$, *the set of integers satisfying simultaneously*

$$
\left| e^{2\pi i p_1(n)} - e^{2\pi i p_1(0)} \right| < \varepsilon, \ldots, \left| e^{2\pi i p_k(n)} - e^{2\pi i p_k(0)} \right| < \varepsilon
$$

is syndetic.[5]

5. Substitution Minimal Sets and Uniform Recurrence in Bebutov Systems

Let Λ be finite set. We shall present some constructions of uniformly recurrent points in $\Lambda^{\mathbf{N}}$. Everything we do could be carried out for $\Lambda^{\mathbf{Z}}$ as well. The following criterion for uniform recurrence in $\Lambda^{\mathbf{N}}$ or $\Lambda^{\mathbf{Z}}$ follows directly from the definition:

PROPOSITION 1.22.   *With* Λ *a finite alphabet, a point* ω *in* $\Lambda^{\mathbf{N}}$ *or* $\Lambda^{\mathbf{Z}}$ *is uniformly recurrent for the corresponding Bebutov system if and only if every word that occurs in* ω *occurs along syndetic set.*

[5] Theorem 1.21 may be applied as follows. Form the uniform closure of the linear combinations of functions on $\mathbf{Z}$ of the form $e^{2\pi i p(n)}$, $p(x)$ a polynomial. This is a translation invariant algebra $\mathcal{W}$ (for Weyl) containing the algebra of almost periodic functions. For any function $f \in \mathcal{W}$ the set $\{n : |f(n) - f(0)| < \varepsilon\}$ is syndetic, and so is each set $\{n : |f(n) - f(n_0)| < \varepsilon\}$. Now let B be a thick subset of $\mathbf{Z}$ (Definition 1.10). Suppose two functions in W agree on B. Their difference is in $\mathcal{W}$ and since every syndetic set interests B, the difference must vanish identically. This proves:

Each function in $\mathcal{W}$ is determined if it is known on a thick set. This result will be generalized in Chapter 9, Section 3.

Since a uniformly recurrent point of Λ^N is also recurrent, such a point must have the form previously prescribed:

$$(1.6) \qquad \omega = \left[(aw^{(1)}a)w^{(2)}(aw^{(1)}a)\right]w^{(3)}\left[(aw^{(1)}a)w^{(2)}(aw^{(1)}a)\right] \cdots$$

where the $w^{(n)}$ are words composed of elements in Λ. A sufficient condition for ω to satisfy the condition of Proposition 1.22 is that the $w^{(n)}$ are of bounded length. Thus the sequence (1.7) is uniformly recurrent. The freedom in the choice of $w^{(n)}$ here makes evident the following fact that was not evident to the earliest investigators of symbolic dynamics:

There exist nonperiodic sequences that represent uniformly recurrent points of the symbolic system Λ^N (Robbins [1]).

One of the earliest examples of such a sequence was constructed by M. Morse. There are various means of presenting this example, but for purposes of generalizing the procedure, the following method is instructive. We construct successively words of length $1, 2, 2^2, \ldots, 2^n$ based on the letters $\Lambda = \{a, b\}$ where each word constitutes the first half of the next word. The procedure for defining the next word at each stage is to perform the replacements $a \to ab$ and $b \to ba$ in the given word. We now begin:

> *a*
> *ab*
> *abba*
> *abbabaab*
> · · · ·

the limit sequence

> *abbabaabbaababbabaababbaabbabaab* . . .

is the Morse sequence. Let us see why the criterion of Proposition 1.22 is satisfied. First of all, the symbol a occurs syndetically since the sequence is made up of pairs ab and ba. Since the sequence reproduces itself by the substitutions $a \to ab$, $b \to ba$, it follows that the word ab occurs syndetically. But then so does the word $abba$, and so on. Since these initial words include all words, all words occur syndetically and the sequence defines a uniformly recurrent point.

More generally we proceed as follows. Rather than describe a sequence in Λ^N, we shall describe a closed translation invariant subset, and we shall determine when this consists of uniformly recurrent points. Let us call a countable list of words a *vocabulary* if it satisfies the following conditions:

(a) every subword of a word in the list is in the list,
(b) every word is a subword of a longer word.

Given a vocabulary we take the set of all sequences in $\Lambda^{\mathbf{N}}$ in which only words appearing in the vocabulary occur. This is in any-case a closed translation invariant set of sequences. Suppose the vocabulary satisfies the following condition: For every l there exists L such that each word of the vocabulary of length l occurs in every word of length L. Clearly, by Proposition 1.22 the corresponding invariant set will consist entirely of uniformly recurrent points.

We now generalize Morse's construction by showing how to construct a vocabulary satisfying the foregoing condition. Namely suppose $\Lambda = \{a_1, a_2, \ldots, a_r\}$ and that $w_1, w_2, \ldots, w_r$ are r words each of which contains all of the letters of Λ. Form a partial vocabulary by beginning with the symbols of Λ and then adding the words $w_1, \ldots, w_r$ and then proceeding by adding for each word in the vocabulary the word that is obtained by substituting

(1.8) $\qquad a_1 \rightarrow w_1, a_2 \rightarrow w_2, \ldots, a_r \rightarrow w_r$

in the given word, together with all its subwords. It is clear that each symbol will occur syndetically in each sequence, and, as before, we conclude that each word occurs syndetically. This describes the general *substitution minimal set*. We refer the reader to Martin [1] for more details.

We conclude this section with one more procedure for constructing uniformly recurrent points in Bebutov systems. This time we shall consider systems $\Lambda^{\mathbf{Z}}$.

Let $d_1, d_2, d_3, \ldots$ be a sequence of natural numbers with $d_n | d_{n+1}$. One can express $\mathbf{Z}$ as a disjoint union of the infinite arithmetic progressions $\{d_k \mathbf{Z} + a_k\}$ where the a_k are appropriately chosen:

$$\mathbf{Z} = \bigcup_{k=1}^{\infty} (d_k \mathbf{Z} + a_k).$$

For example, having chosen $a_1, \ldots, a_k$ we could let a_{k+1} be the number with smallest absolute value not covered by $d_1 \mathbf{Z} + a_1, \ldots, d_k \mathbf{Z} + a_k$.

Now let Λ be any compact space and $\{\lambda_k\}$ any sequence in Λ. Define $\omega \in \Lambda^{\mathbf{Z}}$ by

$$\omega(n) = \lambda_k \quad \text{for} \quad n \in d_k Z + a_k.$$

It is clear that if

$$[-N, N] \subset \bigcup_{k=1}^{l} (d_k \mathbf{Z} + a_k),$$

then if $n \in [-N,N]$, $\omega(n) = \omega(n + d_l)$. From this we see that every word in ω occurs along a periodic sequence. By Proposition 1.22, ω is uniformly recurrent.[6]

6. Combinatorial Applications

The following notion is dual to that of a syndetic set.

DEFINITION 1.10.   *A thick subset R of $\mathbf{N}$ or of $\mathbf{Z}$ is one that contains arbitrarily long intervals $(a_n, a_n + n)$, $n \to \infty$.*

REMARK.   In Gottschalk and Hedlund [1] this is called *replete*.

A set S is syndetic if it intersects nontrivially each thick set, and a set R is thick if it intersects nontrivially each syndetic set.

DEFINITION 1.11.   *A subset of $\mathbf{N}$ or $\mathbf{Z}$ is piecewise syndetic if it is the intersection of a thick set with a syndetic set.*

Equivalently, a set is *piecewise syndetic* if for some fixed l, the set contains arbitrarily long strings $a_1 < a_2 < a_3 < \cdots < a_n$ satisfying $a_{i+1} - a_i \le l$. We can now state the following result.

THEOREM 1.23.   *Let $\mathbf{N} = B_1 \cup B_2 \cup \cdots \cup B_q$ be a partition of $\mathbf{N}$ into finitely many sets. Then one of the sets B_j is piecewise syndetic. A similar statement holds for partitions of $\mathbf{Z}$.*

Proof: Take $\Lambda = \{1,2,\ldots,q\}$ and let $\omega \in \Lambda^{\mathbf{N}}$ be defined by

$$\omega(n) = i \Leftrightarrow n \in B_i.$$

Consider the Bebutov system generated by ω; that is, let

$$X = \{\overline{T^n \omega : n \ge 0}\},$$

the orbit closure of ω. Now (X,T) is a compact dynamical system and by Theorem 1.16, X contains a uniformly recurrent point. Suppose ξ is such a point and suppose that j is a value taken on by some $\zeta(n)$. The value j occurs syndetically in ζ, say with gaps $\le l$ between successive recurrences. Since $\xi \in X$, there are translates $T^m \omega$ arbitrarily close to ξ. This means that for some m,

$$T^n \omega(1) = \xi(1), \qquad T^m \omega(2) = \xi(2), \ldots, \qquad T^m \omega(n) = \xi(n).$$

---

[6] Using the construction at the end of the section, we find it is easy to obtain examples of the following:

(a)   a uniformly recurrent real valued sequence $\omega(n)$ for which

$$\lim_{N \to \infty} \frac{1}{2N+1} \sum_{-N}^{N} \omega(n)$$

does not exist.

(b)   a minimal dynamical system with positive topological entropy.

But then j occurs in $\omega(m + 1)$, $\omega(m + 2)$, ..., $\omega(m + n)$ with gaps $\leq l$ between successive recurrences. Hence B_j is piecewise syndetic.

More generally we have the following theorem.

THEOREM 1.24. *Suppose B is a piecewise syndetic subset of* **N** *or* **Z**. *If* $B = B_1 \cup B_2 \cup \cdots \cup B_q$, *then one of the* B_j *is itself piecewise syndetic.*

Proof: We consider the case $B \subset$ **N**. Form the characteristic function 1_B of B, $1_B(n) = 1$ if $n \in B$, $1_B(n) = 0$ if $n \notin B$. Regarding 1_B as a point of $\{0,1\}^{\mathbf{N}}$, consider its orbit closure in the corresponding Bebutov system. It is easy to see that if B is piecewise syndetic, 1_B has in its orbit closure a point $1_{B'}$ with B' actually syndetic.

Now define $\omega(n) = i \Leftrightarrow n \in B_i$, $i = 1,2,\ldots,q$; with $\omega(n) = 0$ for $n \notin B_i$. We can find a sequence n_k with $T^{n_k}\omega$ convergent in $\{0,1,2,\ldots,q\}^{\mathbf{N}}$ to a point ω' where $\omega'(n) \geq 1 \Leftrightarrow n \in B'$. Next we can find a sequence m_k such that $T^{m_k}\omega'$ converges to a uniformly recurrent point ξ of $\{0,1,2,\ldots,q\}^{\mathbf{N}}$. Since $\omega' \geq 1_{B'}$ and B' is syndetic, it is easy to verify that no limit of translates of ω' can vanish. Hence $\xi \neq 0$. Let j be a nonzero value occurring in ξ. It occurs syndetically, and since ξ is also a limit of translates of ω, it follows that j occurs along arbitrarily long intervals in $\omega(n)$ with gaps between successive occurrences bounded by a fixed l. Hence B_j is piecewise syndetic.

It is sometimes useful to regard a syndetic set as a discrete analogue of a set with nonempty interior in a compact metric space. Then the dual notion of a thick set corresponds to that of a dense subset. A piecewise syndetic set is the analogue of a set dense in some open set, and Theorem 1.24 is a discrete analogue of Baire's theorem on unions of nowhere dense sets.

7. More Diophantine Approximation

Using the theory of minimal dynamical system, we can prove an extension of Theorem 1.6 relating to diophantine inequalities. A more powerful result will be proved in Chapter 3 (Weyl's equidistribution theorem) using measure theory, but it is instructive to see what results can be achieved by topological means.

Let $\mathbf{T}^d$ denote the d-dimensional torus with coordinates $(\theta_1,\theta_2,\ldots,\theta_d)$ where the θ_i take values in the additive group $\mathbf{R}/\mathbf{Z} = $ reals mod 1. We consider homotopy classes of functions from $\mathbf{T}^d$ to $\mathbf{T}$. It is known that in each homotopy class there is a unique linear function $f(\theta_1,\theta_2,\ldots,\theta_d) = a_1\theta_1 + a_2\theta_2 + \cdots + a_d\theta_d$, $a_1,a_2,\ldots,a_d \in \mathbf{Z}$. Let us denote the coefficient a_i which is uniquely determined by the homotopy class of f as $a_i = A_i[f]$.

LEMMA 1.25.   *Let* $T : \mathbf{T}^d \to \mathbf{T}^d$ *have the form*

$$T(\theta_1, \theta_2, \ldots, \theta_d) = (\theta_1 + \alpha_1, \theta_2 + f_1(\theta_1), \theta_2 + f_2(\theta_1, \theta_2), \ldots,$$
$$\theta_d + \theta_{d-1}(\theta_1, \theta_2, \ldots, \theta_{d-1}))$$

where α_1 *is irrational and furthermore for each* $i = 1, \ldots, d - 1, A_i[f_i] \neq 0$. *Then* T *defines a minimal dynamical system on* $\mathbf{T}^d$; *i.e., no proper closed subset of* $\mathbf{T}^d$ *is invariant under* T.

Proof: This follows by induction on d. For $d = 1$ it is valid by Kronecker's theorem. Assume the lemma true for $d - 1$, so that $(\mathbf{T}^{d-1}, T)$ is a minimal system. Let $u = (\theta_1, \ldots, \theta_{d-1}) \in \mathbf{T}^{d-1}$, and denote the points of $\mathbf{T}^d$ as pairs (u, θ) then

$$T(u, \theta) = (Tu, \theta + f_{d-1}(u)),$$

and $(\mathbf{T}^d, T)$ is a group extension of $(\mathbf{T}^{d-1}, T)$. Define the (system) automorphisms $R_\phi : \mathbf{T}^d \to \mathbf{T}^d$ by $R_\phi(u, \theta) = (u, \theta + \phi)$. We verify that $R_\phi T = T R_\phi$. Assume that T does not act minimally on $\mathbf{T}^d$ and let Z be a minimal T-invariant closed subset. Since the projection of Z to $\mathbf{T}^{d-1}$ is invariant and compact and by hypothesis $(\mathbf{T}^{d-1}, T)$ is minimal, Z must project onto all of $\mathbf{T}^{d-1}$. Consequently, for each $u \in \mathbf{T}^{d-1}$, the set $Z_u = \{\theta : (u, \theta) \in Z\}$ is nonempty. We shall show that there exists a number $h \in \mathbf{N}$ such that each set Z_u consists of a coset of the group $\mathbf{T}$ modulo the subgroup $\{0, 1/h, 2/h, \ldots, (h-1)/h\}$. Namely, let $H = \{\phi : R_\phi Z \cap Z \neq \varnothing\}$. Since R_ϕ commutes with T, $R_\phi Z$ is again minimal and so if $R_\phi Z \cap Z \neq \varnothing$, we must have $R_\phi Z = Z$. It follows that H is a closed subgroup, and, moreover, that unless $Z = \mathbf{T}^d$, $H \neq \mathbf{T}$. Hence $H = \{0, 1/h, 2/h, \ldots, (h-1)/h\}$ for some $h \in \mathbf{N}$. In view of the two characterizations of H:

$$H = \{\phi : R_\phi Z \cap Z \neq \varnothing\} = \{\phi : R_\phi Z = Z\},$$

we see that the sets Z_u do indeed have the form ascribed to them.

Now map $\mathbf{T}^d \to \mathbf{T}^d$ by $S_h(u, \theta) = (u, h\theta)$, and define $T_h : \mathbf{T}^d \to \mathbf{T}^d$ by $T_h(u, \theta) = (Tu, \theta + hf_{d-1}(u))$. We check that

$$T_h S_h = S_h T$$

so that S_h is a homomorphism of the system $(\mathbf{T}^d, T)$ onto $(\mathbf{T}^d, T_h)$. It follows that $S_h(Z)$ is a T_h-invariant subset of $\mathbf{T}^d$. Moreover, the fibers $S_h(Z)_u$ consist of singletons, since $S_h(Z)_u = h(Z_u)$. This means that $S_h(Z)$ is the graph of a continuous function $g : \mathbf{T}^{d-1} \to \mathbf{T}$. To say that $S_h(Z)$ is T_h-invariant is to say that

(1.9) $g(Tu) = g(u) + hf_{d-1}(u)$.

Let us show that (1.9) is impossible. Considering the homotopy classes of the various maps of $\mathbf{T}^{d-1} \to \mathbf{T}$, we obtain

(1.10) $\quad A_{d-1}[g \circ T] - A_{d-1}[g] = hA_{d-1}[f_{d-1}] \neq 0.$

But suppose g is homotopic to

$$b_1\theta_1 + b_2\theta_2 + \cdots + b_{d-1}\theta_{d-1}.$$

Then $g \circ T$ is homotopic to

$$b_1\theta_1 + b_2(\theta_2 + f_1(\theta_1)) + \cdots + b_{d-1}(\theta_{d-1} + f_{d-2}(\theta_1, \ldots, \theta_{d-2})).$$

Replacing each function by the linear function in its homotopy class, we find that $A_{d-1}[g \circ T] = b_{d-1} = A_{d-1}[T]$. This contradicts (1.10), and so we must have $Z = \mathbf{T}^d$. This proves the lemma.

This can now be used to obtain the following extension of Theorem 1.6.

THEOREM 1.26. *If $p(x)$ is a real polynomial with at least one coefficient other than the constant term irrational, then for any $\varepsilon > 0$ we can solve the inequality*

$$|p(n) - m| < \varepsilon$$

in integers m,n.

Proof: Let us first assume that the leading coefficient of $p(x)$ is irrational; say $p(x) = \dfrac{\alpha}{d!} x^d + c_1 x^{d-1} + \cdots$, α irrational. Define $T : \mathbf{T}^d \to \mathbf{T}^d$ by

$$T(\theta_1, \ldots, \theta_d) = (\theta_1 + \alpha, \theta_2 + \theta_1, \ldots, \theta_d + \theta_{d-1}).$$

The conditions of Lemma 1.25 are met so that T defines a minimal system in $\mathbf{T}^d$. Let $p_d(x) = p(x)$, and define successively

$$p_{d-1}(x) = p_d(x+1) - p_d(x)$$
$$p_{d-2}(x) = p_{d-1}(x+1) - p_{d-1}(x)$$

$$- \; - \; -$$

$$p_1(x) = p_2(x+1) - p_2(x) = \alpha x + \beta.$$

Then

$$T(p_1(n), p_2(n), \ldots, p_d(n)) = (p_1(n+1), p_2(n+1), \ldots, p_d(n+1)).$$

Since the orbit under T of $(p_1(0), p_2(0), \ldots, p_d(0))$ is dense in $\mathbf{T}^d$, it follows that the set of values of $p(n) = p_d(n)$ is dense in $\mathbf{T}$. This establishes the theorem in this case.

In the general case write $p(x) = q(x) + r(x)$ where $r(x)$ has its leading coefficient irrational and where all the coefficients of $q(x)$ are rational and $q(0) = 0$. Let a be a common denominator of the coefficients of $q(x)$, and consider the polynomial $r(ax)$. This has leading coefficient irrational, and so we can solve $|r(an) - m| < \varepsilon$ in integers n,m. But then

$$|p(an) - q(an) - m| < \varepsilon$$

and since $q(an) + m$ is an integer, this gives the desired inequality.

8. Nonwandering Transformations and Recurrence

Let T be a continuous map of a compact metric space X to itself. A subset $A \subset X$ will be said to be *wandering* if all the transforms $A, T^{-1}A$, $T^{-2}A, \ldots, T^{-n}A, \ldots$ are disjoint from A (and therefore from each other). Let us say that a system (X,T) is *nonwandering* if no nonempty open set is wandering. A sufficient condition that (X,T) be nonwandering is that X be the support of an invariant measure with finite total mass. This is implied by Poincaré's theorem (Introduction, §4).

THEOREM 1.27. *If (X,T) is nonwandering, then the set of ~~non~~recurrent points of (X,T) is residual.*
   Proof: Let $d(,)$ denote the metric on X, and define

$$F(x) = \inf_{n \geq 1} d(x, T^n x).$$

The function $F(x)$ is an upper semicontinuous function. For if $F(x_0) = u$, then for any $\varepsilon > 0$, some $d(x_0, T^n x_0) < u + \varepsilon$, and the same will remain true in a neighborhood of x_0. In other words, $F(x) < F(x_0) + \varepsilon$ in a neighborhood of x_0; so $F(x)$ is upper semicontinuous. Now a semicontinuous function on a compact metric space possesses a residual set of points of continuity. Suppose x_0 is a point of continuity of $F(x)$. If $F(x_0) = 0$, then x_0 is a recurrent point. If $F(x_0) > 0$, then for some $a > 0$ and some neighborhood of x_0, $F(x) > a$ throughout this neighborhood. Let V be an open set of diameter $< a$ for which this inequality prevails. Now V is not a wandering set, so some $T^{-n}V \cap V \neq \varnothing$, or for some $x \in V$, $T^n x \in V$. But then $F(x) < a$, which is a contradiction. We conclude that at points of continuity, $F(x_0) = 0$, so that these are all recurrent points.

For completeness we include a proof of the fact alluded to in this proof that on a complete metric space, a semicontinuous function possesses a residual set of points of continuity.

Nota: If $F(x_0) = 0$ then
F is continuous at x_0.

LEMMA 1.28. *If $F(x)$ is a semicontinuous function, its points of discontinuity lie in the union of countably many closed nowhere dense sets.*

Proof: Say $F(x)$ is upper semicontinuous. Fix $\varepsilon > 0$ and let A_ε be the set of points x for which there exist sequences $x'_n \to x$, $x''_n \to x$ with $|F(x'_n) - F(x''_n)| \geq \varepsilon$. It is clear that A_ε is closed. Let us show that it has empty interior. We do this under the additional hypothesis that $F(x)$ is bounded from below. Suppose now that x is an interior point of A_ε. By upper semicontinuity, if x_n is sufficiently close to x, $F(x'_n) \leq F(x) + \varepsilon/2$ whence $F(x''_n) < F(x) - \varepsilon/2$. But for n large x''_n is again an interior point of A_ε and the process may be repeated. But we cannot repeat it indefinitely if $F(x)$ is bounded from below. This proves the lemma when $F(x)$ is bounded from below. If $F(x)$ is unbounded, replace it by $e^{F(x)}$.

... for suff. large n, x'_n and x''_n are interior to A_ε and

$F(x'_n) < F(x) + \frac{\varepsilon}{2}$, $F(x''_n) < F(x) + \frac{\varepsilon}{2}$; since $\varepsilon <$

$|F(x'_n) - F(x''_n)|$, either $F(x'_n) < F(x) - \frac{\varepsilon}{2}$ or $F(x''_n)$

$< F(x) - \frac{\varepsilon}{2}$.

(Note: $\bigcup_k A_{1/k}$ includes the set of discontinuity for F.)

CHAPTER 2

Van der Waerden's Theorem

In this chapter we shall extend Birkhoff's recurrence theorem, Theorem 1.1, to the situation where several commuting transformations act on a compact space X. We shall show that if $T_1, T_2, \ldots, T_l$ are all continuous maps of the compact metric space X to itself, there exists some $x \in X$ and some sequence $n_k \to \infty$, with $T_i^{n_k} x \to x$ simultaneously for $i = 1, 2, \ldots, l$. (Theorem MBR of the Introduction.)

Let us begin by observing that if the condition of commutativity is omitted, the conclusion of this theorem need not hold. One can even arrange that the set of recurrent points for one transformation be disjoint from that of another. The following is a simple example of this. Take $X = [0,1]$, $Tx = \dfrac{x}{2}$, $Sx = \dfrac{x+1}{2}$; the only recurrent point for T is 0, the only one for S is 1. For an example with invertible transformations take $X = \mathbf{R} \cup \{\infty\}$, $Tx = x + 1$, $Sx = \dfrac{x}{x+1}$; T has only ∞ as recurrent point, S has only 0 as recurrent point.

The phenomenon of disjoint sets of points of recurrence can only take place if some of the transformations have wandering sets (Chapter 1, §8). If each T_i is nonwandering, then by Theorem 1.27, its set of recurrent points is residual; consequently there will be common points of recurrence for all the T_i. Nevertheless, without commutativity it may still happen that the return times of any point to a neighborhood of the point are disjoint for the various transformations. The following is an example of this.

Let $X = \{-1,1\}^{\mathbf{Z}}$ and let T be the shift map $T\omega(n) = \omega(n + 1)$. Let $R: X \to X$ be defined by

$$(2.1) \qquad R\omega(n) = \begin{cases} \omega(n) & \text{for} \quad n = 0 \\ -\omega(n) & \text{for} \quad n \neq 0, \end{cases}$$

and let $S = RTR$. $R = R^{-1}$, so $S^n = RT^nR$. Now $T^n\omega$ close to ω implies that $\omega(n) = \omega(0)$. On the other hand, $S^n\omega$ close to ω means that $RT^nR\omega$ is close to ω and so $T^nR\omega$ is close to $R\omega$; i.e., $R\omega(n) = R\omega(0)$. But by (2.1) this requires $-\omega(n) = \omega(0)$ if $n \neq 0$, and so $T^n\omega$ and $S^n\omega$ cannot both be close to ω.

1. Bowen's Lemma and Homogeneous Sets

The first step in proving Theorem MBR is to establish the following lemma whose simple proof given below is due to Rufus Bowen.

LEMMA 2.1   *Let T be a continuous map of a compact metric space X to itself. Let $A \subset X$ be a subset with the property that for every $x \in A$, and $\varepsilon > 0$, there exists $y \in A$ and $n \geq 1$ with $d(T^ny,x) < \varepsilon$. Then for every $\varepsilon > 0$, there exists a point $z \in A$ and an $n \geq 1$ with $d(T^nz,z) < \varepsilon$.*

Proof: Let $\varepsilon > 0$ be given. We shall define inductively a sequence of points $z_0, z_1, z_2, \ldots$ one of which will satisfy $d(T^nz,z) < \varepsilon$. Set $\varepsilon_1 = \varepsilon/2$. Choose z_0 arbitrarily in A and let $z_1 \in A$ satisfy

$$d(T^{n_1}z_1,z_0) < \varepsilon_1$$

for some n_1. Now let ε_2 with $0 < \varepsilon_2 \leq \varepsilon_1$ be such that whenever $d(z,z_1) < \varepsilon_2$,

$$d(T^{n_1}z,z_0) < \varepsilon_1.$$

With ε_2 determined find $z_2 \in A$ and $n_2 \geq 1$ with

$$d(T^{n_2}z_2,z_1) < \varepsilon_2.$$

Next let ε_3 with $0 < \varepsilon_3 \leq \varepsilon_1$ be such that whenever $d(z,z_2) < \varepsilon_3$

$$d(T^{n_2}z,z_1) < \varepsilon_2.$$

Continue in this way defining $z_0, z_1, \ldots, z_k$ in A, $n_1, n_2, \ldots, n_k$ in **N**, $\varepsilon_1, \varepsilon_2, \ldots, \varepsilon_k$ in $\left(0, \dfrac{\varepsilon}{2}\right)$, where, these having been determined so that

(2.2) $d(T^{n_i}z_i, z_{i-1}) < \varepsilon_i, \qquad i = 1,2,\ldots,k,$

we define ε_{k+1}, $0 < \varepsilon_{k+1} \leq \varepsilon_1$ so that

(2.3) $d(z,z_k) < \varepsilon_{k+1} \Rightarrow d(T^{n_k}z,z_{k-1}) < \varepsilon_k.$

We then determine z_{k+1} and n_{k+1} by

$$d(T^{n_{k+1}}z_{k+1},z_k) < \varepsilon_{k+1}.$$

Matters have now been arranged so that whenever $i < j$,

(2.4) $d(T^{n_j + n_{j-1} + \cdots + n_{i+1}}z_j, z_i) < \varepsilon_{i+1} \leq \varepsilon_1 = \dfrac{\varepsilon}{2}.$

Since X is compact, we shall find some pair $i, j, i < j$, for which $d(z_i, z_j) < \dfrac{\varepsilon}{2}$. This combined with (2.4) gives

$$d(T^{n_j + n_{j-1} + \cdots + n_{i+1}} z_j, z_j) < \frac{\varepsilon}{2} + \frac{\varepsilon}{2} = \varepsilon.$$

This proves the lemma.

We shall now define the notion of a *homogeneous* subset of a dynamical system (X, T). We shall see that when the subset A in Lemma 2.1 is a homogeneous set, then we can weaken the hypothesis of that lemma and also strengthen the conclusion.

DEFINITION 2.1.   *Let T be a continuous map of a compact space X itself and let A be a closed subset of X. The set A is said to be homogeneous with respect to T if there exists a group G of homeomorphisms of X, each of which commutes with T and leaves A invariant, and such that the dynamical system (A, G) is minimal.*

A homeomorphism of X that commutes with T constitutes an automorphism of the system (X, T). If (A, G) is minimal, the G-orbits in A are dense in A. Thus if A is a homogeneous set with respect to T, the automorphisms of the system (X, T) will move any given point of A arbitrarily close to any other.

An example of this occurs in the case of a group extension of a system. Let (Y, T) be a cyclic dynamical system, $\psi: Y \to G$ a continuous map of Y to a compact group G. Set $X = Y \times G$ and define $T: X \to X$ by $T(y, g) = (Ty, \psi(y)g)$. Then if $R_{g'}: X \to X$ is defined by $R_{g'}(y, g) = (y, gg')$, the group G acts on X via the transformation $R_{g'}$. It follows that the fibers $y \times G$ are all homogeneous sets of (X, T).

DEFINITION 2.2.   *A closed subset A of a compact metric space X is said to be recurrent for a transformation $T: X \to X$, if for any $\varepsilon > 0$ and any point $x \in A$, there exists $y \in A$ and $n \geq 1$ with $d(T^n y, x) < \varepsilon$.*

LEMMA 2.2.   *If A is a homogeneous set in the compact metric space X with respect to a transformation T, and for any $\varepsilon > 0$, we can find $x, y \in A$ and $n \geq 1$ so that $d(T^n y, x) < \varepsilon$, then A is recurrent.*

Proof: Let G be a group of homeomorphisms commuting with T, leaving A invariant and for which (A, G) is minimal. We claim that for any $\varepsilon > 0$ we can find a finite subset $G_0 \subset G$ such that for any pair $x, y \in A$,

(2.5) $\min\limits_{g \in G_0} d(gx, y) < \varepsilon.$

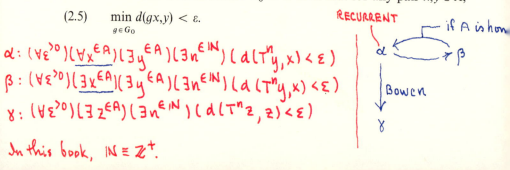

To see this, let $\{V_i\}$ be a finite covering of A by open sets of diameter $< \varepsilon$. For each V_i we can find a finite set $\{g_{ij}\}$ such that

$$\bigcup_j g_{ij}^{-1} V_i = A$$

(Lemma 1.14). It is now clear that the set $G_0 = \{g_{ij}\}$ satisfies (2.5) for any $x, y \in A$. Given the set $G_0 \subset G$, we find $\delta > 0$ such that whenever $d(x_1, x_2) < \delta$, we obtain $d(gx_1, gx_2) < \varepsilon$ for all $g \in G_0$. Apply the hypothesis of the lemma with ε replaced by δ; for some $x, y \in A$, $n \geq 1$ we have $d(T^n y, x) < \delta$. Now let z be any point of A, and find $g \in G_0$ with $d(gx, z) < \varepsilon$. The transformation g commutes with T so that

$$d(T^n gy, gx) = d(gT^n y, gx) < \varepsilon$$

since $d(T^n y, x) < \delta$; and combining this with $d(gx, z) < \varepsilon$, we obtain

$$d(T^n gy, z) < 2\varepsilon.$$

It follows that A is recurrent.

We can illustrate this lemma by considering again the example of a group extension. If y is a recurrent point for (Y, T), then the homogeneous set $y \times G$ clearly satisfies the hypothesis of Lemma 2.2. It follows that in a group extension the fiber above a recurrent point is a recurrent set.

The next lemma says that a recurrent homogeneous set in (X, T) contains a recurrent point.

LEMMA 2.3. *Let A be a recurrent homogeneous set in X with respect to a transformation T. Then A contains a recurrent point for (X, T).*

Proof: Let

$$F(x) = \inf_{n \geq 1} d(T^n x, x).$$

As in the proof of Theorem 1.27, we see that $F(x)$ is upper semicontinuous. Since A is recurrent, the hypothesis of Lemma 2.1 is fulfilled. From that lemma we conclude that $F(x)$ comes arbitrarily close to 0 on A. Now let $x_0 \in A$ be a point of continuity of the restriction of F to A. (See Lemma 1.28 for the existence of points of continuity of the semicontinuous function F.)

Suppose $F(x_0) > 0$. Then $F(x) > \delta > 0$ in an open set $V \subset A$. Since A is homogeneous, we have by Lemma 1.14 applied to the minimal system (A, G),

$$(2.6) \qquad A \subset \bigcup_{g \in G_0} g^{-1} V,$$

where G_0 is some finite subset of G. Let $\eta > 0$ be such that $d(x_1,x_2) < \eta$ implies $d(gx_1,gx_2) < \delta$ for all $g \in G_0$. Then on each $g^{-1}V$, $F(x) \geq \eta$. For if $F(x) < \eta$, we could find n with $d(x,T^n x) < \eta$ which would then give $d(gT^n x,gx) < \delta$ or $d(T^n gx,gx) < \delta$, whence $F(gx) < \delta$ for all $g \in G_0$. But this cannot happen since for some $g \in G_0$, $gx \in V$. Thus in each $g^{-1}V$, $F(x) \geq \eta$ and by (2.6), $F(x) \geq \eta$ throughout A. This contradicts Lemma 2.1 and so we must have had $F(x_0) = 0$, so that x_0 is a recurrent point. This proves the lemma.

Returning once again to the example of a group extension with

$$A = y \times G,$$

with y a recurrent point for (Y,T), we conclude that since A is a homogeneous recurrent set it must contain a recurrent point. In this case it follows by homogeneity that every point is recurrent. In this way we have obtained an alternative proof of Theorem 1.4.

Coming back to the general case, we combine Lemma 2.2 and Lemma 2.3 to obtain the following.

PROPOSITION 2.4. *Let T be a continuous map of a compact metric space X to itself, and let $A \subset X$ be a homogeneous closed subset of X with respect to T. If for any $\varepsilon > 0$, we can find $x,y \in A$ and $n \geq 1$ with $d(T^n y,x) < \varepsilon$, then A contains a recurrent point for T.*

2. The Multiple Birkhoff Recurrence Theorem

We assume $T_1, T_2, \ldots, T_l$ are continuous maps of a compact metric space X to itself, and we wish to find a point in the space that returns close to itself under the action of the same power of $T_1, T_2, \ldots,$ and T_l. We shall assume, to begin with, that the transformations T_i are invertible, and we subsequently reduce the general case to this one. The main idea that we shall use is that x is simultaneously recurrent for $T_1, T_2, \ldots, T_l$ if $(x,x,\ldots,x) \in X^l$ is recurrent for $T_1 \times T_2 \times \cdots \times T_l$. Thus the problem is to determine if the diagonal of X^l contains a recurrent point for the transformation $T_1 \times T_2 \times \cdots \times T_l$. In Section 1 we have developed machinery for ascertaining whether certain subsets in a dynamical system contain recurrent points and we shall put this to use now.

PROPOSITION 2.5. *Let X be a compact metric space and $T_1, T_2, \ldots, T_l$ commuting homeomorphisms of X. Then there exists a point $x \in X$ and a sequence $n_k \to \infty$ with $T_i^{n_k} x \to x$ simultaneously for $i = 1,2,\ldots,l$.*

Proof: Let G be the group of homeomorphisms of X generated by $T_1, T_2, \ldots, T_l$. In proving the proposition, we find it will suffice to restrict our discussion to any closed G-invariant subset of X and to locate the point x in that subset. In particular we can assume that we are dealing with a minimal closed G-invariant set, and, replacing X by this set, we may assume without loss of generality that (X,G) is minimal. We proceed by induction on l. The case $l = 1$ is Birkhoff's theorem; in fact, by Theorem 1.15 every point of a minimal closed T_1-invariant subset is uniformly recurrent. Assume the lemma has been established for any set of $l - 1$ commuting homeomorphisms, and consider the l homeomorphisms $T_1, T_2, \ldots, T_l$. Form the l-fold product $X^{(l)} = X \times X \times \cdots \times X$ and let $\Delta^{(l)}$ denote the diagonal subset consisting of l-tuples $(x,x,\ldots,x)$. Set $T = T_1 \times T_2 \times \cdots \times T_l$ and let G act on $X^{(l)}$ by identifying each $S \in G$ with $S \times S \times \cdots \times S$. Note that $\Delta^{(l)}$ is a homogeneous set for $(X^{(l)}, T)$ since T commutes with the action of G. We claim the diagonal $\Delta^{(l)}$ also satisfies the remaining hypothesis of proposition 2.4. For let $R_i = T_i T_l^{-1}$, $i = 1, 2, \ldots, l - 1$, and let y be a point in X with $R_i^{n_k} y \to y$ for $i = 1, 2, \ldots, l - 1$ and some sequence $n_k \to \infty$. Set

$$x^* = (y, y, \ldots, y) \in \Delta^{(l)}$$
$$y_k^* = (T_l^{-n_k} y, T_l^{-n_k} y, \ldots, T_l^{-n_k} y) \in \Delta^{(l)}.$$

Then for large k, $T^{n_k} y_k^* = (R_1^{n_k} y, R_2^{n_k} y, \ldots, R_{l-1}^{n_k} y, y)$ is close to $(y, y, \ldots, y) = x^*$ with $x^*, y_k^* \in \Delta^{(l)}$. Thus the hypotheses of Proposition 2.4 are fulfilled and this proves the proposition.

We are now in a position to prove the result in general.

THEOREM 2.6. (MBR). *Let X be a compact metric space and $T_1, T_2, \ldots, T_l$ commuting continuous maps of X to itself. Then there exists a point $x \in X$ and a sequence $n_k \to \infty$ with $T_i^{n_k} x \to x$ simultaneously for $i = 1, 2, \ldots, l$.*

Proof: We shall use a standard device for replacing a semigroup action by a group action. Let $\Omega = X^{\mathbf{Z}^l}$ with

$$S_i \omega(n_1, \ldots, n_i, \ldots, n) = \omega(n_1, \ldots, n_i + 1, \ldots, n_l), \qquad i = 1, \ldots, l.$$

Let $\tilde{X} \subset \Omega$ be the subset satisfying

(2.7) $S_i \omega(n_1, \ldots, n_l) = T_i(\omega(n_1, \ldots, n_l)), \qquad i = 1, \ldots, l,$

for each lattice point $(n_1, \ldots, n_l) \in \mathbf{Z}^l$. The set $\tilde{X}$ is nonempty. To see this take any $x \in X$ and for $n \in \mathbf{N}$ set

$$\omega_n(n_1, \ldots, n_l) = T_1^{n_1 + n} T_2^{n_2 + n} \cdots T_l^{n_l + n} x$$

for $n_i \geq -n$. Define ω_n arbitrarily for other lattice points. The point ω_n satisfies (2.7) for $(n_1, \ldots, n_l)$ with all $n_i \geq -n$. Thus a limit point of ω_n in Ω will belong to $\tilde{X}$. So $\tilde{X}$ is nonempty. It is also invariant under S_i and S_i^{-1}.

[handwritten marginalia:] x may vary with n. If the T_j's are surjective one can assume that, for all n, $\omega_n(0)$ has a given value in X. So ev_0 is surjective

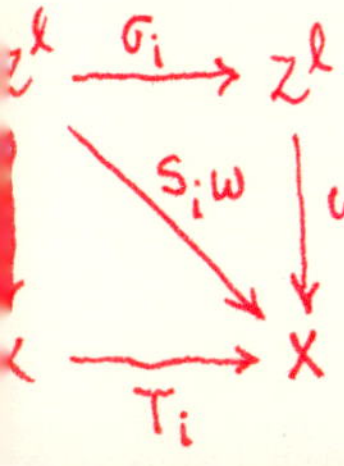

[handwritten diagrams in margins:]

$$\mathbf{Z}^l \xrightarrow{\sigma_i} \mathbf{Z}^l$$
$$S_i\omega \searrow \quad \downarrow \omega$$
$$X \xrightarrow{T_i} X$$

$$\tilde{X} = \{\omega \in \Omega : T_i \cdot \omega = \omega \cdot \sigma_i\}$$

$$\tilde{X} \xrightarrow{S_i} \tilde{X}$$
$$ev_y \downarrow \qquad \qquad \downarrow ev_y$$
$$X \xrightarrow{T_i} X$$

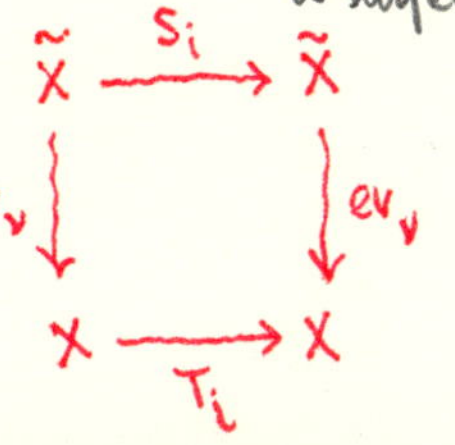

Proposition 2.5 is applicable to $\tilde{X}$ and the homeomorphisms $S_1, \ldots, S_l$, so let $\tilde{x}$ be a multiply recurrent point for $S_1, \ldots, S_l$. By virtue of (2.7) this implies that each component of $\tilde{x}$ is a multiply recurrent point in X for $T_1, T_2, \ldots, T_l$. This proves Theorem 2.6.

3. The Multidimensional van der Waerden Theorem

We have already seen in the Introduction how Theorem 2.6 implies van der Waerden's theorem on arithmetic progressions. Grünwald (also referred to in the literature by the name Gallai) extended this theorem to higher dimensions, and we now proceed to deduce this version from Theorem 2.6. We can formulate this theorem either in terms of partitions of the full lattice $\mathbf{Z}^m$ or of the positive lattice $\mathbf{N}^m$ or partitions of sufficiently large chunks of $\mathbf{N}^m$. We choose the second alternative and discuss variants in the next section.

Note that if $B \subset \mathbf{N}$ is a set that contains arbitrarily long arithmetic progressions then B also contains a homothetic copy of any finite configuration in $\mathbf{N}$. That is, if F is a finite subset of $\mathbf{N}$, $F \subset \{1, 2, \ldots, N\}$ and any arithmetic progression of length N, say, $\{a + b, a + 2b, \ldots, a + Nb\}$, contains the configuration $bF + a$. We use this to formulate Grünwald's theorem.

THEOREM 2.7.   *Let $\mathbf{N}^m = B_1 \cup B_2 \cup \cdots \cup B_q$ be a partition of the m-dimensional positive lattice $\mathbf{N}^m$. One of the sets B_j has the property that if F is any finite subset of $\mathbf{N}^m$, then for some $a \in \mathbf{N}^m$ and $b \in \mathbf{N}$, $bF + a \subset B_j$.*

Proof: It is easily seen that it suffices to produce the set B_j for a given configuration F. For since then are only finitely many possibilities for B_j and since a sequence F_n may be chosen where each contains all the preceding ones and any F is contained in one of them, a set B_j that occurs for infinitely many F_n will work for all F. So we shall assume that a finite set Γ is given.

Let $\Lambda = \{1, 2, \ldots, q\}$ and form $\Omega = \Lambda^{\mathbf{N}^m}$. Ω becomes a compact metric space if we define a metric by

$$(2.8) \qquad d(\omega, \omega') = \inf \left\{ \frac{1}{k} : \omega(i_1, i_2, \ldots, i_m) \right.$$

$$= \omega'(i_1, i_2, \ldots, i_m) \text{ for } 1 \leq i_1, i_2, \ldots, i_m < k \left. \right\}$$

Let $T_i : \Omega \to \Omega$ be defined by $T_i \omega(n) = \omega(n + e_i)$, $i = 1, 2, \ldots, l$, and $n \in \mathbf{N}^m$, where $F = \{e_1, \ldots, e_l\}$. Define a specific point $\omega \in \Omega$ by

$$\omega(n) = i \Leftrightarrow n \in B_i,$$

and let X be the smallest $(T_1, T_2, \ldots, T_l)$-invariant closed subset of Ω containing ω. By Theorem 2.6 there exists some $\xi \in X$ and some $n \geq 1$ for which

$$(2.9) \qquad d(T_1^n \xi, \xi) < 1, d(T_2^n \xi, \xi) < 1, \ldots, d(T_l^n \xi, \xi) < 1.$$

Denote by e the vector $(1, 1, \ldots, 1)$; then (2.9) is equivalent to

$$(2.10) \qquad \xi(e) = \xi(e + ne_1) = \xi(e + ne_2) = \cdots = \xi(e + ne_l).$$

Now $\xi \in X$ is either a translate of ω by some $T_1^{n_1} T_2^{n_2} \cdots T_l^{n_l}$ or a limit of such points. Either way these translates come arbitrarily close so that (2.10) implies

$$(2.11) \qquad \omega(a) = \omega(a + ne_1) = \omega(a + ne_2) = \cdots = \omega(a + ne_l)$$

with $a = e + n_1 e_1 + n_2 e_2 + \cdots + n_l e_l$, for some $n_1, n_2, \ldots, n_l$. It follows that if $j = \omega(a)$, then $a + nF \subset B_j$.

4. Reformulations and Applications: Diophantine Inequalities

Let us say that a subset $B \subset \mathbf{Z}^m$ is a VDW-*set* if for every finite $F \subset \mathbf{Z}^m$ we have some $bF + a \subset B$, $b \in N$, $a \in \mathbf{Z}^m$.

PROPOSITION 2.8.　*If S is a syndetic subset* (Definition 1.7) *of $\mathbf{N}^m$ or of $\mathbf{Z}^m$, then S is a VDW set.*

Proof: Say S is a syndetic subset of $\mathbf{N}^m$. Finitely many translates of S will cover $\mathbf{N}^m$; $\mathbf{N}^m \subset \bigcup(S + a_i)$, $a_i \in \mathbf{Z}^m$. It follows from Theorem 2.7 that some $S + a_j$ is VDW. But this property is translation invariant so that S itself must be a VDW-set.

A finite partition of $\mathbf{N}^m$ amounts to defining a function from $\mathbf{N}^m$ to a finite set. Grünwald's theorem asserts that any such function is constant on a homothetic copy $bF + a$ of any finite F. We can generalize this to functions with values in a compact metric space.

THEOREM 2.9.　*Let Λ be a compact metric space and let $f : \mathbf{N}^m \to \Lambda$ be an arbitrary function with values in Λ. For any $\varepsilon > 0$ and finite set $F \subset \mathbf{N}^m$, we can find a homothetic copy $bF + a$ for which the image under f, $f(bF + a)$, is a set of diameter $< \varepsilon$ in Λ.*

Proof: Simply let

$$\Lambda = \bigcup_{i=1}^{q} V_i$$

where diam $V_i < \varepsilon$. Let $\mathbf{N}^m = \bigcup B_i$ be a partition where $f(B_i) \subset V_i$. Applying Grünwald's theorem to this partition, we obtain the desired result.

Theorems 2.7 and 2.9 have implications for functions on large but finite chunks of $\mathbf{N}^m$. For example, Theorem 2.7 has the following corollary.

THEOREM 2.10. *Let F be a finite subset of* $\mathbf{N}^m$ *and let q be given. There exists a number* $T(m,q,F)$ *such that if* $\mathbf{N}_n^m$ *denotes the set of vectors in* $\mathbf{N}^m$ *with components* $\leq n$, *then whenever* $n \geq T(m,q,F)$ *and* $\mathbf{N}_n^m = B_1 \cup B_2 \cup \cdots \cup B_q$ *is a partition into q sets, one of these will contain a homothetic copy of F*, $bF + a$.

Proof: Regard the partition $\mathbf{N}_n^m = B_1 \cup B_2 \cup \cdots \cup B_q$ as a function from $\mathbf{N}_n^m$ to $\{1,2,\ldots,q\}$. Suppose with $n \to \infty$ we can find partitions for which no homothetic copy of F is contained in any B_j. Consider the corresponding functions from $\mathbf{N}_n^m$ to $\Lambda = \{1,2,\ldots,q\}$ and extend it arbitrarily to $\mathbf{N}^m$ to obtain a point ω_n in $\Lambda^{\mathbf{N}^m}$. Take any limit point of ω_n, say ω, and apply Theorem 2.6 to the corresponding partition. It follows that ω is constant on some $bF + a$. This set $bF + a$ is contained in $\mathbf{N}_n^m$ as soon as n is large, and moreover $\omega_{n'}$ agrees with ω on $\mathbf{N}_n^m$ as soon as n' is large. But this clearly leads to a contradiction that proves the theorem.

Theorem 2.9 can be used as a tool in diophantine approximation. We illustrate this first by a simple example. Theorem 2.9 implies that if $f(n)$ is any bounded function on the integers and $\varepsilon > 0$, there will be three numbers in arithmetic progression n, $n + h$, $n + 2h$ with $|f(n) - f(n + h)| < \varepsilon$, $|f(n) - f(n + 2h)| < \varepsilon$. Choose $f(n) = e^{in^2\alpha}$. We can then solve

(2.12)
$$\left|e^{in^2\alpha} - e^{i(n+h)^2}\right| < \varepsilon$$
$$\left|e^{in^2\alpha} - e^{i(n+2h)^2\alpha}\right| < \varepsilon.$$

This gives

$$\left|1 - e^{i(2nh+h^2)\alpha}\right| < \varepsilon$$
$$\left|1 - e^{i(4nh+4h^2)\alpha}\right| < \varepsilon,$$

and since $|1 - e^{2i\theta}| < 2|1 - e^{i\theta}|$, we find

$$\left|1 - e^{2ih^2\alpha}\right| < 3\varepsilon.$$

This implies that for every $\alpha \in R$ we can solve

(2.13) $$\left|\alpha - \frac{m}{h^2}\right| < \frac{\varepsilon}{h^2}$$

for any $\varepsilon > 0$. This was also proved in Proposition 1.5. We proceed to a second proof of Theorem 1.6 based on van der Waerden's theorem. It will be useful first to extend Theorem 2.9.

LEMMA 2.11. *Let $f_1: \mathbf{N}^{m_1} \to \Lambda_1, \ldots, f_s: \mathbf{N}^{m_s} \to \Lambda_s$ be s functions on lattices of varying dimension to various compact metric spaces Λ_i, and let $\varepsilon > 0$. If $F_1 \subset \mathbf{N}^{m_1}, \ldots, F_s \subset \mathbf{N}^{m_s}$ are s finite configurations, there exists a single constant of homothety b and vectors $a_i \in \mathbf{N}^{m_i}$ such that for each i, diam $f_i(bF_i + a_i) < \varepsilon$.*

Proof: Simply form $f_1 \times \cdots \times f_s: \mathbf{N}^{m_1 + \cdots + m_s} \to \Lambda_1 \times \cdots \times \Lambda_s$ and $F = F_1 \times F_2 \times \cdots \times F_s$, and apply Theorem 2.9.

We shall apply the foregoing to functions taking values in $\mathbf{R}/\mathbf{Z}$, the reals modulo 1. It will be useful to introduce a notation for the distance to the nearest integer.

$$\|x\| = \inf_{n \in \mathbf{Z}} |x - n|.$$

This function is defined on $\mathbf{R}/\mathbf{Z}$ and satisfies $\|x_1 + x_2\| \le \|x_1\| + \|x_2\|$.

We begin with the identity

$$(2.14) \qquad \sum_{j=0}^{r} (-1)^j \binom{r}{j}(x + jy)^r = A_r y^r, \qquad A_r = \sum_{j=0}^{r} (-1)^j \binom{r}{j} j^r$$

which we leave to the reader to verify. For each $r = 1, 2, \ldots, d$ set $f_r(n) = \alpha_r n^r \in \mathbf{R}/\mathbf{Z}$. By Lemma 2.11 we can find d arithmetic progressions with a common difference b, and of lengths $2, 3, \ldots, d + 1$ such that on the rth progression the function $f_r(n)$ oscillates by less than ε. Write

$$\|f_r(a_r + jb) - f_r(a_r)\| < \varepsilon, \qquad j = 1, \ldots, r.$$

Apply (2.14) with $y = b$ and $x = a_r$:

$$\sum_{j=0}^{r} (-1)^j \binom{r}{j} f_r(a_r + jb) = A_r f_r(b),$$

and the fact that

$$\sum_{j=0}^{r} (-1)^j \binom{r}{j} f_r(a_r) = 0,$$

to conclude that

$$(2.15) \quad \|A_r f_r(b)\| \le 2^r \varepsilon, \qquad r = 1, 2, \ldots, d.$$

Finally if

$$p(x) = \sum_{1}^{d} \alpha_r A_r x^r,$$

we obtain

$$(2.16) \quad \|p(b)\| \le \left(\sum |\alpha_r| 2^r \right) \varepsilon.$$

This gives us once again the result of Theorem 1.6 to the effect that for any real polynomial $p(x)$ with $p(0) = 0$ we can find integers m,n with

$$|p(n) - m| < \varepsilon.$$

The following result appears to be new.

THEOREM 2.12. *Let θ_n be an arbitrary sequence in $\mathbf{R}/\mathbf{Z}$, and let $\varepsilon > 0$. Then there exist m,n, $m \neq n$ with*

$$(2.17) \quad \|(n - m)\theta_n\| < \varepsilon \quad \text{and} \quad \|(n - m)\theta_m\| < \varepsilon.$$

Proof: Consider the function $f:\mathbf{N}^2 \to \Lambda$ defined by $f(n,m) = n\theta_m$. Let $F = \{(1,1),(1,2),(2,1),(2,2)\}$; then $bF + a$ consists of four points that we can denote $\{(p + m, m), (p + m, n), (p + n, m), (p + n, n)\}$. Assuming that f oscillates at these four points by less than ε, we find

$$\|(p + m)\theta_m - (p + n)\theta_m\| < \varepsilon$$
$$\|(p + m)\theta_n - (p + n)\theta_n\| < \varepsilon,$$

which proves the theorem.

As an illustration take $\theta_n \equiv \lambda^m$ where λ is a real number > 1. Combining the two inequalities of (2.17), we obtain the following.

COROLLARY. *For any real $\lambda > 1$ and $\varepsilon > 0$ we can solve the inequality*

$$|m(\lambda^m - 1)\lambda^n - p| < \varepsilon$$

in integers, m,n,p with $m,n \geq 1$.

We conclude this section with the following application of van der Waerden's theorem.

THEOREM 2.13. *Let X be an arbitrary space, $T_1, T_2, \ldots, T_l$ commuting transformations of X to itself, and let $\phi_1, \phi_2, \ldots, \phi_l$ be functions from X to the unit circle: $|\phi_i(x)| = 1$. Then for any $\varepsilon > 0$, we can find $x \in X$ and $n \geq 1$ to satisfy the inequalities*

$$|\phi_1(T_1^n x)^n - 1| < \varepsilon, \qquad |\phi_2(T_2^n x)^n - 1| < \varepsilon, \ldots, |\phi_l(T_l^n x)^n - 1| < \varepsilon.$$

Proof: Let x_0 be any point of X. Form the vector valued function $\Phi:\mathbf{N}^{l+1} \to \mathbf{C}^l$ by setting

$$\Phi(n_0, n_1, \ldots, n_l) = \begin{pmatrix} \phi_1(T_1^{n_1} T_2^{n_2} \cdots T_l^{n_l} x_0)^{n_0} \\ \phi_2(T_1^{m_1} T_2^{n_2} \cdots T_l^{n_l} x_0)^{n_0} \\ \vdots \\ \phi_l(T_1^{n_1} T_2^{n_2} \cdots T_l^{n_l} x_0)^{n_0} \end{pmatrix}.$$

Choosing a metric in $\mathbf{C}^l$ so that two vectors of distance $<\varepsilon$ from one another have all their respective components at distance $<\varepsilon$ from one another, we proceed by Theorem 2.9 to find a "cube" in $\mathbf{N}^{l+1}$ on which the function Φ oscillates by less than ε. The cube with vertex $(n_0, n_1, \ldots, n_l)$ and edge n includes the points $(n_0, n_1 + n, n_2, \ldots, n_l)$ and $(n_0 + n, n_1 + n, n_2, \ldots, n_l)$ as well as $(n_0, n_2, n_2 + n, \ldots, n_l)$ and $(n_0 + n, n_1, n_2 + n, \ldots, n_l)$, etc. If we now set

$$x = T_1^{n_1} T_2^{n_2} \cdots T_l^{n_l} x_0$$

and compare the values of Φ at the vertices of the cube in question, we find

$$\left| \phi_1(T_1^n x)^{n_0} - \phi_1(T_1^n x)^{n_0 + n} \right| < \varepsilon, \qquad \text{or} \qquad \left| \phi_1(T_1^n x)^n - 1 \right| < \varepsilon$$

as well as

$$\left| \phi_2(T_2^n x)^{n_0} - \phi_2(T_2^n x)^{n_0 + n} \right| < \varepsilon, \qquad \text{or} \qquad \left| \phi_2(T_2^n x)^n - 1 \right| < \varepsilon$$

and so on for all the components. This proves the theorem.

5. Further Refinements: IP-Sets

In this section we shall see that the phenomena of recurrence and multiple recurrence, $T_i^{n_k} x \to x$, take place even if some restriction is imposed on the iterates n_k that are to appear. The basic idea in order to obtain this refinement is to combine a system (X, T) with another system (Z, T) containing a point z_0 that recurs at known times. We study the product system $(X \times Z, T)$ and look for recurrent points in $X \times z_0$. Clearly the recurrence times will be a subset of those for z_0. The notion of an "*IP*-set" arises naturally when we study the structure of sets of integers that can serve as the set of recurrence times for some point in some system (Z, T). The conclusion will be that our recurrence theorems are valid when the sequence n_k is restricted to an *IP*-set.

Let (Z, T) be a fixed dynamical system on a compact metric space Z, let $z_0 \in Z$, and let $R = \{n : n \geq 1, d(T^n z_0, z_0) < 1\}$.

PROPOSITION 2.14. *Let T be a homeomorphisms of a compact metric space X. There exists a recurrent point $x \in X$ for which a sequence $n_k \to \infty$ can be chosen in R with $T^{n_k} x \to x$.*

Proof: Replacing X by a minimal subset of (X, T), we may assume that (X, T) is minimal. Form $\tilde{X} = X \times Z$ with $\tilde{T}(x, z) = (Tx, Tz)$. We notice that $X \times z_0$ is a recurrent set (Definition 2.2) in $(\tilde{X}, \tilde{T})$ since z_0 is a recurrent point of (Z, T). Moreover, $X \times z_0$ is a homogeneous set since $S : \tilde{X} \to \tilde{X}$ given by $S(x, z) = (Tx, z)$ commutes with $\tilde{T}$ and $X \times z_0$ is minimal with respect to S. By Lemma 2.3, $A = X \times z_0$ contains a recurrent point for $(\tilde{X}, \tilde{T})$. Our proposition follows directly.

We proceed to a similar extension of Proposition 2.5.

PROPOSITION 2.15. *Let X be a compact metric space and $T_1, T_2, \ldots, T_l$ commuting homeomorphisms of X. Then there exists a point $x \in X$ and a sequence $n_k \to \infty$, $n_k \in R$, such that $T_i^{n_k} x \to x$ simultaneously for $i = 1, 2, \ldots, l$.*

Proof: The proof is essentially a repetition of the proof of Proposition 2.5 with the spaces $X \times X \times \cdots \times X$ replaced by $X \times X \times \cdots \times X \times Z$ and the diagonal $\Delta^{(l)}$ replaced by $\Delta^{(l)} \times z_0$. Let G be the group of homeomorphisms of X generated by $T_1, T_2, \ldots, T_l$ and let G act on $X^{(l)} \times Z$ by $S(x_1, x_2, \ldots, x_l, z) = (Sx_1, Sx_2, \ldots, Sx_l, z)$. We may assume that G acts minimally on X so that $\Delta^{(l)} \times z_0$ is homogeneous for $(X^{(l)} \times Z, \tilde{T})$ where $\tilde{T} = T_1 \times T_2 \times \cdots \times T_l \times T$. To prove the proposition we need to know that $\Delta^{(l)} \times z_0$ is a recurrent set, or that the hypotheses of Proposition 2.4 are fulfilled. This will be the case if we can find a point in $\Delta^{(l)} \times z_0$ recurrent for $T_1 T_l^{-1} \times T_2 T_l^{-1} \times \cdots \times I \times T$, $I = T_l T_l^{-1} =$ identity. We thereby reduce the problem to that of finding a recurrent point in $\Delta^{(l-1)} \times z_0$ for $T_1 T_l^{-1} \times T_2 T_l^{-1} \times \cdots \times T_{l-1} T_l^{-1}$. The procedure for proving the proposition is therefore to prove the following assertion by induction on i: For each $i = 1, 2, \ldots$, if $T'_1, T'_2, \ldots, T'_i$ are transformations in G, the set $\Delta^{(i)} \times z_0$ contains a recurrent point for the transformation $T'_1 \times T'_2 \times \cdots \times T'_i \times T$ acting on $X^{(i)} \times Z$. The case $i = 1$ is the case of Proposition 2.14 and the passage from the case i to $i + 1$ is implemented by means of Proposition 2.4. This completes the proof.

Finally the passage from the case of invertible transformations to arbitrary continuous transformations is carried out just as in the proof of Theorem 2.6. We thus obtain the following result.

THEOREM 2.16. *Let X and Z be compact metric spaces, $T_1, T_2, \ldots, T_l$ commuting continuous maps of X to itself and T a continuous map of Z to itself. Suppose z_0 is a recurrent point of (Z, T) and let $R = \{n : n \geq 1, d(T^n z_0, z) < 1\}$. Then there exists a point $x \in X$ and a sequence $n_k \to \infty$, $n_k \in R$, with $T_i^{n_k} x \to x$ simultaneously for $i = 1, 2, \ldots, l$.*

We now wish to characterize the sets of integers that can occur as the set R in the foregoing theorem. These sets turn out to be essentially those that occur in Hindman's theorem (Introduction, §6).

DEFINITION 2.3. *A set of positive integers is called an IP-set if there exists a sequence $p_1, p_2, p_3, \ldots \in \mathbf{N}$ such that the set in question consists of* *the numbers p_i together with all finite sums $p_{i_1} + p_{i_2} + \cdots + p_{i_k}$ with $i_1 < i_2 < \cdots < i_k$.*

Note that the elements p_i need not be distinct, but in the expressions $p_{i_1} + p_{i_2} + \cdots + p_{i_k}$ the indices are distinct. An additive semigroup in $\mathbf{N}$ will be an *IP*-set, but generally an *IP*-set will not be a semigroup unless the elements p_i are repeated infinitely often. In Chapter 7 we shall see the connection between *IP*-sets and idempotence. This connection is the basis of the expression *IP*. We also remark that if $p_1, p_2, \ldots, p_n$ denote vectors in a vector space, then 0 together with the sums $p_{i_1} + p_{i_2} + \cdots + p_{i_k}$, $i_1 < i_2 < \cdots < i_k \leq n$ form the vertices of an n-dimensional parallelopiped (possibly degenerate). Thus an *IP*-set might be thought of as an Infinite-dimensional Parallelopiped. Turning now to the dynamical significance of *IP*-sets, we prove

THEOREM 2.17. *If z_0 is a recurrent point of a dynamical system (Z,T) with Z metric, then for each $\delta > 0$ the set*

$$(2.18) \quad R_\delta = \{n : d(T^n z_0, z_0) < \delta\}$$

contains an IP-set. Conversely if $R \subset \mathbf{N}$ is an IP-set, then there exists a dynamical system (Z,T) and a recurrent point $z_0 \in Z$ such that $R \supset R_1$, where R_1 is defined by (2.18).

Proof: Suppose z_0 is a recurrent point for (Z,T). Let p_1 satisfy

$$(2.19) \quad d(T^{p_1} z_0, z_0) < \delta.$$

Now find $\delta_2 > 0$ with $\delta_2 \leq \delta$ such that

$$(2.20) \quad d(z, z_0) < \delta_2 \Rightarrow d(T^{p_1} z, z_0) < \delta.$$

With this δ_2 define p_2 by

$$(2.21) \quad d(T^{p_2} z_0, z_0) < \delta_2.$$

The conditions (2.19), (2.20), and (2.21) together ensure that

$$(2.22) \quad d(T^m z_0, z_0) < \delta$$

for $m = p_1, p_2$ and $p_1 + p_2$. We continue inductively. Assume $p_1, p_2, \ldots, p_n$ have been found so that (2.22) is valid for all $m = p_{i_1} + p_{i_2} + \cdots + p_{i_k}$ with $i_1 < i_2 < \cdots < i_k \leq n$. We then find $\delta_{n+1} \leq \delta$ such that whenever $d(z, z_0) < \delta_{n+1}$,

$$(2.23) \quad d(T^m z, z_0) < \delta$$

holds for the same set of m. Then if p_{n+1} is defined so that

$$(2.24) \quad d(T^{p_{n+1}} z_0, z_0) < \delta_{n+1},$$

then (2.22) will be valid with m replaced by $m + p_{n+1}$ or by p_{n+1}. This provides the inductive step. We see that the IP-set

$$(2.25) \quad R = \{p_{i_1} + p_{i_2} + \cdots + p_{i_k}, \qquad i_1 < i_2 < \cdots < i_k\}$$

constructed in this manner is contained in R_δ.

Conversely, let $R \subset \mathbf{N}$ be an IP-set as in (2.25). Notice that whenever $H_1, H_2, \ldots, H_n, \ldots$ is a sequence of disjoint sets of $\mathbf{N}$ and we set

$$(2.26) \quad p'_n = \sum_{i \in H_n} p_i,$$

then the sums $p'_{i_1} + p'_{i_2} + \cdots + p'_{i_k}$, $i_1 < i_2 < \cdots < i_k$, form an IP-set R' which is a subset of R. The H_n can be chosen inductively so that p'_n is as large as we please, and, in particular, they may be chosen with

$$(2.27) \quad p'_{n+1} > p'_1 + p'_2 + \cdots + p'_n$$

We shall show that when (2.27) is valid we, can find a system (Z, T) and a recurrent point $z_0 \in Z$ with $R' = \{n : d(T^n z_0, z_0) < 1\}$. Namely, let $Z = \{0, 1\}^{\mathbf{N} \cup \{0\}}$ with the metric

$$(2.28) \quad d(z, z') = \inf\left\{\frac{1}{k+1} : z(i) = z'(i) \quad \text{for} \quad i < k\right\},$$

where $d(z, z') = 1$ if and only if $z(0) \neq z'(0)$. Let $T : Z \to Z$ be the shift map. $T\omega(n) = \omega(n + 1)$, and let $z_0 \in Z$ be defined by

$$z_0(n) = \begin{cases} 1 & \text{if} \quad n = 0 \\ 1 & \text{if} \quad n \in R' \\ 0 & \text{if} \quad n > 0 \quad \text{and} \quad n \notin R'. \end{cases}$$

We claim that z_0 is a recurrent point of the Rebutov system (Z, T). Namely, if $m = p'_{j+1} + p'_{j+2} + \cdots + p'_k$ for any $k > j$, then $T^m z_0(n) = 1 \Leftrightarrow z_0(m + n) = 1 \Leftrightarrow m + n \in R'$. Now condition (2.27) implies that two expressions $p'_{i_1} + p'_{i_2} + \cdots + p'_{i_k}$ and $p'_{j_1} + p'_{j_2} + \cdots + p'_{j_l}$ in R' can be equal only if $i_1 = j_1, i_2 = j_2$, etc.; thus for $n < p'_{j+1}$, $m + n \in R'$ if $n \in R'$. It follows that

$$d(T^m z_0, z_0) \leq \frac{1}{p'_{j+1}}$$

so that z_0 is a recurrent point of (Z, T). Moreover,

$$d(T^n z_0, z_0) < 1 \Leftrightarrow T^n z_0(0) = 1 \Leftrightarrow z_0(n) = 1 \Leftrightarrow n \in R'.$$

This completes the proof of the theorem.

In view of the equivalence between IP-sets and sets of recurrence times, we may now reformulate Theorem 2.16 with the set R replaced by an IP-set. This result leads to a refinement of the theorems of van der Waerden and Grünwald's

THEOREM 2.18. *Let* $\mathbf{N}^m = B_1 \cup B_2 \cup \cdots \cup B_q$ *be a partition of the m-dimensional positive lattice and let* $R \subset \mathbf{N}$ *be an* IP-*set. One of the sets* B_j *has the property that if* F *is any finite subset of* $\mathbf{N}^m$, *then for some* $a \in \mathbf{N}^m$ *and* $b \in R$, $bF + a \subset B_j$.

The proof is identical to that of Theorem 2.7 except that Theorem 2.16 is invoked instead of Theorem 2.6

Theorem 2.18 in turn leads to a refinement of some of the diophantine results of Section 4. For example we have:

THEOREM 2.19. *Let* $p_1(x), p_2(x), \ldots, p_k(x)$ *be real polynomials. For* $\varepsilon > 0$ *form the set*

$$S = \{n \in \mathbf{N} : \|p_1(n) - p_1(0)\| < \varepsilon, \ldots, \|p_k(n) - p_k(0)\| < \varepsilon\}$$

where $\|t\|$ *denotes the distance to the nearest integer. Then* S *has the property that it intersects every* IP-*set in* $\mathbf{N}$.

We shall see later that if a set of integers intersects every IP-set in $\mathbf{N}$, it must itself contain an IP-set and in addition it must be syndetic. Thus Theorem 2.19 extends Theorem 1.21.

PART II
RECURRENCE IN
MEASURE PRESERVING
SYSTEMS

CHAPTER 3

Invariant Measures on
Compact Spaces

1. Measure Preserving Systems and Poincaré Recurrence

Throughout Part II we shall be concerned with *measure spaces* by which we mean a triple $(X,\mathscr{B},\mu)$ where X is a space, $\mathscr{B}$ a σ-algebra of sets in X, μ a nonnegative σ-additive measure on X with $\mu(X) < \infty$. It will sometimes be convenient to normalize the measure so that $\mu(X) = 1$, and we shall assume this to be the case from the start. The expression $L^p(X,\mathscr{B},\mu)$ denotes the usual Lebesgue space of measurable functions whose pth power is integrable, $1 \le p \le \infty$, and $L^\infty(X,\mathscr{B},\mu)$ denotes the space of essentially bounded functions. We shall not be careful to distinguish between the space of functions $L^p(X,\mathscr{B},\mu)$ and the Banach space obtained from the latter by identifying functions equal almost everywhere (*a.e.*). We shall sometimes write $E(f)$ for $\int f \, d\mu$ when $f \in L^1(X,\mathscr{B},\mu)$.

We say that T is a *measure preserving transformation* of $(X,\mathscr{B},\mu)$ if T is a transformation of X to itself for which $B \in \mathscr{B} \Rightarrow T^{-1}B \in \mathscr{B}$ and $\mu(T^{-1}B) = \mu(B)$. This implies that for all $f \in L^1(X,\mathscr{B},\mu)$

$$E(f \circ T) = E(f).$$

We shall speak of the quadruple $(X,\mathscr{B},\mu,T)$ as a *measure preserving system* (*m.p.s.*)

The analogue of a minimal dynamical system is an *ergodic* measure preserving system.

DEFINITION 3.1. *The system* $(X,\mathscr{B},\mu,T)$ *is an ergodic m.p.s. if* $T^{-1}B = B$ *for* $B \in \mathscr{B}$ *implies* $\mu(B) = 0$ *or* $\mu(B) = 1 \, (=\mu(X))$. *In this case we also speak of* T *as being an ergodic transformation.*

A fundamental result is the *ergodic theorem* (Halmos [1]).

THEOREM 3.1. *If* $f \in L^1(X,\mathscr{B},\mu)$ *and* T *is a measure preserving transformation of* $(X,\mathscr{B},\mu)$, *then for almost all* x *with respect to* μ,

$$(3.1) \qquad \frac{f(x) + f(Tx) + \cdots + f(T^n x)}{n + 1} \to \bar{f}(x)$$

where $\bar{f}(x)$ is a T-invariant function, $\bar{f} = \bar{f} \circ T$. If $f \in L^p(X,\mathscr{B},\mu)$, $1 \leq p < \infty$, then the convergence in (3.1) also takes place in the norm of $L^p(X,\mathscr{B},\mu)$.

If $(X,\mathscr{B},\mu,T)$ is ergodic, than a function satisfying $\bar{f} = \bar{f} \circ T$ must be almost everywhere a constant. Since the convergence in (3.1) is in L^1, we have $E(\bar{f}) = E(f)$. It follows that when T is ergodic, $\bar{f} = E(f)$, so that

$$\frac{f(x) + f(Tx) + \cdots + f(T^n x)}{n + 1} \to E(f) = \int f(y) \, d\mu(y)$$

for almost all x.

We remark that we shall in fact never need the full strength of Theorem 3.1. Rather it will suffice for us to know that if $f \in L^2(X,\mathscr{B},\mu)$, then the convergence in (3.1) takes place in the sense of the weak topology on the Hilbert space $L^2(X,\mathscr{B},\mu)$. In other words, whenever $f,g \in L^2(X,\mathscr{B},\mu)$

$$(3.2) \qquad \frac{1}{N + 1} \sum_{n=0}^{N} \int g(x) f(T^n x) \, d\mu(x) \to \int g(x) \bar{f}(x) \, d\mu(x)$$

as $N \to \infty$. The proof of this version of the ergodic theorem is quite elementary. (See, e.g., Chapter 4, §1.)

The elementary properties of recurrence in measure preserving systems follow from the Poincaré recurrence theorem. We repeat the argument given in the Introduction.

THEOREM 3.2. If $(X,\mathscr{B},\mu,T)$ is a measure preserving system and $A \in \mathscr{B}$ with $\mu(A) > 0$, then for some $n \geq 1$, $\mu(A \cap T^{-n}A) > 0$.

Proof: Assume $\mu(A \cap T^{-n}A) = 0$ for all $n \geq 1$; then for any $i \neq j$, $\mu(T^{-i}A \cap T^{-j}A) = 0$. But then $\{T^{-i}A\}$ would form an essentially disjoint family of sets all of whom have the same nonzero measure, $\mu(T^{-i}A) = \mu(A)$. This is clearly impossible since $\mu(X) < \infty$.

Anticipating the reasoning that will lead to more sophisticated recurrence theorems, let us show how to deduce Theorem 3.2 from the weak form of the ergodic theorem. Namely, let $f = 1_A$ be the characteristic function of the set $A: 1_A(x) = 1$ for $x \in A$, $1_A(x) = 0$ for $x \notin A$. Let $\bar{f}(x)$ be defined by (3.2) and apply the latter successively to $g = f, f \circ T, f \circ T^2, \ldots$. If $\mu(A \cap T^{-n}A) = 0$ for all $n \geq 1$, then $\int f(x) f(T^n x) \, d\mu(x) = 0$, and so $\int f(T^m x) f(T^n x) \, d\mu(x) = 0$ for $n \neq m$. This gives us by (3.2),

$$\int f(T^m x) \bar{f}(x) \, d\mu(x) = 0,$$

and one more application of (3.2) implies that

$$\int \bar{f}(x) \cdot \bar{f}(x) \, d\mu(x) = 0.$$

But then $\bar{f} = 0$ *a.e.* so that $\int \bar{f} \, d\mu = 0$. But we have seen that $\int f \, d\mu = \int \bar{f} \, d\mu$ and by hypothesis

$$\int f \, d\mu = \int 1_A \, d\mu = \mu(A) > 0.$$

This contradiction proves Theorem 3.2.

Now suppose that the underlying space X of a measure preserving system $(X, \mathscr{B}, \mu, T)$ is a metric space whose open sets are measurable. Theorem 3.2 implies the following result:

THEOREM 3.3. *Let $(X, \mathscr{B}, \mu, T)$ be a m.p.s. for which X is a separable metric space whose open sets are measurable. Then almost every point of X is recurrent for T.*

Proof: Since X is separable, we can find for each n a countable cover of X by open sets of diameter $< 1/n$,

$$X = \bigcup_{i=1}^{\infty} B_{ni}.$$

Let

$$X_n = \bigcup_{i=1}^{\infty} \left(B_{ni} - \bigcup_{j=1}^{\infty} T^{-j} B_{ni} \right)$$

A nonrecurrent point belongs to some X_n. We shall show that each X_n has measure 0 and so $\bigcup X_n$ also has measure 0. Let A be any measurable set; form

$$A^1 = A - \bigcup_{g=1}^{\infty} T^{-j} A.$$

The set A^1 has the property that $A^1 \cap T^{-n} A^1 = \varnothing$ for each $n \geq 1$. By Theorem 3.2 $\mu(A^1) = 0$. It follows that for each n and i,

$$\mu \left(B_{ni} - \bigcup_{j=1}^{\infty} T^{-j} B_{ni} \right) = 0$$

and so also $\mu(X_n) = 0$.

2. Invariant Measures on Compact Spaces

Let X be a compact metric space, and let $\mathscr{B}$ denote the σ-algebra of borel subsets of X. $\mathscr{M}(X)$ will denote the set of nonnegative borel measures in X that assign mass 1 to the whole space. We call these *probability measures* in X. It is known (Halmos [2]) that for compact metric spaces these measures are "regular," and so there is a one-to-one correspondence between these measures and bounded linear functionals on the Banach space $\mathscr{C}(X)$ of continuous real-valued functions satisfying $f \geq 0 \Rightarrow L(f) \geq 0$ and $L(1) = 1$, the correspondence being given by integration: $\mu \leftrightarrow L_\mu$ with $L_\mu(f) = \int f \, d\mu$. Thus we can imbed $\mathscr{M}(X)$ in the dual space $\mathscr{C}(X)^*$, and in this way $\mathscr{M}(X)$ is endowed with a topology, specifically the weak*-topology: $\mu_n \to \mu \Leftrightarrow \int f \, d\mu_n \to \int f \, d\mu$. This is again a metric topology, and $\mathscr{M}(X)$ is a compact subset of $\mathscr{C}(X)^*$.

Now let T be a continuous map of X to itself. We can define the image $T\mu$ of a measure $\mu \in \mathscr{M}(X)$ under T by either

(3.3) $T\mu(A) = \mu(T^{-1}A), \qquad A \in \mathscr{B}$

or

(3.4) $\int f(x) \, dT\mu(x) = \int f(Tx) \, d\mu(x), \qquad f \in \mathscr{C}(X).$

Let $\mathscr{S}(T)$ denote the set of probability measures satisfying $T\mu = \mu$. Note that $\mathscr{S}(T)$ is never empty since invariant measures can always be constructed. For example, begin with any probability measure v and let $\mu_n = n^{-1}(v + Tv + \cdots + T^{n-1}v)$. Clearly any limit point in $\mathscr{M}(X)$ of $\{\mu_n\}$ is invariant. Also $\mathscr{S}(T)$ is a compact convex set of measures. By the Krein-Milman theorem it is spanned by extremal points. In particular, extremal invariant measures exist. The significance of this appears in the next proposition.

PROPOSITION 3.4.   *The measure μ is an extremal point of $\mathscr{S}(T)$ if and only if $(X,\mathscr{B},\mu,T)$ is an ergodic system. Consequently, the extremal measures of $\mathscr{S}(T)$ will be called ergodic measures.*

Proof: Suppose $(X,\mathscr{B},\mu,T)$ is not ergodic and that $A \in \mathscr{B}$ satisfies: $T^{-1}A = A, 0 < \mu(A) < 1$. Form μ' and μ'' in $\mathscr{S}(T)$ by

$$\mu'(B) = \mu(B \cap A)/\mu(A), \qquad \mu''(B) = \mu(B-A)/\mu(X-A).$$

Then $\mu = \mu(A)\mu' + \mu(X-A)\mu''$ is a nontrivial decomposition, and $\mu' \neq \mu''$ implies that μ is not extremal.

Now suppose that μ is not extremal: $\mu = \alpha\mu' + (1 - \alpha)\mu''$. $\mu \geq \alpha\mu'$ implies that μ' is absolutely continuous with respect to μ. We can write, by the Radon-Nikodym theorem,

$$\mu'(A) = \int_A f \, d\mu$$

for a unique function $f \in L^1(X,\mathscr{B},\mu)$. Now

$$\mu'(A) = \mu'(T^{-1}A) = \int_{T^{-1}A} f \, d\mu = \int_A f \circ T \, d\mu.$$

By uniqueness, $f = f \circ T$. If $(X,\mathscr{B},\mu,T)$ were ergodic, then this would imply that f is constant *a.e.*; so $\mu' = c\mu$, whence $\mu' = \mu$. Hence μ not extremal implies $(X,\mathscr{B},\mu,T)$ not ergodic.

One consequence of Proposition 3.4 is that if $\mathscr{S}(T)$ should happen to consist of just one measure μ, then T is ergodic with respect to μ.

DEFINITION 3.2. *The dynamical system* (X,T) *(or the transformation* T*) is uniquely ergodic if* T *has a unique invariant probability measure* μ.

The ergodic theorem applies here as it does to every ergodic system. However, in the uniquely ergodic case we can both strengthen the assertion of the theorem and give an elementary proof of the ergodic theorem as it applies to continuous functions.

THEOREM 3.5. *Let* (X,T) *be a uniquely ergodic system,* μ *the* T-invariant *probability measure and* $f \in \mathscr{C}(X)$. *Then*

$$(3.5) \qquad \frac{f(x) + f(Tx) + \cdots + f(T^n x)}{n + 1} \to \int f \, d\mu$$

uniformly in X.

Proof: It is not hard to show that if θ is a signed borel measure on X with $T\theta = \theta$, and $\theta = \theta^+ - \theta^-$ is its Hahn-Jordan decomposition into the difference of nonnegative measures, then $T\theta^+ = \theta^+$ and $T\theta^- = \theta^-$. In the case of unique ergodicity θ^+ and θ^- are proportional to the unique invariant probability measure, and it follows that the set of all invariant signed measures is one-dimensional. This means that in $\mathscr{C}(X)^*$, the subspace of functionals vanishing on all functions of the form $f - f \circ T$ is one-dimensional. Letting V be the closed subspace of $\mathscr{C}(X)$ spanned by $\{f - f \circ T\}$, we find this is tantamount to saying that $\mathscr{C}(X)/V$ has a one-dimensional dual. This implies that V is of codimension 1 in $\mathscr{C}(X)$. Since $\int g \, d\mu = 0$ for all $g \in V$ and $\int 1 \, d\mu = 1$, the function $1 \notin V$. We conclude that any continuous function is a uniform limit of functions of the form $f - f \circ T + c$. Inasmuch as the conclusion of the theorem is obvious for these functions, it follows that the conclusion is valid for all continuous functions.

DEFINITION 3.3. *The support of a nonnegative borel measure on a separable metric space is the complement of the union of all open sets having measure* 0.

Since the space has a countable base of open sets, the union of all open sets having measure 0 is a countable union of such sets and so itself has measure 0. So the measure is indeed supported by its support. The support of an invariant measure is an invariant set. It follows that if (X,T) is a minimal dynamical system, every invariant measure has all of X for its support.

PROPOSITION 3.6. *If* T *is uniquely ergodic on* X *and* $X' \subset X$ *is the support of the invariant measure, then* (X',T) *is minimal.*

Proof: Suppose $X'' \subset X'$ is closed and invariant, and let T'' denote the restriction of T to X''. Since $\mathscr{S}(T'')$ is nonempty, there exists an invariant measure whose support $\subset X''$. This would have to be some other measure than the one having support in X', and so T could not be uniquely ergodic.

DEFINITION 3.4.   *Let T be a continuous transformation of the space X, $x_0 \in X$, and $\mu \in \mathscr{S}(T)$. We say that x_0 is a generic point for μ (or for the m.p.s. $(X,\mathscr{B},\mu,T)$) if*

$$(3.6) \qquad \frac{f(x_0) + f(Tx_0) + \cdots + f(T^n x_0)}{n + 1} \to \int f \, d\mu$$

for every continuous function $f \in \mathscr{C}(X)$.

If $(X,\mathscr{B},\mu,T)$ is ergodic, then (3.6) is valid almost everywhere for any $f \in L^1(X,\mathscr{B},\mu)$; in particular, for $f \in \mathscr{C}(X)$. It will be valid simultaneously for a countable set of functions f almost everywhere, and choosing this set uniformly dense in $\mathscr{C}(X)$, as we may when X is compact metric, we obtain (3.6) for all continuous functions. This gives the following.

PROPOSITION 3.7.   *If T is a continuous map of the compact metric space X to itself and μ is an ergodic measure for (X,T), then almost every point of X (with respect to μ) is generic for μ.*

If (X,T) is uniquely ergodic, it follows from Theorem 3.5 that every point in X is generic for the unique invariant measure. Conversely, suppose that every point in X is generic for some invariant measure μ. We claim that $\mathscr{S}(T)$ must consist of a single measure. Otherwise it would have at least two extreme points, i.e., (X,T) would have distinct ergodic measures. But it is obvious from the definition of generic points that distinct measures have distinct generic points, and since ergodic measures must have generic points, it could not be the case that every point is generic for the same measure. We thus have

PROPOSITION 3.8.   *The system (X,T) is uniquely ergodic if every point of X is generic for the same measure.*

According to Proposition 3.6, in the case of unique ergodicity, the support of the invariant measure is a minimal set for the system. By Theorem 1.15 every point in the support of this measure is uniformly recurrent (Definition 1.7). It will be instructive to see how this result follows from other considerations of this chapter. Suppose (X,T) is uniquely ergodic, μ is the invariant measure, and X' is the support of μ. Let $x_0 \in X'$, and let V be a neighborhood of x_0. Let $f(x)$ be a continuous function vanishing outside of V, $f \geq 0$, with $f(x) > 0$ somewhere in V. Then $\int f \, d\mu > 0$ and by Theorem 3.5 there exists an n_0 so that for all x

and $n \geq n_0$,

$$f(x) + f(Tx) + \cdots + f(T^n x) > 0.$$

Hence for all $x \in X$, some $T^m x \in V$ for $m \leq n_0$. In particular $\{n : T^n x_0 \in V\}$ is syndetic (Definition 1.7), and so x_0 is uniformly recurrent.

A more general notion than that of a generic point is given in the following definition.

DEFINITION 3.5. *Let X be a compact metric space, T a continuous transformation of X, and μ an invariant measure. We say $x_0 \in X$ is quasi-generic for μ if for some sequence of pairs of integers $\{a_k, b_k\}$ with $a_k \leq b_k$ and $b_k - a_k \to \infty$ we have*

$$(3.7) \qquad \frac{1}{b_k - a_k + 1} \sum_{n=a_k}^{b_k} f(T^n x_0) \to \int f \, d\mu$$

as $k \to \infty$, for every continuous function $f \in \mathscr{C}(X)$.

Every point of a compact metric space X is quasi-generic for some invariant measure. More precisely we have

PROPOSITION 3.9. *Let T be a continuous map of the compact metric space X to itself, let $x_0 \in X$ and let X' be the forward orbit closure of x_0, $X' = \overline{\{T^n x_0 : n \geq 0\}}$. If $\mu \in \mathscr{M}(X')$ is an ergodic measure with respect to T, then x_0 is quasi-generic for μ.*

Proof: Since μ is an ergodic measure, by Proposition 3.7, some point, say $x_1 \in X'$, is generic for μ. Then for each $f \in \mathscr{C}(X)$,

$$\frac{1}{N+1} \sum_0^N f(T^n x_1) \to \int f \, d\mu.$$

Let $\{f_k\}$ be a dense set of functions in $\mathscr{C}(X)$, and let N_k be an increasing sequence such that

$$(3.8) \qquad \left| \frac{1}{N_k + 1} \sum_{n=0}^{N_k} f_j(T^n x_1) - \int f_j \, d\mu \right| < \frac{1}{k}$$

for $j = 1, 2, \ldots, k$. Equation (3.8) will remain valid if x_1 is replaced by a sufficiently nearby point; in particular it will be true if x_1 is replaced by some $T^{a_k} x_0$. This will give

$$(3.9) \qquad \left| \frac{1}{N_k + 1} \sum_{n=a_k}^{a_k + N_k} f_j(T^n x_0) - \int f_j \, d\mu \right| < \frac{1}{k}, \qquad j = 1, 2, \ldots, n.$$

Now set $b_k = a_k + N_k$ and we obtain (3.7) for each of the functions in $\{f_k\}$. But since these are dense, (3.7) must be valid for all $f \in \mathscr{C}(X)$.

REMARK.   (1) We have been discussing invariant measures for individual transformations T on a space X. We could develop similar notions for a finite set of commuting transformation $T_1, T_2, \ldots, T_l$ of X. We shall in fact need such a theory, but prefer to develop it *ad hoc* where needed rather than presenting it systematically.

(2) Proposition 3.7 depends on the pointwise ergodic theorem and so, on the face of it, does Proposition 3.9 which depends upon Proposition 3.7. Actually all we needed to know was that there exist quasi-generic points for ergodic measures, and this follows from the mean ergodic theorem. This point is relevant for extending Proposition 3.9 to groups (e.g. amenable groups) for which one has a mean ergodic theorem but (so far) no pointwise ergodic theorem.

3. Group Extensions, Unique Ergodicity, and Equidistribution

Let (Y,T) be a dynamical system and v a T-invariant measure on Y; let G be a compact group, and $\psi : Y \to G$ a continuous map. We have defined the group extension $(Y \times G, T)$ of the dynamical system (Y,T) by $T(y,g) = (Ty, \psi(y)g)$. Now let m_G denote the Haar measure of G, and let $\mu = v \times m_G$ be the product measure on $Y \times G$. This will be a T-invariant measure on $Y \times G$, for

$$\int f \circ T \, d\mu = \iint f(Ty, \psi(y)g) \, dv(y) \, dg = \int \left[\int f(Ty, g') \, dg' \right] dv(y)$$

$$= \int \left[\int f(y, g') \, dg' \right] dv(y) = \int f \, d\mu$$

using the left-invariance of Haar measure. We now define the group extension of the measure preserving system $(Y, \mathscr{B}_Y, v, T)$ to be the system $(Y \times G, \mathscr{B}_{Y \times G}, v \times m_G, T)$.

PROPOSITION 3.10.   *Assume (Y,T) is uniquely ergodic and that*

$$(Y \times G, \mathscr{B}_{Y \times G}, v \times m_G, T)$$

is ergodic. Then $(Y \times G, T)$ is uniquely ergodic.

Proof: Define, as in Chapter 1, the automorphisms of $(Y \times G, T)$ given by $R_h(y,g) = (y, gh)$, $h \in G$. Since R_h is an automorphism of $(Y \times G, T)$, if $x \in Y \times G$ is generic for a measure μ, then $R_h x$ is generic for the measure $R_h \mu$. Now the measure $v \times m_G$ is invariant under each R_h. It follows that if (y, g_1) is generic for $v \times m_G$, then (y,g) is generic for $v \times m_G$ for every $g \in G$. We are given that $v \times m_G$ is an ergodic measure. So by Proposition 3.7 almost every point is generic for it. This means that for almost all y (with respect to v) some (y,g)—and therefore, all (y,g)—are generic for $v \times m_G$. Suppose that $(Y \times G, T)$ were not uniquely ergodic and let μ' be some other ergodic measure. Almost all its points are generic for it, and so they must lie in a set $A \times G$ where $A \subset Y$ is a set of v of measure 0. But the projection of μ' onto Y is an invariant measure on Y and by unique ergodicity of (Y,T) this coincides with v. This leads to a contradiction, and so we must conclude that $(Y \times G, T)$ is uniquely ergodic.

We turn now to an example in which these notions will be illustrated. Once again we let $\mathbf{T} = \mathbf{R}/\mathbf{Z}$, the group of reals modulo 1, and we study certain affine transformations of the torus $\mathbf{T}^d$.

PROPOSITION 3.11. *Let $T:\mathbf{T}^d \to \mathbf{T}^d$ be defined by*

$$T(\theta_1, \theta_2, \ldots, \theta_d) = (\theta_1 + \alpha, \theta_2 + b_{21}\theta_1, \ldots,$$

(3.10)
$$\theta_d + b_{d1}\theta_1 + b_{d2}\theta_2 + \cdots + b_{d,d-1}\theta_{d-1})$$

where α is irrational and the coefficients b_{ij} are integers with $b_{21}, b_{32}, \ldots,$ $b_{d,d-1} \neq 0$. Then T is ergodic with respect to Haar measure on $\mathbf{T}^d$.

Proof: We use harmonic analysis. Suppose $f = f \circ T$ for $f \in L^2(\mathbf{T}^d)$. Expand f in a Fourier series

$$f(\theta) \sim \sum_{n \in \mathbf{Z}^d} c_n e^{2\pi i n \cdot \theta}.$$

Then

$$f(T\theta) \sim \sum_{n \in \mathbf{Z}^d} c_n e^{2\pi i n \cdot T\theta}.$$

If we write $n = (n_1, n_2, \ldots, n_d)$ and $\theta = (\theta_1, \theta_2, \ldots, \theta_d)$, then

$$\begin{aligned}
n \cdot T\theta &= (n_1, n_2, \ldots, n_d) \cdot (\theta_1 + \alpha, \theta_2 + b_2 + b_{21}\theta_1, \ldots, \\
&\qquad \theta_d + b_{d1}\theta_1 + \cdots + b_{d,d-1}\theta_{d-1}) \\
&= n_1\alpha + (n_1 + b_{21}n_2 + \cdots + b_{d1}n_d)\theta_1 + \cdots + n_d\theta_d \\
&= n_1\alpha + (nT') \cdot \theta
\end{aligned}$$

where T' is the matrix

$$\begin{pmatrix}
1 & 0 & \cdot & \cdot & 0 \\
b_{21} & 1 & \cdot & \cdot & 0 \\
b_{31} & b_{32} & \cdot & \cdot & \cdot \\
\cdot & \cdot & \cdot & \cdot & \cdot \\
b_{d1} & b_{d2} & \cdot & \cdot & 1
\end{pmatrix}.$$

Consequently, if $f = f \circ T$, we shall have, comparing Fourier coefficients,

$$(3.11) \quad c_{nT''} = c_n e^{2\pi i n_1 \alpha},$$

whence $|c_{nT'}| = |c_n|$. Since $\sum |c_n|^2 < \infty$, $c_n \neq 0$ only if $nT'' = n$ for some integer $r \geq 1$, and this implies

$$n = (n_1, 0, 0, \ldots, 0).$$

But (3.11) gives, in that case,

$$c_n = c_n e^{2\pi i n_1 \alpha},$$

and since α is irrational, we shall have $c_n = 0$ unless $n = 0$. This proves the ergodicity of T.

Notice that the system defined by (3.10) for dimension d is a group extension of the analogous system for dimension $d - 1$. We can therefore invoke Propositions 3.10 and 3.11 to conclude that if (3.10) defines a uniquely ergodic system for $d - 1$, it does so for d. Since the system consisting of a one-point space is uniquely ergodic, we can deduce by combining Propositions 3.10 and 3.11 that for $d = 1$, (3.10) defines a uniquely ergodic system. We may then proceed by induction to conclude that (3.10) defines a uniquely ergodic system for each d.

THEOREM 3.12. *Let α be irrational, b_{ij} integers for $1 \leq j < i \leq d$ with $b_{i,i-1} \neq 0$. Then the transformation*

$$T(\theta_1, \theta_2, \ldots, \theta_d) = (\theta_1 + \alpha, \theta_2 + b_{21}\theta_1, \ldots,$$
$$\theta_d + b_{d1}\theta_1 + \cdots + b_{d,d-1}\theta_{d-1})$$

defines a uniquely ergodic dynamical system on the d-dimensional torus $\mathbf{T}^d$.

In this case every point of the torus is generic for Haar measure. We take a still more specific example:

$$T(\theta_1, \theta_2, \ldots, \theta_d) = (\theta_1 + \alpha, \theta_2 + \theta_1, \ldots, \theta_d + \theta_{d-1}).$$

Let $p(x)$ be a polynomial of degree d with real coefficients whose leading coefficient is $\alpha/d!$ where α is irrational. Define as in Chapter 1, Section 2, $p_d(x) = p(x)$,

$$p_{d-1}(x) = p_d(x + 1) - p_d(x), \ldots,$$
$$p_1(x) = p_2(x + 1) - p_2(x)$$
$$= \alpha x + \beta.$$

Then the orbit under T of the point $(p_1(0), p_2(0), \ldots, p_d(0))$ is the sequence $\{(p_1(n), p_2(n), \ldots, p_d(n))\}$. So by Theorem 3.5, if f is a continuous function on $\mathbf{T}^d$:

$$\frac{1}{N+1} \sum_0^N f(p_1(n), p_2(n), \ldots, p_d(n))$$

$$\rightarrow \int \cdots \int f(\theta_1, \theta_2, \ldots, \theta_d) \, d\theta_1 d\theta_2, \ldots, d\theta_d.$$

In particular, if $f(\theta_1, \theta_2, \ldots, \theta_d) = g(\theta_d)$, this becomes

$$(3.12) \qquad \frac{1}{N+1} \sum_0^N g(p(n)) \rightarrow \int_0^1 g(\theta) \, d\theta.$$

This phenomenon is referred to as the equidistribution of the sequence $\{p(n)\}$ modulo one (Weyl [1]). We have thus shown that if $p(x)$ is a real polynomial with leading coefficient irrational, the sequence $\{p(n)\}$ is equidistributed. With the fact that $\{p(n)\}$ is equidistributed if for some $r = 1, 2, \ldots$, each of the sequences $\{p(rn + s)\}$ $s = 1, 2, \ldots, r - 1$, is equidistributed, it is easy to reduce the case of an arbitrary real polynomial with some coefficient other than the constant term irrational to the foregoing case. This gives us Weyl's equidistribution theorem:

THEOREM 3.13. (Weyl [1]). *If $p(x)$ is a real polynomial with at least one coefficient other than the constant term irrational, then the sequence $\{p(n)\}$ is equidistributed modulo one.*

If we use the full force of Theorem 3.5, then we can say a bit more. In the uniquely ergodic case the averages

$$\frac{f(x) + f(Tx) + \cdots + f(T^n x)}{n + 1}$$

converge uniformly to the integral of f. In particular, for any point x_0, the averages

$$\frac{1}{N+1} \sum_M^{M+N} f(T^n x_0)$$

converge to $\int f \, d\mu$ uniformly in M. This leads to a sharpened form of equidistribution referred to sometimes as *well-distribution*:

$$\frac{1}{N+1} \sum_M^{M+N} g(p(n)) \rightarrow \int_0^1 g(\theta) \, d\theta$$

as $N \rightarrow \infty$, *uniformly* in M.

4. Applications to Unitary Operators

In this section we shall give an application of Weyl's theorem on equidistribution (Theorem 3.13) to the theory of unitary operators on a Hilbert space. These results have implications for measure preserving systems and ultimately also for number theory.

Suppose $p(t)$ is a polynomial with rational coefficients that takes on integer values at integer values of the argument. Let U be a unitary operator on a Hilbert space $\mathscr{H}$. For $\theta \in [0,1)$ let

$$\mathscr{H}_\theta = \{x \in \mathscr{H} : Ux = e^{2\pi i\theta}x\}$$

and let

$$\mathscr{H}_{rat} = \bigoplus \mathscr{H}_\theta, \qquad \theta \text{ rational.}$$

Finally let P_{rat} denote the orthogonal projection from $\mathscr{H}$ to $\mathscr{H}_{rat}$.

LEMMA 3.14.   If $P_{rat}x = 0$, then

$$\frac{1}{N+1} \sum_{n=0}^{N} U^{p(n)}x \to 0$$

as $N \to \infty$.

Proof: Let $dE(\theta)$ be the spectral measure for U, so that

$$U^n = \int_0^1 e^{2\pi in\theta} \, dE(\theta).$$

Let $\langle , \rangle$ denote the inner product in $\mathscr{H}$ and define the measure ω_x on $[0,1]$ by

$$d\omega_x(\theta) = \langle dE(\theta)x,x \rangle.$$

If $P_{rat}x = 0$, then the spectral measure E—and hence the numerical measure ω_x—assigns measure 0 to the rationals. We now evaluate the norm of the expression in the lemma:

$$\left\| \frac{1}{N+1} \sum_{n=0}^{N} U^{p(n)}x \right\|^2 = \frac{1}{(N+1)^2} \sum_{0 \le n_1,n_2 \le N} \langle U^{p(n_1)-p(n_2)}x,x \rangle$$

$$= \frac{1}{(N+1)^2} \sum_{0 \le n_1,n_2 \le N} \int_0^1 e^{2\pi i(p(n_1)-p(n_2))\theta} \, d\omega_x(\theta)$$

$$(3.13) \qquad = \int_0^1 \left| \frac{1}{N+1} \sum_{n=0}^{N} e^{2\pi ip(n)\theta} \right|^2 \, d\omega_x(\theta).$$

According to Theorem 3.13,

$$\frac{1}{N+1} \sum_{n=0}^{N} e^{2\pi ip(n)\theta} \to 0$$

for θ irrational, so the integral in (3.13) becomes restricted as $N \to \infty$ to the rational values of θ. Since the ω_x-measure of the rationals vanishes we obtain the conclusion of the lemma.

This lemma leads to the following result.

PROPOSITION 3.15. *Let $p(t)$ be a polynomial with rational coefficients taking integer values on the integers and with $p(0) = 0$. Let U be a unitary operator in a Hilbert space and let x be a vector with $\langle U^{p(n)}x, x \rangle = 0$ for all $n \geq 1$. Then x is orthogonal to all eigenvectors of U for eigenvalues that are roots of unity.*

Proof: Write $x = x_1 + x_2$ with $x_1 \in \mathscr{H}_{rat}$ and $x_2 \perp \mathscr{H}_{rat}$. Since $p(t)$ has rational coefficients and no constant term, there exists an integer a such that $p(at)$ has integer coefficients. For any integer m we shall then have $m \mid p(amn)$ for all integer values of n. Let $\varepsilon > 0$ and find m such that for some x_1' satisfying $U^m x_1' = x_1'$ we have $\|x_1 - x_1'\| < \varepsilon$. We shall then have for every n, $\|U^{mn}x_1 - x_1\| < 2\varepsilon$. It follows that

$$\left\| \frac{1}{N+1} \sum_{n=0}^{N} U^{p(amn)}x_1 - x_1 \right\| < 2\varepsilon.$$

By the foregoing lemma

$$\frac{1}{N+1} \sum_{n=0}^{N} U^{p(amn)}x_2 \to 0,$$

and so for large N,

$$\left\| \frac{1}{N+1} \sum_{n=0}^{N} U^{p(amn)}x - x_1 \right\| < 2\varepsilon.$$

Inasmuch as each $U^{p(amn)}x$ is orthogonal to x we obtain:

$$|\langle x_1, x \rangle| < 2\varepsilon \|x\|$$

and since ε is arbitrary, $\langle x_1, x \rangle = 0$, whence $x_1 = 0$. This proves the proposition.

We can apply this proposition to measure preserving systems.

THEOREM 3.16. *Let $(X, \mathscr{B}, \mu, T)$ be a measure preserving system, $A \in \mathscr{B}$ with $\mu(A) > 0$. Let $p(t)$ be a polynomial taking integer values on the integers and with $p(0) = 0$. Then for some $n \geq 1$, $\mu(T^{-p(n)}A \cap A) > 0$.*

Proof: We take $\mathscr{H} = L^2(X, \mathscr{B}, \mu)$ and let U be the unitary operator $f \to f \circ T$. If $\mu(T^{-p(n)}A \cap A) = 0$, then the function

$$1_A(x) = \begin{cases} 1 & \text{if } x \in A \\ 0 & \text{if } x \notin A \end{cases}$$

satisfies $\langle U^{p(n)}1_A, 1_A \rangle = 0$ for $n \geq 1$. But then by the proposition, 1_A is orthogonal to every eigenvector for an eigenvalue that is a root of unity. But $U1 = 1$ and $\langle 1_A, 1 \rangle = \mu(A) \neq 0$. This proves the theorem.

Thus, for example, for any measure preserving system $(X,\mathscr{B},\mu,T)$ and any $A \in \mathscr{B}$ with $\mu(A) > 0$, there is some perfect square n^2 with

$$T^{-n^2}A \cap A \neq \varnothing.$$

We shall later give number-theoretic applications of these results.

The foregoing theorem together with the *Poincaré* recurrence theorem suggest the following definition.

DEFINITION 3.6. *A set $W \subset \mathbf{Z}$ will be called a Poincaré sequence if for any m.p.s. $(X,\mathscr{B},\mu,T)$ and $A \in \mathscr{B}$ with $\mu(A) > 0$, we have $\mu(T^{-n}A \cap A) > 0$ for some $n \in W$, $n \neq 0$.*

It follows from Theorem 3.16 that if $p(t)$ is a polynomial satisfying $0 \in p(\mathbf{Z}) \subset \mathbf{Z}$, then $p(\mathbf{Z})$ is a *Poincaré sequence*.

If W is a Poincaré sequence, then $W \cap m\mathbf{Z}$ is also a Poincaré sequence for any $m \in \mathbf{Z}$, $m \neq 0$. To see this, we merely observe that $(X,\mathscr{B},\mu,T^m)$ is a m.p.s. whenever $(X,\mathscr{B}\mu,T)$ is a m.p.s. In particular $W \cap m\mathbf{Z}$ must contain nonzero elements if W is a Poincaré sequence. It follows that $\{n^2 + 1\}$ is not a Poincaré sequence, and, in fact, if $p(t)$ is a polynomial with $p(\mathbf{Z}) \subset \mathbf{Z}$, the set $p(\mathbf{Z})$ will be a Poincaré sequence only if $0 \in p(\mathbf{Z})$.

5. Invariant Measures for Symbolic Flows

Let Λ be a finite alphabet, $\Omega = \Lambda^{\mathbf{Z}}$, the space of all Λ-sequences, and $T:\Omega \to \Omega$ the shift map $T\omega(n) = \omega(n + 1)$. We shall be considering the relation between invariant measures on Ω and the frequency with which symbols occur in certain sequences in Ω. To make this precise we shall define several notions of density of sets of integers. In the definition we shall denote by $|S|$ the cardinality of a finite set S.

DEFINITION 3.7. *Let S be a subset of either $\mathbf{N}$ or $\mathbf{Z}$. The upper Banach density of S, is*

$$BD^*(S) = \limsup_{|I| \to \infty} \frac{|S \cap I|}{|I|}$$

where I ranges over intervals of $\mathbf{N}$ or $\mathbf{Z}$. That is, for some sequence of intervals $\{I_k\}$, $I_k = (a_k,b_k)$, $b_k - a_k \to \infty$

$$(3.14) \qquad \frac{|S \cap I_k|}{|I_k|} \to BD^*(S),$$

and for any other sequence with $|I_k| \to \infty$,

$$\limsup \frac{|S \cap I_k|}{|I_k|} \leq BD^*(S).$$

The upper density of S, $D^*(S)$ is defined by

$$D^*(S) = \lim_{N \to \infty} \sup \frac{|S \cap [1,N]|}{N}.$$

The lower density of S, $D_*(S)$ is defined similarly, and the set has density $D(S)$ if $D^*(S) = D_*(S) = D(S)$.

REMARK. We could also define *lower Banach density* and consider equality of upper and lower Banach density. This, however, is a much rarer phenomenon than the equality of upper and lower density. For convenience we have defined density in terms of the interval $[1,N]$ rather than $[-N,N]$, in order to deal simultaneously with subsets of **N** and **Z**.

Let $\xi \in \Omega$ and $a \in \Lambda$. We say *a occurs in ξ with positive upper Banach density* if $BD^* \{n : \xi(n) = a\} > 0$.

LEMMA 3.17. *Let $\xi \in \Omega$ and let $X = \{T^n\xi : n \in \mathbf{Z}\}$ be its orbit closure with respect to the shift. Let $a \in \Lambda$ and let $A(a) = \{\omega : \omega(0) = a\}$. The symbol a occurs in ξ with positive upper Banach density if and only if there exists an invariant measure μ on X with $\mu(A(a)) > 0$.*

Proof: Suppose there exists an invariant measure on X that assigns positive measure to $A(a)$. By the Krein-Milman theorem applied to the convex compact set of T-invariant measures on X, there exists an extremal measure μ (or, by Proposition 3.4, an ergodic measure) with $\mu(A(a)) > 0$. By Proposition 3.9 (adapted to **Z**-actions instead of **N**-actions), ξ is quasi-generic for μ. This means that for some sequence of intervals $\{I_k\}$, $|I_k| \to \infty$, we have

$$(3.15) \qquad \frac{1}{|I_k|} \sum_{n \in I_k} f(T^n\xi) \to \int f \, d\mu$$

for every continuous function f on X. If we take $f = 1_{A(a)}$, the characteristic function of $A(a)$, then f is continuous on X and (3.15) implies that the upper Banach density of $\{n : \xi(n) = a\}$ is at least $\mu(A(a)) > 0$.

Conversely, suppose the symbol a occurs in ξ with positive upper Banach density, and let $\{I_k\}$ be the sequence of intervals achieving this density in accordance with (3.14). Passing to a subsequence if necessary, we can assume that for any continuous function $f \in \mathscr{C}(X)$,

$$\lim_{k \to \infty} \frac{1}{|I_k|} \sum_{n \in I_k} f(T^n\xi) = L(f)$$

exists. (We can always find a subsequence with the limit in question existing for a given function. Then by a diagonal procedure, we can arrange that $L(f)$ exists for any countable set of functions, and, finally, by the separability of $\mathscr{C}(X)$, we can choose the set uniformly dense in $\mathscr{C}(X)$. The limit

$L(f)$ will then exist for all of $\mathscr{C}(X)$.) The functional $L(f)$ corresponds to a probability measure μ, and since clearly $L(f) = L(f \circ T)$, we have $\mu = T\mu$. But

$$\mu(A(a)) = \int 1_{A(a)} \, d\mu = L(1_{A(a)}) = \lim_{k \to \infty} \frac{1}{|I_k|} \sum 1_{A(a)}(T^n \xi) > 0.$$

This proves the lemma.

We apply this lemma to $\{0,1\}$-sequences, or, equivalently, to subsets of $\mathbf{Z}$. If $S \subset \mathbf{Z}$, we shall denote by $S - S$ the set of differences

$$\{s' - s'' : s', s'' \in S\}.$$

THEOREM 3.18. *If W is a Poincaré sequence and $S \subset \mathbf{Z}$ is a subset of positive upper Banach density, then $S - S \cap W$ contains nonzero integers.*

Proof: Take $\Lambda = \{0,1\}$, $\xi = 1_S \in \Omega = \{0,1\}^{\mathbf{Z}}$. Let X be the orbit closure of ξ and μ a shift-invariant measure on X assigning positive measure to the set $A(1) \subset \Omega$. Such a measure μ exists by Lemma 3.17. Since the support of μ is contained in X, we must have $\mu(X \cap A(1)) > 0$. Suppose $\omega \in X$. Each $\omega(n)$ is either 0 or 1, so $\omega = 1_R$ for some set $R \subset \mathbf{Z}$. We claim that $1_R \in X$ implies $R - R \subset S - S$. For if $m \in R - R$, then $\omega(h) = \omega(h + m) = 1$ for some h, and since ω is a limit of translates of ξ, $\xi(h') = \xi(h' + m)$ for some h', thus $h', h' + m \in S$ and $m \in S - S$. Now since W is a Poincaré sequence and $A(1) \cap X$ has positive measure, there exists $m \in W$ with ω and $T^m \omega$ both belonging to $A(1) \cap X$ for some ω. But that means $\omega(0) = \omega(m) = 1$ for $\omega \in X$, so if $\omega = 1_R$, we have $m \in R - R$.

We mention three examples of Poincaré sequences.

(1)   One is the range $p(\mathbf{Z})$ of a polynomial function for which $p(0) = 0$;
(2)   Another is an *infinite difference set*, i.e., the set $\{n_j - n_i, i < j\}$ for an arbitrary sequence $\{n_k\}$.
      To see this we simply observe that the proof of Theorem 3.2, the Poincaré recurrence theorem, carries over verbatim to this case, replacing the sets $T^{-i}A$ by $T^{-n_i}A$.
(3)   The third is a *thick set* (Definition 1.10), i.e., a set of integers containing arbitrarily long intervals.

In fact every thick set contains an infinite difference set, and so must be a Poincaré sequence. Alternatively we can argue as follows. Let $(X, \mathscr{B}, \mu, T)$ be a *m.p.s.* and $\mu(A) > 0$. There exists an N such that

$$\mu\left(\bigcup_0^N T^{-n}A\right) > \mu\left(\bigcup_0^\infty T^{-n}A\right) - \mu(A).$$

Then for any M,

$$\mu\left(\bigcup_M^{M+N} T^{-n}A\right) > \mu\left(\bigcup_0^\infty T^{-n}A\right) - \mu(A).$$

This implies that $\mu\left(\bigcup_M^{M+N} T^{-n}A \cap A\right) > 0$, for otherwise

$$\mu\left(\bigcup_0^\infty T^{-n}A\right) \geq \mu(A) + \mu\left(\bigcup_M^{M+N} T^{-n}A\right) > \mu\left(\bigcup_0^\infty T^{-n}A\right),$$

a contradiction. Thus each sufficiently long interval of integers includes an n with $\mu(T^{-n}A \cap A) > 0$. This proves that a thick set is a Poincare sequence.

As a corollary to Theorem 3.18 we have the following number-theoretic result.

PROPOSITION 3.19. *Let S be subset of $\mathbf{Z}$ of positive upper Banach density. Then*

(a) $S - S$ *is a syndetic set,*

(b) *if $p(t)$ is a polynomial taking on integer values at the integers and including 0 in its range on the integers, then there exists a solution to the equation $x - y = p(z), x, y \in S, x \neq y, z \in \mathbf{Z}$.*

Part (a) follows since a set that intersects every thick set of integers must be syndetic; otherwise its complement is thick. Part (a) can also be proved directly without great effort. The assertion (b) follows from the fact that under the hypothesis $p(\mathbf{Z})$ is a Poincare sequence. Part (b) was also proved independently by Sárközy.

REMARK. What the proof of Theorem 3.18 actually shows is that whenever $S \subset \mathbf{Z}$ has positive upper Banach density, then $S - S$ contains the return times of a nontrivial set in a *m.p.s.*:

(3.16) $S - S \supset \{n \in \mathbf{N} : \mu(T^{-n}A \cap A) > 0\}$

for some $(X, \mathscr{B}, \mu, T)$ and $A \in \mathscr{B}$ with $\mu(A) > 0$. Let us say that a set $\mathscr{H} \subset \mathbf{N}$ is a *return time set* if it has the form given in (3.16). Notice that the intersection of return time sets is a return time set:

$$\{n \in \mathbf{N} : \mu_1(T_1^{-n}A_1 \cap A_1) > 0\} \cap \{n \in \mathbf{N} : \mu_2(T_2^{-n}A_2 \cap A_2) > 0\}$$
$$= \{n \in \mathbf{N} : \mu_1 \times \mu_2((T_1 \times T_2)^{-n}(A_1 \times A_2) \cap (A_1 \times A_2)) > 0\}.$$

It follows that the assertions made for $S - S$ are also valid for finite intersections of such sets. In particular if $S_1, S_2, \ldots, S_k$ are k sets of positive upper Banach density, then

$$(S_1 - S_1) \cap (S_2 - S_2) \cap \cdots \cap (S_k - S_k)$$

is a syndetic set. This result has also been obtained by Prikry as well as by Stewart, and Tijdeman [1].

6. Density and Upper Density

It is easy to give examples of sets S with $BD^*(S) > 0$ but with $D^*(S) = 0$. For example,

$$S = \bigcup_{n=1}^{\infty} [n^3, n^3 + n]$$

is a set with $BD^*(S) = 1$ and $D^*(S) = 0$. Nevertheless, for certain purposes we may replace sets of positive upper Banach density by sets whose density exists and is positive. The following result is due to R. Ellis (private communication).

THEOREM 3.20.   *Let $S \subset \mathbf{Z}$ be a subset of positive upper Banach density. There exists a set $R \subset \mathbf{Z}$ with the property that if F is any finite subset of R, some translate $F + h \subset S$, $h \in \mathbf{Z}$, and R is a set whose density exists and is positive.*

Proof: Take $\xi = 1_S \in \{0,1\}^{\mathbf{Z}}$ and let X be the orbit closure of ξ with respect to the shift. Recalling the proof of Theorem 3.18, we find an ergodic measure μ on X with $\mu(A(1)) > 0$, where $A(1) = \{\omega : \omega(0) = 1\} \subset \{0,1\}^{\mathbf{Z}}$. Since μ is ergodic, there exists a generic point for μ in X. (Proposition 3.7). Let $\eta = 1_R$ be this point. The density of R is given by

$$\lim \frac{1}{N} \sum_{1}^{N} 1_R(n) = \lim \frac{1}{N} \sum_{1}^{N} 1_{A(1)}(T^n\eta) = \int 1_{A(1)} d\mu = \mu(A(1)) > 0,$$

so that the density exists and is positive. Moreover if $F \subset R$, then $\eta(n) = 1$ for $n \in F$, and since $\eta \in \overline{\{T^n\xi : n \in \mathbf{Z}\}}$, there must be some h with $T^h\xi(n) = 1$ for $n \in F$, i.e., $F + h \subset S$.

In particular we have the following consequence.

COROLLARY   *If $S \subset \mathbf{Z}$ has positive upper Banach density, there exists a set of positive density R with $R - R \subset S - S$.*

The following is open:

QUESTION.   If S is a set of positive (upper Banach) density, does there exist a syndetic set R with $R - R \subset S - S$?

7. Szemerédi's Theorem

In Chapter 7 we shall prove the following extension of Poincaré's recurrence theorem. We refer to it as the *multiple recurrence theorem*.

THEOREM 7.15. *If* $(X,\mathscr{B},\mu)$ *is a measure space and* $T_1, T_2, \ldots, T_l$ *are commuting measure preserving transformation of* $(X,\mathscr{B},\mu)$, *then for any set* $A \in \mathscr{B}$ *with* $\mu(A) > 0$, *there exists an integer* $n \geq 1$ *with*

$$\mu(A \cap T_1^{-n}A \cap T_2^{-n} \cap \cdots \cap T_l^{-n}A) > 0.$$

Using Lemma 3.17, we can show how this theorem implies Szemerédi's theorem on arithmetic progressions.

THEOREM 3.21. (Szemerédi) *If* S *is a subset of* $\mathbf{Z}$ *with positive upper Banach density, then* S *contains arbitrarily long arithmetic progressions.*

Notice that this theorem implies van der Waerden's theorem on arithmetic progressions. For in any partition of N or Z into finitely many sets, at least one of them has positive upper density. To see how Theorem 3.21 follows from Theorem 7.15, suppose $S \subset \mathbf{Z}$ has positive upper Banach density. Once again let $X \subset \{0,1\}^{\mathbf{Z}}$ be the orbit closure of the point $1_S \in \{0,1\}^{\mathbf{Z}}$, and let $A = A(1) = \{\omega : \omega(0) = 1\}$. By Lemma 3.17 there exists a measure μ on X invariant under the shift and with $\mu(A(1)) > 0$. Now let $T_1 = T$, $T_2 = T^2, \ldots, T_l = T^l$, where T is the shift. By Theorem 7.15, there will be some point $\omega \in A(1) \cap X$ with

$$T^n\omega \in A(1) \cap X, \ T^{2n}\omega \in A(1) \cap X, \ldots, T^{ln}\omega \in A(1) \cap X.$$

To say that $\omega, T^n\omega, \ldots, T^{ln}\omega \in A(1)$ is to say that $\omega(0) = \omega(n) = \omega(2n) = \cdots = \omega(ln) = 1$. If, moreover, $\omega \in X$, then ω is a limit of translates of 1_S, so that for some h,

$$1_S(h) = 1_S(h + n) = 1_S(h + 2n) = \cdots = 1_S(h + ln) = 1.$$

That means that $h, h + n, h + 2n, \ldots, h + ln \in S$.

8. Poincaré Sequences and Recurrence

There is some overlap between some of the results of this chapter and the results of Chapter 2, Section 5. Namely we can make the following observation.

PROPOSITION 3.22. *Let* W *be a Poincaré set and let* (X,T) *be a dynamical system with* X *a compact metric space. Then there exists* $x \in X$ *and a sequence* $\{n_k\}$ *of nonzero elements of* W *with* $T^{n_k}x \to x$.

Proof: Let μ be some T-invariant borel measure on X; we shall show that almost every $x \in X$ with respect to μ has the property in question. It suffices to show that *a.e.* x has the property that $d(T^n x, x) < \varepsilon$ for some $n \neq 0$, $n \in W$. Let $B_\varepsilon \subset X$ be the set of points for which this fails to be true. Write $B_\varepsilon = \bigcup A_k$ where the diameter of each A_k is $< \varepsilon$. Then

$T^{-n}A_k \cap A_k = \emptyset$ for each $n \in W$, $n \neq 0$. But then $\mu(A_k) = 0$; so $\mu(B_\varepsilon) = 0$. This proves the proposition.

COROLLARY. *Let $\{s_n\}$ be a nondecreasing sequence of integers, and let (X, T) be a dynamical system with X a compact metric space. Then there exists a point $x \in X$ and a sequence of pairs $\{i_n < j_n\}$ with $T^{s_{j_n} - s_{i_n}}x \to x$.*

This result overlaps Theorem 2.16 which ensures the existence of a point x with $T^{w_n}x \to x$ for $\{w_n\}$ a subsequence of a given IP-set. Namely, observe that any IP-set contains a difference set. For if the IP-set consists of all sums $\{p_{i_1} + p_{i_2} + \cdots + p_{i_k}\}$, then it contains $s_j - s_i$ for $i < j$ where $s_i = p_1 + p_2 + \cdots + p_i$. So if we know that some point recurs along a difference set of times, it also recurs along an IP-set of times.

CHAPTER 4

Some Special Ergodic Theorems

In this chapter we shall present two special cases of the multiple recurrence theorem for commuting measure preserving transformation (Theorem 7.15, quoted in Chapter 3, §7). In these cases we shall be able to compute the average

$$(4.1) \qquad \lim_{N \to \infty} \frac{1}{N + 1} \sum_{n=0}^{N} \mu(A_0 \cap T_1^{-n}A_1 \cap \cdots \cap T_1^{-n}A_l)$$

for any measurable sets $A_0, A_1, \ldots, A_l$ in the measure space $(X, \mathcal{B}, \mu)$. Specializing to the case $A_0 = A_1 = \cdots = A_l = A$ with $\mu(A) > 0$, we shall find that the limit in (4.1) is positive, and this will establish the assertion of the multiple recurrence theorem in these cases.

1. Weakly Mixing Transformations

In what follows we shall frequently consider the product of measure preserving systems. If $(X_1, \mathcal{B}_1, \mu_1, T_1)$ and $(X_2, \mathcal{B}_2, \mu_2, T_2)$ are two *m.p.s.*, the product system, denoted $(X_1 \times X_2, \mathcal{B}_1 \times \mathcal{B}_2, \mu_1 \times \mu_2, T_1 \times T_2)$ consists of the product space endowed with the product measure, with $\mathcal{B}_1 \times \mathcal{B}_2$ denoting the smallest σ-algebra on $X_1 \times X_2$ including the products $A_1 \times A_2$ of measurable sets $A_1 \in \mathcal{B}_1$, $A_2 \in \mathcal{B}_2$ and with $T_1 \times T_2$ defined by $T_1 \times T_2(x_1, x_2) = (T_1 x_1, T_2 x_2)$.

DEFINITION 4.1. *A measure preserving transformation T of a measure space $(X, \mathcal{B}, \mu)$ is weakly mixing if $T \times T$ is an ergodic transformation of $(X \times X, \mathcal{B} \times \mathcal{B}, \mu \times \mu)$. We also speak of $(X, \mathcal{B}, \mu, T)$ as a weak mixing system.*

Naturally, if $(X, \mathcal{B}, \mu, T)$ is not ergodic, T will not be weakly mixing, since if $T^{-1}A = A$, then $T^{-1} \times T^{-1}(A \times X) = A \times X$. We can compare the notions of weak mixing and ergodicity in the following manner. For a system $(X, \mathcal{B}, \mu, T)$, if $A, B \in \mathcal{B}$, set

$$N(A, B) = \{n : \mu(A \cap T^{-n}B) > 0\}.$$

If T is ergodic, then whenever $\mu(A),\mu(B) > 0$, the set $N(A,B) \neq \varnothing$. If T is weakly mixing, then for any A,B,C,D with positive measure in X,

$$N(A \times C,B \times D) \neq \varnothing,$$

and this amounts to saying that $N(A,B) \cap N(C,D) \neq \varnothing$. Thus the sets $N(A,B)$ have to be large enough that they each intersect each other. We shall see later that, in fact, weak mixing implies that each $N(A,B)$ has density 1 in $\mathbf{Z}$. Although it is not obvious, it will develop later that the condition $N(A,B) \cap N(C,D) \neq \varnothing$, for any sets A,B,C,D with positive measure, implies weak mixing.

We begin our discussion of weak mixing by proving a weak version of the ergodic theorem. This will be useful as a model for similar results to be proved subsequently.

THEOREM 4.1.   *If $(X,\mathscr{B},\mu,T)$ is an ergodic m.p.s. and $f, g \in L^2(X,\mathscr{B},\mu)$, then*

$$(4.2) \qquad \frac{1}{N+1} \sum_{n=0}^{N} \int f(x)g(T^n x)\, d\mu(x) \to \left(\int f(x)\, d\mu(x) \right)\left(\int g(x)\, d\mu(x) \right)$$

as $N \to \infty$.

Proof: Since $\phi \to \phi \circ T$ is unitary transformation of $L^2(X,\mathscr{B},\mu)$, the set of averages

$$\left\{ \frac{1}{N+1} \sum_0^N g \circ T^n \right\}$$

belongs to a ball of fixed radius in this Hilbert space. Since a closed ball in Hilbert space is weakly compact, the sequence of averages will converge weakly once it is established that there is a unique limit point to the set in question. But any such limit point is a T-invariant function and hence, by ergodicity, must equal a constant. Since

$$\frac{1}{N+1} \sum_0^N \int g(T^n x)\, d\mu(x) \to \int g(x)\, d\mu(x),$$

the constant in question must be $\int g(x)\, d\mu(x)$. This establishes the theorem.

In the weak mixing case a similar result is true for $T \times T$ acting on $X \times X$. If f_1 and f_2 are functions on spaces X_1 and X_2 respectively, we shall denote by $f_1 \otimes f_2$ the function

$$f_1 \otimes f_2(x_1,x_2) = f_1(x_1)f_2(x_2).$$

If $f_i \in L^2(X_i, \mathcal{B}_i, \mu_i)$, $i = 1,2$, then $f_1 \otimes f_2 \in L^2(X_1 \times X_2, \mathcal{B}_1 \times \mathcal{B}_2, \mu_1 \times \mu_2)$. Notice that

$$\int f_1 \otimes f_2 \cdot g_1 \otimes g_2 \, d\mu_1 \times \mu_2 = \left(\int f_1 g_1 \, d\mu_1 \right) \left(\int f_2 g_2 \, d\mu_2 \right).$$

Now suppose $f, g \in L^2(X, \mathcal{B}, \mu)$, and apply Theorem 4.1 to $T \times T$, $f \otimes f$, $g \otimes g$, where we assume T is weakly mixing. We then obtain

(4.3)
$$\frac{1}{N+1} \sum_{n=0}^{N} \left\{ \int f(x) g(T^n x) \, d\mu(x) \right\}^2$$
$$\to \left\{ \int f(x) \, d\mu(x) \right\}^2 \left\{ \int g(x) \, d\mu(x) \right\}^2,$$

as $N \to \infty$.

Next we invoke the following elementary lemma.

LEMMA 4.2. *If $\{a_n\}$ is a sequence of real numbers with*

$$\frac{1}{N+1} \sum_{0}^{N} a_n \to \alpha, \qquad \frac{1}{N+1} \sum_{0}^{N} a_n^2 \to \alpha^2,$$

then

$$\frac{1}{N+1} \sum_{0}^{N} (a_n - \alpha)^2 \to 0.$$

The proof is by direct computation. The lemma implies that under the given hypothesis on $\{a_n\}$, we shall find that the set

$$\{n : |a_n - \alpha| > \varepsilon\}$$

has 0 density for each $\varepsilon > 0$. In view of (4.2) and (4.3) we now obtain the following.

THEOREM 4.3. *If T is a weakly mixing transformation of the measure space $(X, \mathcal{B}, \mu)$ and $f, g \in L^2(X, \mathcal{B}, \mu)$, then*

(4.4)
$$\frac{1}{N+1} \sum_{0}^{N} \left\{ \int f(x) g(T^n x) \, d\mu(x) - \int f \, d\mu \int g \, d\mu \right\}^2 \to 0.$$

Weak mixing systems can be characterized by a certain multiplier property.

PROPOSITION 4.4. *If $(X, \mathcal{B}, \mu, T)$ is a weak mixing system, and $(Y, \mathcal{D}, \nu, S)$ is an ergodic system, then $(X \times Y, \mathcal{B} \times \mathcal{D}, \mu \times \nu, T \times S)$ is ergodic.*

Proof: To show that $X \times Y$ has no nontrivial invariants sets, it suffices to establish that for $f \in L^2(X \times Y, \mathcal{B} \times \mathcal{D}, \mu \times v)$, the ergodic averages

$$\frac{1}{N+1} \sum_{n=0}^{N} f(T^n x, S^n y)$$

converge weakly to $\int f(x,y) \, d\mu(x) \, dv(y)$ in $L^2(X \times Y, \mathcal{B} \times \mathcal{D}, \mu \times v)$. It suffices to do this for a dense subset of functions, and since linear combinations of functions of the form $f(x,y) = g(x)h(y)$ are dense, we find that the ergodicity in question will follow if we establish that

$$\frac{1}{N+1} \sum_{n=0}^{N} \int g_1(x) g_2(T^n x) \, d\mu(x) \cdot \int h_1(y) h_2(S^n y) \, dv(y)$$

(4.7)

$$\rightarrow \left(\int g_1 \, d\mu \right) \left(\int g_2 \, d\mu \right) \left(\int h_1 \, dv \right) \left(\int h_2 \, dv \right),$$

for $g_1, g_2 \in L^2(X, \mathcal{B}, \mu)$ and $h_1, h_2 \in L^2(Y, \mathcal{D}, v)$. We shall do this first for $g_1 = \text{constant}$; then for g_1 satisfying $\int g_1 \, d\mu = 0$. By linearity (4.7) will follow for the general case and this will prove the proposition. If $g_1 = \text{constant}$, then $\int g_1(x) g_2(T^n x) \, d\mu(x) = (\int g_1 \, d\mu)(\int g_2 \, d\mu)$ and (4.7) reduces to Theorem 4.1 for the ergodic system $(Y, \mathcal{D}, v, S)$. Consider next the case in which $\int g_1 \, d\mu = 0$. We now have

$$\left\{ \frac{1}{N+1} \sum_{n=0}^{N} \int g_1(x) g_2(T^n x) \, d\mu(x) \cdot \int h_1(y) h_2(S^n y) \, dv(y) \right\}^2$$

$$\leq \frac{1}{N+1} \sum_{n=0}^{N} \left\{ \int g_1(x) g_2(T^n x) \, d\mu(x) \right\}^2$$

$$\times \frac{1}{N+1} \sum_{n=0}^{N} \left\{ \int h_1(y) h_2(S^n y) \, dv(y) \right\}^2 \rightarrow 0$$

by (4.4). This completes the proof.

The property of the foregoing proposition can be used as a characterization of weak mixing.

PROPOSITION 4.5.   *A m.p.s. $(X, \mathcal{B}, \mu, T)$ is weak mixing if and only if for every ergodic system $(Y, \mathcal{D}, v, T)$, the product system*

$$(X \times Y, \mathcal{B} \times \mathcal{D}, \mu \times v, T \times S)$$

is ergodic.

Proof: The necessity of the conditions is the content of Proposition 4.4. Conversely, if $(X, \mathcal{B}, \mu, T)$ satisfies the condition in question, then, to begin with, it is ergodic, since it can be regarded as the product of itself with a

1-point ergodic system. Consequently, also $(X \times X, \mathscr{B} \times \mathscr{B}, \mu \times \mu, T \times T)$ will be ergodic and this is the definition of weak mixing.

Suppose now that $(X_i, \mathscr{B}_i, \mu_i, T_i)$, $i = 1, 2, \ldots, n$, are weak mixing systems. Each of them is also ergodic and by Proposition 4.4 the product of any two is ergodic. By the same proposition, the product of any three will be ergodic, and so on. Then $(\Pi X_i, \Pi \mathscr{B}_i, \Pi \mu_i, \Pi T_i)$ is ergodic. We conclude that the product of any number of weak mixing systems is ergodic. But by the same token

$$(\Pi X_i \times \Pi X_i, \Pi \mathscr{B}_i \times \Pi \mathscr{B}_i, \Pi \mu_i \times \Pi \mu_i, \Pi T_i \times \Pi T_i)$$

is ergodic. This means that $(\Pi X_i, \Pi \mathscr{B}_i, \Pi \mu_i, \Pi T_i)$ is weak mixing. This gives us the following result.

PROPOSITION 4.6. *The product of any number of weak mixing systems is weak mixing.*

We also have

PROPOSITION 4.7. *If T is weakly mixing on $(X, \mathscr{B}, \mu)$, so is each power T^m, $m \neq 0$.*

Proof: If T is invertible, then T^{-m} is weakly mixing if and only if T^m is weakly mixing. So we take $m > 0$. Let $(Y, \mathscr{D}, v)$ be the finite measure space consisting of the integers $(0, 1, \ldots, m - 1)$ each carrying measure $1/m$, and let S be the periodic transformation $Si = i + 1 \pmod m$. The transformation S is ergodic, so $T \times S$ is ergodic. We first show that T^m is ergodic. Suppose that $T^m f = f$. Form the function $F(x, i) = f(T^{m-i}x)$ and check that $F(Tx, Si) = F(x, i)$. Then $F(x, i)$ is constant and so $f(x)$ is constant. Hence T^m is ergodic. Now apply this argument to $T \times T$ which is weakly mixing by the foregoing proposition. Then $(T \times T)^m$ is ergodic whence T^m is weakly mixing.

2. Multiple Recurrence for Weakly Mixing Transformations

Our goal in this section is to study expressions of the form

$$\int f_0(x) f_1(T_1^n x) f_2(T_2^n x) \cdots f_l(T_l^n x) \, d\mu(x)$$

where the transformations $T_1, T_2, \ldots, T_l$ belong to a commutative group of measure preserving transformations of the space $(X, \mathscr{B}, \mu)$ and where it is assumed that each nontrivial transformation of the group is weakly mixing. In doing this we shall use systematically the kind of convergence we have encountered in the foregoing section in dealing with the expressions $\mu(A \cap T^{-n}B)$ for T weakly mixing. Namely, we shall speak of *convergence in density* for convergence of a sequence but for a subsequence of 0 density. The following is the precise definition.

DEFINITION 4.2.   *A sequence* $\{x_n, n = 1,2,3, \ldots\}$ *of points in a topological space* X *converges to a point* $x \in X$ *in density if for every neighborhood* V *of* x *in* X, $x_n \in V$ *but for a set of* n *of density* 0. *We write* $D\text{-}\lim_n x_n = x$.

This definition modifies the ordinary definition of convergence by replacing finite sets of n by subsets of 0 density. In Chapter 9 we shall introduce yet another definition of convergence that is based on another notion of nullset. Many properties of ordinary convergence are valid for convergence in density. An exception is the property that subsequences of convergent sequences are convergent. What is valid is that if n_k is a subset of positive upper density of $\mathbf{N}$ and $D\text{-}\lim_n x_n = x$, then

$$D\text{-}\lim_k x_{n_k} = x.$$

We shall need the following lemma.

LEMMA 4.8.   *Let* $Q \subset \mathbf{N}$ *be a subset of density* 1 *and for each* $q \in Q$ *let* R_q *be a subset of* $\mathbf{N}$ *of density* 1. *Let* S *be a subset of* $\mathbf{N}$ *of positive upper density and let* $k \geq 1$ *be given. There exist* k *numbers* $n_1 < n_2 < \cdots < n_k$ *in* S *such that each* $n_j - n_i \in Q$ *for* $i < j$ *and* $n_i \in R_{n_j - n_i}$.

Proof: We shall show by induction that there exists a set S_k of positive upper density, and a set of k integers $m_1 < m_2 < \cdots < m_k$ such that for each $n \in S_k$, the k-tuple $n + m_1, n + m_2, \ldots, n + m_k$ has the properties sought in the lemma. For $k = 1$ there is nothing to prove, so suppose S_k and $\{m_1, m_2, \ldots, m_k\}$ have been found. Assume furthermore that $S_k \subset S$. We shall find $S_{k+1} \subset S_k$ and m_{k+1} satisfying all the requirements. Namely we shall find m_{k+1} such that $m_{k+1} - m_i \in Q$ for $1 \leq i \leq k$, and a subset S_{k+1} of S_k with positive upper density such that for $n \in S_{k+1}$ the additional conditions, $n + m_{k+1} \in S$ and $n + m_i \in R_{m_{k+1} - m_i}$, $1 \leq i \leq k$, are satisfied.

We begin by showing that if S_k has positive upper density, then the set $S_k^* = \{p : (S_k + p) \cap S_k$ has positive upper density$\}$ has positive upper density. (Compare with Chapter 3, §5.) It is easy to see that if for a set of translates $S_k + p_1, S_k + p_2, \ldots, S_k + p_l$, each pair intersects in a set of density 0, then the union has upper density equal to $l \times$ upper density of S_k. That means that for l sufficiently large so that this value exceeds 1, for any choice of $p_1, p_2, \ldots, p_l$, some $(S_k + p_i) \cap (S_k + p_j)$ has positive density, or some $(S_k + p_j - p_i) \cap S_k$ has positive upper density. Now let l be minimal with this property, $l \geq 2$. If $l = 2$, then $S_k^* = \mathbf{N}$ and we are through. If $l > 2$, let $p_1, p_2, \ldots, p_{l-1}$ be such that each $(S_k + p_j - p_i) \cap S$ has density 0, $1 \leq i < j \leq l - 1$. Then for any $p, p - p_i \in S_k^*$ for some i, and so S_k^* is in fact syndetic and certainly has positive upper density.

Having shown that S_k^* has positive upper density, we choose m_{k+1} in $S_k^* \cap \bigcap_{i \leq k} (Q + m_i)$. We then obtain $m_{k+1} - m_i \in Q$ for $1 \leq i \leq k$. Finally we define

$$S_{k+1} = (S_k - m_{k+1}) \cap S_k \cap \bigcap_{i \le k} (R_{m_{k+1}-m_i} - m_i).$$

This set has positive upper density since $(S_k - m_{k+1}) \cap S_k$ does and

$$\bigcap_{i \le k} (R_{m_{k+1}-m_i} - m_i)$$

has density 1. This completes the proof.

We shall use the foregoing lemma to obtain a sufficient condition for the convergence in density of a sequence of vectors in Hilbert space to the 0 vector with respect to the weak topology. It should be noted that the analogous result for ordinary convergence does not hold.

LEMMA 4.9. *Let $\{x_n\}$ be a bounded sequence of vectors in Hilbert space and suppose that*

$$(4.8) \qquad \text{D-}\lim_{m} \left(\text{D-}\lim_{n} \langle x_{n+m}, x_n \rangle \right) = 0.$$

Then with respect to the weak topology, $\text{D-}\lim_{n} x_n = 0$.

Proof: Let x be some vector and suppose that for some $\varepsilon > 0$ the set $S = \{n : \langle x_n, x \rangle > \varepsilon\}$ has positive upper density. We assume for convenience that the Hilbert space is over the reals. We have $x \ne 0$ and with $\delta < \varepsilon^2 / \|x\|^2$, let

$$Q = \left\{ m : \text{D-}\lim_{n} \langle x_{n+m}, x_n \rangle < \frac{\delta}{2} \right\}.$$

Q has density 1 and for each $q \in Q$,

$$R_q = \{n : \langle x_{n+q}, x_n \rangle < \delta\}$$

has density 1. Apply Lemma 4.8 to these sets with k to be specified later. If $n_1, n_2, \dots, n_k$ satisfy the conclusion of Lemma 4.8, then we shall have

(i) $\langle x_{n_i}, x \rangle > \varepsilon, 1 \le i \le k$, and

(ii) $\langle x_{n_i}, x_{n_j} \rangle < \delta, 1 \le i < j \le k$.

Set $y_i = x_{n_i} - \dfrac{\varepsilon x}{\|x\|^2}$. Then

$$\langle y_i, y_j \rangle < \delta - \frac{2\varepsilon^2}{\|x\|^2} + \frac{\varepsilon^2}{\|x\|^2} < \delta - \frac{\varepsilon^2}{\|x\|^2} < 0, \quad 1 \le i < j \le k.$$

But since the y_i are bounded independently of k, and

$$0 \le \left\| \sum y_i \right\|^2 = \sum_{i=1}^{k} \|y_i\|^2 + 2 \sum_{i<j} \langle y_i, y_j \rangle$$

$$\le k \max \|y_i\|^2 - k(k-1)\left(\frac{\varepsilon^2}{\|x\|^2} - \delta\right),$$

we arrive at a contradiction if k is chosen sufficiently large. This shows that the set S in question has density 0 for any x and any $\varepsilon > 0$. This proves the lemma.

DEFINITION 4.3. *Let G be a commutative group of measure preserving transformations of a measure space $(X,\mathscr{B},\mu)$. G is called totally weak mixing if each $T \in G$, $T \neq$ identity, is weakly mixing.*

THEOREM 4.10. *Let G be a totally weak mixing group of transformations of $(X,\mathscr{B},\mu)$ and let $T_1,T_2,\ldots,T_l$ be distinct elements of G, $T_i \neq$ identity. Then for any $f_0,f_1,\ldots,f_l \in L^\infty(X,\mathscr{B},\mu)$,*

(4.9) $\text{D-}\lim_{n} \int f_0(x)f_1(T_1^n x) \cdots f_l(T_l^n x)\, d\mu(x)$

$$= \left(\int f_0\, d\mu\right)\left(\int f_1\, d\mu\right) \cdots \left(\int f_l\, d\mu\right).$$

Proof: We shall prove this by induction on l. The case $l = 1$ is contained in Theorem 4.3. Suppose then that the assertion in question is valid for l; we shall establish it for $l + 1$. So let $S_1,\ldots,S_{l+1}$ be distinct elements of G with $S_i \neq$ identity and let $f_0,f_1,\ldots,f_{l+1} \in L^\infty(X,\mathscr{B},\mu)$. We shall assume at first that $\int f_{l+1}\, d\mu = 0$. Write

(4.10) $g_n(x) = f_1(S_1^n x)f_2(S_2^n x) \cdots f_{l+1}(S_{l+1}^n x).$

We want to show that in the weak topology of $L^2(X,\mathscr{B},\mu)$ we have $\text{D-}\lim_{n} g_n = 0$. For this we apply Lemma 4.9. We set $T_1 = S_1 S_{l+1}^{-1},\ldots,$ $T_l = S_l S_{l+1}^{-1}$. Then the T_i are mutually distinct and distinct from the identity. We have, writing $F_i^{(m)}(x) = f_i(x)f_i(S_i^m x)$,

$\langle g_{n+m}, g_n \rangle = \int f_1(S_1^{n+m} x) \cdots f_{l+1}(S_{l+1}^{n+m} x)f_1(S_1^n x) \cdots f_{l+1}(S_{l+1}^n x)\, d\mu(x)$

(4.11) $= \int F_1^{(m)}(S_1^n x) \cdots F_{l+1}^{(m)}(S_{l+1}^n x)\, d\mu(x)$

$= \int F_{l+1}^{(m)}(x)F_1^{(m)}(T_1^n x) \cdots F_l^{(m)}(T_l^n x)\, d\mu(x).$

We can now apply the induction hypothesis to obtain

(4.12) $\text{D-}\lim_{n} \langle g_{n+m}, g_n \rangle = \left(\int F_{l+1}^{(m)}\, d\mu\right)\left(\int F_1^{(m)}\, d\mu\right) \cdots \left(\int F_l^{(m)}\, d\mu\right).$

Recalling the definition of $F_i^{(m)}$, we find

$$\int F_{l+1}^{(m)} d\mu = \int f_{l+1}(x) f_{l+1}(S_{l+1}^m x) \, d\mu(x),$$

and since $\int f_{l+1} \, d\mu = 0$, we have by Theorem 4.3

$$\underset{m}{D\text{-lim}} \int F_{l+1}^{(m)} d\mu = 0.$$

Inasmuch as the remaining integrals $\int F_i^{(m)} d\mu$ are bounded for all m. we conclude that

$$\underset{m}{D\text{-lim}} \, \underset{n}{D\text{-lim}} \, \langle g_{n+m}, g_n \rangle = 0$$

If follows that $D\text{-lim}_n \, g_n = 0$ in the weak topology of $L^2(X, \mathscr{B}, \mu)$, and this establishes our result under the hypothesis $\int f_{l+1} \, d\mu = 0$. To pass to the general case, we need now only dispose of the case $f_{l+1} = \text{constant}$. But this case follows from the induction hypothesis. This completes the proof of the theorem.

In particular, suppose $(X, \mathscr{B}, \mu, T)$ is a weak mixing system. Then $\{T^n\}$ is a totally weak mixing group of transformations of $(X, \mathscr{B}, \mu)$. Let $0 < e_1 < e_2 < \cdots < e_l$ be distinct integers and apply the theorem to $T^{e_1}, T^{e_2}, \ldots, T^{e_l}$ and functions $f_i = 1_{A_i}$. We then obtain the following result which may be regarded as the statement that T is "weakly mixing of all orders."

THEOREM 4.11. *If T is a weakly mixing transformation of $(X, \mathscr{B}, \mu)$, $A_0, A_1, \ldots, A_l$ are $l + 1$ sets in $\mathscr{B}$, and $0 < e_1 < e_2 < \cdots < e_l$ are distinct integers, then for any $\varepsilon > 0$, the set*

$$\{n : |\mu(A_0 \cap T^{-ne_1} A_1 \cap \cdots \cap T^{-ne_l} A_l) - \mu(A_0)\mu(A_1) \cdots \mu(A_l)| > \varepsilon\}$$

has density 0.

Another corollary is the multiple recurrence theorem for a totally weak mixing group of transformations.

THEOREM 4.12. *If G is a totally weak mixing group of transformations of the measure space $(X, \mathscr{B}, \mu)$ and $T_1, T_2, \ldots, T_l \in G$, then for any $A \in \mathscr{B}$ with $\mu(A) > 0$, there exists $n \geq 1$ with*

$$(4.13) \quad \mu(A \cap T_1^{-n} A \cap \cdots \cap T_l^{-n} A) > 0.$$

Proof: We can assume the T_i are distinct and different from the identity. By Theorem 4.10 we have that for every $\varepsilon > 0$

$$(4.14) \quad \mu(A \cap T_1^{-n} A \cap \cdots \cap T_l^{-n} A) > \mu(A)^{l+1} - \varepsilon$$

for all n but for a set of density 0. Choosing $\varepsilon < \mu(A)^{l+1}$, we obtain the desired result.

3. Generic Measures and a Mean Ergodic Theorem

In this section we generalize the classical mean ergodic theorem to a situation where the underlying measure is not invariant. This will be applied in Section 4 to establish another special case of the multiple recurrence theorem.

We begin with von Neumann's mean ergodic theorem (Riesz and Sz.-Nagy [1]).

THEOREM 4.13.   *Let T be a unitary operator on a Hilbert space $\mathcal{H}$, let $\mathcal{H}_T$ denote the subspace of T-invariant vectors, and let P_T denote the orthogonal projection from $\mathcal{H}$ to $\mathcal{H}_T$. Then for each $u \in \mathcal{H}$*

$$(4.15) \qquad \frac{u + Tu + \cdots + T^n u}{n + 1} \to P_T u$$

in the strong topology of $\mathcal{H}$.

Now let T be a measure preserving transformation of the measure space $(X,\mathcal{B},\mu)$. The transformation T induces a unitary operator on $L^2(X,\mathcal{B},\mu)$ which we shall again denote by T: $Tf = f \circ T$. Theorem 4.13 says that the ergodic average

$$\frac{1}{N + 1} \sum_{n=0}^{N} T^n f$$

converges in the L^2-norm to $P_T f$. We shall extend this to convergence in L^2-norm with respect to other measures.

DEFINITION 4.4.   *Let v be a measure on $(X,\mathcal{B})$ and let $\mathcal{A}$ be an algebra of bounded $\mathcal{B}$-measurable functions closed with respect to complex conjugation. We say that v is generic for $(X,\mathcal{B},\mu,T)$ with respect to $\mathcal{A}$ if for every $f \in \mathcal{A}$*

$$(4.16) \qquad \frac{1}{N + 1} \sum_{n=0}^{N} \int T^n f \, dv \to \int f \, d\mu.$$

Of course, μ itself is generic for $(X,\mathcal{B},\mu,T)$. Also if x_0 is a generic point for μ where X is a compact metric space and T a continuous transformation (Definition 3.4), then the point measure δ_{x_0} is generic for $(X,\mathcal{B},\mu,T)$ with respect to the algebra $\mathcal{C}(X)$, in accordance with Definition 4.4.

Now let us suppose that $\mathcal{A}$ contains a set of T-invariant functions that is dense in $L^2(X,\mathcal{B},\mu)_T$, the subspace of T-invariant functions of $L^2(X,\mathcal{B},\mu)$. The functions of $\mathcal{A}$ can be regarded as belonging to either $L^2(X,\mathcal{B},\mu)$ or $L^2(X,\mathcal{B},v)$, but for T-invariant functions we have

$$\int |f|^2 \, d\mu = \lim_{N \to \infty} \frac{1}{N + 1} \sum_{n=0}^{N} \int T^n |f|^2 \, dv = \int |f|^2 \, dv$$

when v is generic for $(X,\mathscr{B},\mu,T)$. Thus we have a dense subset of $L^2(X,\mathscr{B},\mu)_T$ identified isometrically with functions in $L^2(X,\mathscr{B},v)$ and as a result we can regard $L^2(X,\mathscr{B},\mu)_T$ as a subspace of $L^2(X,\mathscr{B},v)$. We now have the following generalization of Theorem 4.13.

THEOREM 4.14. *Let v be generic for $(X,\mathscr{B},\mu,T)$ with respect to a conjugation-closed algebra $\mathscr{A}$ that contains a set of T-invariant functions dense in $L^2(X,\mathscr{B},\mu)_T$. Then for every $f \in \mathscr{A}$,*

$$\frac{1}{N+1} \sum_{n=0}^{N} T^n f \to P_T f$$

in $L^2(X,\mathscr{B},v)$.

Note that by our introductory remarks $P_T f \in L^2(X,\mathscr{B},\mu)_T$ may be regarded as an element of $L^2(X,\mathscr{B},v)$. The theorem is, of course, of interest only if v is singular with respect to μ, and in this case two bounded functions may define the same element of $L^2(X,\mathscr{B},\mu)$ but different elements of $L^2(X,\mathscr{B},v)$.

Proof: Let $\varepsilon > 0$ be given and find $g \in \mathscr{A} \cap L^2(X,\mathscr{B},\mu)_T$ with

$$(4.17) \quad \|P_T f - g\|_{L^2(X,\mathscr{B},\mu)} < \varepsilon.$$

By Theorem 4.13,

$$\int \left| \frac{(f-g) + T(f-g) + \cdots + T^M(f-g)}{M+1} \right|^2 d\mu < \varepsilon^2$$

for M sufficiently large. Fix M accordingly and choose N large enough so that (4.16) gives

$$\frac{1}{N+1} \sum_{n=0}^{N} \int T^n \left| \frac{1}{M+1} \sum_{m=0}^{M} T^m(f-g) \right|^2 dv < \varepsilon^2.$$

Now

$$\left| \frac{1}{N+1} \sum_{n=0}^{N} T^n \left(\frac{1}{M+1} \sum_{m=0}^{M} T^m(f-g) \right) \right|^2$$

$$\leq \frac{1}{N+1} \sum_{n=0}^{N} T^n \left| \frac{1}{M+1} \sum_{m=0}^{M} T^m(f-g) \right|^2$$

so that

$$(4.18) \quad \int \left| \frac{1}{(N+1)(M+1)} \sum_{n=0}^{N} \sum_{m=0}^{M} T^{n+m}(f-g) \right|^2 dv < \varepsilon^2$$

for fixed M and all sufficiently large N. Since

$$\frac{1}{(N+1)(M+1)} \sum_{n=0}^{N} \sum_{m=0}^{M} a_{n+m} - \frac{1}{N+1} \sum_{n=0}^{N} a_n \to 0$$

for fixed M as $N \to \infty$, assuming $\{a_n\}$ is bounded, and at a rate depending only on sup $|a_n|$, it follows that (4.18) can be replaced by

$$\int \left| \frac{1}{N+1} \sum_{n=0}^{N} T^n f - g \right|^2 dv < \varepsilon^2,$$

and N sufficiently large. Combining this with (4.17) and bearing in mind that $P_T f - g \in L^2(X,\mathscr{B},\mu)_T$ so that its norm is the same in $L^2(X,\mathscr{B},\mu)$ as in $L^2(X,\mathscr{B},v)$, we conclude

$$\left\| \frac{1}{N+1} \sum_{n=0}^{N} T^n f - P_T f \right\|_{L^2(X,\mathscr{B},v)} < 2\varepsilon.$$

This proves the theorem.

4. Roth's Theorem

Consider a measure space $(X,\mathscr{B},\mu)$. If f and g are functions on X, we shall denote by $f \otimes g$ the function on $X \times X$ given by $f \otimes g(x,x') = f(x)g(x')$; $L^\infty(\mathbf{X}) \otimes L^\infty(\mathbf{X})$ will denote the algebra of functions on $X \times X$ that are finite sums of functions $f \otimes g$ with $f, g \in L^\infty(X,\mathscr{B},\mu)$. We shall denote by μ_Δ the diagonal measure on $X \times X$ given by $\int f(x,x') d\mu_\Delta(x,x') = \int f(x,x) d\mu(x)$.

We shall now apply Theorem 4.14 to the following situation. Let T be an ergodic measure preserving transformation of $(X,\mathscr{B},\mu)$, and let $S = T \times T^2$ act on $X \times X$. We notice that μ_Δ is generic for the system $(X \times X, \mathscr{B} \times \mathscr{B}, \mu \times \mu, S)$ with respect to the algebra $\mathscr{A} = L^\infty(\mathbf{X}) \otimes L^\infty(\mathbf{X})$:

$$
\begin{aligned}
\frac{1}{N+1} \sum_{n=0}^{N} \int S^n f \otimes g \, d\mu_\Delta &= \frac{1}{N+1} \sum_{n=0}^{\infty} \int f(T^n x) g(T^{2n} x) \, d\mu(x) \\
&= \frac{1}{N+1} \sum_{n=0}^{\infty} \int f(x) g(T^n x) \, d\mu(x) \\
&\to \left(\int f \, d\mu \right) \left(\int g \, d\mu \right) \\
&= \int f \otimes g \, d\mu \times \mu.
\end{aligned}
$$

(4.19)

The next lemmas will be used to show that a subset of $\mathscr{A}$ is dense in $L^2(X \times X, \mathscr{B} \times \mathscr{B}, \mu \times \mu)_S$.

LEMMA 4.15. *Let U be a unitary operator on a Hilbert space $\mathscr{H}$. The operator U has pure continuous spectrum (i.e., no eigenvectors) if and only* *if for every* $u_1, u_2 \in \mathcal{H}$

$$(4.20) \quad \frac{1}{2N+1} \sum_{n=-N}^{N} |\langle U^n u_1, u_2 \rangle|^2 \to 0$$

as $N \to \infty$.

Proof: Using the spectral theorem, one expresses the inner products in (4.20) as Fourier transforms

$$\langle U^n u_1, u_2 \rangle = \int_0^{2\pi} e^{in\theta} d\pi_{u_1, u_2}(\theta)$$

where π_{u_1, u_2} is a measure on $[0, 2\pi]$. The operator U has continuous spectrum if and only if each measure π_{u_1, u_2} is continuous (assigns 0 mass to single points). The lemma now follows from Wiener's theorem (Wiener [1]).

LEMMA 4.16. *Let U and U' be unitary operators on Hilbert spaces $\mathcal{H}$ and $\mathcal{H}'$, respectively. Let $U \otimes U'$ be the unitary operator defined on the tensor product $\mathcal{H} \otimes \mathcal{H}'$ by $U \otimes U'(u \otimes u') = Uu \otimes U'u'$. If U has pure continuous spectrum, so does $U \otimes U'$.*

Proof: First of all notice that in the foregoing lemma, it suffices to verify (4.20) for u_1, u_2 ranging independently over a basis of $\mathcal{H}$. In the case of $U \otimes U'$ it will suffice to verify that

$$\frac{1}{2N+1} \sum_{-N}^{N} |\langle (U \otimes U')^n u \otimes u', v \otimes v' \rangle|^2 \to 0.$$

But this follows since $\langle (U \otimes U')^n u \otimes u', v \otimes v' \rangle = \langle U^n u, v \rangle \langle U'^n u', v' \rangle$ and $|\langle U'^n u', v' \rangle|$ is bounded.

LEMMA 4.17. *Let U and U' be unitary operators on Hilbert spaces $\mathcal{H}$ and $\mathcal{H}'$ respectively, and let $\mathcal{H}_0$ be the subspace of $\mathcal{H}$ spanned by eigenvectors of U, $\mathcal{H}'_0$ the subspace of $\mathcal{H}'$ spanned by eigenvectors of U'. Then any eigenvectors of $U \otimes U'$ in $\mathcal{H} \otimes \mathcal{H}'$ is in $\mathcal{H}_0 \otimes \mathcal{H}'_0$.*

Proof: Write $\mathcal{H} = \mathcal{H}_0 \oplus \mathcal{H}_1$, $\mathcal{H}' = \mathcal{H}'_0 \oplus \mathcal{H}'_1$ as orthogonal decompositions. $U\mathcal{H}_i = \mathcal{H}_i$ for $i = 0,1$ and $U'\mathcal{H}'_i = \mathcal{H}'_i$ for $i = 0,1$. Thus $\mathcal{H} \otimes \mathcal{H}'$ is the sum of 4 subspaces $\mathcal{H}_i \otimes \mathcal{H}'_j$ invariant under $U \otimes U'$. By the preceding lemma, only $\mathcal{H}_0 \otimes \mathcal{H}'_0$ can have discrete spectrum.

Finally we have

LEMMA 4.18. *Let U and U' be unitary operators on Hilbert spaces $\mathcal{H}$ and $\mathcal{H}'$ respectively and let $w \in \mathcal{H} \otimes \mathcal{H}'$ be an eigenvector of $U \otimes U'$: $U \otimes U'w = \lambda w$. Then $w = \sum c_n u_n \otimes u'_n$ where $Uu_n = \lambda_n u_n$, $U'u'_n = \lambda'_n u'_n$ and $\lambda_n \lambda'_n = \lambda$, and the sequences $\{u_n\}, \{u'_n\}$ are orthonormal.*

Proof: Let $\{v_n\}$ be an orthonormal basis of $\mathcal{H}_0$ consisting of eigenvectors and similarly $\{v'_n\}$ for $\mathcal{H}'_0$. By the foregoing lemma, $w = \sum c_{nm}v_n \otimes v'_m$. Write $Uv_n = \lambda(v_m)v_m$, $U'v'_m = \lambda'(v'_m)v'_n$. We see upon evaluating $U \otimes U'w$ that if $\lambda(v_n)\lambda(v'_m) \neq \lambda$, then $c_{nm} = 0$. This proves the lemma.

Now let us take $\mathcal{H} = \mathcal{H}' = L^2(X,\mathcal{B},\mu)$, $U = T$, $U' = T^2$. The tensor product $\mathcal{H} \otimes \mathcal{H}'$ can be identified with $L^2(X \times X, \mathcal{B} \times \mathcal{B}, \mu \times \mu)$, and $U \otimes U'$ is the operator induced by $S = T \times T^2$ on this space. Let $\mathscr{E}(X,\mathcal{B},\mu,T)$ denote the subspace of $L^2(X,\mathcal{B},\mu)$ spanned by eigenvectors of T. Notice that $\mathscr{E}(X,\mathcal{B},\mu,T) = \mathscr{E}(X,\mathcal{B},\mu,T^2)$. For if $T^2f = \lambda f$, then since $T(Tf \pm \lambda^{1/2}f) = \pm\lambda^{1/2}(Tf \pm \lambda^{1/2}f)$, $Tf \pm \lambda^{1/2}f$ (which may be 0) belongs to $\mathscr{E}(X,\mathcal{B},\mu,T)$, and therefore so does f. The absolute value of an eigenfunction is T-invariant, so each eigenfunction is a limit of bounded eigenfunctions that belong to $L^\infty(X) \cap \mathscr{E}(X,\mathcal{B},\mu,T)$. It follows now from Lemma 4.18 that $L^\infty(X) \otimes L^\infty(X)$ contains a dense subset of $L^2(X \times X, \mathcal{B} \times \mathcal{B}, \mu \times \mu)_S$. Applying Theorem 4.14 to the operator S, we have the following intermediate result:

PROPOSITION 4.19.   *If $g,h \in L^\infty(X,\mathcal{B},\mu)$ and T is ergodic on $(X,\mathcal{B},\mu)$, $S = T \times T^2$, then*

$$\frac{g \otimes h + Tg \otimes T^2h + T^2g \otimes T^4h + \cdots + T^Ng \otimes T^{2N}h}{N + 1} \to P_S(g \otimes h)$$

in $L^2(X \times X, \mathcal{B} \times \mathcal{B}, \mu_\Delta)$, as $N \to \infty$.

Here $P_S(g \otimes h)$ is identified with a function in $L^2(X \times X, \mathcal{B} \times \mathcal{B}, \mu_\Delta)$, and the convergence is in the strong topology of this space. Using just weak convergence, we deduce:

PROPOSITION 4.20.   *If $f,g,h \in L^\infty(X,\mathcal{B},\mu)$ and T is measure preserving on $(X,\mathcal{B},\mu)$, $S = T \times T^2$, then*

$$(4.21) \qquad \frac{1}{N + 1} \sum_{n=0}^{N} \int f(x)g(T^nx)h(T^{2n}x)\,d\mu(x) \to \int f(x)P_S(g \otimes h)(x)\,d\mu(x)$$

as $N \to \infty$. In particular the limit in question always exists.

Here we have identified $L^2(X \times X, \mathcal{B} \times \mathcal{B}, \mu_\Delta)$ with $L^2(X,\mathcal{B},\mu)$. To make use of (4.21), we should determine $P_S(g \otimes h)$ as a function on X. Rather than do this we use the following result.

LEMMA 4.21.   *For each $f \in L^2(X,\mathcal{B},\mu)$ let $\bar{f}$ denote the projection of f onto the subspace $\mathscr{E}(X,\mathcal{B},\mu,T) \subset L^2(X,\mathcal{B},\mu)$. Then for $g,h \in L^\infty(X)$,*

$$P_S(g \otimes h) = P_S(\bar{g} \times \bar{h}) \in \mathscr{E}(X,\mathcal{B},\mu,T).$$

Proof: According to Lemma 4.17,

$$P_S(g\otimes h) \in \mathscr{E}(X,\mathscr{B},\mu,T)\otimes\mathscr{E}(X,\mathscr{B},\mu,T).$$

As a result, if Q denotes the projection of $L^2(X\times X, \mathscr{B}\times\mathscr{B}, \mu\times\mu)$ to $\mathscr{E}(X,\mathscr{B},\mu,T)\otimes\mathscr{E}(X,\mathscr{B},\mu,T)$, then $P_S = QP_S = P_SQ$. But $Qf\otimes g = \bar{f}\otimes\bar{g}$, so $P_S(f\otimes g) = P_S(\bar{f}\otimes\bar{g})$. In identifying $L^2(X\times X, \mathscr{B}\times\mathscr{B}, \mu_\Delta)$ with $L^2(X,\mathscr{B},\mu)$, the function $f_1(x_1)f_2(x_2)$ is identified with $f_1(x)f_2(x)$. Thus the product of eigenfunctions $\phi_1(x_1)\phi_2(x_2)$ is identified with $\phi_1\phi_2(x)$ which belongs to $\mathscr{E}(X,\mathscr{B},\mu,T)$. It follows that $P_S(f\otimes g) \in \mathscr{E}(X,\mathscr{B},\mu,T)$.

Using Lemma 4.21, we see that the right-hand side of (4.21) depends on $f,\bar{g},\bar{h}$. Moreover, since $P_S(f\otimes g) \in \mathscr{E}(X,\mathscr{B},\mu,T)$, we can replace f by $\bar{f}$. Thus the limit in (4.21) is the same for f,g,h as for $\bar{f},\bar{g},\bar{h}$. This gives us the following result.

THEOREM 4.22. *Suppose $(X,\mathscr{B},\mu,T)$ is a measure preserving system, and let $\mathscr{E}(X,\mathscr{B},\mu,T)$ denote the subspace of $L^2(X,\mathscr{B},\mu)$ spanned by eigenvectors of T. For $f \in L^2(X,\mathscr{B},\mu)$ denote by $\bar{f}$ its image under the orthogonal projection to $\mathscr{E}(X,\mathscr{B},\mu,T)$. Then for $f,g,h \in L^\infty(X,\mathscr{B},\mu)$,*

$$(4.22) \quad \lim_{N\to\infty} \frac{1}{N+1} \sum_{n=0}^N \int f(x)g(T^n x)h(T^{2n}x)\,d\mu(x)$$

$$= \lim_{N\to\infty} \frac{1}{N+1} \sum_{n=0}^N \int \bar{f}(x)\bar{g}(T^n x)\bar{h}(T^{2n}x)\,d\mu(x).$$

It can be shown that in case T is ergodic, the limit on the right-hand side of (4.22) can be expressed as an integral over a compact abelian group. Using this, one can show that if $f = g = h = $ nonnegative function that is not *a.e.* zero, then the limit in question is strictly positive. This is carried out in detail in Furstenberg [1]. Here we shall use a slightly different approach.

LEMMA 4.23. *If $f \geq 0$, then $\bar{f} \geq 0$.*

Proof: This follows from the fact that whenever $f,g \in \mathscr{E}(X,\mathscr{B},\mu,T)$, then $f \vee g = \max(f,g) \in \mathscr{E}(X,\mathscr{B},\mu,T)$. For $\bar{f}$ minimizes the distance from $\mathscr{E}(X,\mathscr{B},\mu,T)$ to f and if $f \geq 0$, clearly $\bar{f} \vee 0$ will do at least as well as $\bar{f}$. Hence $\bar{f} \geq 0$. To show that if $f,g \in \mathscr{E}(X,\mathscr{B},\mu,T)$, then $f \vee g \in \mathscr{E}(X,\mathscr{B},\mu,T)$, we notice that $\mathscr{E}(X,\mathscr{B},\mu,T)$ has a dense subspace that is a uniformly closed and conjugation closed algebra. But such an algebra is closed under the max operator and the result follows readily for its L^2-closure.

An immediate corollary is

LEMMA 4.24. *If $f \in L^\infty(X,\mathscr{B},\mu)$, then $\bar{f} \in L^\infty(X,\mathscr{B},\mu)$.*

Proof: Take real and imaginary parts of f and note that if $f \leq c$, then $c - f \geq 0$ so that $c - \bar{f} \geq 0$.

LEMMA 4.25.   *If* $\psi \in \mathscr{E}(X,\mathscr{B},\mu,T)$, *then for any* $\varepsilon > 0$ *there exists a syndetic set of* n *for which* $\|T^n\psi - \psi\|_2 < \varepsilon$, *the norm being that of* $L^2(X,\mathscr{B},\mu)$.

Proof: It suffices to prove this for a dense set of $\psi \in \mathscr{E}(X,\mathscr{B},\mu,T)$. Letting ψ be a finite linear combination of eigenfunctions, we find the lemma follows from the fact that if $\lambda_1, \lambda_2, \ldots, \lambda_m \in$ unit circle in the complex plane, then for any $\varepsilon > 0$ there is a syndetic set of n with all $|\lambda_i^n - 1| < \varepsilon$. This is a special case of Theorem 1.21.

LEMMA 4.26.   *If* $\psi \in L^\infty(X) \cap \mathscr{E}(X,\mathscr{B},\mu,T)$, *then for any* $\varepsilon > 0$ *there exists a syndetic set of* n *with*

$$\int \psi(x)\psi(T^n x)\psi(T^{2n} x)\, d\mu(x) > \int \psi(x)^3\, d\mu(x) - \varepsilon.$$

Proof: We have

$$\left| \int \psi(x)\psi(T^n x)\psi(T^{2n} x)\, d\mu(x) - \int \psi(x)^3\, d\mu(x) \right|$$

$$\leq \int \psi(x)\psi(T^n x)|\psi(T^{2n} x) - \psi(x)|\, d\mu(x)$$

$$+ \int \psi(x)^2 |\psi(T^n x) - \psi(x)|\, d\mu(x)$$

$$\leq \|\psi T^n\psi\|_2 \|T^{2n}\psi - \psi\|_2 + \|\psi^2\|_2 \cdot \|T^n\psi - \psi\|_2$$

$$\leq 3\|\psi\|_\infty^2 \|T^n\psi - \psi\|_2.$$

We now obtain

THEOREM 4.27.   *If* f *is a nonnegative function in* $L^\infty(X,\mathscr{B},\mu)$, f *not a.e.* 0, *and* T *a measure preserving transformation of* $(X,\mathscr{B},\mu)$, *then*

$$\lim_{N \to \infty} \frac{1}{N+1} \sum_{n=0}^{N} \int f(x)f(T^n x)f(T^{2n} x)\, d\mu(x) > 0.$$

In particular, for every $A \in \mathscr{B}$, $\mu(A) > 0$, *there exists* $n \geq 1$ *with*

(4.23) $\mu(A \cap T^{-n}A \cap T^{-2n}A) > 0$.

Proof: The limit in question is the same for f as for $\bar{f} \in L^\infty(X) \cap \mathscr{E}(X,\mathscr{B}\mu,T)$. We now apply Lemmas 4.25 and 4.26 for a small $\varepsilon > 0$. Since $\bar{f} \geq 0$ and $\int f\, d\mu = \int \bar{f}\, d\mu$, it follows that $\bar{f}$ is not *a.e.* 0 and $\int \bar{f}^3\, d\mu > 0$, so that $\int \bar{f}(x)\bar{f}(T^n x)\bar{f}(T^{2n} x)\, d\mu(x) > \delta > 0$ for some δ, for a syndetic set of n. It follows that the right-hand side of (4.22) is positive.

If we now recall the argument in Chapter 3, Section 7 showing how the multiple recurrence theorem, Theorem 7.15, implies Szemerédi's theorem on arithmetic progressions, we see that Theorem 4.28 implies the following restricted version, first proved by K. F. Roth.

THEOREM 4.28. *If S is a subset of **Z** with positive upper Banach density, then S contains three numbers in arithmetic progression.*

5. More on Generic Measures

In this section we shall give another application of Theorem 4.14. Let $\xi \in \{0,1\}^{\mathbf{N}} = \Omega$ be a generic point for some ergodic measure θ on Ω with respect to the shift map $S:\Omega \to \Omega$, $S\omega(n) = \omega(n + 1)$.
In particular

$$\lim_{N \to \infty} \frac{1}{N} \sum_{n=1}^{N} \xi(n) = \alpha$$

exists. Let us assume $\alpha \neq 0$. Arrange the values of n with $\xi(n) = 1$ in increasing order: $n_1 < n_2 < n_3 < \cdots$. Then $n_k/k \to 1/\alpha$. Now let $(X,\mathscr{B},\mu,T)$ be a weak mixing system and form the system

$$(\Omega \times X, \mathscr{B}_\Omega \times \mathscr{B}, \theta \times \mu, S \times T)$$

which is ergodic by Proposition 4.6. ($\mathscr{B}_\Omega$ denotes the borel sets of Ω.) Let $\mathscr{A}$ be the algebra $\mathscr{C}(\Omega) \otimes L^\infty(X)$, and let ν be the measure $\delta_\xi \times \mu$ on $\Omega \times X$. We claim that ν is generic for $\theta \times \mu$ with respect to $\mathscr{A}$. For

$$\frac{1}{N} \sum_{n=1}^{N} (S \times T)^n \delta_\xi \times \mu = \frac{1}{N} \sum_{n=1}^{N} S^n \delta_\xi \times \mu$$

$$= \left(\frac{1}{N} \sum_{n=1}^{N} \delta_{S^n \xi}\right) \times \mu \to \theta \times \mu.$$

The hypotheses of Theorem 4.14 are fulfilled for the measure ν with respect to the algebra $\mathscr{A}$ and the operator $S \times T$, since

$$L^2(\Omega \times X, \mathscr{B}_\Omega \times \mathscr{B}, \theta \times \mu)_{S \times T}$$

consists of constants. We therefore conclude that for $\psi \in \mathscr{C}(\Omega)$, $f \in L^\infty(X)$,

$$\frac{1}{N} \sum_{n=1}^{N} \psi(S^n \xi) f(T^n x) \to \left(\int \psi \, d\theta\right)\left(\int f \, d\mu\right).$$

in $L^2(X,\mathscr{B},\mu)$. In particular let $\psi(\omega) = \omega(1)$ and replace f by Tf:

$$\frac{1}{N} \sum_{n=1}^{N} \xi(n) f(T^n x) \to \alpha \int f \, d\mu$$

or

$$\frac{1}{N} \sum_{k=1}^{N} f(T^{n_k}x) \rightarrow \int f \, d\mu$$

since $N/n_N \rightarrow \alpha$.

A refinement of the foregoing argument replacing the generic point by a quasi-generic one, enables one to prove the following:

THEOREM 4.29.   *Let $\{n_k\}$ be a sequence of natural numbers with $n_k = O(k)$. Then if $(X,\mathscr{B},\mu,T)$ is a weak mixing measure preserving system,*

$$(4.24) \qquad \frac{1}{N} \sum_{k=1}^{N} f(T^{n_k}x) \rightarrow \int f \, d\mu$$

in $L^2(X,\mathscr{B},\mu)$ for every $f \in L^2(X,\mathscr{B},\mu)$.

This theorem was also proved by Jones and Lin [1].

The following questions suggest themselves in connection with the L^2-convergence that has been established using Theorem 4.14:

QUESTION 1.   If $(X,\mathscr{B},\mu,T)$ is a *m.p.s* and $m \neq 0,1$, $f,g \in L^\infty(X,\mathscr{B},\mu)$, does

$$\frac{1}{N+1} \sum_{n=0}^{N} f(T^n x)g(T^{nm}x)$$

converge almost everywhere?

QUESTION 2.   If $(X,\mathscr{B},\mu,T)$ is a weakly mixing *m.p.s.* and $n_1 < n_2 < n_3 < \cdots$ with $n_k = O(k)$, $f \in L^2(X,\mathscr{B},\mu)$, does

$$\frac{1}{N} \sum_{k=1}^{N} f(T^{n_k}x) \rightarrow \int f \, d\mu$$

almost everywhere?

6. Weak Mixing and Eigenfunctions

If $(X,\mathscr{B},\mu,T)$ is a weak mixing system, then

$$(X \times X,\mathscr{B} \times \mathscr{B},\mu \times \mu,T \times T)$$

is ergodic and has no nonconstant invariant functions. From this it is easy to conclude that $\mathscr{E}(X,\mathscr{B},\mu,T)$ consists of only the constants. For suppose $f(Tx) = \lambda f(x)$ with f nonconstant. The fact that $\int |f(Tx)|^2 \, d\mu(x) = \int |f(x)|^2 \, d\mu(x)$ gives $|\lambda| = 1$. This in turn implies that $|f(x)|$ is T-invariant, so, by ergodicity, $|f(x)|$ is a constant that we may suppose $\neq 0$. We then find that $f(x)/f(x')$ is a bounded invariant function on $X \times X$ that can be constant only if f is constant.

Using Lemma 4.17, we obtain the converse as well.

THEOREM 4.30. *A m.p.s. $(X,\mathscr{B},\mu,T)$ is weak mixing if and only if $\mathscr{E}(X,\mathscr{B},\mu,T)$ is one-dimensional, i.e., if the only eigenfunctions of $f \to f \circ T$ are the constants.*

Proof: We have seen that the condition is necessary for weak mixing. Suppose now that $\mathscr{E}(X,\mathscr{B},\mu,T)$ is one-dimensional. Then by Lemma 4.17,

$$L^2(X \times X, \mathscr{B} \times \mathscr{B}, \mu \times \mu)_{T \times T} \subset \mathscr{E}(X,\mathscr{B},\mu,T) \otimes \mathscr{E}(X,\mathscr{B},\mu,T)$$

is one-dimensional; hence $T \times T$ is ergodic.

Equivalently, if $(X,\mathscr{B},\mu,T)$ is not weak mixing, there must exist non-trivial eigenfunctions. This gives us very specific information about nonweak-mixing systems. For example, recalling the notation $N(A,B)$ of Section 1, we have

THEOREM 4.31. *Let $(X,\mathscr{B},\mu,T)$ be a m.p.s. A necessary and sufficient condition for weak mixing is that for every pair of sets $A,B \in \mathscr{B}$, $\mu(A) > 0$, $\mu(B) > 0$, $N(A,B) \cap N(A,A) \neq \varnothing$.*

Proof: The necessity was proved in Section 1. Suppose that the condition is satisfied. Since $N(A,B) \neq \varnothing$ for all nontrivial A,B, the system is ergodic. If it is not weak mixing, there will exist eigenfunctions $f(x)$ with $|f(x)| = 1$, $f(x)$ not constant. We can find two arcs, α,β, on the unit circle of lengths $< \delta$ that are separated by an arc of length $> \delta$ for which $\mu\{x : f(x) \in \alpha\} > 0$, $\mu\{x : f(x) \in \beta\} > 0$. Let

$$A = \{x : f(x) \in \alpha\}, \quad B = \{x : f(x) \in \beta\}$$

and suppose $f(Tx) = \lambda f(x)$. Then

$$N(A,B) \subset \{n : \lambda^n \alpha \cap \beta \neq \varnothing\}, \quad N(A,A) \subset \{n : \lambda^n \alpha \cap \alpha \neq \varnothing\},$$

and since $\lambda^n \alpha$ is an arc of length $< \delta$, the two conditions are mutually exclusive.

CHAPTER 5

Measure Theoretic Preliminaries

In this chapter we have collected the technical measure-theoretic material that will be needed in the next two chapters. Most of this material relates to the notion of a factor of a measure space, a notion that will be used repeatedly in the sequel.

1. Factors of Measure Spaces and Measure Preserving Systems

We recall the definition of a measure space as a triple $(X,\mathscr{B},\mu)$, where X is an arbitrary space, $\mathscr{B}$ is a σ-algebra of subsets of X, μ is a σ-additive non-negative measure on the sets of $\mathscr{B}$ with (for convenience) $\mu(X) = 1$. If $(Y,\mathscr{D},v)$ is another measure space, a map $\phi: X \rightarrow Y$ is *measure preserving* if

  (i)   it is measurable, i.e., for each $A \in \mathscr{D}$, $\phi^{-1}(A) \in \mathscr{B}$, and

  (ii) for each $A \in \mathscr{D}$, $\mu(\phi^{-1}(A)) = v(A)$.

    In much of our discussion the underlying space X of a measure space $(X,\mathscr{B},\mu)$ plays a secondary role, merely providing the means for defining the σ-algebra $\mathscr{B}$ and the measure μ. Moreover, sets of $\mathscr{B}$ that differ by a null set (a set with μ-measure 0) can often be identified. As a result it will often be advantageous to attach to a measure space $(X,\mathscr{B},\mu)$ the associated abstract σ-algebra $\tilde{\mathscr{B}}$ consisting of equivalence classes of sets in $\mathscr{B}$, where $A \sim A'$ if $\mu(A \cup A' - A \cap A') = 0$, together with the measure μ as a function on $\tilde{\mathscr{B}}$. Note, for example, that the spaces $L^p(X,\mathscr{B},\mu)$ are determined by $(\tilde{\mathscr{B}},\mu)$.

    When we have a measure preserving transformation $\phi: X \rightarrow Y$ between two measure spaces $(X,\mathscr{B},\mu)$ and $(Y,\mathscr{D},v)$, we can see from the defining conditions (i) and (ii) that the map $\phi^{-1}: \mathscr{D} \rightarrow \mathscr{B}$ is playing the central role. We now generalize the notion of a measure preserving transformation in the notion of a *homomorphism* $\alpha:(X,\mathscr{B},\mu) \rightarrow (Y,\mathscr{D},v)$ where all that is given is the map, which we shall denote α^{-1}, of the (mod null sets) algebras: $\tilde{\mathscr{D}} \rightarrow \tilde{\mathscr{B}}$. Not all homomorphisms arise from spacial maps $X \rightarrow Y$. Since these, however, furnish the most important examples, we shall regard the

(*) If $\tilde{A} \le \tilde{B}$ and $\mu(\tilde{A}) = \mu(\tilde{B})$
then $\tilde{A} = \tilde{B}$.

direction of an abstract homomorphism α as being from $(X,\mathscr{B},\mu)$ to $(Y,\mathscr{D},v)$ even though we are only given the map $\alpha^{-1}:\tilde{\mathscr{D}} \to \tilde{\mathscr{B}}$.

DEFINITION 5.1. *A homomorphism* $\alpha(X,\mathscr{B},\mu) \to (Y,\mathscr{D},v)$ *between two measure spaces is given by an injection* $\alpha^{-1}:\tilde{\mathscr{D}} \to \tilde{\mathscr{B}}$ *satisfying*

(i) $\alpha^{-1}(\tilde{A}_1 \cup \tilde{A}_2) = \alpha^{-1}(\tilde{A}_1) \cup \alpha^{-1}(A_2),\ \tilde{A}_1,\tilde{A}_2 \in \tilde{\mathscr{D}},$

(ii) $\alpha^{-1}(\tilde{Y}-\tilde{A}) = \tilde{X}-\alpha^{-1}(\tilde{A}),\ \tilde{A} \in \tilde{\mathscr{D}},$

(iii) $\mu(\alpha^{-1}(\tilde{A})) = v(\tilde{A}),\ \tilde{A} \in \tilde{\mathscr{D}}$

We remark that in the definition of homomorphism, the mapping α^{-1} is necessarily one-to-one. For if $\alpha^{-1}(\tilde{A}) = \alpha^{-1}(\tilde{B})$, then $\alpha^{-1}(\tilde{A} \cup \tilde{B}) = \alpha^{-1}(\tilde{A} \cap \tilde{B})$ whence $v(\tilde{A} \cup \tilde{B}) = v(\tilde{A} \cap \tilde{B})$, which implies that $\tilde{A} = \tilde{B}$. Moreover, α^{-1} also preserves σ-algebra operations. If $(*)$

$$\tilde{A} = \bigcup_1^{\infty} \tilde{A}_n,$$

then

$$v\left(\tilde{A}-\bigcup_1^N \tilde{A}_n\right) \to 0,$$

whence

$$\mu\left(\alpha^{-1}(\tilde{A})-\bigcup_1^N \alpha^{-1}(\tilde{A}_n)\right) \to 0,$$

and so $(*)$

$$\alpha^{-1}(\tilde{A}) = \bigcup_1^{\infty} \alpha^{-1}(\tilde{A}_n).$$

Given a homomorphism determined by a spacial map $\phi:X \to Y$, we can lift a measurable function on $(Y,\mathscr{D},v)$ to a measurable function on $(X,\mathscr{B},\mu)$ by $f \to f \circ \phi$. This correspondence maps $L^p(Y,\mathscr{D},v)$ isometrically into $L^p(X,\mathscr{B},\mu)$. For a general homomorphism $\alpha:(X,\mathscr{B},\mu) \to (Y,\mathscr{D},v)$ one can also associate to each equivalence class of measurable functions on $(Y,\mathscr{D},v)$ modulo null function such an equivalence class of functions on $(X,\mathscr{B},\mu)$. Namely to each measurable function $f(y)$ attach the family of sets $A_t(f) = \{y:f(y) < t\}$, $t \in$ rationals. Conversely given a family of measurable sets A_t, t rational, with $t_1 < t_2 \Rightarrow A_{t_1} \subset A_{t_2}$, we define $f(y) = \inf \{t:y \in A_t\}$. Beginning with a measurable function $f(y)$ on $(Y,\mathscr{D},v)$ we form $\{A_t(f)\}$ and map these to $\{\alpha^{-1}(A_t(f)) \subset \tilde{\mathscr{B}}$. By choosing representatives of these sets in $\mathscr{B}$, we construct a function f^α on $(X,\mathscr{B},\mu)$ which satisfies

(5.1) $\alpha^{-1}(f^{-1}(B)) = (f^\alpha)^{-1}(B)$

for each borel set $B \subset \mathbf{R}$. From (5.1) we deduce that if $f \in L^p(Y,\mathscr{D},v)$, then $f^\alpha \in L^p(X,\mathscr{B},\mu)$. It is clear that in the case of a spacial map $\phi: X \to Y$, the function f^ϕ agrees (*a.e.*) with $f \circ \phi$.

DEFINITION 5.2.   *The measure space* $(Y,\mathscr{D},v)$ *is a factor of* $(X,\mathscr{B},\mu)$ *if there is a homomorphism of* $(X,\mathscr{B},\mu)$ *to* $(Y,\mathscr{D},v)$. *In this case we also say that* $(X,\mathscr{B},\mu)$ *is an extension of* $(Y,\mathscr{D},v)$. *Two measure spaces* $(X,\mathscr{B},\mu)$ *and* $(Y,\mathscr{D},v)$ *are equivalent if there is a homomorphism* $\alpha^{-1}:(X,\mathscr{B},\mu) \to (Y,\mathscr{D},v)$ *satisfying* $\alpha^{-1}(\tilde{\mathscr{D}}) = \tilde{\mathscr{B}}$.

Since in any case $\alpha: \tilde{\mathscr{D}} \to \tilde{\mathscr{B}}$ is $1 - 1$, if it is also onto, it can be inverted and the inverse defines a homomorphism $\alpha^{-1}:(Y,\mathscr{D},v) \to (X,\mathscr{B},\mu)$.

We denote by $\tilde{A} \bigtriangleup \tilde{B}$ the symmetric difference of $\tilde{A}$ and $\tilde{B}$: $\tilde{A} \bigtriangleup \tilde{B} = \tilde{A} \cup \tilde{B} - \tilde{A} \cap \tilde{B}$. If $(X,\mathscr{B},\mu)$ is a measure space, then $\tilde{\mathscr{B}}$ is a metric space with metric $d(\tilde{A},\tilde{B}) = \mu(\tilde{A} \bigtriangleup \tilde{B})$. If we identify $\tilde{A}$ with $1_A \in L^1(X,\mathscr{B},\mu)$, the metric in question coincides with the L^1-metric. From this we see that $\tilde{\mathscr{B}}$ is a complete metric space. We shall speak of a dense subset of $\tilde{\mathscr{B}}$ referring to this metric. If $\tilde{\mathscr{B}}_0$ is a subalgebra of $\tilde{\mathscr{B}}$ it can be shown that $\tilde{\mathscr{B}}_0$ is dense in $\tilde{\mathscr{B}}$ if and only if $\tilde{\mathscr{B}}_0$ generates $\tilde{\mathscr{B}}$ as a σ-algebra. The following criterion for equivalence of measure spaces will be useful.

when $\mathscr{B}$ is countably generated, $\tilde{\mathscr{B}}$ is also separable.

PROPOSITION 5.1.   *Two measure spaces* $(X,\mathscr{B},\mu)$ *and* $(Y,\mathscr{D},v)$ *are equivalent if and only if* $\tilde{\mathscr{B}}$ *contains a dense subalgebra* $\tilde{\mathscr{B}}_0$ *and* $\tilde{\mathscr{D}}$ *contains a dense subalgebra* $\tilde{\mathscr{D}}_0$ *such that there exists a measure preserving isomorphism between* $\tilde{\mathscr{B}}_0$ *and* $\tilde{\mathscr{D}}_0$.

Proof: If the two spaces are equivalent, we can take $\tilde{\mathscr{B}}_0 = \tilde{\mathscr{B}}, \tilde{\mathscr{D}}_0 = \tilde{\mathscr{D}}$. Conversely suppose $\tilde{\mathscr{B}}_0 \cong \tilde{\mathscr{D}}_0$. If $\tilde{A} \in \tilde{\mathscr{D}}$ take $\tilde{A}_n \in \tilde{\mathscr{D}}_0$ with $\mu(\tilde{A} \bigtriangleup \tilde{A}_n) \to 0$. $\{\tilde{A}_n\}$ forms a Cauchy sequence. Hence $\{\alpha^{-1}(\tilde{A}_n)\}$ forms a Cauchy sequence, and we can set $\alpha^{-1}(\tilde{A}) = \lim \alpha^{-1}(\tilde{A}_n)$. It is easy to check that α^{-1} satisfies the conditions for a homomorphism and since $\tilde{\mathscr{B}}_0$ is dense in $\tilde{\mathscr{B}}$, α^{-1} will be onto.

We turn now to *measure preserving systems*. It will be convenient for our present purposes to replace the single measure preserving transformation T of the measure space $(X,\mathscr{B},\mu)$ by a group Γ of measure preserving transformations (thereby requiring in addition that each transformation be invertible).

DEFINITION 5.3.   *A measure preserving system (m.p.s.) consists of a measure space* $(X,\mathscr{B},\mu)$ *and a group* Γ *acting on* X *by measure preserving transformations. We denote this system* $(X,\mathscr{B},\mu,\Gamma)$.

If $S \in \Gamma$, we shall denote by the same letter the corresponding transformation $S: X \to X$. We should bear in mind that different group elements may define the same transformation.

Each $S \in \Gamma$ defines an automorphism of $\mathscr{B}$ with $A \to S^{-1}A$, and this induces an automorphism $S^{-1}:\tilde{\mathscr{B}} \to \tilde{\mathscr{B}}$. We use this in defining homomorphisms of measure preserving systems.

DEFINITION 5.4. *A homomorphism of measure preserving systems* $\alpha:(X,\mathscr{B},\mu,\Gamma) \to (Y,\mathscr{D},v,\Gamma)$, *is given by a homomorphism* $\alpha:(X,\mathscr{B},\mu) \to (Y,\mathscr{D},v)$ *satisfying in addition to conditions* (i) *to* (iii) *of Definition 5.1, the condition*

(iv) $\qquad \alpha^{-1}(S^{-1}\tilde{A}) = S^{-1}\alpha^{-1}(\tilde{A}), \qquad \tilde{A} \in \tilde{\mathscr{D}}, \qquad S \in \Gamma.$

In this case we say that $(Y,\mathscr{D},v,\Gamma)$ *is a factor of* $(X,\mathscr{B},\mu,\Gamma)$ *and that* $(X,\mathscr{B},\mu,\Gamma)$ *is an extension of* $(Y,\mathscr{D},v,\Gamma)$. *The two measure preserving systems are equivalent if the homomorphism of one to the other is invertible* (i.e., *if* $\alpha^{-1}(\tilde{\mathscr{D}}) = \tilde{\mathscr{B}}$).

Once again, the standard example of a homomorphism of measure preserving systems is that of a spacial map $\phi:X \to Y$ which in addition to being measure preserving satisfies $S\phi(x) = \phi S(x)$ for $x \in X$ and $S \in \Gamma$. Then $\phi^{-1}S^{-1}(A) = S^{-1}\phi^{-1}(A)$, and so $\alpha^{-1}(\tilde{A}) = \phi^{-1}(A)$ satisfies condition (iv).

We mention two examples of spacial homomorphisms of measure preserving systems. Let $(X,\mathscr{B},\mu,\Gamma)$ be a *m.p.s.*, and let $\mathscr{D} \subset \mathscr{B}$ be a σ-algebra with $S\mathscr{D} \subset \mathscr{D}$ for $S \in \Gamma$. Then the transformations $S \in \Gamma$ are measure preserving transformations of $(X,\mathscr{D},\mu)$, and the system $(X,\mathscr{D},\mu,\Gamma)$ is a factor of $(X,\mathscr{B},\mu,\Gamma)$, with ϕ the identity map of X to itself. Notice that although ϕ is $1 - 1$ and onto, it need not define an equivalence of measure preserving systems. The two systems $(X,\mathscr{B},\mu,\Gamma)$ and $(X,\mathscr{D},v,\Gamma)$ will be equivalent if and only if $\tilde{\mathscr{D}} = \tilde{\mathscr{B}}$, i.e., if $\mathscr{B}$ differs from $\mathscr{D}$ only by null sets.

Another example consists of forming the product of two systems with the same group $\Gamma:(Y,\mathscr{D},v,\Gamma) \times (Z,\mathscr{E},\theta,\Gamma) = (Y \times Z,\mathscr{D} \times \mathscr{E},v \times \theta,\Gamma)$ where $S(y,z) = (Sy,Sz)$. Then $(Y,\mathscr{D},v,\Gamma)$ is a factor of this product system with $\phi(y,z) = y$. An important generalization is that of a *skew product*. Let $(Y,\mathscr{D},v,\Gamma)$ be a *m.p.s.* and $(Z,\mathscr{E},\theta)$ a measure space. Let $\sigma(S,x)$ define a measure preserving transformation of $(Z,\mathscr{E},\theta)$ for each $S \in \Gamma$, $y \in Y$ such that:

(i) for each S, $\sigma(S,y)(z)$ defines a measurable map from $Y \times Z$ to Z,
(ii) $\sigma(S,y)$ is a *cocycle*, i.e.,

(5.2) $\qquad \sigma(S_1 S_2,y) = \sigma(S_1,S_2 y) \, \sigma(S_2,y).$

We now define the *skew product* (determined by σ), $(Y \times Z,\mathscr{D} \times \mathscr{E},v \times \theta,\Gamma)$ by

(5.3) $\qquad S(y,z) = (Sy,\sigma(S,y)z).$

We use Fubini's theorem to verify that (5.3) defines a measure preserving transformation and the cocycle condition (5.2) to show that the transformation $S_1 S_2$ of $Y \times Z$ is the composition of S_1 and S_2. The ordinary product corresponds to the special cocycle: $\sigma(S, y)$ independent of y.

The latter class of examples is very broad. In fact if $(X, \mathscr{B}, \mu, \Gamma)$ is ergodic and satisfies some mild regularity conditions and $(Y, \mathscr{D}, v, \Gamma)$ is a factor, then by a theorem of Rokhlin [1], $(X, \mathscr{B}, \mu, \Gamma)$ is equivalent to a skew product of $(Y, \mathscr{D}, v, \Gamma)$ with some space $(Z, \mathscr{E}, \theta)$. We shall not need this result and we won't give any more details. It will be helpful to the reader, however, for heuristic purposes, to have in mind the case of a skew product when dealing with an arbitrary extension. This is particularly useful when considering the group generated by a single transformation T. Here T may be extended to the product space $Y \times Z$ by letting $\sigma(y)$ be a measurable family of measure preserving transformation of $(Z, \mathscr{E}, \theta)$ and setting

$$T(y,z) = (Ty, \sigma(y)z).$$

Iterating T, we find

$$T^n(y,z) = (T^n y, \sigma_n(y)z)$$

where

(5.4) $\sigma_n(y) = \sigma(T^{n-1}y)\sigma(T^{n-2}y) \cdots \sigma(y).$

This implies $\sigma_{n+m}(y) = \sigma_n(T^m y)\,\sigma_m(y)$ which gives (5.2) if we write $\sigma(T^n, y) = \sigma_n(y)$. (For $n \leq 0$ we use (5.2) to define $\sigma(T^n, y)$.)

By analogy with Proposition 5.1 we have the following criterion for the equivalence of measure preserving systems.

PROPOSITION 5.2.   *Two m.p.s. $(X, \mathscr{B}, \mu, \Gamma)$ and $(Y, \mathscr{D}, v, \Gamma)$ are equivalent if and only if $\tilde{\mathscr{B}}$ contains a dense Γ-invariant algebra $\tilde{\mathscr{B}}_0$ and $\tilde{\mathscr{D}}$ contains a dense Γ-invariant algebra $\tilde{\mathscr{D}}_0$ such that there exists a measure preserving isomorphism between $\tilde{\mathscr{B}}_0$ and $\tilde{\mathscr{D}}_0$ which is compatible with the action of Γ.*

The proof is the same as that of Proposition 5.1.

Before concluding this section, let us given an example of a homomorphism of measure spaces that is not implemented by a spacial map. This example is of a pathological nature; however, it is useful in illustrating the precautions that must be taken in considering general measure spaces.

Let $\Omega \subset [0,1]$ be a nonmeasurable set with outer measure 1 and inner measure 0. Let $\mathscr{D}$ consist of intersections of borel sets with Ω, and if A is a borel set, let $v(A \cap \Omega) = \mu(A)$, μ denoting Lebesgue measure. Note that if $A \cap \Omega = A' \cap \Omega$ for two borel sets, then the symmetric difference of A and A' has inner measure—and therefore measure—zero. Hence the fore-

For each borel subset A
of $[0,1]$, if $A \subseteq \Omega$ or
if $A \subseteq [0,1]\backslash\Omega$ then
$\mu(A) = 0$.

going definition of v is unambiguous. We now take $(X,\mathscr{B},\mu)$ to be the unit interval with Lebesgue measure on borel sets, and $(Y,\mathscr{D},v) = (\Omega,\mathscr{D},v)$. While the map $A \cap \Omega \to A$ is not well defined, it is well defined up to sets of measure 0. Thus $\alpha^{-1}(A \cap \Omega) = \tilde{A}$ is well defined and gives a homomorphism of $(X,\mathscr{B},\mu)$ to $(Y,\mathscr{D},v)$ which does not arise from a map of $X \to Y$.

We close this section with some conventions regarding notation. We shall often find it convenient to denote both a measure space $(X,\mathscr{B},\mu)$ and a measure preserving system $(X,\mathscr{B},\mu,\Gamma)$ by the more concise symbol, boldface $\mathbf{X}$, when there is no danger of confusion. The space $L^p(X,\mathscr{B},\mu)$ will be denoted accordingly by $L^p(\mathbf{X})$. Homomorphisms of systems are also written $\alpha:\mathbf{X} \to \mathbf{Y}$, and when ϕ is a spacial map, $\phi:X \to Y$, we shall also write $\phi:\mathbf{X} \to \mathbf{Y}$.

When α is a homomorphism of systems $\alpha:\mathbf{X} \to \mathbf{Y}$, then we have seen that we can lift measurable functions f on Y to X by $f \to f^\alpha$. This extends the lifting $f \to f \circ \phi$ in the case of a spacial map $\phi:X \to Y$. We shall also denote the composition $f \circ \phi$ by f^ϕ to accord with this notation.

If Γ is the group of transformations of a $m.p.s.$ $(X,\mathscr{B},\mu,\Gamma)$, we denote its members by Latin characters $S,T,\dots$. In this case we shall denote the composition $f \circ T$ by Tf rather than f^T. This notation could lead to difficulties $(S(Tf) = TSf)$ if not for the fact we shall only be dealing with commutative groups Γ.

If ϕ is a *measurable* map of $(X,\mathscr{B})$ to $(Y,\mathscr{D})$, i.e., if $\phi:X \to Y$ with $\phi^{-1}(\mathscr{D}) \subset \mathscr{B}$, and μ is a measure on $(X,\mathscr{B})$, then $\phi\mu$ denotes the measure

$$(5.5) \qquad \phi\mu(A) = \mu(\phi^{-1}A).$$

We can then verify that $f \to f^\phi$ maps $L^1(Y,\mathscr{D},\phi\mu)$ into $L^1(X,\mathscr{D},\mu)$ and

$$(5.6) \qquad \int f^\phi \, d\mu = \int f \, d\phi\mu.$$

In the case of a $m.p.s.$ $(X,\mathscr{B},\mu,\Gamma)$, each $T \in \Gamma$ maps $L^p(\mathbf{X})$ to itself and by (5.6) defines an isometry of $L^p(\mathbf{X})$ with itself.

2. Regular Measure Spaces

For certain of the constructions to be carried out it will be necessary to choose between equivalent measure spaces, confining ones attention to *regular* spaces.

DEFINITION 5.5.   *A measure space $(X,\mathscr{B},\mu)$ is regular if X is a compact metric space and $\mathscr{B}$ consists of all borel sets in X. A m.p.s. $(X,\mathscr{B},\mu,\Gamma)$ is regular if the underlying measure space is regular.*

DEFINITION 5.6.   *A measure space $(X,\mathscr{B},\mu)$ is separable if the associated σ-algebra $\tilde{\mathscr{B}}$ (see §1) is generated by a countable subset. A measure preserving system $(X,\mathscr{B},\mu,\Gamma)$ is separable if $(X,\mathscr{B},\mu)$ is separable and, in addition, Γ is countable.*

An equivalent condition for separability of the measure space $(X,\mathscr{B},\mu)$ is the separability of the metric space $\tilde{\mathscr{B}}$, or any of the spaces $L^p(\mathbf{X}), p < \infty$.

Since a compact metric space has a countable basis for open sets and these generate the borel sets, it is clear that a regular measure space is separable. The following weak converse is also true. We shall defer the proof of this proposition to Section 6 where a stronger result will be proved (Theorem 5.15).

PROPOSITION 5.3.   *Every separable measure space is equivalent to a regular measure space. Every separable m.p.s. is equivalent to a regular m.p.s.*

In our analysis of the structure of a measure preserving system we shall consider together with a given system, all of its factors. We shall always begin with a separable system, and it is easily seen that a factor of a separable system is separable. According to Proposition 5.3 any such factor is equivalent to a regular system that will itself be a factor of the given system. Thus, in studying the structure of a separable m.p.s., we shall be justified in confining our attention to its regular factors.

3. Conditional Expectation

Whenever we have a homomorphism $\alpha:(X,\mathscr{B},\mu) \to (Y,\mathscr{D},v)$, there is defined a natural map $f \to f^\alpha$ lifting measurable functions from Y to X. In this section we shall see how to define an operator in the reverse direction for integrable functions. This operator will map each $L^p(X,\mathscr{B},\mu)$ to $L^p(Y,\mathscr{D},v)$. We shall define it first for $p = 2$ and then extend the operator from $L^2(X,\mathscr{B},\mu)$ to $L^1(X,\mathscr{B},\mu)$. We refer the reader to Doob [1] for an alternative approach to the topic as well as for an explanation of the connection with conditional probabilities.

Let $\alpha:(X,\mathscr{B},\mu) \to (Y,\mathscr{D},v)$ be a homomorphism. The map $f \to f^\alpha$ identifies $L^2(Y,\mathscr{D},v)$ with a closed subspace $L^2(Y,\mathscr{D},v)^\alpha \subset L^2(X,\mathscr{B},\mu)$. If P denotes the orthogonal projection of $L^2(X,\mathscr{B},\mu)$ to $L^2(Y,\mathscr{D},v)^\alpha$, then we define $E(f\,|\,\mathbf{Y})$ for $f \in L^2(X,\mathscr{B},\mu)$ by

(5.7) $E(f\,|\,\mathbf{Y}) \in L^2(Y,\mathscr{D},v), E(f\,|\,\mathbf{Y})^\alpha = Pf.$

We remark that our notation is not the conventional one. In the more conventional approach one considers only homomorphisms of the form $\alpha:(X,\mathscr{B},\mu) \to (X,\mathscr{D},\mu)$ where $\mathscr{D}$ is a sub-σ-algebra of $\mathscr{B}$. Then the conditional expectation is denoted by $E(f\,|\,\mathscr{D})$. Since α is the identity map, $E(f\,|\,\mathscr{D})$ is already a function in $L^2(X,\mathscr{B},\mu)$. We shall find it necessary to consider the more general situation so that our notation $E(f|Y)$ will be more convenient. We shall have to distinguish between $E(f|Y) \in L^2(Y)$ and $E(f|Y)^\alpha \in L^2(X)$. Note that in the notation $E(f|Y)$ we have suppressed the homomorphism $\alpha:X \to Y$ which is implicit in the definition.

PROPOSITION 5.4. *The conditional expectation operator* $f \to E(f|Y)$ *defined for* $f \in L^2(X,\mathscr{B},\mu)$ *by* (5.7) *has the following properties*:

(i) $f \to E(f|Y)$ *is a linear operator of* $L^2(X,\mathscr{B},\mu)$ *to* $L^2(Y,\mathscr{D},v)$.
(ii) *If* $f \geq 0$, $E(f|Y) \geq 0$
(iii) *If* $f \in L^2(Y,\mathscr{D},v)$, $E(f^\alpha|Y) = f$. *In particular* $E(1|Y) = 1$.
(iv) *If* $g \in L^\infty(Y,\mathscr{D},v)$, $E(g^\alpha f|Y) = gE(f|Y)$.
(v) *In particular*, $\int f\, d\mu = \int E(f|Y)\, dv$.

Proof: Property (i) is an immediate consequence of the definition. Property (ii) follows by an argument similar to that of Lemma 4.23. Namely, since the subspace $L^2(Y,\mathscr{D},v) \subset L^2(X,\mathscr{B},\mu)$ is closed with respect to the operations $(f_1,f_2) \to \max(f_1,f_2)$ and since Pf represents the function in the subspace closest to f, a nonnegative function is projected onto a nonnegative function. Property (iii) is immediate from (5.7). To prove properties (iv) and (v) we use the following characterization of P. In general, if V is a closed subspace of a Hilbert space and P_V is the orthogonal projection onto V, then $\|P_V u - u\| = \min\{\|v - u\| : v \in V\}$ determines $P_V u$ uniquely. This is equivalent to

$$\langle u,v \rangle = \langle P_V u, v \rangle \qquad \text{for all} \qquad v \in V.$$

In our case this becomes

$$\int fh^\alpha\, d\mu = \int Pfh^\alpha\, d\mu \qquad \text{for all} \qquad h \in L^2(Y,\mathscr{D},v),$$

and transferring this to $E(f|Y)$ by (5.7),

$$(5.8) \qquad \int fh^\alpha\, d\mu = \int E(f|Y)^\alpha h^\alpha\, d\mu = \int E(f|Y)h\, dv, \qquad h \in L^2(Y,\mathscr{D},v).$$

Now, given that (5.8) is valid for all $h \in L^2(Y,\mathscr{D},v)$, replace h by gh with $g \in L^\infty(Y,\mathscr{D},v)$. Then

$$\int f(gh)^\alpha\, d\mu = \int E(f|Y)gh\, dv$$

or

$$\int (fg^\alpha)h^\alpha\, d\mu = \int (E(f|Y)g)h\, dv, \qquad \text{for all} \qquad h \in L^2(Y,\mathscr{D},v).$$

This proves property (iv), in view of the characterization (5.8). At the same time we have also proved property (v).

It will be useful to record the following corollary of the foregoing proof.

COROLLARY.   *If $f \in L^2(\mathbf{X})$, the conditional expectation $E(f \mid \mathbf{Y})$ is characterized by*

$$\int E(f \mid \mathbf{Y}) h \, dv = \int f h^\alpha \, d\mu$$

for all $h \in L^2(\mathbf{Y})$.

PROPOSITION 5.5.   *Let $\psi(t)$ be a convex nonnegative function of a real variable. If f and $\psi \circ f$ are in $L^2(X,\mathscr{B},\mu)$, then*

(5.9) $\psi(E(f \mid \mathbf{Y})) \le E(\psi \circ f \mid \mathbf{Y})$

Proof: The convex function ψ can be taken as the upper envelope of a countable set of linear functions:

$$\psi(x) = \sup L_n(x), \qquad L_n(x) = a_n x + b_n.$$

By virtue of (i) and (iii) of the foregoing proposition $E(L_n(f) \mid \mathbf{Y}) = L_n(E(f \mid \mathbf{Y}))$. Hence

$$\psi(E(f \mid \mathbf{Y})) = \sup L_n(E(f \mid \mathbf{Y})) = \sup E(L_n(f) \mid \mathbf{Y}).$$

According to (ii), $f_1 \le f_2 \Rightarrow E(f_1 \mid \mathbf{Y}) \le E(f_2 \mid \mathbf{Y}$ and since $L_n(f) \le \psi \circ f$, $E(L_n(f) \mid \mathbf{Y}) \le E(\psi \circ f \mid \mathbf{Y})$. This proves the proposition.

THEOREM 5.6.   *The conditional expectation map, $f \to E(f \mid \mathbf{Y})$, extends to a map of $L^1(X,\mathscr{B},\mu)$ to $L^1(Y,\mathscr{D},v)$ satisfying (i)–(v) of Proposition 5.4, and, in addition, it maps each $L^p(X,\mathscr{B},\mu)$ to $L^p(Y,\mathscr{D},v)$, $1 \le p \le \infty$, with $\|E(f \mid \mathbf{Y})\|_p \le \|f\|_p$.*

Proof: Since $L^2(X,\mathscr{B},\mu)$ is dense in $L^1(X,\mathscr{B},\mu)$ the conditional expectation can be extended to $L^1(X,\mathscr{B},\mu)$ if it defines a uniformly continuous map from $L^2(X,\mathscr{B},\mu)$ to $L^2(Y,\mathscr{D},v)$ in the L^1-norms of both spaces. This follows, however, from the foregoing proposition, taking $\psi(t) = |t|$, together with property (v) of Proposition 5.4. It follows readily that properties (i) to (v) of Proposition 5.4 are valid for this extension. We may now extend Proposition 5.5, replacing the condition $f, \psi \circ f \in L^2$ by $f, \psi \circ f \in L^1$. Now suppose that $f \in L^p(X,\mathscr{B},\mu)$. Apply Proposition 5.5 with $\psi(t) = |t|^p$. We conclude that $E(f \mid \mathbf{Y}) \in L^p(Y,\mathscr{D},v)$ and that the map $f \to E(f \mid \mathbf{Y})$ is a contraction for each $p < \infty$. The case $p = \infty$ follows from properties (ii) and (iii) of Proposition 5.4.

We conclude our discussion of the conditional expectation with the following result pertaining to measure preserving systems.

PROPOSITION 5.7. *Let $(Y,\mathscr{D},v,\Gamma)$ be a factor of $(X,\mathscr{B},\mu,\Gamma)$. Then for each $f \in L^1(X,\mathscr{B},\mu)$ and $S \in \Gamma$, $E(Sf \mid Y) = SE(f \mid Y)$.*

Proof: It suffices to prove this for $f \in L^2(X,\mathscr{B},\mu)$. We use Proposition 5.4, property (v):

$$\int E(Sf \mid Y)h \, dv = \int (Sf)h^\alpha \, d\mu = \int f(S^{-1}h^\alpha) \, d\mu,$$

using the fact that S is a unitary operator on $L^2(X,\mathscr{B},\mu)$. Now turning to the characterization (5.1) of f^α, we see that

$$\alpha^{-1}\{y:Sf(y) \in B\} = \alpha^{-1}S^{-1}\{y:f(y) \in B\} = S^{-1}\alpha^{-1}\{y:f(y) \in B\}$$
$$= S^{-1}\{x:f^\alpha(x) \in B\} = \{x:Sf^\alpha(x) \in B\}$$

from which it follows that

$$(Sf)^\alpha = S(f^\alpha).$$

Hence

$$\int E(Sf \mid Y)h \, dv = \int f(S^{-1}h)^\alpha \, d\mu = \int E(f \mid Y)S^{-1}h \, dv$$
$$= \int SE(f \mid Y)h \, dv$$

using the fact that S is unitary on $L^2(Y,\mathscr{D},v)$. According to the corollary to Proposition 5.4, $E(Sf \mid Y) = SE(f \mid Y)$.

4. Disintegration of Measures

Suppose a map $\phi:X \to Y$ defines a homomorphism with respect to some structure on X and Y. It is natural to examine what takes place above the points $y \in Y$, i.e., on *fibers* of the map, $X_y = \phi^{-1}(y)$. When X and Y are endowed with the structure of measure spaces, we shall want to replace the fibers by measure spaces. In this section we shall show how this can be done for a homomorphism $\alpha:(X,\mathscr{B},\mu) \to (Y,\mathscr{D},v)$. We shall obtain a measure-valued function $y \to \mu_y$ defined almost everywhere on Y with μ_y a measure on $(X,\mathscr{B})$, so that the measure space $(X,\mathscr{B},\mu_y)$ plays the role of the fiber over y. In "good" situations with α induced by a spacial map $\phi:X \to Y$, the measure μ_y will in fact be concentrated on the fiber $\phi^{-1}(y)$. This will not always be so, but it will be useful to picture the situation in this way.

Let $(X,\mathscr{B},\mu)$ be a regular measure space, and let $\alpha:(X,\mathscr{B},\mu) \to (Y,\mathscr{D},v)$ be a homomorphism to another measure space (not necessarily regular). Here X is a compact metric space, and we shall denote by $\mathscr{M}(X)$ the compact metric space of probability measures on X.

THEOREM 5.8.   *There exists a measurable map from Y to $\mathcal{M}(X)$ which we shall denote $y \to \mu_y$ which satisfies*

(i)   *For every $f \in L^1(X,\mathcal{B},\mu), f \in L^1(X,\mathcal{B},\mu_y)$ for a.e. $y \in Y$, and $E(f\,|\,\mathbf{Y})(y) = \int f \, d\mu_y$ for a.e. $y \in Y$.*

(ii) $\int \{\int f \, d\mu_y\} \, dv(y) = \int f \, d\mu$ *for every $f \in L^1(X,\mathcal{B},\mu)$.*

The map $y \to \mu_y$ is characterized by condition (i). We shall write $\mu = \int \mu_y \, dv$ and refer to this as the *disintegration of μ with respect to the factor* $(Y,\mathcal{D},v)$.

Proof: The main idea of the proof is that on the compact metric space X, there is a one-to-one correspondence between borel measures and linear functionals on $\mathcal{C}(X)$, the space of real valued continuous functions on X. We can therefore use property (i) to define the measures μ_y. The delicate point in the argument is handling the fact that $E(f\,|\,\mathbf{Y})$ is well-defined in $L^1(Y,\mathcal{D},v)$; that is to say, it is defined only almost everywhere as function of Y. We proceed as follows. Choose a countable dense set of functions (with respect to uniform convergence) in $\mathcal{C}(X)$, assume that 1 is in the set, and let $\mathcal{A}$ consist of all finite linear combinations of these with rational coefficients. The set $\mathcal{A}$ is a countable subset of $\mathcal{C}(X)$. Now let $L(f)$ denote a functional on $\mathcal{A}$ satisfying (i) $L(af_1 + bf_2) = aL(f_1) + bL(f_2)$ for $f_1, f_2 \in \mathcal{A}$, a,b rational, (ii) $|L(f)| \le \max |f(x)|$, (iii) $f \ge 0 \Rightarrow L(f) \ge 0$, (iv) $L(1) = 1$. We claim that we can extend L to a linear functional on $\mathcal{C}(X)$ satisfying (iii). This is done by uniform continuity; property (ii) guarantees that $L(f)$ is uniformly continuous on the dense subset $\mathcal{A}$ of $\mathcal{C}(X)$. Extending L by continuity, we obtain linearity for rational coefficients directly from property (i) and subsequently for all real coefficients. If $f \ge 0$ and $f_n \in \mathcal{A}$ with

$$\left| f_n - \left(f + \frac{1}{n} \right) \right| < \frac{1}{n},$$

then $f_n \ge 0$ and $f_n \to f$ uniformly. Since $L(f_n) \ge 0$, we shall have $L(f) \ge 0$. Now each linear functional on $\mathcal{C}(X)$ satisfying $L(1) = 1$ and $f \ge 0 \Rightarrow L(f) \ge 0$ corresponds to a measure in $\mathcal{M}(X)$. We conclude that a measure in $\mathcal{M}(X)$ is determined by a functional on $\mathcal{A}$ satisfying properties (i) to (iv). Having made this correspondence, we point out that a subbasis for the open sets in $\mathcal{M}(X)$ consists of sets of the form $U_f = \{L : L(f) > 0\}$ as f ranges over $\mathcal{A}$. These sets generate the family of borel sets in $\mathcal{M}(X)$, and so to check the measurability of a function $Y \to \mathcal{M}(X)$, it suffices to establish the measurability of the counterimage of each U_f.

Now choose, for each $f \in \mathcal{A}$, some everwhere defined version of $E(f\,|\,\mathbf{Y})$. If we set $L_y(f) = E(f\,|\,Y)(y)$, then we see that almost every L_y satisfies properties (i) to (iv), since these amount to countably many conditions.

Moreover, $\{y:L_y(f) > 0\}$ is a measurable set for $(Y,\mathcal{D},v)$; so the map $y \to L_y$ is measurable. Identifying L_y with a measure μ_y by way of $L_y(f) = \int f \, d\mu_y$ defines the map $y \to \mu_y$, and we see that property (i) is automatically satisfied for $f \in \mathcal{A}$. Using uniform convergence, we conclude that property (i) is valid for $f \in \mathscr{C}(X)$. Note that whenever property (i), is valid for a function f, then property (ii) will be valid by virtue of Proposition 5.4, property (v). For $f \in L^1(X,\mathscr{B},\mu)$ note that we have $f = \sum g_n$, a.e., $g_n \in \mathscr{C}(X)$ where $\sum \|g_n\|_1 < \infty$. Then $\sum \|E(|g_n| \, \|Y)\|_1 < \infty$, so that $\sum E(|g_n| \, |Y) < \infty$ a.e. At a point y for which this series converges, $\sum |g_n| \in L^1(X,\mathscr{B},\mu_y)$, and so

$$\int (\sum g_n) \, d\mu_y = \sum \int g_n \, d\mu_y.$$

We also have a.e., $\int g_n \, d\mu_y = E(g_n|Y)(y)$, so that

$$\int (\sum g_n) \, d\mu_y = \sum E(g_n|Y)(y) \quad a.e.$$

Since $f \to E(f|Y)$ is a contraction in L^1, we have

$$E(\sum g_n|Y) = L^1\text{-lim} \sum_n E(g_n|Y).$$

But the a.e.-limit, if it exists, must coincide with the L^1-limit, and so

$$\int (\sum g_n) \, d\mu_y = E(\sum g_n|Y) \quad a.e.$$

To complete the proof of (i), we must show that the null function $f - \sum g_n$ remains a null function for almost every μ_y. Let $\mu(A) = 0$ and let A_n be open with $A_n \supset A$, $\mu(A_n) \to 0$. Let h_n be a continuous function, $h_n = 0$ outside of A_n, and $0 < h_n \leq 1$ on A_n. For each $\varepsilon > 0$, the power h_n^ε is continuous and

$$\int \left\{ \int h_n^\varepsilon \, d\mu_y \right\} dv(y) = \int h_n^\varepsilon \, d\mu \leq \mu(A_n).$$

Letting $\varepsilon \to 0$, we find

$$\int \mu_y(A_n) \, dv(y) \leq \mu(A_n),$$

and so

$$\int \mu_y(A) \, dv(y) \leq \mu(A_n) \to 0.$$

This completes the proof of property (i), and as we have already observed property (ii) follows from (i) by Proposition 5.4 property (v).

To complete the proof of the theorem, we remark that a measure on $(X,\mathscr{B})$ is determined by the integrals of continuous function with respect to it, so property (i) clearly determines μ_y a.e.

Theorem 5.8 is a generalization of Fubini's theorem to the context of arbitrary extensions of measure spaces. Naturally if $(X,\mathcal{B},\mu) = (Y,\mathcal{D},v) \times (Z,\mathcal{E},\theta)$, then $\mu_y = \delta_y \times \theta$, which is the copy of θ lying over the point y.

Finally, let us consider disintegration of measures for measure preserving systems.

PROPOSITION 5.9.   *Let $(Y,\mathcal{D},v,\Gamma)$ be a factor of $(X,\mathcal{B},\mu,\Gamma)$, and let $\mu = \int \mu_y \, dv(y)$ be the disintegration of μ with respect to this factor. For each $S \in \Gamma$ and for almost every $y \in Y$,*

$$(5.10) \quad \mu_{Sy} = S\mu_y$$

Proof: The property (i) of Theorem 5.8 characterizes the measure valued function μ_y. Use Proposition 5.7 to write

$$\int Sf \, d\mu_y = \int f \, d\mu_{Sy}.$$

Replace f by $S^{-1}f$:

$$\int f \, d\mu_y = \int S^{-1}f \, d\mu_{Sy} = \int f \, d(S^{-1}\mu_{Sy})$$

by virtue of (5.6). Thus $\mu_y = S^{-1}\mu_{Sy}$ *a.e.*, and this is equivalent to (5.10).

5. Relative Products of Measure Spaces

Let $(X_1,\mathcal{B}_1,\mu_1)$ and $(X_2,\mathcal{B}_2,\mu_2)$ be two regular measure spaces that are extensions of the same space $(Y,\mathcal{D},v)$:

$$\alpha_1:(X_1,\mathcal{B}_1,\mu_1) \to (Y,\mathcal{D},v), \qquad \alpha_2:(X_2,\mathcal{B}_2,\mu_1) \to (Y,\mathcal{D},v).$$

In this section we shall define the product of the two measure spaces relative to their common factor. This will play the same role measure theoretically as the fiber-product plays set-theoretically. (If $F_1 \cdot X_1 \to Y$, $F_2:X_2 \to Y$ are arbitrary set maps the *fiber-product* consists of

$$\{(x_1,x_2):F_1(x_1) = F_2(x_2)\} = \bigcup_{y \in Y} F_1^{-1}(y) \times F_2^{-1}(y).$$

When Y reduces to a single point, the fiber-product is the ordinary product). From the probabilistic point of view, the relative product provides a probability space on which both σ-algebras of events $\mathcal{B}_1$ and $\mathcal{B}_2$ are represented and are "independent conditioned upon $\mathcal{D}$".

DEFINITION 5.7.   *Denote by $\mu_1 \times_Y \mu_2$ the measure on $X_1 \times X_2$ defined by*

$$(5.11) \quad \mu_1 \times_Y \mu_2(A) = \int \mu_{1,y} \times \mu_{2,y}(A) \, dv(y)$$

for $A \in \mathscr{B}_1 \times \mathscr{B}_2$, *where* $\mu_i = \int \mu_{i,y} \, dv(y)$ *are the disintegrations of* μ_i *with respect to the factor* $(Y, \mathscr{D}, v)$. *The measure space* $(X_1 \times X_2, \mathscr{B}_1 \times \mathscr{B}_2,$ $\mu_1 \times_Y \mu_2)$ *is called the product of* $(X_1, \mathscr{B}_1, \mu_1)$ *and* $(X_2, \mathscr{B}_2, \mu_2)$ *relative to* $(Y \mathscr{D}, v)$ *and is denoted* $\mathbf{X}_1 \times_Y \mathbf{X}_2$.

To justify the measurability of the integrand in (5.11), let us note that whenever $\mu = \int \mu_y \, dv(y)$ is the disintegration of μ with respect to $(Y, \mathscr{D}, v)$, with μ a borel measure on the compact metric space X, then $\int f \, d\mu_y$ is measurable for each continuous $f \in \mathscr{C}(X)$, by the definition of measurability of μ_y. Since the set of f for which $\int f \, d\mu_y$ is measurable is closed under monotone limits, the same is true for f borel measurable. We now conclude that $\mu_{1,y} \times \mu_{2,y}(A)$ is measurable for $A = A_1 \times A_2$, $A_i \in \mathscr{B}_i$, $i = 1, 2$. Again by passage to the limit, the result is valid for all $A \in \mathscr{B}_1 \times \mathscr{B}_2$. Once the expression (5.11) is seen as meaningful, it is easily verified that it defines a σ-additive probability measure on $(X_1 \times X_2, \mathscr{B}_1 \times \mathscr{B}_2)$.

If f_1 is a function on a space X_1 and f_2 is a function on X_2, we continue the convention of denoting by $f_1 \otimes f_2$ the function

$$f_1 \otimes f_2(x_1, x_2) = f_1(x_1) f_2(x_2).$$

We now have the following result.

PROPOSITION 5.10. *The measure* $\mu_1 \times_Y \mu_2$ *is characterized by the equality*

$$(5.12) \quad \int f_1 \otimes f_2 \, d\mu_1 \times_Y \mu_2 = \int E(f_1 \mid \mathbf{Y}) E(f_2 \mid \mathbf{Y}) \, dv$$

satisfied whenever $f_1 \in L^2(\mathbf{X}_1)$, $f_2 \in L^2(\mathbf{X}_2)$.

In this expression the two conditional expectations refer to the corresponding operators from $L^2(X_i, \mathscr{B}_i, \mu_i)$ to $L^2(Y, \mathscr{D}, v)$.

Proof: We have

$$\int f_1(x_1) f_2(x_2) \, d\mu_1 \times_Y \mu_2(x_1, x_2)$$

$$= \int \{ f_1(x_1) f_2(x_2) \, d\mu_{1,y} \times \mu_{2,y}(x_1, x_2) \} \, dv(y)$$

$$= \int \left\{ \int f_1(x_1) \, d\mu_{1,y}(x_1) \right\} \left\{ \int f_2(x_2) \, d\mu_{2,y}(x_2) \right\} dv(y)$$

$$= \int E(f_1 \mid \mathbf{Y})(y) E(f_2 \mid \mathbf{Y})(y) \, dv(y).$$

Let π_1, π_2 be the respective projections of $X_1 \times X_2$ onto its components X_1 and X_2. We verify that these are measure preserving maps of

$$(X_1 \times X_2, \mathscr{B}_1 \times \mathscr{B}_2, \mu_1 \times_Y \mu_2) \to (X_i, \mathscr{B}_i, \mu_i).$$

For example,

$$\mu_1 \times_{\mathbf{Y}} \mu_2(\pi_1^{-1}(A_1)) = \mu_1 \times_{\mathbf{Y}} \mu_2(A_1 \times X_2)$$
$$= \int \mu_{1,y} \times \mu_{2,y}(A_1 \times X_2)\, dv(y)$$
$$= \int \mu_{1,y}(A_1)\, dv(y) = \mu_1(A_1).$$

Thus $(X_1 \times X_2, \mathscr{B}_1 \times \mathscr{B}_2, \mu_1 \times_{\mathbf{Y}} \mu_2)$ is an extension of both $(X_i, \mathscr{B}_i, \mu_i)$ and these are both extensions of $(Y, \mathscr{D}, v)$. We thus obtain two homomorphisms of the relative product to $(Y, \mathscr{D}, v)$ as indicated in the diagram:

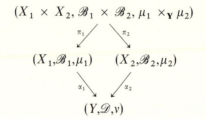

PROPOSITION 5.11.   *The above diagram is commutative. In other words, the compositions* $\alpha_1 \pi_1$ *and* $\alpha_2 \pi_2$ *define the same homomorphism of* $(X_1 \times X_2, \mathscr{B}_1 \times \mathscr{B}_2, \mu_1 \times_{\mathbf{Y}} \mu_2)$ *to* $(Y, \mathscr{D}, v)$. *Thus the relative product is unambiguously an extension of* $(Y, \mathscr{D}, v)$.

Proof: We have to consider the two maps of $\tilde{\mathscr{D}} \to \tilde{\mathscr{B}}_1 \times \tilde{\mathscr{B}}_2$:

$$\tilde{A} \to \pi_1^{-1}\alpha_1^{-1}(\tilde{A}), \pi_2^{-1}\alpha_2^{-1}(\tilde{A}).$$

To show that these maps are identical, it will be convenient to consider more generally the two injections of

$$L^2(Y, \mathscr{D}, v) \to L^2(X_1 \times X_2, \mathscr{B}_1 \times \mathscr{B}_2, \mu_1 \times_{\mathbf{Y}} \mu_2).$$

Using the notation of Section 1, we have: $f \to f^{\alpha_1 \pi_1}$, $f \to f^{\alpha_2 \pi_2}$. We shall show that $f^{\alpha_1 \pi_1} - f^{\alpha_2 \pi_2}$ is a null function on $X_1 \times X_2$:

$$\int (f^{\alpha_1 \pi_1} - f^{\alpha_2 \pi_2})^2 \, d\mu_1 \times_{\mathbf{Y}} \mu_2$$
$$= \int (f^{\alpha_1 \pi_1})^2 \, d\mu_1 \times_{\mathbf{Y}} \mu_2 - 2 \int f^{\alpha_1 \pi_1} f^{\alpha_2 \pi_2} \, d\mu_1 \times_{\mathbf{Y}} \mu_2$$
$$+ \int (f^{\alpha_2 \pi_2})^2 \, d\mu_1 \times_{\mathbf{Y}} \mu_2.$$

In these integrals the integrand is either a function of x_1 or of x_2 or a product of two such functions. Applying Proposition 5.10 and Proposition

5.4, property (iii), we find

$$\int (f^{\alpha_1\pi_1})^2 \, d\mu_1 \times_Y \mu_2 = \int (f^{\alpha_1})^2 \, d\mu_1 = \int f^2 \, dv,$$

$$\int f^{\alpha_1\pi_1} f^{\alpha_2\pi_2} \, d\mu \times_Y \mu_2 = \int f^{\alpha_1} \otimes f^{\alpha_2} \, d\mu_1 \times_Y \mu_2$$

$$= \int E(f^{\alpha_1}|Y) E(f^{\alpha_2}|Y) \, dv = \int f^2 \, dv,$$

$$\int (f^{\alpha_2\pi_2})^2 \, d\mu_1 \times_Y \mu_2 = \int (f^{\alpha_2})^2 \, d\mu_2 = \int f^2 \, dv,$$

so that

$$\int (f^{\alpha_1\pi_1} - f^{\alpha_2\pi_2})^2 \, d\mu_1 \times_Y \mu_2 = 0.$$

This proves the proposition.

Since $(X_1 \times X_2, \mathscr{B}_1 \times \mathscr{B}_2, \mu_1 \times_Y \mu_2)$ is an extension of $(Y, \mathscr{D}, v)$, we can disintegrate $\mu_1 \times_Y \mu_2$ with respect to $(Y, \mathscr{D}, v)$.

PROPOSITION 5.12. *We have for* $f_1 \in L^2(X_1)$, $f_2 \in L^2(X_2)$,

(5.13) $E(f_1 \otimes f_2 | Y) = E(f_1 | Y) E(f_2 | Y)$

and the disintegration of $\mu_1 \times_Y \mu_2$ *with respect to* $(Y, \mathscr{D}, v)$ *is*

$$(\mu_1 \times_Y \mu_2)_y = \mu_{1,y} \times \mu_{2,y}.$$

Proof: It is easy to see from the definition of conditional expectation that $E(f | Y) = E(E(f | X_1) | Y) = E(E(f | X_2) | Y)$. We proceed to evaluate $E(f_1 \otimes f_2 | X_1)$. We claim $E(f_1 \otimes f_2 | X_1) = f_1 E(f_2 | Y)^{\alpha_1}$, and to prove it we shall use the corollary to Proposition 5.4. We must show that

$$\int f_1(x_1) f_2(x_2) h(x_1) \, d\mu_1 \times_Y \mu_2 = \int f_1(x_1) E(f_2 | Y)^{\alpha_1} h(x_1) \, d\mu_1(x_1).$$

If Proposition 5.4, property (v) is applied to the homomorphism α_1, the right-hand side becomes

$$\int E(f_1 h | Y) E(f_2 | Y) \, dv = \int f_1(x_1) h(x_1) f_2(x_2) \, d\mu_1 \times_Y \mu_2(x_1, x_2)$$

according to Proposition 5.10. This establishes our assertion. We now use Proposition 5.4, property (iv) to obtain

$$E(f_1 \otimes f_2 | Y) = E(E(f_1 \otimes f_2 | X_1) | Y) = E(f_1 E(f_2 | Y)^{\alpha_1} | Y)$$

$$= E(f_2 | Y) E(f_1 | Y) = \int f_1 \otimes f_2 \, d\mu_{1,y} \times \mu_{2,y}.$$

This proves both statements of the proposition.

Suppose now we are given three measure spaces, $(\mathbf{X}_i, \mathscr{B}_i, \mu_i)$, $i = 1,2,3$, that are extensions of $(Y, \mathscr{D}, v)$. We can form the relative product in three ways corresponding to the three measures in $X_1 \times X_2 \times X_3$:

$$(\mu_1 \times_\mathbf{Y} \mu_2) \times_\mathbf{Y} \mu_3, \qquad \mu_1 \times_\mathbf{Y} (\mu_2 \times_\mathbf{Y} \mu_3), \qquad \mu_1 \times_\mathbf{Y} \mu_2 \times_\mathbf{Y} \mu_3$$

where the last of these is defined symmetrically by generalizing (5.11):

$$\mu_1 \times_\mathbf{Y} \mu_2 \times_\mathbf{Y} \mu_3(A) = \int \mu_{1,y} \times \mu_{2,y} \times \mu_{3,y}(A) \, dv(y).$$

Proposition 5.12 shows that these are all identical and the triple relative product is defined unambiguously.
In particular

$$(\mathbf{X}_1 \times_\mathbf{Y} \mathbf{X}_2) \times_\mathbf{Y} \mathbf{X}_3 = \mathbf{X}_1 \times_\mathbf{Y} (\mathbf{X}_2 \times_\mathbf{Y} \mathbf{X}_3).$$

Similarly multiple relative products are defined unambigously.
Another technical result is the following.

PROPOSITION 5.13.   *If* $\mathbf{X}_1, \mathbf{X}_2$ *are extensions of* $\mathbf{Y}$ *and* $\mathbf{X}'_1$ *is equivalent to* $\mathbf{X}_1$ *and* $\mathbf{X}'_2$ *is equivalent to* $\mathbf{X}_2$, *then in a natural way,* $\mathbf{X}'_1$ *and* $\mathbf{X}'_2$ *are extensions of* $\mathbf{Y}$ *and* $\mathbf{X}'_1 \times_\mathbf{Y} \mathbf{X}'_2$ *is equivalent to* $\mathbf{X}_1 \times_\mathbf{Y} \mathbf{X}_2$.

Proof: We shall use the criterion of Proposition 5.1 for equivalence of measure space. Namely we establish an isomorphism between the (mod null sets) algebras generated by product sets in $\mathbf{X}_1 \times \mathbf{X}_2$ and in $\mathbf{X}'_1 \times \mathbf{X}'_2$. We have by hypothesis a correspondence between $L^2(\mathbf{X}_i)$ and $L^2(\mathbf{X}'_i)$ for for $i = 1,2$, so it will suffice to show that if $f'_1 \in L^2(X'_1, \mathscr{B}'_1, \mu'_1)$ corresponds to $f_1 \in L^2(X_1, \mathscr{B}_1, \mu_1)$ and $f'_2 \in L^2(X'_2, \mathscr{B}'_2, \mu'_2)$ corresponds to $f_2 \in L^2(X_2, \mathscr{B}_2, \mu_2)$, then

$$\int f'_1 \otimes f'_2 \, d\mu'_1 \times_\mathbf{Y} \mu'_2 = \int f_1 \otimes f_2 \, d\mu_1 \times_\mathbf{Y} \mu_2.$$

But this follows from Proposition 5.10 and the fact that $E(f'_i | \mathbf{Y}) = E(f_i | \mathbf{Y})$.

This proposition shows that although in the category of measure spaces the relative product is defined only for regular spaces, when we consider equivalence classes of measure spaces, the relative product is defined for all separable spaces. In the next chapter we shall define relative properties of extensions of measure preserving systems in terms of relative products. The present proposition shows that these properties may be defined even if the systems in question are not regular.
We now turn to relative products of measure preserving systems.

PROPOSITION 5.14.   *Suppose* $(X_i, \mathscr{B}_i, \mu_i, \Gamma)$, $i = 1,2$, *are extensions of the measuring preserving system* $(Y, \mathscr{D}, v, \Gamma)$. *Then the measure* $\mu_1 \times_\mathbf{Y} \mu_2$ *is* Γ-

invariant where the action of Γ *on* $X_1 \times X_2$ *is given by* $S(x_1,x_2) = (Sx_1,Sx_2)$.
Hence $(X_1 \times X_2,\mathscr{B}_1 \times \mathscr{B}_2,\mu_1 \times_Y \mu_2,\Gamma)$ *is a m.p.s. which is the relative product*
$(X_1,\mathscr{B}_1,\mu_1,\Gamma) \times_Y (X_2,\mathscr{B}_2,\mu_2,\Gamma)$.

Proof: We have to verify that

$$\mu_1 \times_Y \mu_2(S^{-1}A) = \mu_1 \times_Y \mu_2(A)$$

for $A \in \mathscr{B}_1 \times \mathscr{B}_2$, and $S \in \Gamma$. But according to equation (5.11),

$$\mu_1 \times_Y \mu_2(S^{-1}A) = \int \mu_{1,y} \times \mu_{2,y}(S^{-1}A)\,dv(y)$$

$$= \int S\mu_{1,y} \times S\mu_{2,y}(A)\,dv(y).$$

The latter identity is established for all $A \in \mathscr{B}_1 \times \mathscr{B}_2$ by checking it for
$A = A_1 \times A_2$, $A_i \in \mathscr{B}_i$. Now we use Proposition 5.9,

$$\mu_1 \times_Y \mu_2(S^{-1}A) = \int \mu_{1,Sy} \times \mu_{2,Sy}(A)\,dv(y)$$

$$= \int \mu_{1,y} \times \mu_{2,y}(A)\,dSv(y),$$

using (5.6). But $Sv = v$ for $S \in \Gamma$ and this proves the proposition.

6. Regular Homomorphisms

We collect a number of desirable properties of a homomorphism between
measure spaces in the following definition. We state it directly for measure
preserving systems since these are our principal concern. In any case a
measure space is a measure preserving system with trivial group Γ.

DEFINITION 5.8 *A homomorphism* $\alpha:(X,\mathscr{B},\mu,\Gamma) \to (Y,\mathscr{D},v,\Gamma)$ *is regular
if* $(X,\mathscr{B},\mu,\Gamma)$ *is a regular m.p.s. and* α *is a spacial map,* $\alpha:X \to Y$.

THEOREM 5.15. *Let* $\alpha:\mathbf{X} \to \mathbf{Y}$ *be a homomorphism of separable measure
preserving systems. There exists a system* $\mathbf{X}'$ *equivalent to* $\mathbf{X}$ *and a system*
$\mathbf{Y}'$ *equivalent to* $\mathbf{Y}$ *such that the corresponding homomorphism* $\alpha':\mathbf{X}' \to \mathbf{Y}'$
is regular.

Proof: Write $\mathbf{X} = (X,\mathscr{B},\mu,\Gamma)$, $\mathbf{Y} = (Y,\mathscr{D},v,\Gamma)$. Let $\mathscr{D}_0$ be a countable
algebra of sets in $\mathscr{D}$, with $\mathscr{D}_0$ invariant under Γ and with $\tilde{\mathscr{D}}_0$ dense in $\tilde{\mathscr{D}}$,
and write $\mathscr{D}_0 = \{A_m\}$. Let $Y' = \{0,1\}^N$, the space of all 0,1-sequences and
$\eta_m:Y' \to \{0,1\}$ be the m'th coordinate function. Let

$$A'_m = \{y' \in Y':\eta_m(y') = 1\}.$$

We claim that there exists a measure v' on Y' with

$$v'(A'_{i_1} \cap A'_{i_2} \cap \cdots \cap A'_{i_k}) = v(A_{i_1} \cap A_{i_2} \cap \cdots \cap A_{i_k})$$

for every k-tuple $i_1 < i_2 < \cdots < i_k$. To prove this we first find a measure v'_N that satisfies this condition for $i_k \leq N$. Since these conditions are consistent and involve a finite algebra of sets, it is easy to see that v'_N exists. Taking a limit of some subsequence $v' = \lim v'_N$, we obtain v' as required. Let $\mathcal{D}'$ be the σ-algebra of borel sets on Y'. Finally we wish to define an action of Γ on Y'. We assume the sets $\{A_m\}$ are all distinct. Each $S \in \Gamma$ determines a permutation $\sigma_S : \mathbf{N} \to \mathbf{N}$ with $S^{-1}A_m = A_{\sigma_s(m)}$. We now define $S : Y' \to Y'$ by $Sy'(m) = y'(\sigma_S(m))$. This defines an action of Γ on Y' as one may check using $\sigma_{ST}(m) = \sigma_T(\sigma_S(m))$. Moreover, $S^{-1}A'_m = A'_{\sigma_s(m)}$. It follows that Γ is measure preserving on $(Y',\mathcal{D}',v')$. The system $\mathbf{Y}'$ is regular, and using Proposition 5.1, we conclude that $(Y',\mathcal{D}',v',\Gamma)$ is equivalent to $(Y,\mathcal{D},v,\Gamma)$.

By a similar construction, we begin with an algebra $\mathcal{B}_0 \subset \mathcal{B}$ that is Γ-invariant and with $\tilde{\mathcal{B}}_0$ dense in $\tilde{\mathcal{B}}$ to obtain $(X',\mathcal{B}',\mu',\Gamma)$ regular and equivalent to $(X,\mathcal{B},\mu,\Gamma)$. $\mathcal{B}_0 = \{B_m\}$ and the action of Γ on $\mathcal{B}_0$ induces permutations $S^{-1}B_m = B_{\tau(m)}$. We now assume in addition, that $\alpha^{-1}(\tilde{\mathcal{D}}_0) \subset \tilde{\mathcal{B}}_0$. More specifically, we assume that we have a map $\alpha^{-1} : \mathcal{D}_0 \to \mathcal{B}_0$ with $\overline{\alpha^{-1}(A_m)} = \alpha^{-1}(\tilde{A}_m)$. We define $\beta : \mathbf{N} \to \mathbf{N}$ by $\alpha^{-1}(A_m) = B_{\beta(m)}$ and we define $\phi : X' = \{0,1\}^{\mathbf{N}} \to Y'$ by $\phi(x')(m) = x'(\beta(m))$. Then

$$S\phi(x')(m) = \phi(x')(\sigma(m)) = x'(\beta \circ \sigma(m)),$$
$$\phi S(x')(m) = Sx'(\beta(m)) = x'(\tau \circ \beta(m)).$$

These expressions are equal, as one verifies on account of $\alpha^{-1}(S^{-1}A_m) = S^{-1}\alpha^{-1}(A_m)$. This completes the proof of the theorem.

CHAPTER 6

Structure of Measure Preserving Systems

The objective in the next two chapters is to prove the multiple recurrence theorem for commuting measure preserving transformations of a measure space. In Chapter 4 we established two special cases of this theorem, Theorem 4.12 and Theorem 4.27. If we examine the proofs of these special cases, we see that two distinct phenomena are exploited to obtain recurrence. In the case of a totally weak mixing group of transformations the various terms $T_i^{-n}A$ in the multiple intersection tend to become independent of one another, and $\mu(A \cap T_1^{-n}A \cap \cdots \cap T_l^{-n}A)$ behaves, on the average, like $\mu(A)^{l+1}$. In the proof of Theorem 4.27, the general case is reduced to one in which the space of functions is spanned by eigenfunctions. In this case the function $f(T^n x)$ returns "almost periodically" to the function $f(x)$ and the integral $\int f(x)f(T^n x)f(T^{2n} x)\,d\mu(x)$ behaves repeatedly like $\int f(x)^3\,d\mu(x)$. Unlike the proof of Theorem 2.6, the topological analogue of multiple recurrence, in which we could handle all dynamical systems in one stroke, we shall now have to decompose the general situation into situations where the mixing phenomena and the periodicity phenomena can be called upon separately. In the present chapter, we provide the apparatus for this analysis in the form of a structure theorem for a measure space on which a (finitely generated, commutative) group of measure preserving transformations acts.

The model for the type of structure theorem we shall need is the structure theorem (proved in Furstenberg [2]) for minimal distal systems. A *distal dynamical system* (X,G) is one for which distinct points $x,y \in X$ remain at a distance bounded away from 0, $d(gx,gy) > \varepsilon(x,y) > 0$, for all $g \in G$. It is easy to check that if a given system is distal, this property is maintained by any isometric extension. (See Definition 1.5 for the definition of an isometric extension of a cyclic system.) This enables one to construct a family of distal systems, beginning with the trivial 1-point system and extending it successively by isometric extensions. The main structure theorem claims that if one interprets isometric extensions sufficiently broadly, and if one admits transfinite chains of successive isometric extensions, then one can, by this procedure, describe all (metric) distal systems.

The structure theorem for distal systems suggests the possibility of forming a "composition series" for various classes of dynamical systems—or measure preserving systems—in which the given system $\mathbf{X} = (X, \mathscr{B}, \mu, \Gamma)$ appears in a chain $\mathbf{X} \to \cdots \to \mathbf{X}_{\alpha+1} \to \mathbf{X}_\alpha \to \cdots \mathbf{X}_1 \to \mathbf{X}_0$, possibly transfinite, for which the individual links $\mathbf{X}_{\alpha+1} \to \mathbf{X}_\alpha$ can be explicitly described, or are, at least, of a more elementary nature. Relativizing the notion of a weak mixing system, we shall describe in the present chapter the notion of a *weak mixing extension* $\mathbf{X}_{\alpha+1} \to \mathbf{X}_\alpha$ relative to a transformation $T \in \Gamma$ as well as the complementary notion of a *compact extension* $\mathbf{X}_{\alpha+1} \to \mathbf{X}_\alpha$ relative to a transformation $T \in \Gamma$. The two forms of behavior are combined in the notion of a *primitive extension* $\mathbf{X}_{\alpha+1} \to \mathbf{X}_\alpha$, for which the group Γ splits into the product of two subgroups $\Gamma_1 \times \Gamma_2$ with $\mathbf{X}_{\alpha+1} \to \mathbf{X}_\alpha$ weak mixing relative to $T \in \Gamma_1$ and compact relative to $T \in \Gamma_2$. The structure theorem to be proved in this chapter asserts that by a succession of primitive extensions we can arrive at an arbitrary (separable) measure preserving system. This will then be used in the next chapter to provide the inductive step in our proof of the multiple recurrence theorem by transfinite induction.

A significant feature of this structure theory is the reconstruction of a dynamical system in terms of its factors, i.e., its homomorphic images, and not only in terms of its subsystems which present themselves more naturally as the building blocks of the system. The idea of studying the factors of a dynamical system seems to have occurred first in topological dynamics in connection with minimal systems, which, of course, have no subsystems (Ellis [1]). With the development of the theory of Bernoulli automorphisms and entropy, factors of measure preserving systems have also come into prominence. Thus Sinai's theorem [1] asserts that any system of positive entropy has Bernoulli factors. Also, a K system can be characterized as one that has no nontrivial 0-entropy factors. The notions Bernoulli K-system, and entropy do not occur in the structure theory presented here, but there is a parallelism between the two theories, with random-like behavior represented in one case by Bernoulli or K-systems and in the other by weak mixing systems, and deterministic behavior represented in one theory by 0-entropy and in the other, by compactness.

Throughout this chapter Γ will be a multiplicative group isomorphic to $\mathbf{Z}^r$. The group Γ will be acting, not necessarily faithfully, on the measure spaces under discussion. All the measure spaces that occur will be assumed to be separable. We shall adhere to the notational conventions of the preceding chapter. If S is a transformation of a space X and f is a function on X, Sf continues to denote the function defined by $Sf(x) = f(Sx)$.

Most of the substance of this chapter and the next is based on joint work with Y. Katznelson appearing in Furstenberg and Katznelson [1]. Results relating to the structure theorem can also be found in the work of Zimmer [1].

1. Ergodic and Weak Mixing Extensions

Let $\mathbf{X} = (X, \mathscr{B}, \mu, \Gamma)$ be a measure preserving system and $\mathbf{Y} = (Y, \mathscr{D}, v, \Gamma)$ a factor of $\mathbf{X}$. We write $\alpha : \mathbf{X} \to \mathbf{Y}$ where α corresponds to a map

$$\alpha^{-1} : \tilde{\mathscr{D}} \to \tilde{\mathscr{B}};$$

that is, α^{-1} takes sets of $\mathscr{D}$ modulo null sets to sets of $\mathscr{B}$ modulo null sets.

DEFINITION 6.1.   *The system* $\mathbf{X}$ *is an ergodic extension of* $\mathbf{Y}$ *relative to* $T \in \Gamma$ *if the only T-invariant sets of* $\mathscr{B}$ *are images (modulo null sets) under* α^{-1} *of T-invariant sets of* $\mathscr{D}$. *We write* $\mathbf{X} \to \mathbf{Y}$ *ergodic rel. T.*

If we denote the trivial one-point system by $\mathbf{X}_0$, then for any $\mathbf{X}$ we have a homomorphism $\mathbf{X} \to \mathbf{X}_0$. It is clear that the usual notion of ergodicity of a transformation T on $(X, \mathscr{B}, \mu)$ now becomes the assertion that $\mathbf{X} \to \mathbf{X}_0$ is ergodic rel. T.

DEFINITION 6.2.   *We say* $\mathbf{X}$ *is a weak mixing extension of* $\mathbf{Y}$ *relative to* $T \in \Gamma$ *if* $\mathbf{X} \times_{\mathbf{Y}} \mathbf{X} \to \mathbf{Y}$ *is an ergodic extension of* $\mathbf{Y}$ *relative to T. We write* $\mathbf{X} \to \mathbf{Y}$ *w.m.rel.T.*

Since we have homomorphisms $\mathbf{X} \times_{\mathbf{Y}} \mathbf{X} \xrightarrow{\pi_i} \mathbf{X} \xrightarrow{\alpha} \mathbf{Y}$, with π_i either the projection onto the first component or the second, it follows that if $\mathbf{X} \times_{\mathbf{Y}} \mathbf{X} \to \mathbf{Y}$ is ergodic relative to $T \in \Gamma$, so is $\mathbf{X} \to \mathbf{Y}$. Thus a weak mixing extension is also an ergodic extension.

Given a homomorphism $\alpha : \mathbf{X} \to \mathbf{Y}$, we have an imbedding $L^p(\mathbf{Y}) \xrightarrow{\sim} L^p(\mathbf{Y})^\alpha \subset L^p(\mathbf{X})$, $1 \le p \le \infty$. It is evident that if $\mathbf{X} \to \mathbf{Y}$ is an ergodic extension relative to T, then a T-invariant function of $L^p(\mathbf{X})$ must belong to $L^p(\mathbf{Y})^\alpha$.

LEMMA 6.1.   *Suppose* $\mathbf{X} \to \mathbf{Y}$ *is ergodic relative to Y and suppose* $\phi, \psi \in L^2(\mathbf{X})$ *with* $E(\phi \mid \mathbf{Y}) = 0$. *Then*

$$(6.1) \qquad \lim_{N \to \infty} \frac{1}{N+1} \sum_{n=0}^{N} \int E(\psi T^n \phi \mid \mathbf{Y}) \, dv = 0.$$

Proof: By Proposition 5.4, property (v), this is the same as asserting that

$$\lim_{N \to \infty} \frac{1}{N+1} \sum_{n=0}^{N} \int \psi T^n \phi \, d\mu = 0.$$

Consider the sequence

$$\frac{1}{N+1}\sum_{n=0}^{N} T^n\phi \quad \text{in} \quad L^2(\mathbf{X}).$$

We shall prove that it converges weakly to 0. It suffices to show that if $\Phi \in L^2(\mathbf{X})$ is a weak limit point of this sequence, then $\Phi = 0$. But since

$$T\frac{1}{N+1}\sum_{n=0}^{N} T^n\phi - \frac{1}{N+1}\sum_{n=0}^{N} T^n\phi \to 0$$

as $N \to \infty$, we shall have $T\Phi = \Phi$. By hypothesis $\alpha: X \to Y$ is ergodic, so that $\Phi \in L^2(\mathbf{Y})^\alpha$. It follows by Proposition 5.4, property (iii) that $E(\Phi|\mathbf{Y})^\alpha = \Phi$. Now the map $f \to E(f|\mathbf{Y})$ from $L^2(\mathbf{X})$ to $L^2(\mathbf{Y})$ is continuous in the weak topology. It follows that if

$$\frac{1}{N_k+1}\sum_{n=0}^{N_k} T^n\phi \to \Phi$$

weakly, then

$$\frac{1}{N_k+1}\sum_{n=0}^{N} T^n E(\phi|\mathbf{Y}) = \frac{1}{N_k+1}\sum_{n=0}^{N_k} E(T^n\phi|\mathbf{Y}) \to E(\Phi|\mathbf{Y}).$$

But then $E(\Phi|\mathbf{Y}) = 0$, so that $\Phi = 0$. This proves the lemma.

We now apply the lemma to the extension $\mathbf{X} \times_\mathbf{Y} \mathbf{X} \to \mathbf{Y}$ where $\mathbf{X} \to \mathbf{Y}$ is $w.m.rel.T$. Take $\phi, \psi \in L^\infty(\mathbf{X})$, so that $\phi \otimes \phi, \psi \otimes \psi \in L^2(\mathbf{X} \times_\mathbf{Y} \mathbf{X})$. By Proposition 5.12,

$$E((\psi \otimes \psi)T^n(\phi \otimes \phi)|\mathbf{Y}) = E(\psi T^n\phi \otimes \psi T^n\phi|\mathbf{Y}) = E(\psi T^n\phi|\mathbf{Y})^2.$$

Assume that $E(\phi|\mathbf{Y}) = 0$; then $E(\phi \otimes \phi|\mathbf{Y}) = E(\phi|\mathbf{Y})^2 = 0$. By the lemma we obtain

$$(6.2) \qquad \lim_{N \to \infty} \frac{1}{N+1}\sum_{n=0}^{N} \int E(\psi T^n\phi|\mathbf{Y})^2 \, dv = 0$$

Now let ϕ be an arbitrary bounded function. The function $\phi - E(\phi|\mathbf{Y})^\alpha$ satisfies $E(\phi - E(\phi|\mathbf{Y})^\alpha|\mathbf{Y}) = 0$ by Proposition 5.4, property (iii). Replace ϕ by $\phi - E(\phi|\mathbf{Y})^\alpha$ in (6.2) and use Proposition 5.4 and 5.7 to conclude that

$$E(\psi T^n\phi - \psi T^n E(\phi|\mathbf{Y})^\alpha|\mathbf{Y}) = E(\psi T^n\phi|\mathbf{Y}) - E(\psi|\mathbf{Y})T^n E(\phi|\mathbf{Y}).$$

We thus obtain

PROPOSITION 6.2.   *If* $\mathbf{X} \to \mathbf{Y}$ *is* $w.m.rel.T$ *and* $\phi, \psi \in L^\infty(\mathbf{X})$, *then*

$$\lim_{N \to \infty} \frac{1}{N+1}\sum_{n=0}^{N} \int \{E(\psi T^n\phi|\mathbf{Y}) - E(\psi|\mathbf{Y})T^n E(\phi|\mathbf{Y})\}^2 \, dv = 0.$$

PROPOSITION 6.3.   *If $\mathbf{X}_1 \to \mathbf{Y}$ is w.m.rel.T and $\mathbf{X}_2 \to \mathbf{Y}$ is ergodic rel.T, then $\mathbf{X}_1 \times_{\mathbf{Y}} \mathbf{X}_2 \to \mathbf{Y}$ is ergodic rel.T.*

Proof: Write $\alpha_1 : \mathbf{X}_1 \to \mathbf{Y}$, $\alpha_2 : \mathbf{X}_2 \to \mathbf{Y}$, and $\beta : \mathbf{X}_1 \times_{\mathbf{Y}} \mathbf{X}_2 \to \mathbf{Y}$. We want to show that if $f \in L^2(\mathbf{X}_1 \times_{\mathbf{Y}} \mathbf{X}_2)$ is T-invariant, then $f \in L^2(\mathbf{Y})^{\beta}$, or equivalently, $f = E(f \mid \mathbf{Y})^{\beta}$. We do this by showing that for all $f \in L^2(\mathbf{X}_1 \times_{\mathbf{Y}} \mathbf{X}_2)$,

$$(6.3) \qquad \frac{1}{N+1} \sum_{n=0}^{N} T^n(f - E(f \mid \mathbf{Y})^{\beta}) \to 0$$

weakly in the Hilbert space. Consider first functions of the form $f = \phi_1 \otimes \phi_2$ where furthermore we suppose that $E(\phi_1 \mid \mathbf{Y}) = 0$. Form the averages as in (6.3) and take inner products with functions of the form $\psi_1 \otimes \psi_2$. We obtain

$$\frac{1}{N+1} \sum_{n=0}^{N} \langle T^n(\phi_1 \otimes \phi_2), \psi_1 \otimes \psi_2 \rangle$$

$$= \frac{1}{N+1} \sum_{n=0}^{N} \int \psi_1 T^n \phi_1 \otimes \psi_2 T^n \phi_2 \, d\mu_1 \times_{\mathbf{Y}} \mu_2$$

$$= \frac{1}{N+1} \sum_{n=0}^{N} \int E(\psi_1 T^n \phi_1 \mid \mathbf{Y}) E(\psi_2 T^n \phi_2 \mid \mathbf{Y}) \, dv$$

by Proposition 5.10. We are assuming that $\phi_1, \phi_2, \psi_1, \psi_2$ are bounded, so $E(\psi_2 T^n \phi_2 \mid \mathbf{Y})$ is certainly bounded in $L^2(\mathbf{Y})$. According to equation (6.2) the average of $\| E(\psi_1 T^n \phi_1 \mid \mathbf{Y}) \|^2$ converges to 0 and consequently the average of $\| E(\psi_1 T^n \phi_1 \mid \mathbf{Y}) \|$ converges to 0. Since $E(\phi_1 \otimes \phi_2 \mid \mathbf{Y}) = 0$, this establishes (6.3) for this class of functions. A similar argument establishes (6.3) for $f = \phi_1 \otimes \phi_2$ with $E(\phi_2 \mid \mathbf{Y}) = 0$. In general we have

$$\phi_1 \otimes \phi_2 = (\phi_1 - E(\phi_1 \mid \mathbf{Y})^{\alpha_1}) \otimes \phi_2 + E(\phi_1 \mid \mathbf{Y})^{\alpha_1}$$
$$\otimes (\phi_2 - E(\phi_2 \mid \mathbf{Y})^{\alpha_2}) + E(\phi_1 \mid \mathbf{Y})^{\alpha_1} \otimes E(\phi_2 \mid \mathbf{Y})^{\alpha_2}.$$

The last summand is the same as $\{ E(\phi_1 \mid \mathbf{Y}) E(\phi_2 \mid \mathbf{Y}) \}^{\beta}$ so that (6.3) is clearly valid for it. Moreover, $E(\phi_i - E(\phi_i \mid \mathbf{Y})^{\alpha_1} \mid \mathbf{Y}) = 0$ by Proposition 5.4, property (iii). This establishes (6.3) for all functions of the form $\phi_1 \otimes \phi_2$, and since linear combinations of these are dense in $L^2(\mathbf{X}_1 \times_{\mathbf{Y}} \mathbf{X}_2)$, the proposition is proven.

PROPOSITION 6.4.   *If $\mathbf{X} \to \mathbf{Y}$ is w.m.rel.T, then $\mathbf{X} \times_{\mathbf{Y}} \mathbf{X} \to \mathbf{Y}$ is also w.m.rel.T.*

Proof: This follows from the foregoing proposition using the associativity of relative products:

$$(\mathbf{X} \times_{\mathbf{Y}} \mathbf{X}) \times_{\mathbf{Y}} (\mathbf{X} \times_{\mathbf{Y}} \mathbf{X}) = \mathbf{X} \times_{\mathbf{Y}} (\mathbf{X} \times_{\mathbf{Y}} (\mathbf{X} \times_{\mathbf{Y}} \mathbf{X})).$$

We conclude this section with the following definition.

DEFINITION 6.3.   *We say that the extension $\alpha:(X,\mathscr{B},\mu,\Gamma) \to (Y,\mathscr{D},\nu,\Gamma)$ is weak mixing relative to a subgroup $\Gamma' \subset \Gamma$ if $\alpha:X \to Y$ is weak mixing relative to T for every $T \in \Gamma'$, $T \neq identity$.*

Note that an extension $X \to Y$ fails to be weak mixing relative to Γ' if some set in $X \times_Y X$ that is not the preimage of a set in Y is invariant under some T, $T \neq 1$, $T \in \Gamma'$. Ordinarily when one speaks of ergodicity of a group action, one is concerned about invariance of sets under every transformation in the group.

2. Some Examples of Compactness

Before embarking on a formal discussion of the notion complementary to weak mixing, we shall discusss some examples that I hope will illuminate the definition.

The first example will be one of a *m.p.s.* that is compact relative to the trivial flow. Let G be a compact metrizable abelian group, let $S \to g_S \in G$ be a homomorphism of $\Gamma \to G$ and define the action of Γ on G by $Sx = g_S x$ for $x \in G$. Now take $X = G$, $\mathscr{B} = $ borel sets of G, $\mu = $ haar measure on G, and $X = (X,\mathscr{B},\mu,\Gamma)$ with this action of Γ on G.

Let us say that a subset M of a topology space N is *precompact* (in N) if the closure $\overline{M}$ is compact. If N is a complete metric space, then a subset M is precompact if and only if for each $\varepsilon > 0$ there exist finitely many elements $x_1, \ldots, x_k \in N$ such that for each $x \in M$

$$\min_{1 \leq j \leq k} \ d(x,x_j) \leq \varepsilon.$$

In what follows we shall be interested in subsets of $\mathscr{H} = L^2(X)$ with respect to the norm topology of the Hilbert space $\mathscr{H}$, with X the system described above.

PROPOSITION 6.5.   *If $f \in \mathscr{H}$, then $\Gamma f = \{Sf, S \in \Gamma\}$ is a precompact subset of $\mathscr{H}$.*

Proof: It suffices to show that for $f \in \mathscr{H}$, the set of all translates of f by elements in G, $\{f^g, g \in G\}$ where $f^g(x) = f(gx)$, is precompact. Consider the map $g \to f^g$ of $G \to \mathscr{H}$. We claim this is continuous. If f is continuous, this is obvious, since in that case $g \to f^g \in L^\infty(X)$ is continuous. For arbitrary f, take $f_n \to f$ in $\mathscr{H}$ with f_n continuous. Then the function f^g is the uniform limit of continuous functions f_n^g and therefore is continuous. It follows that $\{f^g, g \in G\}$ is the continuous image of a compact set and so is even compact.

We are now interested in averages over Γ that correspond to expressions of the form

$$\lim \frac{1}{N + 1} \sum_{n=0}^{N} f(n)$$

for $\mathbf{Z}$. Recall that $\Gamma \cong \mathbf{Z}^r$. It is easy to see that one can find a sequence of subsets of Γ, $I_k \subset \Gamma$, $k = 1,2,3, \ldots$, such that the cardinality $|I_k| \to \infty$ and $|SI_k \bigtriangleup I_k|/|I_k| \to 0$ as $k \to \infty$ for each $S \in \Gamma$, where $SI_k = \{ST, T \in I_k\}$. Such a sequence $\{I_k\}$ can be called a *Følner sequence*.

PROPOSITION 6.6. *Let $\{I_k\}$ be a Følner sequence and let $f \in \mathcal{H}$. Let $G' \subset G$ be the closed subgroup of G generated by the elements $g_S, S \in \Gamma$. Then if $\bar{f}(x) = \int_{G'} f(g'x)\, dg'$, we have*

$$(6.4) \qquad \frac{1}{|I_k|} \sum_{S \in I_k} Sf \to \bar{f}$$

as $k \to \infty$, in the weak topology of $\mathcal{H}$.

Proof: Note that $\bar{\bar{f}} = \bar{f}$ so that $\overline{(f - \bar{f})} = 0$. Hence every $f \in \mathcal{H}$ is a sum $f = f_1 + f_2$ with $\bar{f}_1 = f_1$ and $\bar{f}_2 = 0$. It suffices to establish the theorem in these two cases. Suppose $f_1 = \bar{f}_1$ so that $f_1(x) = \int_{G'} f_1(g'x)\, dg'$. Then

$$Sf_1(x) = f_1(g_S x) = \int_{G'} f_1(g'g_S x)\, dg' = \int_{G'} f_1(g'x)\, dg' = f_1(x)$$

since $g_S \in G'$ and the measure dg' is invariant. But this means that the left-hand side of (6.4) yields $f_1 = \bar{f}_1$ for every k and so the assertion is obvious.

Now suppose that $\bar{f} = 0$. Inasmuch as the averages in question belong to a weakly compact subset of $\mathcal{H}$, it suffices to show that if a subsequence of (6.4) converges, the limit is 0. Changing notation, we assume

$$(6.5) \qquad \frac{1}{|I_k|} \sum_{S \in I_k} Sf \to f_0$$

as $k \to \infty$. Using the property of a Følner sequence, we conclude that

$$\frac{1}{|I_k|} \sum_{S \in I_k} S'Sf \to f_0$$

for each $S' \in \Gamma$. On the other hand, by linearity and continuity, the foregoing limit should be $S'f_0$. Thus $S'f_0 = f_0$ for each $S' \in \Gamma$. Using the continuity of $g \to f^g$ (see proof of Proposition 6.5), we conclude that $f_0 = f_0^{g'}$ for all $g' \in G'$. Now clearly if $f_n \to f_0$, then $f_n^g = f_0^g$ (weakly). Hence

$$\frac{1}{|I_k|} \sum_{S \in I_k} Sf^{g'} \to f_0$$

for all $g' \in G'$. Next using Fubini's theorem, we find it is easy to justify integrating these expressions for $g' \in G'$:

$$\frac{1}{|I_k|} \sum_{S \in I_k} S\bar{f} \to f_0.$$

But since $\bar{f} = 0$, f_0 must vanish. This completes the proof of the proposition.

Actually we shall apply the foregoing not to the group G itself but to the group $G \times G$ with the diagonal action $S(x,x') = (Sx,Sx')$. Let us define the operation P on $L^2(\mathbf{X} \times \mathbf{X})$ by

$$(6.6) \qquad Pf(x,x') = \int_{G'} f(g'x,g'x')\,dg'.$$

PROPOSITION 6.7.   Let $f \in L^2(\mathbf{X})$, f not a.e. 0. Then for any Følner sequence $\{I_k\}$,

$$(6.7) \qquad \lim_{k > \infty} \frac{1}{|I_k|} \sum_{S \in I_k} Sf(x)\overline{Sf(x')} = P(f \otimes \bar{f})(x,x')$$

is not a.e. 0.

We shall give two proofs. In the first we make full use of the special form of X. In the second proof we shall use only the precompactness property established in Proposition 6.5.

Proof 1:   We use (6.6) to calculate $P(f \otimes \bar{f})$ taking advantage of the development of $f(x)$ into a Fourier series in terms of the characters on G. If we write $f \sim \sum c_\chi \chi$, then $f \otimes \bar{f} \sim \sum c_\chi \bar{c}_\lambda \chi \otimes \bar{\lambda}$, and $P(f \otimes \bar{f}) \sim \sum \delta(\chi,\lambda)c_\chi \bar{c}_\lambda \chi \otimes \bar{\lambda}$ where

$$\delta(\chi,\lambda) = \int_{G'} \chi(g')\overline{\lambda(g')}\,dg'.$$

In particular the coefficient of $\chi \otimes \bar{\chi}$ in $P(f \otimes \bar{f})$ is $|c_\chi|^2$, and so if $P(f \otimes f') = 0$, we must have $f = 0$.

Proof 2:   Suppose the limit in (6.7) vanishes. The convergence is in the weak topology, so we can multiply both sides of (6.7) by a function in $L^2(\mathbf{X} \times \mathbf{X})$ and integrate. In particular

$$\frac{1}{|I_k|} \sum_{S \in I_k} \iint h(x)\overline{h(x')}Sf(x)\overline{Sf(x')}\,d\mu(x)\,d\mu(x') \to 0$$

or

$$(6.8) \qquad \frac{1}{|I_k|} \sum_{S \in I_k} \left| \int h(x)Sf(x)\,d\mu(x) \right|^2 \to 0$$

This should be understood as saying that for any $h \in L^2(\mathbf{X})$, "most" of the expressions $\int h(x)Sf(x)\,d\mu(x)$, for $S \in \Gamma$, are small, We now use the characterization of precompactness in a complete metric space and Proposition 6.5. Namely for any $\varepsilon > 0$, there is a set of functions $h_1, h_2, \ldots, h_k \in L^2(\mathbf{X})$ with

$$\min_{1 \le j \le k} \|Sf - h_j\| \le \varepsilon$$

for each S.

Write

$$\|Sf - h_j\|^2 \ge \|Sf\|^2 - 2\left\{\int \overline{h_j(x)}Sf(x)\,d\mu(x)\right\}$$

$$\ge \|f\|^2 - 2\left|\int \overline{h_j(x)}Sf(x)\,d\mu(x)\right|,$$

$$\min_{1 \le j \le h} \{\|Sf - h_j\|^2\} \ge \|f\|^2 - 2 \max_{1 \le j \le k}\left\{\left|\int \overline{h_j(x)}Sf(x)\,d\mu(x)\right|\right\}$$

$$\ge \|f\|^2 - 2\sum_{j=1}^{k}\left|\int \overline{h_j(x)}Sf(x)\,d\mu(x)\right|.$$

Write $a_j(S) = \left|\int \overline{h_j(x)}Sf(x)\,d\mu(x)\right|$. By equation (6.8),

$$\frac{1}{|I_k|}\sum_{S \in I_k} a_j(S)^2 \to 0 \Rightarrow \frac{1}{|I_K|}\sum_{S \in I_k} a_j(S) \to 0$$

for each j, and summing over j, we also have

$$\frac{1}{|I_k|}\sum_{S \in I_k}\left(\sum_{j=1}^{k} a_j(S)\right) \to 0$$

Putting this together, we conclude $\|f\|^2 \le \varepsilon^2$, and since ε was arbitrary, $f = 0$.

Proposition 6.7 insures us an ample supply of Γ-invariant functions on $X \times X$. For each function $P(f \otimes \bar{f})$ is Γ-invariant (see proof of Proposition 6.6). These are certainly not all constant since if we choose $f \in L^2(\mathbf{X})$ with $\int f\,d\mu = 0$, then $\int P(f \otimes \bar{f})\,d\mu \times \mu = 0$ and by the foregoing proposition, $P(f \otimes \bar{f}) \ne 0$, and so it is not constant. The next proposition defines more precisely what is meant by an "ample supply" of Γ-invariant functions on $X \times X$. Note that whenever $H(x,x')$ is in $L^2(\mathbf{X} \times \mathbf{X})$, then for a.e. x', $H(\cdot,x') \in L^2(\mathbf{X})$.

PROPOSITION 6.8.   *If   H   ranges   over   the Γ-invariant   functions   of*
$L^2(\mathbf{X} \times \mathbf{X})$, *then the system of equations*

$$\int H(x,x')\psi(x)\,d\mu(x) = 0$$

for a function $\psi \in L^2(X)$ implies $\psi = 0$.

Another version of the same statement is that as H ranges over the Γ-invariant functions of $L^2(\mathbf{X} \times \mathbf{X})$ and x' ranges over any subset of full measure in X', then the set of sections $H(\cdot,x')$, spans all of $L^2(\mathbf{X})$. In this form the proof of the proposition is immediate since for each $f \in L^2(\mathbf{X})$, the function

$$H(x,x') = f(x'x^{-1})$$

is Γ-invariant, $x'x^{-1}$ referring to the group operation. Once again we give an alternative proof that does not use the group G explicitly and depends only on properties of X already established.

Proof: Suppose ψ has the property described in the proposition. Then, in particular

$$\int \psi(x)P(\bar{\psi}\otimes\psi)(x,x')\,d\mu(x) = 0$$

and so

$$\iint \psi(x)\overline{\psi(x')}P(\bar{\psi}\otimes\psi)(x,x')\,d\mu(x)\,d\mu(x') = 0$$

$$\iint \psi\otimes\bar{\psi}\,\overline{P(\psi\otimes\bar{\psi})}\,d\mu \times \mu = 0.$$

Since $P(\psi\otimes\bar{\psi})$ is S-invariant for $S \in \Gamma$, we obtain

$$\iint (S\psi\otimes S\bar{\psi})\overline{P(\psi\otimes\bar{\psi})}\,d\mu \times \mu = 0$$

for each $S \in \Gamma$. If we average (6.9) over $S \in I_\lambda$ and let $\lambda \to \infty$, we find

$$\iint P(\psi\otimes\bar{\psi})\overline{P(\psi\otimes\bar{\psi})}\,d\mu \times \mu = 0$$

or $P(\psi\otimes\bar{\psi}) = 0$ *a.e.* But by the foregoing proposition this implies $\psi = 0$ *a.e.* This completes the proof.

Let us mention another formulation of the property expressed in Proposition 6.8.

PROPOSITION 6.8′.   *As H ranges over Γ-invariant functions in $L^2(\mathbf{X} \times \mathbf{X})$ and ϕ ranges over $L^2(\mathbf{X})$, the functions*

$$\int H(x,x')\phi(x')\,d\mu(x')$$

span $L^2(\mathbf{X})$.

Proof: Assume to the contrary that $\psi \in L^2(\mathbf{X})$ is orthogonal to all of these functions. This implies that

$$\int H(x,x')\psi(x)\,d\mu(x) = 0,$$

and by the previous proposition $\psi = 0$.

We have described three properties of the system $(X,\mathscr{B},\mu,\Gamma)$ under discussion: (1) precompactness of Γf, $f \in L^2(\mathbf{X})$, (2) nonvanishing of the averages

$$\lim \frac{1}{|I_k|} \sum_{S \in I_k} Sf \times S\overline{f},$$

and (3) the existence of an ample supply of Γ-invariant functions in $L^2(\mathbf{X} \times \mathbf{X})$. Moreover, we have also shown that (1) $\Rightarrow$ (2) $\Rightarrow$ (3). We leave it to the reader to complete the circle and show that (3) $\Rightarrow$ (1). (We shall do this is in a more general context.) In fact, if we add the condition that Γ acts ergodically on $(X,\mathscr{B},\mu)$, then any of the conditions (1), (2), or (3) implies that $(X,\mathscr{B},\mu,\Gamma)$ is a system based on a compact abelian group as described here. We shall not pursue this, and we shall satisfy ourselves with regarding the group system as an example of the phenomenon we shall call *compactness*.

We turn now to our second example in which we shall illustrate the phenomenon of compactness relative to a nontrivial factor. Let $\mathbf{Y} = (Y,\mathscr{D},\nu,\Gamma)$ be an arbitrary system, let M be a compact metric space, and let G be the compact group of isometries of M. Suppose we can extend the action of Γ to $Y \times M$ by

(6.9) $\quad S(y,u) = (Sy,\sigma_S(y)u)$

where $\sigma_S(y) \in G$. The group G preserves some measure θ on M and if $\mathscr{E}$ denotes the σ-algebra of borel sets on M, we can form the skew product system $X = (Y \times M,\mathscr{D} \times \mathscr{E},\nu \times \theta,\Gamma)$, where now (compare Chapter 5 §1) the measure preserving maps $\sigma_S(y)$ are also isometries of M.

If we write $\mu = \nu \times \theta$, then the disintegration of μ with respect to the factor $\mathbf{Y}$ is given by $\mu_y = \delta_y \times \theta$. Then

$$\mu \times_{\mathbf{Y}} \mu = \int \delta_y \times \delta_y \times \theta \times \theta \, d\nu(y)$$

and $\mathbf{X} \times_{\mathbf{Y}} \mathbf{X}$ is equivalent to the skew product $(Y \times M \times M,\mathscr{D} \times \mathscr{E} \times \mathscr{E}, \nu \times \theta \times \theta,\Gamma)$ where

(6.10) $\quad S(y,u,u') = (Sy,\sigma_S(y)u,\sigma_S(y)u').$

From this we see immediately that there exist nontrivial Γ-invariant functions on $X \times_Y X$. Namely, if $d(u,u')$ is the metric on M, then $H(y,u,u') = d(u,u')$ is an invariant function. Now it is not hard to show that on the metric space M, the function $F_{n,u'}(u) = d(u,u')^n$ span a dense subset of $L^2(M,\mathscr{E},\theta)$. Using this, we obtain the following property of the extension $\alpha: X \to Y$:

PROPOSITION 6.9.   Let Ω be a set of Γ-invariant functions in $L^2(X \times_Y X)$ that spans the set of all of these. Then for almost every $y \in Y$, the functions $H(\cdot,x')$, for $H \in \Omega$ and x' ranging over a set with full μ_y-measure in X, span $L^2(X,\mathscr{B},\mu_y)$.

We can also formulate this property in a manner analogous to the properties of Propositions 6.8 and 6.8', but we leave this to the next section. What is significant for now is that we have a relativized version of property (3) described earlier. The difference between this example and the first example is that $X \times_Y X$ takes the place of $X \times X$, and the Hilbert spaces $L^2(X,\mathscr{B},\mu_y)$ replace the single space $L^2(X,\mathscr{B},\mu)$.

We can also obtain an analogue of the property of precompactness. To begin with, let $\phi \in L^2(M,\mathscr{E},\theta)$ and consider the set of functions $\{\phi^g, g \in G\}$ where $\phi^g(u) = \phi(gu)$. Once again the map $g \to \phi^g$ taking $G \to L^2(M,\mathscr{E},\theta)$ is continuous as we see by first taking ϕ continuous and then taking ϕ to be a limit in $L^2(M,\mathscr{E},\theta)$ of continuous functions. It follows that $\{\phi^g, g \in G\}$ is compact in $L^2(M,\mathscr{E},\theta)$. Now consider a function $f \in L^2(X)$ having the form

$$f(y,u) = \sum_{n=1}^{N} \psi_n(y)\phi_n(u)$$

with $\psi_n \in L^\infty(Y)$ and $\phi_n \in L^2(M)$. Then

$$(6.11) \quad Sf(y,u) = f(Sy,\sigma_S(y)u) = \sum_{n=1}^{N} \psi_n(Sy)\phi_n(\sigma_S(y)u)$$

Now fix y and let S range over Γ. It follows from the foregoing that the set of functions of u in (6.11) is precompact. Transferring this to X, we can say that for almost every $y \in Y$, the set of functions $\Gamma f = \{Sf, S \in \Gamma\}$ is precompact in $L^2(X,\mathscr{B},\mu_y)$ with this Hilbert space corresponding to $L^2(M,\mathscr{E},\theta)$ where we regard functions of $x = (y,u)$ as functions of u for fixed y. Since the family of functions under consideration is dense in $L^2(X)$, we have:

PROPOSITION 6.10.   There is a dense subset of functions $f \in L^2(X)$ with the property that for a.e. $y \in Y$, the set $\Gamma f = \{Sf, S \in \Gamma\}$ is precompact in $L^2(X,\mathscr{B},\mu_y)$.

Unlike the result in Proposition 6.5, we cannot make the present assertion for all $f \in L^2(\mathbf{X})$. For example, if $\psi \notin L^\infty(\mathbf{Y})$, then $f(y,u) = \psi(y)\phi(u)$ will not have the property in question. Later on we shall see that a modified property is valid for all $f \in L^2(X,\mathscr{B},\mu)$. In any event the analogue of the precompactness of Γf is valid, the main difference between the present example and the earlier one being that the single Hilbert space $L^2(X,\mathscr{B},\mu)$ is replaced by the family of Hilbert spaces $L^2(X,\mathscr{B},\mu_y)$, $y \in Y$.

3. Characterizing Compact Extensions

We begin by generalizing Proposition 6.6 to a (weak) ergodic theorem for unitary representations of Γ.

PROPOSITION 6.11.   *Let Γ act on a Hilbert space $\mathscr{H}$ by unitary operators, $f \to Sf$, $f \in \mathscr{H}$, $S \in \Gamma$, and let $\{I_k\}$ be a Følner sequence for Γ. Then for any $f \in \mathscr{H}$,*

$$\frac{1}{|I_k|} \sum_{S \in I_k} Sf \to Pf$$

weakly as $k \to \infty$, where P is the orthogonal projection of $\mathscr{H}$ onto the subspace of Γ-invariant vectors.

Proof: Let $\mathscr{H}_0 = \{f \in \mathscr{H} : Pf = 0\} = (P\mathscr{H})^\perp$. Now $P\mathscr{H}$ consists of all Γ-invariant vectors, so clearly $SP\mathscr{H} \subset P\mathscr{H}$ for $S \in \Gamma$. Therefore, the S being unitary, $S\mathscr{H}_0 \subset \mathscr{H}_0$. We shall now show that if $f \in \mathscr{H}_0$, then

$$(6.12) \qquad \frac{1}{|I_k|} \sum_{S \in I_k} Sf \to 0$$

weakly. As usual it suffices to show that any weak limit of a subsequence of (6.12) must be 0. Passing to a subsequence, we can suppose

$$\frac{1}{|I_k|} \sum_{S \in I_k} Sf \to f_0.$$

The Følner property again assures us that $Sf_0 = f_0$ for each $S \in \Gamma$. Hence $f_0 \in \mathscr{H}_0^\perp$. On the other hand, all the $Sf \in \mathscr{H}_0$, so $f_0 \in \mathscr{H}_0$. Hence $f_0 = 0$. This proves (6.12). We now have for an arbitrary $f \in \mathscr{H}$,

$$\frac{1}{|I_k|} \sum_{S \in I_k} S(f - Pf) \to 0,$$

and this proves the proposition.

We apply the foregoing to the case $\mathcal{H} = L^2(\mathbf{Z})$ for some measure preserving system $\mathbf{Z} = (Z,\mathcal{E},\theta,\Gamma)$ to conclude that

$$\frac{1}{|I_k|} \sum_{S \in I_k} Sf$$

converges weakly in $L^2(\mathbf{Z})$ as $k \to \infty$ for $f \in L^2(\mathbf{Z})$. We can extend this to weak convergence in $L^1(\mathbf{Z})$ for $f \in L^1(\mathbf{Z})$ where, as usual, $f_n \to f$ weakly in $L^1(\mathbf{Z})$ if $\int f_n g \, d\theta \to \int fg \, d\theta$ for $g \in L^\infty(\mathbf{Z})$. This is easily checked using the fact that $L^\infty(\mathbf{Z}) \subset L^2(\mathbf{Z})$ and the density of $L^2(\mathbf{Z})$ in $L^1(\mathbf{Z})$ and noting that

$$f \to \frac{1}{|I_k|} \sum_{S \in I_k} Sf$$

is a norm-decreasing operator. We thus have the following ergodic theorem.

PROPOSITION 6.12.   *If $(Z,\mathcal{E},\theta,\Gamma)$ is a m.p.s. and $\{I_k\}$ is a Folner sequence for Γ, then there exists an operator $P:L^1(\mathbf{Z}) \to L^1(\mathbf{Z})$ defined by*

$$Pf = \underset{k \to \infty}{\text{wk. lim}} \frac{1}{|I_k|} \sum_{S \in I_k} Sf.$$

Let $\alpha:\mathbf{X} \to \mathbf{Y}$ be a regular homomorphism (Chapter 5, §6) and $\beta:\mathbf{X} \times_\mathbf{Y} \mathbf{X} \to \mathbf{Y}$ the corresponding homomorphism of the relative product. If $\mathbf{X} = (X,\mathcal{B},\mu,\Gamma)$, we have the disintegration $\mu = \int \mu_y \, dv(y)$ as well as $\mu \times_\mathbf{Y} \mu = \int \mu_y \times \mu_y \, dv(y)$. Together with the Hilbert space $\mathcal{H} = L^2(X,\mathcal{B},\mu)$, we shall consider the space $L^2(X \times X, \mathcal{B} \times \mathcal{B}, \mu \times_\mathbf{Y} \mu)$ which we denote by $\mathcal{H} \otimes_\mathbf{Y} \mathcal{H}$ and also the fiber spaces $\mathcal{H}_y = L^2(X,\mathcal{B},\mu_y)$, $\mathcal{H}_y \otimes \mathcal{H}_y = L^2(X,\mathcal{B},\mu_y \times \mu_y)$. If $f \in \mathcal{H}$ or $\otimes_\mathbf{Y} \mathcal{H}$, we denote its norm by $\|f\|$; if $f \in \mathcal{H}_y$ or $\mathcal{H}_y \otimes \mathcal{H}_y$, we denote its norm by $\|f\|_y$. If $f \in \mathcal{H}$, then for almost every y, $f \in \mathcal{H}_y$, with a similar statement for $f \in \mathcal{H} \otimes_\mathbf{Y} \mathcal{H}$. We shall say that $f \in \mathcal{H}$ is *fiberwise bounded* if $\|f\|_y$ is bounded as a function of y.

If $H \in \mathcal{H} \otimes_\mathbf{Y} \mathcal{H}$ and $\phi \in \mathcal{H}$, we can define the convolution $H*\phi$ by

(6.13) $$H*\phi(x) = \int H(x,x')\phi(x') \, d\mu_{\alpha(x)}(x').$$

For *a.e.* y, $H(x,x')$ is a function in $L^2(X \times X, \mathcal{B} \times \mathcal{B}, \mu_y \times \mu_y)$, so that for *a.e.* x(with respect to μ_y) the integrand in (6.13) is the product of two functions in $L^2(X,\mathcal{B},\mu_y)$, and so the integral exists. Thus (6.13) defines a function for *a.e.* x with respect to $\mu = \int \mu_y \, dv(y)$. Moreover, we have

(6.14) $\|H*\phi\|_y \leq \|H\|_y \|\phi\|_y$

for *a.e.* $y \in Y$. From this it follows that if H is bounded, $\|H\|_y \le M$, then

$$\|H*\phi\|^2 = \int \|H*\phi\|_y^2 \, dv(y) \le M^2 \int \|\phi\|_y^2 \, dv(y) = M^2 \|\phi\|^2$$

so that $H*\phi \in \mathcal{H}$ and $\phi \to H*\phi$ is a bounded operator.

Suppose $S \in \Gamma$, so that S is a measure preserving map of $(X,\mathcal{B},\mu)$ and of $(Y,\mathcal{D},v)$. We have according to Proposition 5.9, $\mu_{Sy} = S\mu_y$ for $y \in Y$. If $f \in \mathcal{H}_{Sy}$, then

$$(6.15) \quad \int |Sf|^2 \, d\mu_y = \int |f|^2 \, dS\mu_y = \int |f|^2 \, d\mu_{Sy} < \infty,$$

so that $Sf \in \mathcal{H}_y$. We also see from equation (6.15) that $f \to Sf$ defines an isometry from $\mathcal{H}_{Sy}$ to $\mathcal{H}_y$. This fact will be used repeatedly in the sequel.

Finally we note that if $f,g \in L^2(X)$ then $f \otimes g \in L^1(X \times_Y X)$. For

$$\int |f(x)g(x')| \, d\mu \times_Y \mu(x,x')$$

$$= \int \left\{ \int |f(x)g(x')| \, d\mu_y(x) \, d\mu_y(x') \right\} dv(y)$$

$$= \int \left\{ \int |f| \, d\mu_y \int |g| \, d\mu_y \right\} dv(y) \le \int \|f\|_y \|g\|_y \, dv(y)$$

$$\le \left\{ \int \|f\|_y^2 \, dv(y) \right\}^{1/2} \left\{ \int \|g\|_y^2 \, dv(y) \right\}^{1/2} = \|f\| \cdot \|g\|.$$

Consider now the following properties that an extension $\alpha:(X,\mathcal{B},\mu,\Gamma) \to (Y,\mathcal{D},v,\Gamma)$ may have.

C_1: The functions $\{H*\phi\}$ span a dense subset of $L^2(X)$ as H ranges over bounded Γ-invariant functions in $L^2(X \times_Y X)$ and ϕ ranges over $\mathcal{H}$.

C_2: There exists a dense set of functions $\mathcal{F} \subset \mathcal{H}$ with the following property. If $f \in \mathcal{F}$ and $\delta > 0$, there exists a finite set of functions $g_1, g_2, \ldots, g_k \in \mathcal{H}$ such that for each $S \in \Gamma$, $\min_{1 \le j \le k} \|Sf - g_j\|_y < \delta$ for *a.e.* $y \in Y$.

C_3: For every $f \in \mathcal{H}$ the following holds. For every $\varepsilon, \delta > 0$ there exists a set $B \subset Y$, $B \in \mathcal{D}$, with $v(B) > 1 - \varepsilon$ and a finite set of functions $g_1, g_2, \ldots, g_k \in \mathcal{H}$ such that if we denote by f_B, the function $f \cdot 1_{\alpha^{-1}(B)}$, then for each $S \in \Gamma$, $\min_{1 \le j \le k} \|Sf_B - g_j\|_y < \delta$ for *a.e.* $y \in Y$.

C_4: For each $f \in \mathcal{H}$ the following holds. For every $\varepsilon, \delta > 0$ there exists a finite set of functions $g_1, g_2, \ldots, g_k \in \mathcal{H}$ such that for each $S \in \Gamma$, $\min_{1 \le j \le k} \|Sf - g_j\|_y < \delta$ but for a set of $y \in Y$ of measure $< \varepsilon$.

C_5:   For each $f \in L^2(\mathbf{X})$ let $\tilde{f}$ denote the function $\tilde{f}(x,x') = f(x)f(x')$ in $L^1(\mathbf{X} \times_{\mathbf{Y}} \mathbf{X})$. Form the function

$$P\tilde{f}(x,x') = \underset{l \to \infty}{\text{wk. lim}} \frac{1}{|I_l|} \sum_{S \in I_l} f(Sx)\overline{f(Sx')}$$

where the limit exists in the weak topology of $L^1(\mathbf{X} \times_{\mathbf{Y}} \mathbf{X})$ by Proposition 6.12. Then $P\tilde{f} = 0$ only if $f = 0$.

THEOREM 6.13.   *The five properties $C_1 - C_5$ of an extension $\alpha: \mathbf{X} \to \mathbf{Y}$ are equivalent. When these properties hold, we say $\alpha: \mathbf{X} \to \mathbf{Y}$ is a compact extension.*

Proof: Property $C_1 \Rightarrow C_2$. Let us say that a function $f \in \mathscr{H}$ is AP (almost periodic) if it has the property described in C_2; that is, for every $\delta > 0$, there exist $g_1, g_2, \dots, g_k \in \mathscr{H}$ with $\underset{1 \leq j \leq k}{\min} \|Sf - g_j\|_y < \delta$ for a.e. $y \in Y$. Clearly any linear combination of AP functions is AP. To prove that $C_1 \Rightarrow C_2$ it will suffice to show that by an arbitrarily small modification of a function of the form $H*\phi$ we can obtain an AP function. Since $\phi \to H*\phi$ is bounded, we can restrict to a dense set of ϕ; in particular we can assume ϕ is fiberwise bounded, say $\|\phi\|_y \leq M$. With this assumption we shall show that for any $\eta > 0$, if $H*\phi$ is replaced by 0 on a set of measure $\leq \eta$, the resulting function will be AP.

Since the σ-algebra $\mathscr{B} \times \mathscr{B}$ is generated by product sets $A \times A'$, $A, A' \in \mathscr{B}$, it follows that the set of finite linear combinations $\sum \psi_i(x)\psi_i'(x')$, $\psi_i, \psi_i' \in L^\infty(\mathbf{X})$, is dense in $L^2(X \times X, \mathscr{B} \times \mathscr{B}, \mu \times_{\mathbf{Y}} \mu)$. Let $H_n \to H$ where the H_n have this special form, and the convergence is in $L^2(\mathbf{X} \times_{\mathbf{Y}} \mathbf{X})$. That means that $\|H - H_n\|_y \to 0$ as a function of $y \in Y$ in $L^2(\mathbf{Y})$. Passing to a subsequence, we can assume that this convergence is a.e. on Y. We can then find a subset $E \subset Y$ with $\nu(E) < \eta$ such that $\|H - H_n\|_y \to 0$ uniformly for $y \notin E$. Now set $F = \bigcap_{S \in \Gamma} S^{-1}E$, and let

$$(6.16) \quad f(x) = \begin{cases} H*\phi(x) & \text{if} \quad x \notin \alpha^{-1}(F). \\ 0 & \text{if} \quad x \in \alpha^{-1}(F). \end{cases}$$

Since $F \subset E$, $\nu(F) < \eta$ so that f differs from $H*\phi$ only on a set of measure $< \eta$. We shall show that f is AP.

Let us say that a set of functions $g_1, g_2, \dots, g_k \in \mathscr{H}$ is δ-*spanning for* f *on the set* $B \subset Y$, if for each $y \in B$ and $S \in \Gamma$, $\underset{1 \leq j \leq k}{\min} \|Sf - g_j\|_y < \delta$. Observe that if we have a δ-spanning set on B, we can extend it to a δ-spanning set on

$$\bigcup_{S \in \Gamma} S^{-1}B.$$

Namely,

$$\|S_1 Sf - S_1 g_j\|_{S_1^{-1}y} = \|Sf - g_j\|_y$$

so that if $\{g_i\}$ is δ-spanning on B, $\{S_1 g_i\}$ is δ-spanning on $S_1^{-1}B$. We now express

$$\bigcup_{S \in \Gamma} S^{-1}B$$

as a disjoint union $\bigcup B_n$, where $B_n \subset S_n^{-1}B$, $S_n \in \Gamma$, and define $\tilde{g}_i = S_n g_i$ on B_n. Then $\{\tilde{g}_i\}$ is δ-spanning for f on $\bigcup_{S \in \Gamma} S^{-1}B$. To prove that f defined by (6.16) is AP, we shall show that it has a δ-spanning set of functions on Y–E. By the foregoing procedure, we can extend this to

$$Y\text{–}F = \bigcup_{S \in \Gamma} S^{-1}(Y\text{–}E).$$

Now on F, f and all its translates Sf vanish, so by adjoining 0 to the δ-spanning set on Y–F, we obtain a δ-spanning set on Y.

We proceed to show that $f(x)$ defined by (6.16) has a δ-spanning set on Y–E. The crucial fact needed for this is the identity

$$(6.17) \quad S(H*\psi) = H*(S\psi)$$

which holds for a Γ-invariant H and any $\psi \in \mathscr{H}$. Namely

$$S(H*\psi)(x) = H*\psi(Sx) = \int H(Sx,x')\psi(x')\, d\mu_{\alpha(Sx)}(x')$$

$$= \int H(Sx,x')\psi(x')\, dS\mu_{\alpha(x)}(x')$$

$$= \int H(Sx,Sx')\psi(Sx')\, d\mu_{\alpha(x)}(x')$$

$$= \int H(x,x')\psi(Sx')\, d\mu_{\alpha(x)}(x') = H*(S\psi)(x).$$

Since $\psi \to S\psi$ is an isometry of $\mathscr{H}_{Sy} \to \mathscr{H}_y$ and our function ϕ—and therefore f—is fiberwise bounded in norm by M, the same is true of all the functions Sf, $S \in \Gamma$. It follows that $g_1, g_2, \ldots, g_k$ will be a δ-spanning set for f on Y–E if for any $\psi \in \mathscr{H}$ satisfying $\|\psi\|_y \le M$ we have $\min_{1 \le j \le k} \|H*\psi - g_j\|_y < \delta$ for $y \notin E$. But since $\|H - H_n\|_y \to 0$ uniformly on Y–E, we can find n with $\|H - H_n\|_y < \delta/2M$ for $y \notin E$. According to (6.14)

$$(6.18) \quad \|H*\psi - H_n*\psi\|_y < \delta/2M \cdot M = \delta/2,$$

If

$$H_n = \sum_1^N \psi_i(x)\psi_i'(x')$$

with ψ_i, ψ_i' bounded and $\|\psi\|_y \leq M$,

$$H_n * \psi = \sum_1^N c_i \psi_i$$

with $c_i = c_i(\psi)$ bounded. Now for this set of functions it is obvious that a finite subset $g_1, g_2, \ldots, g_k$ can be chosen so that

$$\min_{1 \leq j \leq k} \left\| \sum_1^N c_i \psi_i - g_j \right\| < \delta/2.$$

Combined with (6.18) this gives the desired result.

Property $C_2 \Rightarrow C_3$. Let $f \in \mathcal{H}$ and let $f' \in \mathcal{F}$, the set of AP functions, with $\|f - f'\| < \frac{1}{2}\delta\sqrt{\varepsilon}$. Let $g_1, g_2, \ldots, g_{k-1}$ be a $\delta/2$-spanning set for f' on Y and set $g_k = 0$. Take $B = \{y : \|f - f'\|_y < \delta/2\}$. Using the identity $\|f - f'\|^2 = \int \|f - f'\|_y^2 \, dv(y)$, we find that $v(B) > 1 - \varepsilon$. Form $f_B = f \cdot 1_{\alpha^{-1}(B)}$, and let $S \in \Gamma$. If $y \in S^{-1}B$,

$$\|Sf_B - Sf'\|_y = \|f_B - f'\|_{Sy} < \delta/2$$

so that $\min_{1 \leq j \leq k-1} \|Sf_B - g_j\|_y < \delta/2 + \delta/2 = \delta$. If $y \notin S^{-1}B$, then $f_B = 0$ as an element of $\mathcal{H}_{Sy}$ so that $Sf_B = 0$ as an element of $\mathcal{H}_y$. But then $\|Sf_B - g_k\|_y = 0$, and this verifies the property C_3.

Property $C_3 \Rightarrow C_4$. This is immediate since $\|Sf - g_j\|_y = \|Sf_B - g_j\|_y$ as long as $y \in S^{-1}B$, i.e., but for a set of measure $< \varepsilon$.

Property $C_4 \Rightarrow C_5$. Let us say that $g_1, g_2, \ldots, g_k$ is an ε, δ-spanning set for $f \in \mathcal{H}$ if C_4 holds as stated. Without loss of generality we can suppose that the g_i are bounded functions since we may approximate the given set by bounded functions without destroying the property C_4. Then $\bar{g}_j \otimes g_j \in L^\infty(X \times_Y X)$. Assume then that this is the case and suppose that $Pf = 0$. For each j, $\int \bar{g}_j \otimes g_j P\tilde{f} \, d\mu \times_Y \mu = 0$. Using the definition of $P\tilde{f}$, we find

$$\lim_{l \to \infty} \frac{1}{|I_l|} \sum_{S \in I_l} \int^* \overline{g_j(x)} g_j(x') Sf(x) \overline{Sf(x')} \, d\mu \times_Y \mu(x, x') = 0,$$

or,

$$\lim_{l \to \infty} \frac{1}{|I_l|} \sum_{S \in I_l} \int \left| \int \overline{g_j(x)} Sf(x) \, d\mu_y(x) \right|^2 dv(y) = 0$$

and combining these for $j = 1, 2, \ldots, k$, we find

$$\lim_{l \to \infty} \frac{1}{|I_l|} \sum_{S \in I_l} \left\{ \sum_{j=1}^k \int \left| \int \overline{g_j(x)} Sf(x) \, d\mu_y(x) \right|^2 dv(y) \right\} = 0.$$

It follows that for l large there will exist an $S \in I_l$ for which the expression inside the braces { } is arbitrarily small. For that S, but for a set of y of measure $<\varepsilon$ we shall have for all j,

$$\left|\int \overline{g_j(x)} Sf(x) \, d\mu_y(x)\right|^2 < \delta^2.$$

But since $g_1, g_2, \ldots, g_k$ is an ε, δ-spanning set for f, there will be some j with $\|Sf - g_j\|_y^2 < \delta^2$ but for a set of y of measure $<\varepsilon$. It follows that but for a set of y of measure $<2\varepsilon$ we have $\|Sf\|_y^2 < 3\delta^2$. The same is then true of f and since ε, δ were arbitrary it follows that $f = 0$.

Property $C_5 \Rightarrow C_1$. Suppose the functions of the form $H*\phi$ were not dense in $\mathcal{H}$ as H ranges over Γ-invariant bounded functions in $L^2(X \times_Y X)$ and ϕ ranges over $L^2(X)$. Let $f \in \mathcal{H}$ be orthogonal to all of these and form the function

$$H(x,x') = \lim_{l \to \infty} \frac{1}{|I_l|} \sum_{S \in I_l} Sf(x)Sf(x') = P\tilde{f}(x,x')$$

for some Følner sequence $\{I_l\}$.

This function is Γ-invariant and so are the functions

$$H_M(x,x') = \begin{cases} H(x,x') & \text{for} & |H(x,x')| \leq M \\ 0 & \text{for} & |H(x,x')| > M. \end{cases}$$

By hypothesis $f \perp H_M*f$, so that

$$\int \overline{f(x)} \left\{ \int H_M(x,x')f(x') \, d\mu_{\alpha(x)}(x') \right\} d\mu(x) = 0.$$

Using the disintegration $\mu = \int \mu_y \, d\nu(y)$, we can write this as follows:

$$0 = \iint f(x) \left\{ \int H_M(x,x')f(x') \, d\mu_y(x') \right\} d\mu_y(x) \, d\nu(y)$$

$$= \iiint \overline{f(x)}f(x')H_M(x,x') \, d\mu_y(x) \, d\mu_y(x') \, d\nu(y)$$

$$= \int \tilde{f} \otimes f H_M \, d\mu \times_Y \mu.$$

Thus $H_M \in L^1(X \times_Y X)$ is orthogonal to $\tilde{f} = f \otimes \tilde{f} \in L^\infty(X \times_Y X)$. Since H_M is Γ-invariant, we obtain $S\tilde{f} \perp H_M$ for all $S \in \Gamma$. From this it follows that $P\tilde{f}$ which is the average of $S\tilde{f}$ is orthogonal to H_M. But $P\tilde{f} = H$; so $H \perp H_M$. But $\int \bar{H}H_M = \int |H_M|^2$ so that $H_M = 0$ a.e. This being the case for all M, $H = 0$. Finally by C_5, $f = 0$. This concludes the proof.

DEFINITION 6.4. *If $\alpha:(X,\mathcal{B},\mu,\Gamma) \to (Y,\mathcal{D},\nu,\Gamma)$ is a homomorphism and $\Gamma' \subset \Gamma$, then we say $\alpha:X \to Y$ is compact relative to Γ' if the extension $\alpha:(X,\mathcal{B},\mu,\Gamma') \to (Y,\mathcal{D},\nu,\Gamma')$ is compact, i.e., has one of the properties $C_1 - C_5$.*

Using C_1, we see that *if* $\alpha: X \to Y$ is compact relative to Γ' and $\Gamma'' \subset \Gamma'$, then $\alpha: X \to Y$ is compact relative to Γ''. Using C_2, C_3 or C_4, we see that if $\Gamma'' \subset \Gamma'$ with Γ'' of finite index in Γ' and $\alpha: X \to Y$ is compact relative to Γ'', then $\alpha: X \to Y$ is compact relative to Γ'. Using C_5, we see that the notion of compactness does not depend on the regularity of the homomorphism $\alpha: X \to Y$ and is defined for any homomorphism.

PROPOSITION 6.14. *If $\alpha: X \to Y$ is compact relative to two subgroups, $\Gamma', \Gamma'' \subset \Gamma$, then it is compact relative to $\Gamma' \Gamma''$.*

Proof: We use the characterization C_4 of compactness. Let $f \in \mathcal{H}$ and let $\varepsilon, \delta > 0$ be given. Choose $g_1 g_2, \ldots, g_k \in \mathcal{H}$ such that for each $T \in \Gamma'$, $\min_{1 \le j \le k} \|Tf - g_j\|_y < \delta/2$ for $y \notin E(T)$ where $v(E(T)) < \delta/2$. For each g_j choose $h_{j1}, h_{j2}, \ldots, h_{jq_j} \in \mathcal{H}$ so that for each $S \in \Gamma''$,

$$\min_{1 \le i \le q_j} \|Sg_j - h_{ji}\|_y < \delta/2$$

for $y \notin F_j(S)$ where $v(F_j(S)) < \varepsilon/2k$. For $T \in \Gamma'$, $S \in \Gamma''$ and $y \notin S^{-1}E(T)$,

$$\min_{1 < j < k} \|Tf - g_j\|_{Sy} < \delta/2,$$

or

$$\min_{1 < j < k} \|STf - Sg_j\|_y < \delta/2.$$

If, in addition, $y \notin F_j(S)$ for any j, then

$$\min_{i,j} \|STf - h_{ji}\|_y < \delta/2 + \delta/2 = \delta.$$

Thus outside of

$$S^{-1}E(T) \cup \bigcup_j F_j(S),$$

$$\min_{i,j} \|STf - h_{ji}\|_v < \delta.$$

Since

$$v\left(S^{-1}E(T) \cup \bigcup_j F_j(S)\right) < \varepsilon,$$

this completes the proof.

4. Existence of Compact Extensions

Suppose we have a chain of extensions $\alpha: X \to Y$, $\alpha': Y \to Z$ and suppose that the latter is compact relative to some subgroup $\Gamma' \subset \Gamma$. This means that there exist Γ'-invariant functions on $Y \times_Z Y$ which do not come from

Z. (This follows from property C_1.) In particular $\alpha':\mathbf{Y} \to \mathbf{Z}$ is not a weakly mixing extension. Nor is the composition $\alpha'\alpha:\mathbf{X} \to \mathbf{Z}$ since functions on $\mathbf{Y} \times_{\mathbf{Z}} \mathbf{Y}$ lift naturally by way of α to $\mathbf{X} \times_{\mathbf{Z}} \mathbf{X}$. The next theorem shows that, conversely, whenever $\gamma:\mathbf{X} \to \mathbf{Z}$ is an extension which is not weak mixing, we can interpolate a system $\mathbf{Y}$ between $\mathbf{X}$ and $\mathbf{Z}$, $\mathbf{X} \to \mathbf{Y} \to \mathbf{Z}$ for which the second step is compact for some subgroup $\Gamma' \subset \Gamma$. This is the relativized version of the statement that a system fails to be weak mixing only if it has nontrivial eigenfunctions (Chapter 4, §6). In the relative case we shall speak of a nontrivial extension $\alpha':\mathbf{Y} \to \mathbf{Z}$ which means that $L^p(\mathbf{Z})^{\alpha'} \subsetneqq L^p(\mathbf{Y})$.

THEOREM 6.15.   *If* $\gamma:(X,\mathscr{B},\mu,\Gamma) \to (Z,\mathscr{E},\theta,\Gamma)$ *is a nontrivial extension that is not weak mixing, i.e., for some* $T \in \Gamma$, $T \neq I$, $\gamma:\mathbf{X} \to \mathbf{Z}$ *is not weak mixing relative to* T, *then there exists a system* $(Y,\mathscr{D},\nu,\Gamma)$ *and homomorphisms* $\alpha:\mathbf{X} \to \mathbf{Y}$, $\alpha':\mathbf{Y} \to \mathbf{Z}$ *such that* $\gamma = \alpha'\alpha$ *with the property that* $\alpha':\mathbf{Y} \to \mathbf{Z}$ *is a nontrivial extension and is compact relative to a subgroup* $\Gamma' \subset \Gamma$, $\Gamma' \neq \{I\}$.

Proof: We suppose that $T \in \Gamma$, $T \neq I$ and $H \in L^2(X \times_Z X)$ is a T-invariant function that does not come from a function on **Z**. It is easily seen that this implies that the functions $H*\phi$ for $\phi \in L^2(\mathbf{X})$ are not all in $L^2(\mathbf{Z})^\gamma \subset L^2(\mathbf{X})$. Repeat the argument in the proof of the implication $C_1 \Rightarrow C_2$ of Theorem 6.12, and we find that there exist functions $f \in L^2(\mathbf{X})$, $f \notin L^2(\mathbf{Z})^\gamma$ with the following property: for each $\delta > 0$ there exists a finite set of functions $g_1, g_2, \ldots, g_k \in L^2(\mathbf{X})$ such that for each n,

$$\min_{1 < j < k} \left\| T^n f - g_j \right\|_z < \delta$$

for *a.e.* $z \in Z$. Let us call a function $f \in L^2(\mathbf{X})$ *compact* (with respect to $T \in \Gamma$ and the factor **Z**) if it satisfies the weaker property: for every $\varepsilon, \delta > 0$ there exist $g_1, g_2, \ldots, g_k \in L^2(\mathbf{X})$ such that for each n,

$$\min_{1 \leq j \leq k} \left\| T^n f - g_j \right\|_z < \delta$$

for $z \in Z$ except for a set of z with measure $< \varepsilon$. The property has the advantage that it persists under passage to the limit in $L^2(\mathbf{X})$. Moreover, the bounded compact functions form an algebra that is dense in the space of all compact functions since, clearly, if f is compact so is the function obtained by truncating f at large values. From this it may be seen that if $\mathscr{D} = \{A \in \mathscr{B}: 1_A \text{ is compact}\}$, then $\mathscr{D}$ is a σ-algebra and the subspace of compact functions of $L^2(X,\mathscr{B},\mu)$ consists precisely of the functions measurable with respect to $\mathscr{D}$. We claim that $\mathscr{D}$ is Γ-invariant. More generally, if $f \in L^2(\mathbf{X})$ is compact so is Sf, $S \in \Gamma$. For

$$\left\| T^n S f - S g_j \right\|_z = \left\| S T^n f - S g_j \right\|_z = \left\| T^n f - g_j \right\|_{Sz}.$$

Now set $Y = (X,\mathscr{D},\mu,\Gamma)$. Then Y is a factor of X; moreover, since $f \in$ $L^2(Z)^\gamma$ is clearly compact, $\mathscr{D} \supset \gamma^{-1}(\mathscr{E})$ and Z is a factor of Y. It is a nontrivial factor since there exist compact functions $\notin L^2(Z)^\gamma$. Finally $Y \to Z$ is compact relative to the subgroup $\Gamma' = \{T^n, n \in Z\}$ since C_4 is satisfied. This completes the proof.

5. Primitive Extensions and the Structure Theorem

DEFINITION 6.5.   *We say that* $\alpha:(X,\mathscr{B},\mu,\Gamma) \to (Y,\mathscr{D},v,\Gamma)$ *is primitive if* Γ *is the direct product of two subgroups* $\Gamma = \Gamma_c \times \Gamma_w$ *with* $\alpha:X \to Y$ *compact relative to* Γ_c *and* $\alpha:X \to Y$ *weak mixing relative to* Γ_w.

THEOREM 6.16.   *If* $\gamma:X \to Z$ *is a nontrivial extension, we can find a system* Y *and homomorphisms* $\alpha:X \to Y$, $\alpha':Y \to Z$ *with* $\gamma = \alpha'\alpha$ *and such that* Y *is a nontrivial primitive extension of* Z.

Proof: Let Γ_c be a maximal subgroup of Γ with the following property: there exists a system Y and homomorphisms $\alpha:X \to Y$, $\alpha':Y \to Z$ with $\gamma = \alpha'\alpha$ and such that Y is a nontrivial compact extension of Z relative to Γ_c. (Γ_c exists because $\{I\}$ has this property and because any increasing chain of subgroups of Z^r terminates.) Now Γ_c is not of finite index in a larger group, for otherwise the larger group would have the compactness property. Hence Γ/Γ_c has no torsion and is a free abelian group on a finite set of generators. Find preimages for these generators in Γ, and let Γ_w be the group generated by these preimages. Clearly $\Gamma = \Gamma_c \times \Gamma_w$.

We claim that Y is a primitive extension of Z; more precisely, we contend that $\alpha:Y \to Z$ is weak mixing relative to Γ_w. For suppose the extension failed to be weak mixing relative to some $T \neq I$ in Γ_w. According to Theorem 6.15 we could interpolate a system Y' between Y and Z such that $Y' \to Z$ is compact relative to some $\Gamma' \subset \Gamma_w$. Since $Y \to Z$ is compact relative to Γ_c, the same is true for $Y' \to Z$ and by Proposition 6.14 we would have $Y' \to Z$ compact relative to $\Gamma' \cdot \Gamma_c \gneqq \Gamma_c$. This contradicts the maximality of Γ_c and this proves the theorem.

The foregoing theorem provides the main ingredient in the proof of the structure theorem. To formulate the theorem, we need in addition to the above the notion of a *limit* of factors. If $X = (X,\mathscr{B},\mu,\Gamma)$ is a *m.p.s.* and $\pi_\alpha:X \to X_\alpha$, $\alpha \in A$, is a system of factors of X, we shall say that X *is a limit of the factors* $\{X_\alpha, \alpha \in A\}$, if the σ-algebra $\mathscr{B}$ is generated by $\{\alpha^{-1}(\mathscr{B}_\alpha)\}$.

THEOREM 6.17.   *Let* $X = (X,\mathscr{B},\mu,\Gamma)$ *be a separable measure preserving system with* Γ *a finite rank free abelian group. There is an ordinal* η *and a system of factors* X_ξ *of* X, $\pi_\xi:X \to X_\xi$, *for each ordinal* $\xi \leq \eta$ *such that the following are satisfied*:

(i) $\mathbf{X}_0$ *is the trivial 1-point system and* $\mathbf{X}_\eta = \mathbf{X}$.
(ii) *If* $\xi < \xi'$, *then there is a homomorphism* $\pi_\xi^{\xi'} : \mathbf{X}_{\xi'} \to \mathbf{X}_\xi$ *with* $\pi_\xi = \pi_\xi^{\xi'} \pi_{\xi'}$.
(iii) *For each* $\xi < \eta$, $\mathbf{X}_{\xi+1} \to \mathbf{X}_\xi$ *is primitive.*
(iv) *If* ξ *is a limit ordinal* $\leq \eta$, *then* $\mathbf{X}_\xi$ *is a limit of the systems* $\{\mathbf{X}_{\xi'}, \xi' < \xi\}$.

We refer to this system of factors as the *composition series* of $\mathbf{X}$.

Proof: Let Σ denote the family of systems of factors of $\mathbf{X}$ satisfying a modified version of the conditions above. If $\sigma \in \Sigma$, we have for each ordinal $\xi \leq \eta(\sigma)$ factor $\mathbf{X}_\xi^\sigma$ and the following are satisfied: (i) $\mathbf{X}_0^\sigma$ is trivial, (ii) for $\xi < \xi'$ we have the factorization $\mathbf{X} \to \mathbf{X}_{\xi'}^\sigma \to \mathbf{X}_\xi^\sigma$, (iii) for each $\xi < \eta(\sigma)$, $\mathbf{X}_{\xi+1}^\sigma \to \mathbf{X}_\xi^\sigma$ is nontrivial and primitive, (iv) if ξ is a limit ordinal $\leq \eta(\sigma)$, then $\mathbf{X}_\xi^\sigma$ is a limit of $\{X_{\xi'}^\sigma, \xi' < \xi\}$.

Note that we require $X_{\xi+1}^\sigma$ to be a nontrivial extension of $\mathbf{X}_\xi^\sigma$. This implies that the ordinals $\eta(\sigma)$ occurring correspond to cardinals not exceeding the cardinality of $\tilde{\mathscr{B}}$. It therefore is meaningful to talk of all such systems. We shall order Σ by setting $\sigma_1 \leq \sigma_2$ if $\eta(\sigma_1) \leq \eta(\sigma_2)$ and if, for $\xi \leq \eta(\sigma_1)$, $\mathbf{X}_\xi^{\sigma_1} = \mathbf{X}_\xi^{\sigma_2}$. We claim that Σ has a maximal element σ_0 and that $\mathbf{X}_{\eta(\sigma_0)}^{\sigma_0} = \mathbf{X}$. To show that Σ has a maximal element, we show that a totally ordered subset of Σ has an upper bound. Let Σ' be a totally ordered subset of Σ and consider $\{\eta(\sigma), \sigma \in \Sigma'\}$. Let η' be the first ordinal larger than all of these. If $\eta' = \eta'' + 1$, then $\eta'' = \eta(\sigma'')$ for some $\sigma'' \in \Sigma'$ and clearly σ'' is an upper bound of Σ'. Therefore η' is a limit ordinal. We now define σ' by $\mathbf{X}_\xi^{\sigma'} = \mathbf{X}_\xi^\sigma$ for $\xi < \eta'$ and $\sigma \in \Sigma', \eta(\sigma') = \eta'$ and $\mathbf{X}_{\eta(\sigma')}^{\sigma'} = (X, \mathscr{B}', \mu, \Gamma)$ where $\mathscr{B}'$ is the least σ-algebra such that all the $\mathbf{X}_\xi^\sigma$, $\sigma \in \Sigma'$, $\xi \leq \eta(\sigma)$, are factors of $\mathbf{X}_{\eta(\sigma')}^{\sigma'}$. This defines $\sigma' \in \Sigma$ and σ' is an upper bound of Σ'.

According to Zorn's lemma, Σ has a maximal element σ_0, and the main part of the proof consists in showing that $\mathbf{X}_{\eta(\sigma_0)}^{\sigma_0} = \mathbf{X}$. Here we use Theorem 6.16. If $\mathbf{X}$ were a nontrivial extension of $\mathbf{X}_{\eta(\sigma_0)}^{\sigma_0}$, we could interpolate a factor of $\mathbf{X}$ that would be a primitive extension of $\mathbf{X}_{\eta(\sigma_0)}^\sigma$. We could then find $\sigma \in \Sigma$ with $\eta(\sigma) = \eta(\sigma_0) + 1$ and $\mathbf{X}_{\eta(\sigma)}^\sigma = $ this primitive extension of $\mathbf{X}_{\eta(\sigma_0)}^{\sigma_0}$, $\mathbf{X}_\xi^\sigma = \mathbf{X}_\xi^{\sigma_0}$ for $\xi \leq \eta(\sigma_0)$. This contradicts the maximality of σ_0, and so we must have $\mathbf{X}_{\eta(\sigma_0)}^{\sigma_0} = \mathbf{X}$.

When $\Gamma = \{T^n, n \in \mathbf{Z}\}$, then each extension $\mathbf{X}_{\xi+1} \to \mathbf{X}_\xi$ is either compact or weak mixing. In fact it is easy to show that one can take all the extensions compact with the possible exception of the last link: $\mathbf{X} = \mathbf{X}_\eta \to \mathbf{X}_{\eta'}$ where (necessarily $\eta = \eta' + 1$). For if $\mathbf{X}_{\eta'}$ is the "maximal factor" of $\mathbf{X}$ with the property that all extensions in its composition series are compact, then by Theorem 6.14, $\mathbf{X} \to \mathbf{X}_{\eta'}$ is either trivial or weak mixing.

CHAPTER 7

The Multiple Recurrence Theorem

In this chapter we shall show how the multiple recurrence property can be deduced for an arbitrary system using the structure theorem of the last chapter. We use the transfinite composition series for the system and move up the ladder by arguing that the property in question is preserved both under passage to the limit and under primitive extensions. Most of our work—and this is the heart of the proof—is devoted to the latter, namely, to showing that a primitive extension of a system with the multiple recurrence property also has the multiple recurrence property.

1. Limits of SZ-Systems

DEFINITION 7.1. *Let* $\mathbf{X} = (X, \mathcal{B}, \mu, \Gamma)$ *be a m.p.s. with* Γ *an abelian group. We say that* $\mathbf{X}$ *is an SZ-system (SZ for Szemerédi) if whenever* $A \in \mathcal{B}$ *with* $\mu(A) > 0$ *and* $T_1, T_2, \ldots, T_l \in \Gamma$

$$(7.1) \quad \liminf_{N \to \infty} \frac{1}{N} \sum \mu(T_1^{-n}A \cap T_2^{-n}A \cap \cdots \cap T_l^{-n}A) > 0.$$

Equation (7.1) is a strengthened version of what we want to prove for an arbitrary system (with an abelian group acting). It is this version, however, that we shall be able to push through by transfinite induction.

An equivalent formulation of this property is that for $A \in \mathcal{B}$ with $\mu(A) > 0$ there exist $\delta, \varepsilon > 0$, a set $P \subset \mathbf{N}$ with lower density $> \delta$, and for each $n \in P$ a subset $A_n \in \mathcal{B}$ with $\mu(A_n) > \varepsilon$, and with

$$T_1^n A_n \subset A, \qquad T_2^n A_n \subset A, \ldots, \qquad T_l^n A_n \subset A.$$

Clearly if $\mathbf{X}$ is an SZ-system, so is any system equivalent to $\mathbf{X}$. Since a factor of $(X, \mathcal{B}, \mu, \Gamma,)$ is equivalent to $(X, \mathcal{B}', \mu, \Gamma)$ for some sub-σ-algebra $\mathcal{B}' \subset \mathcal{B}$, determining the factors of $\mathbf{X}$ that are SZ-systems amounts to determining the sub-σ-algebras $\mathcal{B}' \subset \mathcal{B}$ for which $(X, \mathcal{B}', \mu, \Gamma)$ is an SZ-system. We now have:

PROPOSITION 7.1. *Let* $\{\mathcal{B}_t\}$ *be a totally ordered family of sub-σ-algebras of* $\mathcal{B}$, *i.e., for* t_1, t_2 *either* $\mathcal{B}_{t_1} \subset \mathcal{B}_{t_2}$ *or* $\mathcal{B}_{t_2} \subset \mathcal{B}_{t_1}$. *Let* $\mathcal{B}$ *be the σ-algebra* *generated by the union* $\bigcup \mathscr{B}_t$. *If each* $(X, \mathscr{B}_t, \mu, \Gamma)$ *is an SZ-system, so is* $(X, \mathscr{B}, \mu, \Gamma)$.

Proof: Let $A \in \mathscr{B}$ and let $T_1, T_2, \ldots, T_l \in \Gamma$. Let $\varepsilon > 0$. Since $\bigcup \mathscr{B}_t$ is an algebra generating $\mathscr{B}$ as σ-algebra, we can find $A' \in \mathscr{B}_{t'}$ for some t' with $\mu(A \triangle A') < \varepsilon$. Then $\|1_A - 1_{A'}\| < \sqrt{\varepsilon}$ in $L^2(\mathbf{X})$. If $\mathbf{X}_{t'} = (X, \mathscr{B}_{t'}, \mu, \Gamma)$, then $E(1_A | \mathbf{X}_{t'})$ is the projection of 1_A on the subspace of functions measurable over $\mathscr{B}_t$. Since $1_{A'}$ is in this subspace, we shall have

$$\|1_A - E(1_A | \mathbf{X}_{t'})\| < \sqrt{\varepsilon}.$$

From this we can deduce that if ε is sufficiently small, $E(1_A | \mathbf{X}_{t'}) > 1 - (1/2l)$ on a set of positive measure. Otherwise $E(1_A | \mathbf{X}_{t'}) \leq 1 - (1/2l)$ a.e. and $1_A - E(1_A | \mathbf{X}_{t'}) \geq (1/2l)$ on a set of measure $\mu(A)$ whence

$$\|1_A - E(1_A | \mathbf{X}_{t'})\| \geq \frac{\sqrt{\mu(A)}}{2l}.$$

So having chosen $\varepsilon < \mu(A)/4l^2$, we can suppose that $E(1_A | X_{t'}) > 1 - (1/2l)$ in B where $B \in \mathscr{B}_{t'}$. Now let $\mu = \int \mu'_x \, d\mu(x)$ be the disintegration of μ with respect to the factor $\mathbf{X}_{t'}$. Then for $x \in B$, $\mu'_x(A) > 1 - (1/2l)$. We now put to use the hypothesis that $\mathbf{X}_{t'}$ is an *SZ*-system. Let P be a subset of $\mathbf{N}$ of positive lower density such that for $n \in P$ there exists a set B_n in $\mathscr{B}_{t'}$ with $T_i^n B_n \subset B$ for $i = 1, 2, \ldots, l$ with $\mu(B_n) > \eta$. Let $x \in B_n$ and consider $\mu'_x(T_1^{-n}A \cap T_2^{-n}A \cap \cdots \cap T_l^{-n}A)$. We have

$$\mu'_x(T_1^{-n}A \cap T_2^{-n}A \cap \cdots \cap T_l^{-n}A)$$
$$= 1 - \mu'_x(X - (T_1^{-n}A \cap T_2^{-n}A \cap \cdots \cap T_l^{-n}A))$$
$$= 1 - \mu'_x\left(\bigcup_{i=1}^{l} (X - T_i^{-n}A)\right) \geq 1 - \sum_{i=1}^{l} (1 - \mu'_x(T_i^{-n}A))$$
$$= 1 - \sum_{i=1}^{l} (1 - \mu'_{T_i^n x}(A)).$$

We have used Proposition 5.9 to equate $T_i^n \mu'_x$ with $\mu'_{T_i^n x}$. Since $x \in B_n$, $T_i^n x \in B$ and $\mu'_{T_i^n x}(A) > 1 - (1/2l)$, this gives

$$\mu'_x(T_1^{-n}A \cap T_2^{-n}A \cap \cdots \cap T_l^{-n}A) > \frac{1}{2}$$

and finally

$$\mu(T_1^{-n}A \cap T_2^{-n}A \cap \cdots \cap T_l^{-n}A) \geq \frac{1}{2}\mu(B_n) > \frac{\eta}{2}$$

for $n \in P$. This proves the proposition.

This result immediately gives the following.

PROPOSITION 7.2. *Let* $\{\mathbf{X}_{\xi}\}$ *be the composition series for a system in accordance with Theorem 6.16. Assume that the ordinal ξ is a limit ordinal and that for each $\xi' < \xi$, $\mathbf{X}_{\xi'}$ is an SZ-system. Then $\mathbf{X}_{\xi}$ is an SZ-system.*

2. Weak Mixing Extensions

The defining property of an SZ-system can be formulated in terms of the behavior of integrals $\int T_1^n f T_2^n f \cdots T_l^n f \, d\mu$ where f is a bounded non-negative function. In this section we shall show that if $\mathbf{X} \to \mathbf{Y}$ is a weak mixing extension, then we can relate the average behavior of integrals of the above form on X to that of the corresponding expressions where f is replaced by $E(f \mid \mathbf{Y})$ and the integration is over Y. The results we obtain are entirely analogous to those obtained in Chapter 4, Section 2, for totally weak mixing groups of transformations. The results there correspond to $\mathbf{Y}$ the trivial system on one point, so that the functions $E(f \mid \mathbf{Y})$ are constants. As in Chapter 4, we shall use the notion of convergence in density.

We begin with the following property of weak mixing extensions relative to a transformation.

LEMMA 7.3. *If $\mathbf{X} \to \mathbf{Y}$ is weak mixing relative to T and $\phi \in L^{\infty}(\mathbf{X})$ with $E(\phi \mid \mathbf{Y}) = 0$, then in the weak topology of $L^2(\mathbf{X})$, $T^n \phi$ converges to 0 in density.*

Proof: We want to show that for any $\psi \in L^2(\mathbf{X})$ and any $\varepsilon > 0$ the set of n for which $|\int \psi T^n \phi \, d\mu| > \varepsilon$ has density 0. It suffices to do this for a dense set of ψ, so we assume $\psi \in L^{\infty}(\mathbf{X})$. According to Proposition 6.2 we have

$$\frac{1}{N+1} \sum_{n=0}^{N} \int E(\psi T^n \phi \mid \mathbf{Y})^2 \, dv \to 0$$

so that

$$\frac{1}{N+1} \sum_{n=0}^{N} \left\{ \int E(\psi T^n \phi \mid \mathbf{Y}) \, dv \right\}^2 \to 0.$$

However, $\int E(\psi T^n \phi \mid \mathbf{Y}) \, dv = \int \psi T^n \phi \, d\mu$. This proves the lemma.

For an arbitrary $\phi \in L^{\infty}(\mathbf{X})$ we shall have $T^n(\phi - E(\phi \mid \mathbf{Y})^{\alpha}) \to 0$ weakly in density, where α is the homomorphism of $\mathbf{X} \to \mathbf{Y}$. We now generalize this result.

PROPOSITION 7.4. *Let $T_1, T_2, \ldots, T_l$ be distinct elements of Γ, $T_i \neq I$, where $\alpha:(X,\mathscr{B},\mu,\Gamma) \to (Y,\mathscr{D},\nu,\Gamma)$ is a weak mixing extension relative to Γ. If $f_1, f_2, \ldots, f_l \in L^\infty(X)$, then*

$$D\text{-}\lim_n \left(\prod_{i=1}^l T_i^n f_i - \prod_{i=1}^l T_i^n E(f_i|Y)^\alpha \right) = 0$$

in the weak topology of $L^2(X)$.

In proving this proposition we shall need a variant of Lemma 4.9 that provides a sufficient condition for convergence in density to 0 of a sequence of vectors in a Hilbert space. We first define

$$D\text{-}\lim_n \sup a_n$$

for a sequence of real numbers $\{a_n\}$ as the g.l.b. of those values λ such that the density of n with $a_n > \lambda$ is 0. If $\{a_n\}$ is bounded, then $D\text{-}\lim_n \sup a_n$ is finite. The proof of Lemma 4.9 now gives the following.

LEMMA 7.5. *Let $\{x_n\}$ be a sequence of vectors in a Hilbert space and suppose that*

$$D\text{-}\lim_m D\text{-}\lim_n \sup |\langle x_{n+m}, x_n \rangle| = 0.$$

Then with respect to the weak topology, $D\text{-}\lim_n x_n = 0$.

We shall also use the following consequence of Proposition 6.2.

LEMMA 7.6. *If $X \to Y$ is w.m. rel. T and $\phi, \psi \in L^\infty(X)$ with $E(\phi|Y) = 0$, then $\|E(\psi T^n \phi|Y)\|_2 \to 0$ in density.*

Proof of Proposition 7.4: We shall proceed by induction on l, the case $l = 1$ being a consequence of Lemma 7.3. Assume the proposition is valid for a particular value of l. Let $S_1, S_2, \ldots, S_{l+1}$ be distinct elements of Γ with $S_i \neq I$ and let $f_1, f_2, \ldots, f_{l+1} \in L^\infty(X)$. We want to show that

$$(7.2) \qquad D\text{-}\lim_n \left(\prod_{i=1}^{l+1} S_i^n f_i - \prod_{i=1}^{l+1} S_i^n E(f_i|Y)^\alpha \right) = 0$$

in the weak topology of $L^2(X)$. Apply to the foregoing expression the identity

$$(7.3) \qquad \prod_{i=1}^{l+1} a_i - \prod_{i=1}^{l+1} b_i = \sum_{j=1}^{l+1} \left(\prod_{i=1}^{j-1} a_i \right)(a_j - b_j)\left(\prod_{i=j+1}^{l+1} b_j \right)$$

where an empty product is interpreted as 1. Using the identity, we reduce the proof of (7.2) to the proof of

$$D\text{-}\lim_n \prod_{i=1}^{l+1} S_i^n f_1 = 0$$

under the hypothesis that for some j, $E(f_j|\mathbf{Y}) = 0$ (since

$$E(f_j - E(f_j|\mathbf{Y})^\alpha|\mathbf{Y}) = 0).$$

Assume $E(f_{l+1}|\mathbf{Y}) = 0$ and apply Lemma 7.5 to the sequence of vectors

$$g_n = \prod_{i=1}^{l+1} S_i^n f_i.$$

Setting $F_i^{(m)} = f_i S_i^m f_i$, we have

$$\langle g_{n+m}, g_n \rangle = \int \prod_{i=1}^{l+1} S_i^n F_i^{(m)}\, d\mu.$$

Now let $T_1 = S_1 S_{l+1}^{-1}$, $T_2 = S_2 S_{l+1}^{-1}, \ldots, T_l = S_l S_{l+1}^{-1}$. Then we can write

$$\langle g_{n+m}, g_n \rangle = \int F_{l+1}^{(m)} \prod_{i=1}^{l} T_i^n F_i^{(m)}\, d\mu.$$

Since $T_1, T_2, \ldots, T_l$ are distinct and different from the identity, we can apply the induction hypothesis to obtain

$$D\text{-}\lim_n \left\{ \int F_{l+1}^{(m)} \prod_{i=1}^{l} T_i^n F_i^{(m)}\, d\mu - \int F_{l+1}^{(m)} \prod_{i=1}^{l} T_i^n E(F_i^{(m)}|\mathbf{Y})^\alpha\, d\mu \right\} = 0$$

so that

$$D\text{-}\lim_n \sup |\langle g_{n+m}, g_n \rangle| = D\text{-}\lim_n \sup \left| \int F_{l+1}^{(m)} \prod_{i=1}^{l} T_i^n E(F_i^{(m)}|\mathbf{Y})^\alpha\, d\mu \right|.$$

Since $\prod_{i=1}^{l} T_i^n E(F_i^{(m)}|\mathbf{Y})^\alpha \in L^2(\mathbf{Y})^\alpha \subset L^2(\mathbf{X})$, we have

$$\int F_{l+1}^{(m)} \left(\prod_{i=1}^{l} T_i^n E(F_i^{(m)}|\mathbf{Y}) \right)^\alpha d\mu$$

$$= \int E\left(F_{l+1}^{(m)} \left(\prod_{i=1}^{l} T_i^n E(F_i^{(m)}|\mathbf{Y}) \right)^\alpha \middle| \mathbf{Y} \right) d\nu$$

$$= \int E(F_{l+1}^{(m)}|\mathbf{Y}) \prod_{i=1}^{l} T_i^n E(F_i^{(m)}|\mathbf{Y})\, d\nu.$$

Letting c be a bound for the L^2-norms of

$$\prod_{i=1}^{l} T_i^n E(F_i^{(m)}|\mathbf{Y})$$

(all functions in sight are bounded), we find

$$D\text{-}\lim_n \sup |\langle g_{n+m}, g_n \rangle| \leq c \|E(F_{l+1}^{(m)}|\mathbf{Y})\|_2.$$

Recalling that $F_{l+1}^{(m)} = f_{l+1}S_{l+1}^m f_{l+1}$ and invoking Lemma 7.6, we obtain

$$D\text{-}\lim_{m} D\text{-}\lim_{n} \sup |\langle g_{n+m}, g_n \rangle| = 0$$

and so, by Lemma 7.5, $D\text{-}\lim g_n = 0$. This proves the proposition.

As a corollary we have

PROPOSITION 7.7. *If* $\alpha:(X,\mathcal{B},\mu,\Gamma) \to (Y,\mathcal{D},v,\Gamma)$ *is weak mixing relative to* Γ, $T_1, T_2, \ldots, T_l$ *are distinct elements of* Γ (*one of them may be the identity*) *and* $f_1, f_2, \ldots, f_l \in L^\infty(X)$, *then*

$$D\text{-}\lim_{n} \left\{ \int \prod_{i=1}^{l} T_i^n f_i \, d\mu - \int \prod_{i=1}^{l} T_i^n E(f_i|\mathbf{Y}) \, dv \right\} = 0.$$

Proof: We rewrite this as

$$(7.4) \qquad D\text{-}\lim \int \left[\prod_{i=1}^{l} T_i^n f_i - \prod_{i=1}^{l} T_i^n E(f_i|\mathbf{Y})^\alpha \right] d\mu = 0.$$

If all the T_i are different from the identity, this follows immediately from Proposition 7.4. If not, replace all the T_i by some $T_0 T_i$ where $T_0 \neq T_j^{-1}$ for any j. This will not affect (7.4) since T_0^n is measure preserving.

This proposition makes it evident that a weak mixing extension of an SZ-system is an SZ-system. For $(X,\mathcal{B},\mu,\Gamma)$ is an SZ-system if for any $f \in L^\infty(X)$, $f \geq 0$ and f not a.e. 0,

$$\liminf \frac{1}{N} \sum_{n=1}^{N} \int T_1^n f T_2^n f \cdots T_l^n f \, d\mu > 0,$$

where $T_1, T_2, \ldots, T_l \in \Gamma$. In this expression we can assume all the T_i are distinct and then the foregoing proposition enables us to replace the integrals here by the corresponding integrals for $\mathbf{Y}$ with f replaced by $E(f|\mathbf{Y})$.

Recall that if $\mathbf{X} \to \mathbf{Y}$ is weak mixing relative to a transformation, so is $\mathbf{X} \times_{\mathbf{Y}} \mathbf{X} \to \mathbf{Y}$. Let us apply Proposition 7.7 to this extension with the f_i replaced by $f_i \otimes f_i$. Since

$$\int f \otimes f \, d\mu \times_{\mathbf{Y}} \mu = \int E(f|\mathbf{Y})^2 \, dv$$

we obtain

$$D\text{-}\lim_{n} \left\{ \int E\left(\prod_{i=1}^{l} T_i^n f_i \Big| \mathbf{Y} \right)^2 dv - \int \prod_{i=1}^{l} T_i^n E(f_i|\mathbf{Y})^2 \, dv \right\} = 0.$$

In particular, if some $E(f_j|\mathbf{Y}) = 0$, then

$$D\text{-}\lim_n \int E\left(\prod_{i=1}^l T_i^n f_i \middle| \mathbf{Y}\right)^2 dv = 0.$$

Thus the functions

$$E\left(\prod_{i=1}^l T_i^n f_i \middle| \mathbf{Y}\right) \to 0$$

in density with respect to the strong topology in $L^2(\mathbf{Y})$. Returning to the general case, we consider the function

$$E\left(\prod_{i=1}^l T_i^n f_i \middle| \mathbf{Y}\right) - \prod T_i^n E(f_i|\mathbf{Y})$$

$$= E\left(\prod_{i=1}^l T_i^n f_i - \prod_{i=1}^l T_i^n E(f_i|\mathbf{Y}) \middle| \mathbf{Y}\right).$$

Applying the identity (7.3) to the latter expression, we see that the function in question is a sum of functions

$$E\left(\prod_{i=1}^l T_i^n g_i \middle| \mathbf{Y}\right)$$

where some $E(g_i|\mathbf{Y}) = 0$. It follows therefore that

$$D\text{-}\lim\left\{E\left(\prod_{i=1}^l T_i^n f_i \middle| \mathbf{Y}\right) - \prod T_i^n E(f_i|\mathbf{Y})\right\} = 0$$

in the strong topology of $L^2(\mathbf{Y})$. We have proved the following:

PROPOSITION 7.8. *If* $(X,\mathscr{B},\mu,\Gamma) \to (Y,\mathscr{D},v,\Gamma)$ *is weak mixing relative to* Γ *and* $T_1,T_2,\ldots,T_l$ *are distinct elements of* Γ, *then for any functions* $f_1,f_2,\ldots,f_l \in L^\infty(\mathbf{X})$

$$\frac{1}{N}\sum_{n=1}^N \int\left\{E\left(\prod_{i=1}^l T_i^n f_i \middle| \mathbf{Y}\right) - \prod_{i=1}^l T_i^n E(f_i|\mathbf{Y})\right\}^2 dv \to 0$$

as $N \to \infty$.

3. Primitive Extensions

We now assume that $\alpha:(X,\mathscr{B},\mu,\Gamma) \to (Y,\mathscr{D},v,\Gamma)$ is a primitive extension, so that $\Gamma = \Gamma_c \times \Gamma_w$ with $\alpha:X \to Y$ compact relative to Γ_c and $\alpha:X \to Y$ weak mixing relative to Γ_w. In this discussion Γ is a finitely generated free abelian group. We want to show that if $\mathbf{Y}$ is an SZ-system, then so is $\mathbf{X}$. Once again, we shall assume that $\mathbf{X}$ is a regular system, and so we may decompose the measure $\mu:\mu = \int \mu_y \, dv$. As in Chapter 6, Section 3, we let $\mathcal{H} = L^2(X,\mathcal{B},\mu)$ and $\mathcal{H}_y = L^2(X,\mathcal{B},\mu_y)$. Also $\|f\|_y$ denotes the norm of a function in $\mathcal{H}_y$. The results of the foregoing section imply the following:

LEMMA 7.9. *Let* $S_1, S_2, \ldots, S_s$ *be distinct elements of* Γ_w, *and let* $f \in L^\infty(\mathbf{X})$. *Let* $\varepsilon, \delta > 0$. *For each* y *let* $\psi(y) = \int f \, d\mu_y$. *Then the set of* $n \geq 0$ *for which*

$$(7.5) \quad v\left\{ y \in Y : \left| \int \prod S_i^n f \, d\mu_y - \prod \psi(S_i^n y) \right| > \varepsilon \right\} > \delta$$

has density 0.

Proof: We have $\psi = E(f \mid Y)$ and by Proposition 7.8,

$$\lim_{N \to \infty} \frac{1}{N} \sum_{n=1}^{N} \int [E(\prod S_i^n f \mid Y) - \prod S_i^n E(f \mid Y)]^2 \, dv = 0.$$

The implications of compactness are summarized in the next lemma.

LEMMA 7.10. *Let* $A \in \mathcal{B}$ *with* $\mu(A) > 0$. *We can find* $A' \subset A$ *with* $\mu(A')$ *as close as we like to* $\mu(A)$ *having the following property: For any* $\varepsilon > 0$ *there exists a finite set of functions* $g_1, \ldots, g_K \in \mathcal{H}$ *and a function* $k: Y \times \Gamma_c \to \{1, 2, \ldots, K\}$ *such that for a.e.* $y \in Y$ *and every* $R \in \Gamma_c$, $\|R 1_{A'} - g_{k(y,R)}\|_y < \varepsilon$.

Proof: We have used property C_3 of compactness with $A' = A \cap \alpha^{-1}(B)$, where B is a set arbitrarily close to Y.

We shall need the following result which is a consequence of the multi-dimensional van der Waerden (Grünwald's) theorem.

LEMMA 7.11. *Let the number* K *be given and let* $T_1, T_2, \ldots, T_H$ *be elements of* Γ. *There is a finite subset* $Q \subset \Gamma$ *and a number* $M < \infty$ *such that for any map* $k: \Gamma \to \{1, 2, \ldots, K\}$ *there exists* $T' \in Q$ *and* $m \in \mathbf{N}$, $1 \leq m \leq M$, *such that* $k(T'T_1^m) = k(T'T_2^m) = \cdots = k(T'T_H^m)$.

Proof: Consider the corresponding situation for $\mathbf{Z}^r$ with $k: \mathbf{Z}^r \to \{1, 2, \ldots, K\}$. According to Theorem 2.10, if $v_1, v_2, \ldots, v_H \in \mathbf{Z}^r$, there exists a finite subset $L \subset \mathbf{Z}^r$ such that for any map $k: L \to \{1, 2, \ldots, K\}$, k will be constant on a subset of L of the form $\{v' + mv_1, v' + mv_2, \ldots, v' + mv_H\}$. But there are only finitely many possibilities for m and v' with $v' + mv_1 \in L$ and $v' + mv_2 \in L$, assuming, as we may, that $v_1 \neq v_2$. Carrying this over to Γ, we obtain the lemma.

We are now ready to prove the following result.

PROPOSITION 7.12. *If* $\alpha: \mathbf{X} \to \mathbf{Y}$ *is a primitive extension and* Y *is an SZ-system, then* $\mathbf{X}$ *is an SZ-system.*

Proof: Let $A \in \mathcal{B}$ with $\mu(A) > 0$, *and let* $T_1, T_2, \ldots, T_l \in \Gamma$. Replacing A by a slightly smaller set, we can assume that 1_A has the compactness property described in Lemma 7.10. Writing $\mu(A) = \int \mu_y(A) \, dv(y)$, we see that if we set $a = \mu(A)/2$, there exists a subset $B \subset Y$, $v(B) > 0$ with $\mu_y(A) \geq a$ for $y \in B$. We express the T_j as products of elements in Γ_c and in Γ_w. We can then assume that $\{T_j, j = 1, 2, \ldots, l\} \subset \{R_i S_j, i = 1, 2, \ldots, r, j = 1, 2, \ldots, s\}$ where $R_1, R_2, \ldots, R_r \in \Gamma_c$ and $S_1, S_2, \ldots, S_s$ are distinct elements of Γ_w. We also take $R_1 = I$, and we replace the set $\{T_j\}$ by the possibly larger set $\{R_i S_j\}$. Let $a_1 < a^s$. We shall show that there exists a set $P \subset \mathbf{N}$ with positive lower density and $\eta > 0$ such that for each $n \in P$ there is a set $B_n \subset Y$, $B_n \in \mathcal{D}$, with $v(B_n) > \eta$, such that if $y \in B_n$,

$$(7.6) \qquad \mu_y\left(\bigcap_{i,j} R_i^{-n} S_j^{-n} A\right) > a_1.$$

This will prove the proposition since it implies that

$$\mu\left(\bigcap_i T_i^{-n} A\right) > a_1 \eta,$$

for $n \in P$. The set B_n will be determined by two requirements. For $a_1 < a_2 < a^s$ we shall require

$$(7.7) \qquad \mu_y\left(\bigcap_j S_j^{-n} A\right) > a_2$$

when $n \in P$ and $y \in B_n$. Secondly we prescribe $\varepsilon_1 > 0$ such that if $\mu_y(S_j^{-n} R_i^{-n} A \triangle S_j^{-n} A) < \varepsilon_1$ for $1 \leq i \leq r$, $1 \leq j \leq s$, then (7.7) implies (7.6). Then we require

$$(7.8) \qquad \mu_y(S_j^{-n} R_i^{-n} A \triangle S_j^{-n} A) < \varepsilon_1, \qquad i \leq i \leq r, \quad 1 \leq j \leq s,$$

when $n \in P$ and $y \in B_n$.

Suppose now that P and $\{B_n, n \in P\}$ have been found so that (7.8) is satisfied and, in addition,

$$(7.9) \qquad S_j^n y \in B, \qquad \text{for } y \in B_n, \quad 1 \leq j \leq s.$$

Apply Lemma 7.9 with $f = 1_A$, $\varepsilon < a^s - a_2$ and $\delta < 1/2 \lim \inf v(B_n)$. Then

$$\mu_y\left(\bigcap_i S_j^{-n} A\right) = \int \prod S_j^n f \, d\mu_y > \prod \psi(S_j^n y) - \varepsilon \geq a^s - \varepsilon > a_2$$

for $y \in B_n$ but for a set of y of measure $< (1/2) v(B_n)$ and for n outside a set of density 0. Modifying P and B_n accordingly, we will be left with a set—call it again P—with positive lower density, and for each $n \in P$ a set—call it again B_n—with lim inf $v(B_n) > 0$ such that for these n and y, (7.7) and (7.8) are valid. As we saw this gives (7.6) as needed. The problem is reduced to finding P and $\{B_n, n \in P\}$ such that (7.8) and (7.9) are satisfied.

Next we replace (7.8) by the requirement that there exist $g \in \mathcal{H}_y$ with

$$(7.10) \qquad \left\| S_j^n R_i^n 1_A - S_j^n g \right\|_y < \varepsilon_2, \qquad 1 \leq i \leq r, \quad 1 \leq j \leq s,$$

with $\varepsilon_2 < \sqrt{\varepsilon_1}/2$. Since $R_1 = I$, we shall have $\left\| S_j^n R_i^n 1_A - S_j^n 1_A \right\|_y < 2\varepsilon_2 < \sqrt{\varepsilon_1}$ which gives (7.8).

Now recall that A was chosen to comply with the requirements of Lemma 7.10. We can therefore find $g_1, g_2, \ldots, g_K \in L^2(\mathbf{X})$ and a function $k: Y \times \Gamma_c \to \{1, 2, \ldots, K\}$, so that $\left\| R 1_A - g_{k(y,R)} \right\|_y < \varepsilon_2$ for every $R \in \Gamma_c$ and $a.e.$ y. We now define a sequence of functions $k_p: Y \times \Gamma \to \{1, 2, \ldots, K\}$ by $k_p(y, RS) = k(S^p y, R^p)$, $p = 1, 2, \ldots$, where $R \in \Gamma_c$, $S \in \Gamma_w$. This is well defined since $\Gamma = \Gamma_c \times \Gamma_w$ is a direct decomposition. We shall then have

$$(7.11) \qquad \left\| S^p R^p 1_A - S^p g_{k_p(y, RS)} \right\|_y = \left\| R^p 1_A - g_{k(S^p y, R^p)} \right\|_{S^p y} < \varepsilon_2.$$

Fix p and y and apply Lemma 7.8 to the function $k_p(y, \cdot)$ on Γ. Independently of p and y there is a finite subset $Q \subset \Gamma$ and a number M such that $k_p(y, T' R_i^m S_j^m)$ takes on the same value for $1 \leq i \leq r$, $1 \leq j \leq s$, for some $T' \in Q$ and some m with $1 \leq m \leq M$. If k is this value, write $g_{(p,y)}$ for the corresponding g_k. Then if $T' = R'S'$,

$$(7.12) \qquad \left\| S_j^{pm} R_i^{pm} 1_A - S_j^{pm}(R'^{-p} g_{(p,y)}) \right\|_{T'^p y}$$
$$= \left\| T'^p S_j^{pm} R_i^{pm} 1_A - T'^p R'^{-p} S_j^{pm} g_{(p,y)} \right\|_y$$
$$= \left\| S'^p S_j^{pm} R'^p R_i^{pm} 1_A - S'^p S_j^{pm} g_{(p,y)} \right\|_y < \varepsilon_2$$

for $1 \leq i \leq r$, $1 \leq j \leq s$, by (7.11). What we have shown is that for every $p = 1, 2, \ldots$ and $a.e.$ $y \in Y$, there exist m and T', both having a finite range of possibilities, such that (7.10) is satisfied with $n = pm$ for $T'^p y$.

We are now ready to produce the set P and the sets B_n, $n \in P$, such that both (7.9) and (7.10) are satisfied for (y, n), $y \in B_n$. For each p form the set

$$C_p = \bigcap_{j, m, T'} S_j^{-mp} T'^{-p} B \subset Y$$

where the intersection is taken over j, m, T' with $1 \leq j \leq s$, $1 \leq m \leq M$, $T' \in Q$. Here we use the fact that $\mathbf{Y}$ is an SZ-system. It follows that for a set P' of positive lower density, if $p \in P'$, $v(C_p) > \eta' > 0$. Now let $y \in C_p$ for $p \in P'$. There exists $m = m(p, y)$ and $T' = T'(p, y)$ such that $T'^p y$ satisfies (7.10) for $n = pm$. $T'^p y$ also satisfies (7.9) for this T' and m, since by the definition of C^p, this condition is satisfied by all $T'^p y$ and for all m, $T' \in Q$, $1 \leq m \leq M$. Let J be the total number of possibilities for (m, T'). Then for a subset $D_p \subset C_p$ with $v(D_p) > \eta'/J$, $m(p, y)$ and $T'(p, y)$ take on a constant value, say, $m(p)$, $T'(p)$. We now define $n(p) = pm(p)$, and set $P = \{n(p), p \in P'\}$ and

(7.13) $B_{n(p)} = T'(p)^p D_p.$

Then $v(B_{n(p)}) = v(D_p) > \eta'/J$, $S_j^{n(p)} B_{n(p)} \subset B, j = 1, 2, \ldots, s,$ and

$$\left\| S_j^{n(p)} R_i^{n(p)} 1_A - S_j^{n(p)} g'_{(p,y)} \right\|_y < \varepsilon_2$$

for $y \in B_{n(p)}$, $1 \le i \le r$, $1 \le j \le s$ for an appropriately defined $g'_{(p,y)}$. Finally the lower density of $\{pm(p)\}$ is at least $1/M \times$ the lower density of p. We have fulfilled all our requirements, so this concludes the proof of the proposition.

4. The Multiple Recurrence Theorem

Combining the structure theorem (Theorem 6.16) and Propositions 7.1 and 7.12, we can lift the SZ-property from the trivial 1-point system to an arbitrary system with a finitely generated abelian group Γ. Since any finite set of commuting invertible measure preserving transformations of a measure space generate such a group, we have proved the following theorem.

THEOREM 7.13.   *If* $T_1, T_2, \ldots, T_l$ *are commuting invertible measure preserving transformations of a measure space* $(X, \mathscr{B}, \mu)$ *and* $A \in \mathscr{B}$ *with* $\mu(A) > 0$, *then*

(7.14) $\displaystyle \liminf_{N \to \infty} \frac{1}{N} \sum_{n=1}^{N} \mu(T_1^{-n} A \cap T_2^{-n} A \cap \cdots \cap T_l^{-n} A) > 0.$

To complete the picture, we should show that the same result is valid for measure preserving transformations that are not necessarily invertible. We do this as follows. We consider first a special case of this situation and show that if the result in question is established here, it will be valid generally. Then we show that in the special case we can pass from non-invertible transformations to invertible ones.

We take $X_0 = \{0,1\}^{\mathbf{N}_0^l}$ where $\mathbf{N}_0 = \{0,1,2,\ldots\}$, so that X_0 is a compact metric space, and let $\mathscr{B}_0$ be the borel σ-algebra on X_0. The elements of $\mathbf{N}_0^l$ are l-tuples of integers and we denote by $e_1, e_2, \ldots, e_l$ the basis vectors, $e_1 = (1, 0, \ldots, 0), e_2 = (0, 1, \ldots, 0), \ldots, e_l = (0, 0, \ldots, 1)$. Define $S_i: X_0 \to X_0$ by $S_i \omega(u) = \omega(u + e_i), i = 1, 2, \ldots, l$, and let μ_0 be a measure on $(X_0, \mathscr{B}_0)$ invariant under $S_1, S_2, \ldots, S_l$. Finally let $A_0 = \{\omega : \omega(0) = 1\}$. We shall show that if $\mu_0(A_0) > 0$, then

(7.15) $\displaystyle \liminf_{N \to \infty} \frac{1}{N} \sum_{n=1}^{N} \mu_0(S_1^{-n} A_0 \cap S_2^{-n} A_0 \cap \cdots \cap S_l^{-n} A_0) > 0.$

Suppose this has been done and let $(X,\mathscr{B},\mu)$ be any measure space, $T_1, T_2, \ldots, T_l$ commuting measure preserving transformations of $(X,\mathscr{B},\mu)$. If $A \in \mathscr{B}$, we define a map $\alpha:(X,\mathscr{B},\mu) \to (X_0,\mathscr{B}_0,\mu_0)$ by

$$\alpha(x)(n_1,n_2, \ldots,n_l) = 1_A(T_1^{n_1} T_2^{n_2} \cdots T_l^{n_l}x).$$

Let μ_0 be the image of μ under this map, $\mu_0(B) = \mu(\alpha^{-1}(B))$. We have $S_i\alpha(x) = \alpha(T_ix)$, so that μ_0 is S_i-invariant. $\alpha^{-1}(A_0) = A$ so that if $\mu(A) > 0$, then $\mu_0(A_0) > 0$. Thus (7.15) for the special measure space $(X_0,\mathscr{B}_0,\mu_0)$ with the transformations S_i implies the analogous result in general.

We now establish (7.15) by constructing an m.p.s. with invertible transformations. Take $\tilde{X}_0 = \{0,1\}^{\mathbf{Z}^l}$, $\tilde{\mathscr{B}}_0$ the corresponding borel σ-algebra, $\tilde{S}:\tilde{X}_0 \to \tilde{X}_0$ defined by $\tilde{S}_i\omega(u) = \omega(u + e_i)$, and $\tilde{A}_0 = \{\omega:\omega(0) = 1\}$. There is a natural map $\pi:\tilde{X}_0 \to X_0$ defined by $\pi(\omega)(u) = \omega(u)$ for $u \in \mathbf{N}_0^l \subset \mathbf{Z}^l$. We shall show that there exists a measure $\tilde{\mu}_0$ on $(\tilde{X}_0,\tilde{\mathscr{B}}_0)$ satisfying $\pi\tilde{\mu}_0 = \mu_0$, and which is $\tilde{S}_i$-invariant, $i = 1,2, \ldots,l$. Let $\tilde{\mu}$ be any measure on $\tilde{X}_0$ with $\pi\tilde{\mu} = \mu_0$. It is easy to find such a measure by writing $\tilde{X}_0 = X_0 \times Y_0$ for an appropriate Y_0 and setting $\tilde{\mu} = \mu_0 \times v_0$ for any v_0 on Y_0. Now since $\pi\tilde{S}_i = S_i\pi$ and $S_i\mu_0 = \mu_0$, $\pi(\tilde{S}_i\tilde{\mu}) = \mu$, and similarly $\pi(\tilde{S}_1^{n_1}\tilde{S}_2^{n_2} \cdots \tilde{S}_l^{nl}\tilde{\mu}) = \mu$. So if $\tilde{\mu}_0$ is any limit of

$$\frac{1}{N^l} \sum_{n_1 = 1}^N \sum_{n_2 = 1}^N \cdots \sum_{n_l = 1}^N \tilde{S}_1^{n_1}\tilde{S}_2^{n_2} \cdots \tilde{S}_l^{n_l}\tilde{\mu},$$

then $\tilde{\mu}_0$ is $\tilde{S}_i$-invariant and $\pi(\tilde{\mu}_0) = \mu_0$. We can now apply Theorem 7.13 to $(\tilde{X}_0,\tilde{\mathscr{B}}_0,\tilde{\mu}_0; \tilde{S}_1, \ldots,\tilde{S}_l)$ and the set $\tilde{A}_0$. Mapping down to $(X_0,\mathscr{B}_0,\mu_0)$, we obtain (7.15). We have established the following.

THEOREM 7.14. *If $T_1,T_2, \ldots,T_l$ are commuting measure preserving transformations of a measure space $(X,\mathscr{B},\mu)$ and $A \in \mathscr{B}$ with $\mu(A) > 0$, then*

$$\liminf_{N \to \infty} \frac{1}{N} \sum_{n=1}^N \mu(T_1^{-n}A \cap T_2^{-n}A \cap \cdots \cap T_l^{-n}A) > 0.$$

As a corollary we have the "multiple recurrence theorem".

THEOREM 7.15. *If $T_1,T_2, \ldots,T_l$ are commuting measure preserving transformations of a measure space $(X,\mathscr{B},\mu)$ and $A \in \mathscr{B}$ with $\mu(A) > 0$, then for some $n \geq 1$, $\mu(T_1^{-n}A \cap T_2^{-n}A \cap \cdots \cap T_l^{-n}A) > 0$.*

By a procedure analogous to that described in Chapter 3, Section 7, we can deduce from Theorem 7.15 a multidimensional analogue of Szemerédi's theorem. To formulate this we need the notion of positive upper Banach density for higher dimensional sets. By a *block* in $\mathbf{Z}^r$ or $\mathbf{R}^r$, we shall mean the product of intervals. The *width* of a block will mean the length of its shortest edge. A set $S \subset \mathbf{Z}^r$ will be said to have positive upper Banach density if for some sequence of blocks B_n whose widths $\to \infty$,

$$\frac{|S \cap B_n|}{|B_n|} > \delta > 0$$

for some δ. Here $|\cdot|$ denotes the number of elements in the finite set in question. We shall also speak of a measurable set $S \subset \mathbf{R}^r$ as having *positive upper Banch density* (with respects to $\mathbf{R}^r$) if for a sequence of blocks B_n whose widths $\to \infty$,

$$\frac{\lambda(S \cap B_n)}{\lambda(B_n)} > \delta > 0$$

for some δ, where λ denotes Lebesgue measure on $\mathbf{R}^r$. We can now state the following two results:

THEOREM 7.16. *If $S \subset \mathbf{Z}^r$ has positive upper Banach density and F is a finite subset of $\mathbf{Z}^r$, then for some vector $a \in \mathbf{Z}^r$ and integer $b \geq 1$, $a + bF \subset S$.*

THEOREM 7.17. *If $S \subset \mathbf{R}^r$ has positive upper Banach density with respect to $\mathbf{R}^r$ and F is a finite subset of $\mathbf{R}^r$, then for some vector $a \in \mathbf{R}$ and integer $b \geq 1$, $a + bF \subset S$.*

Proof of Theorem 7.16. Let $X = \{0,1\}^{\mathbf{Z}^r}$ with $\mathscr{B}$ its borel σ-algebra. $\mathbf{Z}^r$ acts on X in the usual manner: if $u = (n_1, n_2, \ldots, n_r) \in \mathbf{Z}^r$, then define $T(u) : X \to X$ by

$$T(u)(\omega)(n'_1, n'_2, \ldots, n'_r) = \omega(n'_1 + n_1, n'_2 + n_2, \ldots, n'_r + n_r).$$

Let $F = \{u_1, u_2, \ldots, u_l\}$ be a finite subset of $\mathbf{Z}^r$ and S a subset of positive upper Banach density. We regard the characteristic functions of S, 1_S, as a point of X. Let $\{B_n\}$ be a sequence of blocks of increasing widths for which $(|S \cap B_n|/|B_n|) > \delta > 0$. Define measures μ_n on X by

$$\int f \, d\mu_n = \frac{1}{|B_n|} \sum_{u \in B} f(T(u) 1_S).$$

$\{B_n\}$ is a Følner sequence (Chapter 6, §2), and so if μ is any limit point of $\{\mu_n\}$, we will have $T(u)\mu = \mu$, $u \in \mathbf{Z}^r$. Let $A = \{\omega : \omega(0) = 1\}$. Since A is a closed and open set, 1_A is a continuous function. We have $1_A(T(u)1_S) = 1 \Leftrightarrow T(u)1_S(0) = 1 \Leftrightarrow u \in S$, so that $\mu_n(A) = (|S \cap B_n|/|B_n|) > \delta$. Hence $\mu(A) \geq \delta > 0$. We now apply Theorem 7.15 to the transformations $T(u_1), T(u_2), \ldots, T(u_l)$ and the set $A \subset X$. We conclude that for a subset of X of positive μ measure and some $b \geq 1$, we shall have $T(u_1)^b \omega \in A$, $T(u_2)^b \omega \in A, \ldots, T(u_l)^b \omega \in Z$. Since A is open, this will be valid for a neighborhood of ω. Now it is easily seen that the support of μ is contained in the closure of translates of 1_S. We conclude that for some vector $a \in \mathbf{Z}^r$, $T(u_i)^b T(a) 1_S \in A$, $i = 1, 2, \ldots, l$, or $T(a + bu_i) 1_S \in A \Leftrightarrow a + bu_i \in S$ for $i = 1, 2, \ldots, l$. This proves the theorem.

Proof of Theorem 7.17:   We shall use Theorem 7.16 as applied to a lattice $\mathbf{Z}^q$ whose dimension q equals the number of points in the configuration $F \subset \mathbf{R}^r$. Let $F = \{u_1, u_2, \ldots, u_q\}$, and for each $v \in \mathbf{R}^r$ define a map $\phi_v : \mathbf{Z}^q \to \mathbf{R}^r$ by

$$\phi_v(n_1, n_2, \ldots, n_q) = \sum_{i=1}^{q} n_i u_i + v.$$

We leave it to the reader to verify that if $S \subset \mathbf{R}^r$ has positive upper Banach density with respect to $\mathbf{R}^r$, then for a set of v of positive measure, $\phi_v^{-1}(S) \subset \mathbf{Z}^q$ has positive upper Banach density in $\mathbf{Z}^q$. Thus for some $v \in \mathbf{R}^r, (a_1, a_2, \ldots, a_q) \in \mathbf{Z}^q$ and integer $b \geq 1$, $be_i + (a_1, a_2, \ldots, a_q) \in \phi_v^{-1}(S)$ for $i = 1, 2, \ldots, q$ where $\{e_i\}$ is the usual basis of $\mathbf{Z}^q$. In other words,

$$(v + \sum_{1}^{q} a_j u_j) + bu_i \in S$$

for $i = 1, 2, \ldots, q$. This completes the proof.

PART III
DYNAMICS AND LARGE SETS
OF INTEGERS

CHAPTER 8

Proximality in Dynamical Systems and the Theorems of Hindman and Rado

In this chapter we return to topological dynamics with a discussion of the notion of proximality that we shall tie together with that of recurrence and then apply to combinatorial questions. The main combinatorial consequences are the theorems of Hindman and Rado mentioned in the Introduction. The combinatorial aspects of certain large sets of integers of which Hindman's theorem is an example will be seen in the next chapter to have implications for topological dynamics and ergodic theory. Our goal has been to focus on the interaction between the two areas, and there will be little by way of new results here. The material in these two chapters is based on work done jointly with B. Weiss and some of it appears in Furstenberg and Weiss [1].

1. Proximality

We shall be considering dynamical systems (X,G) where X is a compact metric space and G a group or semigroup of continuous maps of X to itself. $d(,)$ denotes the metric on X.

DEFINITION 8.1. *Two points $x_1, x_2 \in X$ are proximal if for some sequence $\{g_n\} \subset G$, we have $d(g_n x_1, g_n x_2) \to 0$. Two points that are not proximal are said to be distal.*

We shall be particularly interested in the phenomenon of proximality for $G = \mathbf{N}$, the natural numbers. We denote the system by (X,T) where T is a continuous map of $X \to X$. Two points are proximal if for some $n_k \to \infty$, $d(T^{n_k}x_1, T^{n_k}x_2) \to 0$.

Recall the definition of a *thick subset* of $\mathbf{N}$ (Definition 1.10). We shall say that R is a thick subset of $\mathbf{N}$ if R contains arbitrarily long consecutive blocks of integers. We now have

LEMMA 8.1. *If x_1, x_2 are proximal points for (X,T), then for every $\varepsilon > 0$ the set $\{n : d(T^n x_1, T^n x_2) < \varepsilon\}$ is thick.*

Proof: For any l there exists $\delta(l)$ with

$$d(x_1,x_2) < \delta(l) \Rightarrow d(T^i x_1, T^i x_2) < \varepsilon$$

for $i = 1, 2, \ldots, l$. So if $d(T^{n_l} x_1, T^{n_l} x_2) < \delta(l)$, the interval $(n_l + 1, n_l + 2, \ldots, n_l + l) \subset \{n : d(T^n x_1, T^n x_2) < \varepsilon\}$.

This lemma can be readily applied to the symbolic dynamical system $(\Lambda^{\mathbf{N}}, T)$ where Λ is a finite set and $T: \Lambda^{\mathbf{N}} \to \Lambda^{\mathbf{N}}$ is the shift $T\omega(n) = \omega(n + 1)$. (See Chapter 1, Section 3 for details.)

LEMMA 8.2.   *If* $\omega_1, \omega_2 \in \Lambda^{\mathbf{N}}$, *then* ω_1 *and* ω_2 *are proximal if and only if* $\{n : \omega_1(n) = \omega_2(n)\}$ *is thick.*

Proof: The necessity is clear from the foregoing lemma. Conversely, suppose the condition is satisfied. If $\omega_1(n) = \omega_2(n)$ for $n \in (n_l + 1, n_l + 2, \ldots, n_l + l)$ then $T^{n_l}\omega_1, T^{n_l}\omega_2$ agree on the interval $(1, 2, \ldots, l)$ and so $d(T^{n_l}\omega_1, T^{n_l}\omega_2) \to 0$.

The phenomenon of proximality is by no means rare. For example, if X is any infinite, closed, T-invariant subset of $\Lambda^{\mathbf{N}}$, then there necessarily exist pairs of proximal points for (X, T). In fact there exists a pair of points $\omega_1, \omega_2 \in X$ with $\omega_1(1) \neq \omega_2(1)$ and $\omega_1(n) = \omega_2(n)$ for all $n > 1$. If this were not the case, then for any pair of distinct values $a, b \in \Lambda, a \neq b$, there exists $N = N_{a,b}$ such that if $\omega_1(1) = a$, $\omega_2(1) = b$, then $\omega_1(n) \neq \omega_2(n)$ for some n with $2 \leq n \leq N$. If such an N did not exist, then we could find for each N a pair ω_1^N, ω_2^N with $\omega_1^N(1) = a, \omega_2^N(1) = b$ and $\omega_1^N(n) = \omega_2^N(n)$, for $n \leq N$. Passing to the limit of a subsequence, we obtain a pair ω_1, ω_2 as asserted. In the absence of such a pair $N = N_{a,b}$ exists as claimed, and we let

$$N' = \max_{a,b} N_{a,b}.$$

Then each N'-tuple of consecutive values, $\omega(2), \omega(3), \ldots, \omega(N' + 1)$ determines $\omega(1)$. Since X is T-invariant, $\omega(m + 1), \omega(m + 3), \ldots, \omega(m + N')$ determines $\omega(m)$. This means that each sequence ω is determined by N'-tuples arbitrarily far out in the sequence. From this it follows that X is finite.

2. The Enveloping Semigroup and Idempotents

The phenomenon of proximality is best understood when one attaches to a dynamical system (X, G) its "enveloping semigroup." The notion of an enveloping semigroup was developed by Ellis [1], [2]. Let (X, G) be a dynamical system with G a group or a semigroup and let X be compact.

Form the space X^X of all mappings, continuous or not, of X to itself and endow it with the product topology. The sets $\mathscr{U}(x_1, \ldots, x_k; U_1, \ldots, U_k) = \{f \in X^X : f(x_1) \in U_1, \ldots, f(x_k) \in U_k\}$, $U_1, \ldots, U_k$ open sets of X, form a basis for this topology on X^X, and X^X is compact in this topology. The space X^X is also a semigroup and the map $f \to f f_0$ is continuous for all f_0; on the other hand, the map $f \to f_0 f$ is continuous only if f_0 is a continuous function.

Given the dynamical system (X, G) we may map $G \to X^X$, To simplify notation let us identify G with its image in X^X. Let E be the closure of G in X^X.

PROPOSITION 8.3. *The set E is a compact subsemigroup of X^X.*

Proof: If $g \in G$, then $f \to gf$ is continuous. So the image of the closure is contained in the closure of the image and $g\bar{G} \subset \overline{gG}$, or $gE \subset E$. Hence $GE \subset E$. If $f \in E$, then $f' \to f'f$ is continuous so $Ef \subset \overline{Gf} \subset E$, whence $E \cdot E \subset E$. E is a closed subset of X^X and so is compact.

The set E is a topological semigroup for which the one-sided multiplication $f \to f f_0$ is always continuous and the other multiplication $f \to f_0 f$ is continuous whenever f_0 is continuous. We also have the multiplication $E \times X \to X$ where clearly $f \to fx$ is continuous for all x, but $x \to fx$ is continuous only for continuous f.

The key to the present analysis is in the next lemma which generalizes a well-known fact for finite semigroups. This lemma seems to have been first proved by Namakura [1]. Recall that an idempotent of a semigroup is an element u satisfying $u^2 = u$.

LEMMA 8.4. *If E is a compact semigroup for which one-sided multiplication $x \to xx_0$ is continuous, then E contains an idempotent.*

Proof: Let A be a minimal subset of E satisfying (i) $AA \subset A$, (ii) A is compact. The family of these is nonempty since E itself has these properties, and one can check that Zorn's lemma applies to guarantee the existence of a minimal set of this kind. Take any $u \in A$. Au is compact and, moreover, $AuAu \subset Au$ whence $Au = A$. So for some $v \in A$, $vu = u$. Let $A' = \{v : vu = v\}$. The set A' is nonempty, compact, and also a semigroup. So $A' = A$, and so $u \in A'$, and $u^2 = u$.

Proposition 8.4 ensures that an enveloping semigroup always contains idempotents. While this is clear when G is a group, it is not obvious for G a semigroup. The next proposition relates idempotents to proximality. We shall take X to be metric for convenience.

PROPOSITION 8.5. *If $u \in E$ is an idempotent, then for any $x \in X$, x and ux are proximal. In case (X, G) is minimal, then the converse is true: x_1 and x_2 are proximal only if there is an idempotent u with $x_2 = ux_1$.*

Proof: For $\varepsilon > 0$ let U_ε be the ball of radius ε about ux. Let $\mathcal{U}_\varepsilon = \{f \in E : f(x) \in U_\varepsilon$ and $f(ux) \in U_\varepsilon\}$. Note that $u \in \mathcal{U}_\varepsilon$ and so, since $u \in E$ which is the closure of G, we must have $G \cap \mathcal{U}_\varepsilon \neq \varnothing$. If $g \in G \cap \mathcal{U}_\varepsilon$, then $d(gx, gux) < 2\varepsilon$. This gives one direction of the proposition. Assume now that (X, G) is minimal and that x_1, x_2 are proximal points in X. The map $f \to (fx_1, fx_2)$ of E into $X \times X$ is continuous and its image is closed and invariant under the action $(x', x'') \to (gx', gx'')$ of G. Since (x_1, x_2) are proximal, $E(x_1, x_2)$ contains points arbitrarily close to the diagonal; hence it contains the diagonal. In particular, $(x_2, x_2) \in E(x_1, x_2)$. Let $J = \{j \in E : jx_1 = x_2, jx_2 = x_2\}$. Notice that J is a closed subsemigroup of E. If u is an idempotent in J, $ux_1 = x_2$ as claimed.

We can consider also the notion of proximality of a point to a set. Let $A \subset X$ be a closed subset and let $d(x, A) = \inf_{y \in A} d(x, y)$. We shall say that x and A are proximal if for some $\{g_n\} \subset G$, $d(g_n x, g_n A) \to 0$. It is possible for a point x and a set A to be proximal without x being proximal to any $y \in A$. The next proposition, however, deals with a case for which proximality to a set implies proximality to a point.

PROPOSITION 8.6.   *Let (X, G) be a dynamical system with A a G-invariant, closed subset of X. If $x \in X$ is proximal to A, then there exists $y \in A$ with x proximal to y.*

Proof: Notice that Ex is closed and comes arbitrarily close to A; hence $Ex \cap A \neq \varnothing$. Let $F = \{f \in E : fx \in A\}$. Since A is invariant, $EF \subset F$. In particular F is a semigroup and since it is closed, F contains an idempotent u. Let $y = ux$; then $y \in A$ and x is proximal to y.

Now let $x \in X$ and let A be any minimal subset of $\overline{Gx}$. Then x is clearly proximal to A, and so by the foregoing proposition, x is proximal to some point of A. Recall that every point belonging to a minimal subset is uniformly recurrent (Definition 1.8). We have then proved the following result which is the main result of this section.

THEOREM 8.7.   *In a dynamical system on a compact metric space, every point is proximal to a uniformly recurrent point in its orbit closure.*

This result is due to Auslander [1] and Ellis [2].

DEFINITION 8.2.   *A point of a dynamical system is said to be a distal point if it is proximal only to itself. A system is called distal if every point is a distal point.*

Theorem 8.7 has the following corollary.

COROLLARY.   *A distal point is uniformly recurrent. In a distal system, every point is uniformly recurrent, and the space decomposes into the union of minimal sets.*

The last part of the statement follows from Theorem 1.17.

3. Central Sets and Hindman's Theorem

In this section we shall use Theorem 8.7 to deduce Hindman's theorem on *IP*-sets. (See Introduction.) Another proof based on the use of ultra-filters has been given by Glazer [1]. Glazer's proof of Hindman's theorem runs along very similar lines to the present proof but without the dynamical appurtenances.

The following definition is motivated by Theorem 8.7. From this point on we take the semigroup G to be the natural numbers N. The subsets to be defined will play an important part in our combinatorial considerations.

DEFINITION 8.3. *A subset $S \subset N$ is a central subset if there exists a system (X,T), a point $x \in X$ and a uniformly recurrent point y proximal to x, and a neighborhood U_y of y such that*

$$S = \{n: T^n x \in U_y\}.$$

If x is itself uniformly recurrent, then the return times to a neighborhood U_x of x, $\{n: T^n x \in U_x\}$ form a central set. By definition of uniform recurrence, this set is a syndetic set. In general, a central set need not be syndetic but it may be shown that it must be *piecewise syndetic*, i.e., the intersection of a syndetic set and a thick set.

THEOREM 8.8. *In a finite partition of N, $N = B_1 \cup B_2 \cup \cdots \cup B_q$, one of the sets B_j contains a central set.*

Proof: Form the dynamical system (Ω,T) where $\Omega = \{1,2,\ldots,q\}^Z$, and T is the shift, $T\omega(n) = \omega(n + 1)$. Let $\xi \in \Omega$ be any point with $\xi(n) = i \Rightarrow n \in B_i$ for $n \geq 1$. Let $\eta \in \Omega$ be a uniformly recurrent point proximal to ξ, and let $\eta(0) = j$. Set $U_\eta = \{\omega: \omega(0) = j\}$, and let $S = \{n: T^n \xi \in U_\eta\}$. Clearly S is a central set and if $n \in S$ then $\xi(n) = T^n \xi(0) = j$, so that $n \in B_j$. Thus $S \subset B_j$.

We shall see that central sets have many desirable combinatorial properties.

PROPOSITION 8.9. *Central sets contain arbitrarily long arithmetic progressions.*

Proof: Let $S = \{n: T^n x \in U_y\}$ with x proximal to y and y uniformly recurrent, $y \in U_y$. Let $Y \subset X$ be the orbit closure of y, so that Y is a minimal set, and $Y \cap U_y \neq \emptyset$. Suppose

$$\bigcup_1^N T^{-i} U_Y \supset Y,$$

and let $\varepsilon > 0$ be so small that any set in Y of diameter $< \varepsilon$ is contained in some $T^{-i} U_y$, $1 \leq i \leq N$. Let l be given, $l \geq 1$. According to Theorem

2.6, there exists in Y a point y' such that for some n, $T^n y', T^{2n} y', \ldots, T^{ln} y'$ are within ε of one another. Then $T^{i+n} y', T^{i+2n} y', \ldots, T^{i+ln} y' \in U_y$ for some i, $1 \le i \le N$. Now the fact that x is proximal to y implies that the orbit closure of x meets Y and since Y is minimal, the orbit closure of x contains Y. We can then find $T^j x$ so close to y' that $T^{i+j+n} x, T^{i+j+2n} x, \ldots,$ $T^{i+j+ln} x$ are still in U_y. This gives

$$(i + j) + n, (i + j) + 2n, \ldots, (i + j) + ln \in S.$$

Recall that an IP-set in $\mathbf{N}$ consists of a sequence $p_1, p_2, \ldots$ together with sums of terms with distinct indices. Hindman's theorem asserts that in any partition $\mathbf{N} = B_1 \cup B_2 \cup \cdots \cup B_q$, some B_j contains an IP-set. This now follows from

PROPOSITION 8.10. *Every central set contains an IP-set.*

Proof: Once again let $S = \{n : T^n x \in U_y\}$, $y \in U_y$, y proximal to x, and y uniformly recurrent. Let Y be the orbit closure of y. We define inductively a sequence of open sets $\{U_n\}$ containing y, at the same time constructing a sequence $\{p_n\}$ with $p_{i_1} + p_{i_2} + \cdots + p_{i_r}$, $i_1 < i_2 < \cdots < i_r$, belonging to S. First note that if x is proximal to y, then for any neighborhood V of y, there exists p with $T^p x \in V$, $T^p y \in V$. To see this, choose V' with $V' \subset V$, $y \in V'$ and let $\varepsilon > 0$ be such that any point at distance $\le \varepsilon$ from V' is still in V. Let

$$\bigcup_1^N T^{-i} V' \supset Y$$

and let $\delta > 0$ be such that $d(x', x'') < \delta$ implies $d(T^i x', T^i x'') < \varepsilon$ for $1 \le i \le N$. We have $d(T^r x, T^r y) < \delta$ for some r and so $d(T^{i+r} x, T^{i+r} y) < \varepsilon$ for $1 \le i \le N$. But some $T^{i+r} y \in V'$ as i runs from 1 to N and then $T^{i+r} x \in V$, so that $p = i + r$ has the required property.

Now let $U_1 = U_y$. Let p_1 be such that $T^{p_1} x, T^{p_1} y \in U_1$ and set $U_2 = U_1 \cap T^{-p_1} U_1$. Repeat this for $U_2 : p_2$ is chosen so that $T^{p_2} x, T^{p_2} y \in U_2$ and set $U_3 = U_2 \cap T^{-p_2} U_2$. Continue in this manner so that $U_{k+1} \subset U_k$ and

$$T^{p_k} U_{k+1} \subset U_k$$

and both

$$T^{p_k} x, T^{p_k} y \in U_k.$$

Then

$$T^{p_{i_1} + p_{i_2} + \cdots + p_{i_r}} x \in T^{p_{i_1} + \cdots + p_{i_{r-1}}}(U_{i_r})$$
$$\subset T^{p_{i_1} + \cdots + p_{i_{r-2}}}(U_{i_{r-1}}) \subset \cdots \subset T^{p_{i_1}} U_{i_2} \subset U_{i_1} \subset U_1.$$

Hence $p_{i_1} + p_{i_2} + \cdots + p_{i_r} \in S$.

This gives us Hindman's theorem.

THEOREM 8.11. *In a finite partition of* $\mathbf{N}$, $\mathbf{N} = B_1 \cup B_2 \cup \cdots \cup B_q$, one of the *sets* B_j *contains an IP-set.*

REMARK. We mention without proof the following generalization of Theorem 8.8 which can be proved along the same lines: If S is a central set and $S = B_1 \cup B_2 \cup \cdots \cup B_q$, then some B_j contains a central set.

Before discussing further combinatorial properties of central sets, we shall extend the notion of *IP*-sets.

4. *IP*-Systems

One can define *IP*-sets in any semigroup and Hindman's theorem remains valid. In fact we can replace semigroups by *IP*-sets, which themselves form an approximate semigroup. Hindman's theorem still remains true; namely, in a finite partition of an *IP*-set, one of the subsets contains an *IP*-set. All this follows from yet another version of Hindman's theorem—also due to Hindman—which we prove in this section. In this version, the relation to Ramsey's theorem becomes evident.

Let $\mathscr{F}$ denote the set of all finite nonempty subsets of N. We denote the elements of $\mathscr{F}$ by α, β, γ, etc. The family $\mathscr{F}$ is closed under unions and nonempty intersections, and we shall consider lattice isomorphisms of $\mathscr{F}$ into $\mathscr{F}$, which we call homomorphisms. Explicitly, a homomorphism $\phi : \mathscr{F} \to \mathscr{F}$ is a map such that $\alpha \cap \beta = \varnothing \Rightarrow \phi(\alpha) \cap \phi(\beta) = \varnothing$, $\phi(\alpha \cup \beta) = \phi(\alpha) \cup \phi(\beta)$. Evidently such a homomorphism is determined by $\phi\{i\}$ for all i, and these can be an arbitrary family of disjoint finite sets α_i. Then $\phi\{i_1, i_2, \ldots, i_k\} = \alpha_{i_1} \cup \alpha_{i_2} \cup \cdots \cup \alpha_{i_k}$.

DEFINITION 8.4. *An IP-subset of* $\mathscr{F}$ *is the image* $\phi(\mathscr{F})$ *of a homomorphism* $\phi : \mathscr{F} \to \mathscr{F}$.

We now have the following version of Hindman's theorem.

THEOREM 8.12. *If* $\mathscr{F} = B_1 \cup B_2 \cup \cdots \cup B_q$, *then some* B_j *contains an IP-set.*

Proof: Map $\mathscr{F} \to \mathbf{N}$ by $\{i_1, i_2, \ldots, i_k\} \to 2^{i_1} + 2^{i_2} + \cdots + 2^{i_k}$. This is a 1-1 map and the partition of $\mathscr{F}$ induces a partition $\mathbf{N} = B_1' \cup B_2' \cup \cdots \cup B_q'$. According to Theorem 8.11 some B_j' contains an *IP*-set in $\mathbf{N}$. Write this *IP*-set as $\{p_\alpha, \alpha \in \mathscr{F}\}$ where $P_{\{i_1, i_2, \ldots, i_k\}} = p_{i_1} + p_{i_2} + \cdots + p_{i_k}$. To each $\alpha \in \mathscr{F}$ attach the set of exponents in the binary expansion of p_α:

$$p_\alpha = \sum_{i \in \psi(\alpha)} 2^i.$$

We then find that $p_\alpha \in B_j'$ implies that $\psi(\alpha) \in B_j$. If $\alpha \cap \beta = \varnothing$ would imply $\psi(\alpha) \cap \psi(\beta) = \varnothing$, then $\alpha \to \psi(\alpha)$ would define a homomorphism and $\psi(\mathscr{F})$ would be an *IP*-set in B_j. This is not necessarily the case, but we shall find a homomorphism $\psi':\mathscr{F} \to \mathscr{F}$ such that $\alpha \cap \beta = \varnothing$ implies $\psi(\psi'(\alpha)) \cap \psi(\psi'(\beta)) = 0$. Then $\psi \circ \psi'$ is the desired homomorphism.

We shall define $\psi'(\{1\}) = \alpha_1$, $\psi'(\{2\}) = \alpha_2, \ldots$ inductively and extend ψ' to all of $\mathscr{F}$. We use the following property of *IP*-sets in $\mathbf{N}$. If S is an *IP*-set in $\mathbf{N}$ and m is an arbitrary positive integer, then there exists $s \in S$ with s divisible by 2^m (or any other number, for that matter). This amounts to showing that in any finite group an *IP*-set must contain the identity. This is easily seen, and we leave it as an exercise. Now let α_1 be arbitrary in $\mathscr{F}$, and let $2^{m_1} > p_{\alpha_1}$. By the foregoing remark there exists α_2 disjoint from α_1 with p_{α_2} divisible by 2^m. Then $\psi(\alpha_2) \cap \psi(\alpha_i) = \varnothing$. Let $2^{m_2} > p_{\alpha_1} + p_{\alpha_2}$ and choose α_3 disjoint from $\alpha_1 \cup \alpha_2$ with p_{α_3} divisible by 2^{m_2}. Continuing in this manner, we obtain a sequence of disjoint α_i with $\psi(\alpha_i)$ disjoint. Set $\psi'(\{i_1, i_2, \ldots, i_k\}) = \alpha_{i_1} \cup \alpha_{i_2} \cup \cdots \cup \alpha_{i_k}$, then ψ' has the desired property: $\alpha \cup \beta = \varnothing \Rightarrow \psi(\psi'(\alpha)) \cap \psi(\psi'(\beta)) = \varnothing$.

Theorem 8.12 implies a corresponding theorem for any semigroup. First we define *IP-systems* and *IP-sets* for arbitrary semigroups.

DEFINITION 8.5. *An $\mathscr{F}$-sequence of elements in an arbitrary space X is a sequence $\{x_\alpha\}$ indexed by elements $\alpha \in \mathscr{F}$. If X is a semigroup we say that an $\mathscr{F}$-sequence defines a IP-system if $x_{\{i_1, i_2, \ldots, i_k\}} = x_{i_1} x_{i_2} \cdots x_{i_k}$ where $i_1 < i_2 < \cdots < i_k$. The set $\{x_\alpha\}$ is then called an IP-set.*

In general we shall be dealing with commutative semigroups, and then, if $\{x_\alpha\}$ is an *IP*-system, $x_{\alpha \cup \beta} = x_\alpha x_\beta$ for any disjoint sets $\alpha, \beta \in \mathscr{F}$. In general this will be the case if all the indices of α are less than all those of β. We write $\alpha < \beta$ in this case.

PROPOSITION 8.13. *If an IP-set in a semigroup is the union of finitely many subsets, one of these again contains an IP-set.*

Proof: Let $\Sigma = \{\sigma(\alpha)\}$ be the *IP*-set in question and suppose $\Sigma = B_1 \cup B_2 \cup \cdots \cup B_q$. Then $\mathscr{F} = \sigma^{-1}(B_1) \cup \sigma^{-1}(B_2) \cup \cdots \cup \sigma^{-1}(B_q)$. For some j, $\sigma^{-1}(B_j)$ contains an *IP*-set in $\mathscr{F}$. Suppose this is generated by the disjoint sets $\{\alpha_1, \alpha_2, \alpha_3, \ldots\}$. Refine this sequence to a sequence of sets $\{\alpha'_1, \alpha'_2, \alpha'_3, \ldots\}$ where $\alpha'_{n+1} > \alpha'_n$. Then

$$\sigma(\alpha'_{i_1} \cup \alpha'_{i_2} \cup \cdots \cup \alpha'_{i_k}) = \sigma(\alpha'_{i_1})\sigma(\alpha'_{i_2}) \cdots \sigma(\alpha'_{i_k}),$$

and it is clear that the set of these forms an *IP*-subset of B_j.

Given an $\mathscr{F}$-sequence $\{x_\alpha\}$, an $\mathscr{F}$-*subsequence* is defined by a homomorphism $\phi:\mathscr{F} \to \mathscr{F}$ and forming $\{x_{\phi(\alpha)}\} \subset \{x_\alpha\}$. We shall also find it useful to introduce a notion of convergence for $\mathscr{F}$-sequences.

DEFINITION 8.6. *Let* $\{x_\alpha\}$ *be an* $\mathcal{F}$-*sequence in a topological space* X *and let* $x \in X$. *We say that* $x_\alpha \to x$ *as an* $\mathcal{F}$-*sequence, if for every neighborhood* V *of* x *there is some* $\alpha_V \in \mathcal{F}$ *such that* $x_\alpha \in V$ *for all* $\alpha > \alpha_V$.

We shall usually write simply $x_\alpha \to x$, the index α indicating that we are dealing with convergence as an $\mathcal{F}$-sequence. Note that if $x_\alpha \to x$ as an $\mathcal{F}$-sequence and V is a neighborhood of x, x_α may fail to belong to V for infinitely many α.

Note that Theorem 8.12 can be reformulated as stating that if $\{x_\alpha\}$ is an $\mathcal{F}$-sequence with values in a finite space then there exists an $\mathcal{F}$-subsequence which is constant. We generalize this in the next theorem.

THEOREM 8.14. *If* $\{x_\alpha\}$ *is an* $\mathcal{F}$-*sequence with values in a compact metric space, then there exists an* $\mathcal{F}$-*subsequence* $\{x_\alpha\}$ *which converges as an* $\mathcal{F}$-*sequence.*

Proof: Partitioning X into a finite number of sets of diameter $< \varepsilon$, we can use Theorem 8.12 to extract an $\mathcal{F}$-subsequence $\{x_{\phi(\alpha)}\}$ all of whose terms are within a distance $< \varepsilon$ of one another. Now form successive $\mathcal{F}$-subsequences: $\cdots \subset \{x_{\phi_k(\alpha)}\} \subset \{x_{\phi_{k-1}(\alpha)}\} \subset \cdots \subset \{x_\alpha\}$, with

$$\operatorname{diam} \{x_{\phi_k(\alpha)}\} < 1/k.$$

We now carry out the analogue of a diagonal procedure. Define sets $\alpha_1, \alpha_2, \ldots, \alpha_k, \ldots$ inductively by choosing $\alpha_{k+1} \in \phi_{k+1}(\mathcal{F})$ disjoint from $\alpha_1, \alpha_2, \ldots, \alpha_k$. Then set $\phi(\{i_1, i_2, \ldots, i_k\}) = \alpha_{i_1} \cup \alpha_{i_2} \cup \cdots \cup \alpha_{i_k}$. All the $\phi(\alpha)$ for $\alpha \cap \{1, 2, \ldots, j\} = \varnothing$ are subsets of $\phi_{j+1}(\mathcal{F})$, and the corresponding $x_{\phi(\alpha)}$ form a set of diameter $< 1/(j+1)$. It follows that $\{x_{\phi(\alpha)}\}$ is convergent.

Consider now an *IP*-system of continuous maps of a space X to itself. That is, $\{T^\alpha\}$ is a system of maps indexed by $\alpha \in \mathcal{F}$, and for $\alpha < \beta$, $T^{\alpha \cup \beta} = T^\alpha T^\beta$.

LEMMA 8.15. *If* $\{T^\alpha\}$ *is an IP-system of maps of a compact metric space* X *to itself and* $x \in X$, *there exists a subsystem* $\{T^{\phi(\alpha)}\}$ *such that* $T^{\phi(\alpha)}x$ *converges as an* $\mathcal{F}$-*sequence to a point* $y \in X$ *and at the same time* $T^{\phi(\alpha)}y \to y$.

Proof: The fact that for some subsequence of $T^\alpha x$, we have convergence $T^{\phi(\alpha)}x \to y$ follows from Theorem 8.14. Now let $\varepsilon > 0$ be given and choose α_0, so that for $\alpha > \alpha_0$, $d(T^{\phi(\alpha)}x, y) < \varepsilon/2$. Fix $\alpha > \alpha_0$ and find $\delta > 0$, so that $d(x', x'') < \delta \Rightarrow d(T^{\phi(\alpha)}x', T^{\phi(\alpha)}x'') < \varepsilon/2$. Choose $\beta > \alpha$ such that $d(T^{\phi(\beta)}x, y) < \delta$. Then $d(T^{\phi(\alpha \cup \beta)}x, T^{\phi(\alpha)}y) < \varepsilon/2$, and since $\alpha \cup \beta > \alpha_0$, $d(T^{\phi(\alpha \cup \beta)}x, y) < \varepsilon/2$. Putting the two inequalities together, we obtain $d(T^{\phi(\alpha)}y, y) < \varepsilon$.

5. *IP*-Systems of Transformations and Birkhoff Recurrence

Lemma 8.15 can be used as a means for proving the existence of recurrent points. It gives the following general result.

PROPOSITION 8.16.   *Given any IP-system of continuous transformation of a compact metric space, $T^\alpha: X \to X$, there exists an IP-subsystem $T^{\phi(\alpha)}$ and a point $x_0 \in X$ such that $T^{\phi(\alpha)}x_0 \to x_0$.*

Proof: Let x be any point in X and let $x_0 = \lim T^{\phi(\alpha)}x$ for an appropriate subsequence. Then by Lemma 8.15, $T^{\phi(\alpha)}x_0 \to x_0$.

This result implies Birkhoff's recurrence theorem for a single transformation. Given $T: X \to X$, we can form many *IP*-systems $\{T^\alpha\}$ by setting

$$T^{\{i_1, i_2, \ldots, i_k\}} = T^{n_{i_1} + n_{i_2} + \cdots + n_{i_k}},$$

for $\{n_i\}$ an arbitrary sequence in **N**. If $T^\alpha x_0 \to x_0$, then x_0 is clearly a recurrent point. The proof of Proposition 8.16 also show how to find recurrent points; namely, take any limit of a subsequence of

$$T^{n_{i_1} + \cdots + n_{i_k}}x$$

for any point $x \in X$.

In the remainder of this section we shall prove an "IP" analogue of of the multiple (Birkhoff) recurrence theorem. We shall start with an analogue of Lemma 2.1.

LEMMA 8.17.   *Let $\{T^\alpha\}$ be an IP-system of continuous transformations of X, let A be a subset of X and suppose that for any $y \in A$ and any $\beta \in \mathscr{F}$ and $\varepsilon > 0$, there exists $x \in A$ and $\alpha \in \mathscr{F}$ with $\alpha > \beta$ such that $d(T^\alpha x, y) < \varepsilon$. Then for any $\varepsilon > 0$ and $\beta \in \mathscr{F}$ there exists $z \in A$, and $\alpha \in \mathscr{F}$ with $\alpha > \beta$, such that $(T^\alpha z, z) < \varepsilon$.*

Proof: Let $\varepsilon > 0$ be given. We define inductively a sequence $\{z_n\} \subset A$, one of whose points will satisfy $d(T^\alpha z, z) < \varepsilon$. Set $\varepsilon_1 = \varepsilon/2$. Choose z_0 arbitrarily and let $z_1 \in A$ satisfy

$$d(T^{\alpha_1}z_1, z_0) < \varepsilon_1$$

for some $\alpha_1 > \beta$. Now let ε_2 with $0 < \varepsilon_2 \leq \varepsilon_1$ be such that whenever $d(z, z_1) < \varepsilon_2$,

$$d(T^{\alpha_1}z, z_0) < \varepsilon_1.$$

With ε_2 determined find $z_2 \in A$ and $\alpha_2 > \alpha_1$ with

$$d(T^{\alpha_2}z_2, z_1) < \varepsilon_2.$$

Continue in this way defining $z_0, z_1, \ldots, z_k \in Z$, $\alpha_1 < \alpha_2 < \cdots < \alpha_k \in \mathscr{F}$, $\varepsilon_1, \varepsilon_2, \ldots, \varepsilon_k$ in $(0, \varepsilon/2)$, where, these having been determined so that

$$d(T^{\alpha_i} z_i, z_{i-1}) < \varepsilon_i, \qquad i = 1, 2, \ldots, k,$$

we define $\varepsilon_{k+1}, 0 < \varepsilon_{k+1} \le \varepsilon$, so that

$$d(z, z_k) < \varepsilon_{k+1} \Rightarrow d(T^{\alpha_k} z, z_{k-1}) < \varepsilon_k.$$

We then determine z_{k+1} and α_{k+1} by

$$d(T^{\alpha_{k+1}} z_{k+1}, z_k) < \varepsilon_{k+1}.$$

It now follows that whenever $i < j$,

$$d(T^{\alpha_{i+1} \cup \cdots \cup \alpha_j} z_j, z_i) = d(T^{\alpha_{i+1}} \cdots T^{\alpha_j} z_j, z_i) < \varepsilon_{i+1} \le \varepsilon_1 = \varepsilon/2.$$

Since X is compact, we shall find some pair i, j, $i < j$ for which $d(z_i, z_j) < \varepsilon/2$. This combined with the foregoing gives

$$d(T^{\alpha} z_j, z_j) < \frac{\varepsilon}{2} + \frac{\varepsilon}{2}$$

for $\alpha = \alpha_{i+1} \cup \cdots \cup \alpha_j$. This proves the lemma.

In analogy with Definition 2.1 we say that a subset $A \subset X$ is homogeneous with respect to a set of transformations $\{T_i\}$ acting on X if there exists a group of homomorphisms G of X each of which commutes with each T_i and such that G leaves A invariant and (A, G) is minimal.

LEMMA 8.18. *Let $\{T^\alpha\}$ be an IP-system of transformations of X and let A be a homogeneous set for $\{T^\alpha\}$. If for every $\varepsilon > 0$ and $\beta \in \mathscr{F}$, there exists some pair of points $x, y \in A$ with $d(T^\alpha x, y) < \varepsilon$ with $\alpha > \beta$, then we can find a point $z \in A$ and an IP-subsystem $\{T^{\phi(\alpha)}\}$, so that $T^{\phi(\alpha)} z \to z$.*

Proof: Using the homogeneity, as in the proof of Lemma 2.2, it is easy to see that for any $y \in A$ there exists $x \in A$ with $d(T^\alpha x, y) < \varepsilon$. We can then invoke the foregoing lemma to conclude that for any $\varepsilon > 0$, $\beta \in \mathscr{F}$, there exists $z \in A$ with $d(T^\alpha z, z) < \varepsilon$, $\alpha > \beta$. Now let V be an open set in X that intersects A. Let V' be a smaller open set meeting A and such that any point at distance δ from V' is still inside V. There is a finite set G_0 such that

$$\bigcup_{g \in G_0} g^{-1} V' \supset A.$$

Choose ε so that if $d(x', x'') < \varepsilon$, then $d(gx', gx'') < \delta$ for any $g \in G_0$, $x', x'' \in X$. Now let z' be such that $d(T^\alpha z', z') < \varepsilon$, $z' \in A$. For some $g \in G_0$, $gz' \in V'$ and $d(T^\alpha gz', gz') < \delta$. Since $gz' \in V'$, $T^\alpha gz' \in V$. Thus we see that each open set V meeting A contains a point $z'' \in A$ with $T^\alpha z'' \in V$ with $\alpha > \beta$, β

preassigned. Finally we can conclude that for each open set V meeting A there is an open subset V_1 with $T^\alpha V_1 \subset V$ for some $\alpha > \beta$.

We now proceed inductively to define a sequence of open sets and a sequence $\alpha_1 < \alpha_2 < \cdots < \alpha_n < \cdots$ in $\mathscr{F}$ by requiring that $\bar{V}_{n+1} \subset V_n$, $T^{\alpha_{n+1}} V_{n+1} \subset V_n$ and $V_{n+1} \cap A \neq \varnothing$. In addition we may require that the diameter of V_n tends to 0. Then there will be a (unique) point $z \in A$ common to all the A_i. Now set $\phi(\{i_1,i_2,\ldots,i_k\}) = \alpha_{i_1} \cup \alpha_{i_2} \cup \cdots \cup \alpha_{i_k}$. If $i_1 < i_2 < \cdots < i_k$, then

$$T^{\alpha_{i_1} \cup \alpha_{i_2} \cup \cdots \cup \alpha_{i_k}} V_{i_k} = T^{\alpha_{i_1}} T^{\alpha_{i_2}} \cdots T^{\alpha_{i_k}} V_{i_k} \subset V_{i_1 - 1}.$$

Hence $T^{\phi(\alpha)} z \in V_i$ for $\alpha > \{1,2,\ldots,i\}$. It follows that $T^{\phi(\alpha)} z \to z$.

THEOREM 8.19.   Let $\{S_1^\alpha\}, \{S_2^\alpha\}, \ldots \{S_l^\alpha\}$ be l IP-systems of maps of a compact metric space X, all contained in a commutative group of homomorphisms of X. Then there exists a point $x_0 \in X$ and a homomorphism $\phi : \mathscr{F} \to \mathscr{F}$ such that

$$S_1^{\phi(\alpha)} x_0 \to x_0, \, S_2^{\phi(\alpha)} x_0 \to x_0, \ldots, S_l^{\phi(\alpha)} x_0 \to x_0.$$

Proof: Let G be the commutative group containing all the S_i^α. Passing to a subset of X if necessary, we can assume that G acts minimally on X. If we apply the foregoing lemma with $A = X$ and $T^\alpha = S_1^\alpha$, we obtain the assertion of the theorem for the case $l = 1$. We proceed by induction. Suppose the theorem established for l and suppose $\{S_1^\alpha\}, \{S_2^\alpha\}, \ldots \{S_{l+1}^\alpha\}$ are $l + 1$ commuting systems. Form $T_i^\alpha = S_i^\alpha (S_{l+1}^\alpha)^{-1}$. If $\beta > \alpha$,

$$T_i^{\alpha \cup \beta} = S_i^\alpha S_i^\beta (S_{l+1}^\beta)^{-1} (S_{l+1}^\alpha)^{-1} = T_i^\alpha T_i^\beta,$$

since all the maps commute. By the induction hypothesis $T_i^{\phi(\alpha)} z \to z$, $i = 1, 2, \ldots, l$ for some $z \in X$. Consider the $(l + 1)$-fold product $X^{(l+1)}$ on which $S_1^\alpha \times S_2^\alpha \times \cdots \times S_{l+1}^\alpha$ acts. The diagonal $\Delta^{(l+1)}$ is a homogoneous subset and to prove the theorem it will suffice to show that the diagonal contains a point $(x_0, x_0, \ldots, x_0) = x_0$ with $S_1^{\psi(\alpha)} \times S_2^{\psi(\alpha)} \times \cdots \times S_{l+1}^{\psi(\alpha)} x_0^* \to x_0^*$. By the lemma it suffices to find a pair of points $x^*, y^* \in \Delta^{(l+1)}$ and some $\alpha > \beta$ with $d(S_1^\alpha \times S_2^\alpha \times \cdots \times S_{l+1}^\alpha x^*, y^*) < \varepsilon$. But

$$d((S_1^{\phi(\alpha)} \times \cdots \times S_{l+1}^{\phi(\alpha)})(S_{l+1}^{-\phi(\alpha)} \times \cdots \times S_{l+1}^{-\phi(\alpha)})(z, \ldots, z), (z, \ldots, z))$$

converges to 0 and this shows that the lemma is applicable. This completes the proof.

We shall be interested in the following consequence.

PROPOSITION 8.20.   Let $\{S_1^\alpha\}, \ldots \{S_l^\alpha\}$ be IP-systems of maps of X to itself belonging to a commutative group G that we assume acts minimally *on X. If U is any open set in X, there exists an IP-subset $\{\phi(\alpha)\}$ of $\mathscr{F}$ such that*

$$\bigcap_{i=1}^{l} S_i^{-\phi(\alpha)} U \neq \varnothing$$

Proof: Let G_0 be a finite subset of G with

$$\bigcup_{g \in G_0} g^{-1} U = X.$$

Let $\delta > 0$ be such that any set of diameter $< \delta$ is contained in some $g^{-1}U$, $g \in G_0$. Find a point x_0 with $S_i^{\phi(\alpha)} x_0 \to x_0$, $i = 1, \ldots, l$, and refine the IP-subset $\{\phi(\alpha)\}$ so that $d(S_i^{\phi(\alpha)} x_0, x_0) < \delta/2$ for all α, $i = 1, 2, \ldots, l$. We then find that for any α, the points $S_1^{\phi(\alpha)} x_0, S_2^{\phi(\alpha)} x_0, \ldots, S_l^{\phi(\alpha)} x_0$ are all inside some $g^{-1}U$. But then

$$g x_0 \in \bigcap_{i=1}^{l} S_i^{-\phi(\alpha)} U.$$

6. A Basic Property of Central Sets

In Section 3 we obtained two results regarding central sets: they contain arbitrarily long arithmetic progressions, and they contain IP-sets. We shall now formulate another property of central sets that will imply both the others. In what follows S will denote a subset of $\mathbf{N}$ and $S^m = S \times S \times \cdots \times S$ consists of those vectors in $\mathbf{N}^m$, all of whose components lie in S. We look upon $\mathbf{N}^m$ as an additive semigroup, and so we may consider IP-systems $\{v_\alpha\}$ in $\mathbf{N}^m$. If $h \in \mathbf{N}$, we shall denote by $\bar{h}^{(m)}$ the vector $(h, h, \ldots, h) \in \mathbf{N}^m$.

PROPOSITION 8.21. *Let S be a central set in $\mathbf{N}$, and for any $m > 1$, let $\{v_\alpha\}$ be any IP-system in $\mathbf{Z}^m$. We can find an IP-subsystem $\{v_{\phi(\alpha)}\}$ and an IP-system $\{h_\alpha\} \subset \mathbf{N}$ such that the vector $\bar{h}_\alpha^{(m)} + v_{\phi(\alpha)} \in S^m$ for each $\alpha \in \mathscr{F}$.*

If we let $m = 1$, this proposition implies in particular that S contains an IP-set. Moreover, if we choose $v_\alpha = (n_\alpha, 2n_\alpha, \ldots, mn_\alpha)$, where $\{n_\alpha\}$ is any IP-system in $\mathbf{N}$, the conclusion of the proposition becomes the statement that S contains arithmetic progressions of length m with differences in a preassigned IP-set.

Proof: We begin by remarking that in defining central sets in terms of dynamical systems, we can confine ourselves to the case of systems (X, T) with T invertible. This is seen by comparing an arbitrary system (X, T) with its invertible cover $(\tilde{X}, \tilde{T})$ where, as in Chapter 2, Section 2, we take $\tilde{X} \subset X^{\mathbf{Z}}$ as the set of all sequences $\{\tilde{x}(n)\}$ with $\tilde{x}(n + 1) = T(\tilde{x}(n))$, and $\tilde{T}\tilde{x}(n) = \tilde{x}(n + 1)$ for any $\tilde{x} \in X^{\mathbf{Z}}$. Let $\pi: \tilde{X} \to X$ be defined by $\pi(\tilde{x}) = \tilde{x}(0)$; then

π defines a homomorphism of systems. One can verify that $\tilde{x}$ and $\tilde{y}$ in $\tilde{X}$ are proximal if and only if $\pi(\tilde{x})$ and $\pi(\tilde{y})$ are proximal. Moreover, if $y \in X$ is uniformly recurrent, there exists $\tilde{y} \in \tilde{X}$ with $\pi(\tilde{y}) = y$ and $\tilde{y}$ uniformly recurrent. From this it follows that the set of n for which $T^n x$ enters a neighborhood of a uniformly recurrent point y proximal to x, is the same as the corresponding set for $\tilde{x}$ with $\pi(\tilde{x}) = x$. We omit the details.

Suppose now that (X,T) is a dynamical system with T invertible and that $S = \{n: T^n x_0 \in V_{y_0}\}$ where y_0 is a uniformly recurrent point proximal to x, V_{y_0} a neighborhood of y_0. We let Y be the orbit closure of y_0, so that (Y,T) is minimal. We shall define inductively a sequence

$$\alpha_1 < \alpha_2 < \cdots < \alpha_n, \ldots$$

in $\mathscr{F}$ and a sequence $h_1, h_2, \ldots, h_n, \ldots$ in $\mathbf{N}$. Write $v_\alpha = (v_\alpha^{(1)}, v_\alpha^{(2)}, \ldots, v_\alpha^{(m)})$, and consider the IP-systems $T^{v_\alpha^{(i)}}$, $i = 1, 2, \ldots, m$, acting on X. If we restrict these to Y, we can apply Proposition 8.20, and so there exists $\alpha_1 \in \mathscr{F}$ with

$$U_1 = \bigcap_i T^{-v_{\alpha_1}^{(i)}} V_{y_0} \neq \emptyset,$$

U_1 meets Y which is contained in the orbit closure of x_0, and so there exists $h_1 \in \mathbf{N}$ with $T^{h_1} x_0 \in U_1$.

Now suppose $\alpha_1 < \alpha_2 < \cdots < \alpha_l$ and $h_1, h_2, \ldots, h_l$ have been determined and satisfy the following conditions. For each $\beta \subset \{1, 2, \ldots, l\}$ let

$$\phi(\beta) = \bigcup_{i \in \beta} \alpha_i$$

and

$$h_\beta = \sum_{i \in \beta} h_i.$$

We require that for each $\beta \subset \{1, 2, \ldots, l\}$,

$$(8.1) \qquad U_\beta = \bigcap_i T^{-v_{\phi(\beta)}^{(i)}} V_{y_0} \neq \emptyset$$

and that

$$(8.2) \qquad T^{h_\beta} x_0 \in U_\beta, \quad T^{h_\beta} y_0 \in U_\beta.$$

For each $\beta \in \mathscr{F}$ let X_β denote a copy of X and form

$$X_l = \prod_{\beta \subset \{1, 2, \ldots, l\}} X_\beta$$

and $Y_l \subset X_l$ defined by

$$Y_l = \{(T^{h_\beta} y)_{\beta \subset \{1, 2, \ldots, l\}}, y \in Y\}.$$

T acts on X_l componentwise, and Y_l is a minimal subset of (X_l,T), being a homomorphic image of Y under the homomorphism

$$\pi_l(x) = (T^{h_\beta}x)_{\beta \subset \{1,2,\ldots,l\}}.$$

Let $\tilde{U}_l \subset X_l$ be the product open set:

$$\tilde{U}_l = \prod_{\beta \subset \{1,2,\ldots,l\}} U_\beta.$$

$\pi_l(y_0) \in \tilde{U}_l$ and so $\tilde{U}_l$ meets Y_l. Since x_0 is proximal to y_0, $\pi_l(x_0)$ is proximal to $\pi_l(y_0)$. Applying Proposition 8.20, we can find α_{l+1} such that

$$(8.3) \qquad Y_l \cap \left(\bigcap_i T^{-v_{\alpha_{l+1}}^{(i)}} \tilde{U}_l \right) \neq \varnothing,$$

Since $\pi_l(x_0)$ is proximal to $\pi_l(y_0)$, some common power of T brings these points arbitrarily close together, and using the uniform recurrence of $\pi_l(y_0)$, we find the two points can be brought simultaneously into a pre-assigned open set meeting Y_l. We can therefore define h_{l+1} by the requirement

$$(8.4) \qquad \begin{aligned} T^{h_{l+1}}\pi_l(x_0) &\in \bigcap_i T^{-v_{\alpha_{l+1}}^{(i)}}\tilde{U}_l, \\ T^{h_{l+1}}\pi_l(y_0) &\in \bigcap_i T^{-v_{\alpha_{l+1}}^{(i)}}\tilde{U}_l. \end{aligned}$$

Now we must show that the conditions (8.1), (8.2) are satisfied with β replaced by $\beta' = \beta \cup \{l+1\}$. This in fact follows from (8.3) and (8.4). First of all

$$(8.5) \qquad \begin{aligned} U_{\beta'} &= \bigcap_i T^{-v_{\phi(\beta)}^{(i)} - v_{\alpha_{l+1}}^{(i)}} V_{y_0} \supset \bigcap_i T^{-v_{\alpha_{l+1}}^{(i)}} \bigcap_j T^{-v_{\phi(\beta)}^{(j)}} V_{y_0} \\ &= \bigcap_i T^{-v_{\alpha_{l+1}}^{(i)}} U_\beta. \end{aligned}$$

This is seen to be nonempty by considering the β-component in (8.3). Similarly, considering the β-component in (8.4) we obtain

$$T^{h_{\beta'}}(x_0) = T^{h_{l+1} + h_\beta}(x_0) \in \bigcap_i T^{-v_{\alpha_{l+1}}^{(i)}} U_\beta \subset \bigcap_i T^{-v_{\phi(\beta')}^{(i)}} V_{y_0}$$

as in (8.5). Similarly for $T^{h_{\beta'}}(y_0)$. This shows that our inductive procedure may be continued indefinitely so now assume that (8.1) and (8.2) are valid for all $\beta \in \mathscr{F}$. We now have

$$T^{h_\beta}x_0 \in \bigcap_i T^{-v_{\phi(\beta)}^{(i)}} V_{y_0},$$

and so $h_\beta + v_{\phi(\beta)}^{(i)} \in S$ for each $i = 1,2,\ldots,m$. This proves the proposition.

7. Rado's Theorem

Rado characterized in (Rado [1]) those homogeneous systems of linear equations $\sum a_{ij}x_j = 0$ for which a "monochromatic" solution exists for any finite coloring of $\mathbf{N}$. He called these "regular" systems. Rado's explicit description of these systems is the following. Let us say that a $p \times q$ matrix (a_{ij}) is of *level l* if the index set $\{1,2,\ldots,q\}$ can be divided into l disjoint subsets $I_1,I_2,\ldots,I_l$ and rational numbers c^r_j may be found for $1 \leq r \leq l$ and $j \in I_1 \cup I_2 \cup \cdots \cup I_r$ such that the following relationships are satisfied:

$$\sum_{j \in I_1} a_{ij} = 0$$

$$\sum_{j \in I_2} a_{ij} = \sum_{j \in I_1} c^1_j a_{ij}$$

$$\sum_{j \in I_3} a_{ij} = \sum_{j \in I_1 \cup I_2} c^2_j a_{ij}$$

$$\cdots$$

$$\sum_{j \in I_l} a_{ij} = \sum_{j \in I_1 \cup I_2 \cup \cdots \cup I_{l-1}} c^{l-1}_j a_i$$

for $i = 1,2,\ldots,p$.

Rado's theorem then states that if for some l the matrix (a_{ij}) is of level l, then the system

$$(8.7) \qquad \sum_{j=1}^{q} a_{ij}x_j = 0, \qquad i = 1,2,\ldots,p$$

is regular.

We shall prove a stronger statement.

THEOREM 8.22. *If (a_{ij}) is a matrix of level l for some l, then for any central set $S \subset \mathbf{N}$, the system (8.7) has a solution $(x_1,x_2,\ldots,x_q)$ all of whose components lie in S.*

As a preliminary to the proof, we introduce the following definition.

DEFINITION 8.7. *If (a_{ij}) and (b_{ij}) are two $p \times q$ matrices, we shall say that (a_{ij}) is an R-modification of (b_{ij}) if there is a subset $J \subset \{1,2,\ldots,q\}$ and a set of rationals $\lambda_j, j \in J$ such that for all $i, 1 \leq i \leq p$,*

(i) $\displaystyle\sum_{j \in J} a_{ij} = 0,$

(ii) $a_{ij} = b_{ij}$ for $j \notin J,$

(iii) $\lambda_j a_{ij} = b_{ij}$ for $j \in J.$

LEMMA 8.23. *If a matrix (a_{ij}) is of level l, it can be obtained by a series of l R-modifications beginning with the 0 matrix.*

Proof: Let (a_{ij}) be of level l with $\{1,2,\ldots,q\} = I_1 \cup I_2 \cup \cdots \cup I_l$, and coefficients c_j^r as before. Without any loss of generality we may suppose that $c_j^1 \neq 0$, for we see from (8.6) that a constant may always be added to the $c_j^1, j \in I_1$. Now define (b_{ij}) by

$$b_{ij} = a_{ij}, j \notin I_1, \quad \text{and} \quad b_{ij} = -c_j^1 a_{ij}, j \in I_1.$$

Then

$$\sum_{j \in I_1 \cup I_2} b_{ij} = 0,$$

and the remaining equations of (8.6) may be rewritten in terms of (b_{ij}) so that (b_{ij}) is seen to be of level $l - 1$ with respect to the partition of $\{1,2,\ldots,q\}$ into the sets $I_1 \cup I_2, I_3, I_4, \ldots, I_l$. Moreover, it is clear that (a_{ij}) is an R-modification of (b_{ij}). Finally if $l = 1$, it is evident that (a_{ij}) is an R-modification of the 0 matrix. This proves the lemma.

To make matters concrete, let us give one illustration. The matrix

$$B = \begin{pmatrix} 1 & 0 & 0 & -1 \\ 0 & 0 & 1 & -1 \end{pmatrix}$$

is of level 1, with $I_1 = \{1,2,3,4\}$. It is an R-modification of the 0 matrix with $J = \{1,3,4\}, \lambda_1 = \lambda_3 = \lambda_4 = 0$. This matrix can be modified to

$$A = \begin{pmatrix} 1 & -1 & 0 & -1 \\ 0 & 1 & -1 & -1 \end{pmatrix}$$

by taking $J = \{1,2,3\}, \lambda_1 = 1, \lambda_2 = 0, \lambda_3 = -1$. We can also check that A is of level 2. The corresponding system can be written

(8.8)
$$\begin{aligned} x - y &= w \\ y - z &= w, \end{aligned}$$

and the existence of a monochromatic solution to (8.8) means having an arithmetic progression of length 3 inside one of the sets with the difference inside the same set.

LEMMA 8.24. *Let S be a central set and let $d \in \mathbf{N}$. Let $S_d = \{n : dn \in S\}$. Then S_d is a central set.*

Proof: Let x be proximal to y in $(X,T), V_y$ a neighborhood of y, y a uniformly recurrent point. All this remains true if T is replaced by T^d. If $S = \{n : T^n x \in V_j\}$, then $S_d = \{n : (T^d)^n x \in V_j\}$.

LEMMA 8.25.   *Let A and B be $p \times q$ matrices such that A is an R-modification of B. Suppose that for every central set S there exists an IP-set of vectors $\{u_\alpha\} \subset S^q$ with $Bu_\alpha = 0$. Then the same is true for A.*

Proof: Write $A = (a_{ij})$, $B = (b_{ij})$ and suppose the two matrices related as in Definition 8.7. Let d be the common denominator of the rationals λ_j. Let S be a central set and let $\{u_\alpha\}$ be an IP-set of solutions to $Bu = 0$ with components in the central set S_d. We can write

$$\sum_{j \in J} a_{ij} u_\alpha^{(j)} + \sum_{j \in J} \lambda_j a_{ij} u_\alpha^{(j)} = 0$$

and using the fact that

$$\sum_{j \in J} a_{ij} = 0,$$

we find

$$\sum_{j \notin J} a_{ij} du_\alpha^{(j)} + \sum_{j \in J} a_{ij}(d\lambda_j u_\alpha^{(j)} + h) = 0$$

for any h. Then for any h, the vector v with components

$$v^{(j)} = \begin{cases} du_\alpha^{(j)} & \text{for} \quad j \notin J \\ d\lambda_j u_\alpha^{(j)} + h & \text{for} \quad j \in J \end{cases}$$

will satisfy $Av = 0$. We have $du_\alpha^{(j)} \in S$, so $v \in S^q$ if $d\lambda_j u_\alpha^{(j)} + h \in S$ for $j \in J$. Here we invoke Proposition 8.21. The vectors with components $d\lambda_j u_\alpha^{(j)}$ for $j \in J$ define an IP-system in $\mathbf{Z}^m$ with $m = |J|$. So we may find an IP-subsystem and an IP-system $\{h_\alpha\}$ such that

$$d\lambda_j u_{\phi(\alpha)}^{(j)} + h_\alpha \in S.$$

Defining v_α by

$$v_\alpha^{(j)} = \begin{cases} du_{\phi(\alpha)}^{(j)} & \text{for} \quad j \notin J \\ d\lambda_j u_{\phi(\alpha)}^{(j)} + h_\alpha & \text{for} \quad j \in J \end{cases}$$

gives an IP-set of solutions $Av_\alpha = 0$.

An alternative approach to Rado's theorem is given by Deuber [1]. A (c,k,r) *Deuber system* generated by a sequence $\{p_0, p_1, \ldots, p_k\}$ consists of all the numbers of the form $i_0 p_0 + i_1 p_1 + \cdots + i_{m-1} p_{m-1} + c p_m$, $|i_0|, |i_1|, \ldots, |i_{m-1}| \leq r$ and $m \leq k$. Deuber shows that any system $\sum a_{ij} x_j = 0$ with (a_{ij}) of level l has a solution in any (c,k,r) Deuber system provided c, k, r are appropriately chosen. One can complete the proof once again by using Proposition 8.21 to prove (inductively) that a central set contains, in fact, an IP-system of Deuber systems. The details are easily filled in.

CHAPTER 9

The Fine Structure of Recurrence
and Mixing

In this chapter we continue to study the interplay between recurrence properties of dynamical systems and combinatorial properties of sets of integers. Given a notion of a "large" set of integers, we associate with it a form of recurrence, saying that a point x of a dynamical system (X,T) displays this form of recurrence if for each neighborhood V of x, the set of return times of x to V is a large set. Thus the notion of a syndetic set leads to that of a uniformly recurrent point. Starting with two other classes of large sets, we shall arrive at further refinements of ordinary recurrence. These will turn out to correspond to familiar dynamical properties. In this way we shall retrieve the notions of distal and almost automorphic points of dynamical systems. These have been investigated extensively from the dynamical point of view (see in particular Veech [1] and [2]); nonetheless, it will be instructive to relate this theory to combinatorial properties of large sets of integers.

Each mode of recurrence of points of a dynamical system leads in addition to a notion of "almost periodicity" of bounded functions on the integers (or on any group whose actions are being studied). Namely we regard a bounded function $f(n)$ on $\mathbf{Z}$ as being a point of the Bebutov system it generates under the shift. We then say that $f(n)$ is a ( )-function if the corresponding point is ( )-recurrent with respect to the shift. In particular the two notions of recurrence alluded to above give rise to two uniformly closed algebras of functions displaying recurrence properties analogous to those of Bohr almost periodic functions.

In a different vein, we can also describe forms of mixing of a measure preserving system $(X,\mathscr{B},\mu,T)$ in terms of the sets of the integers n for which $\mu(T^{-n}A \cap B)$ is close to $\mu(A)\mu(B)$. Strong mixing corresponds to sets with finite complement and weak mixing to sets with density 1. A variant known as mild mixing which was introduced in Furstenberg and Weiss [2] turns out to correspond to the same kind of "large" set that will play a role in our discussion of recurrence. We close this chapter with a proof of the fact that "mildly mixing" implies "mildly mixing of all orders".

Throughout this chapter our exposition is not intended to be exhaustive but merely illustrative of certain ideas. Some of the material has been developed systematically by S. Glasner and can be found in a forthcoming work [1]. Once again, the ideas of this chapter have been developed largely together with B. Weiss.

1. Dual Families of Sets in Z

Let $\mathscr{S}$ be a collection of subsets of $\mathbf{Z}$. We define the dual family $\mathscr{S}^*$ to consist of all those sets that intersect each set in $\mathscr{S}$ nontrivially. We shall be concerned principally with three examples:

(i) $\mathscr{S}_T = \{$thick subsets of $\mathbf{Z}\}$, where we recall that a thick subset is one that contains arbitrarily long intervals.

(ii) $\mathscr{S}_{IP} = \{IP$-sets of $\mathbf{Z}\}$.

(iii) $\mathscr{S}_\Delta = \{$difference sets of $\mathbf{Z}\}$. A *difference set* or Δ-set $S \in \mathscr{S}_\Delta$ is obtained by taking an arbitrary sequence in $\mathbf{Z}$, $\{s_n\}$, and forming the differences $S = \{s_n - s_m, n > m\}$.

The dual families are $\mathscr{S}_T^*$, $\mathscr{S}_{IP}^*$, and $\mathscr{S}_\Delta^*$. It is clear that $\mathscr{S}_T^*$ is simply the collection of all syndetic sets in $\mathbf{Z}$. For the sets of $\mathscr{S}_{IP}^*$ and $\mathscr{S}_\Delta^*$ we do not have an alternative characterization and we call them IP^*-sets and Δ^*-sets respectively.

LEMMA 9.1. *Every thick set contains an IP-set, and every IP-set contains a Δ-set.*

Proof: Let S be a thick set. We define inductively a sequence $p_1, p_2, \ldots, p_n, \ldots$ such that $p_{i_1} + p_{i_2} + \cdots + p_{i_k} \in S$ whenever $i_1 < i_2 < \cdots < i_k$. Let p_1 be any element in S. Since S contains an interval of length $> |p_1|$, it contains p_2 such that $p_1 + p_2$ also belongs to S. We continue in this fashion. Suppose $p_1, p_2, \ldots, p_n$ have been chosen so that all partial sums $p_{i_1} + p_{i_2} + \cdots + p_{i_k}$, $i_1 < i_2 < \cdots < i_k \le n$ are in S. The set S contains an interval of length $> |p_1| + |p_2| + \cdots + |p_n|$. It follows that there exists $p_{n+1} \in S$ with all sums $p_{i_1} + p_{i_2} + \cdots + p_{i_k} + p_{n+1} \in S$ for $i_1 < i_2 < \cdots < i_k \le n$. So we can adjoin p_{n+1} and maintain the property. Hence S contains an IP-set.

Now let S be an IP-set: $S = \{p_{i_1} + p_{i_2} + \cdots + p_{i_k} : i_1 < i_2 < \cdots < i_k\}$. Set $s_n = p_1 + p_2 + \cdots + p_n$. Clearly $s_n - s_m \in S$ for $n > m$, so S contains a Δ-set.

If $\mathscr{S}_1$ and $\mathscr{S}_2$ are two collections of sets in $\mathbf{Z}$ such that each set of $\mathscr{S}_2$ contains a set of $\mathscr{S}_1$, then it follows that $\mathscr{S}_1^* \subset \mathscr{S}_2^*$. We then find that

LEMMA 9.2. *Every Δ^*-set is an IP^*-set and every IP^*-set is syndetic.*

We shall soon see that these inclusions are proper; not every syndetic set is an IP^*-set, and not every IP^*-set is a Δ^*-set.

An example of a Δ^*-set occurring in the context of ergodic theory is the following. Let $(X, \mathscr{B}, \mu, T)$ be a m.p.s. and $A \in \mathscr{B}$ with $\mu(A) > 0$. Then $N(A) = \{n : \mu(A \cap T^{-n}A) > 0\}$ is a Δ^*-set. Indeed, for any $\{s_n\}$ we must have $\mu(T^{-s_m}A \cap T^{-s_n}A) > 0$ for some $n > m$. Then $s_n - s_m \in N(A)$, so that $N(A)$ intersects every Δ-set.

From this it follows that if A is a nonempty open set in X where (X, T) is a minimal compact dynamical system, then $N_1(A) = \{n : A \cap T^{-n}A \neq \varnothing\}$ is also a Δ^*-set. For X carries a T-invariant measure whose support is a closed T-invariant set. Hence A has positive measure, and $N_1(A) \supset N(A)$.

More generally, for A a nonempty open set in X, (X, T) a minimal compact dynamical system, form

$$(9.1) \qquad N_l(A) = \{n : A \cap T^{-1}A \cap \cdots \cap T^{-ln}A \neq \varnothing\}.$$

PROPOSITION 9.3. *For every nonempty open set A in X, $N_l(A)$ is an IP*-set.*

Proof: Let $\{p_{i_1} + p_{i_2} + \cdots + p_{i_k}\}$ be an *IP*-set. Apply Proposition 8.20 to the *IP*-systems

$$S_j^\alpha = T^{j(p_{i_1} + p_{i_2} + \cdots + p_{i_k})},$$

where $\alpha = \{i_1, i_2, \ldots, i_k\}$, and $j = 1, 2, \ldots, l$. It follows that some $p_{i_1} + p_{i_2} + \cdots p_{i_k} \in N_l(A)$ and so $N_l(A)$ is an *IP**-set.

Now we shall show by example that $N_l(A)$ need not be a Δ^*-set. This will show immediately that not every *IP**-set is a Δ^*-set. We leave it to the reader to provide a straightforward proof of this fact.

Let X be the torus $\mathbf{T}^2$, $\mathbf{T} = \mathbf{R}/\mathbf{Z}$, and let $T : X \to X$ be given by $T(\theta, \phi) = (\theta + \alpha, \phi + \theta)$, α irrational. We have shown in Chapter 1 that this system is minimal (Lemma 1.25). We denote by $\|\theta\|$ the infimum of $|\theta + n|$, $n \in \mathbf{Z}$, and regard $\|\theta\|$ as defined for $\theta \in \mathbf{T}$. Using the fact that the orbit $\{T^n(0,0)\}$ is dense in $\mathbf{T}^2$, we conclude that for any $\beta_1, \beta_2 \in \mathbf{T}$, and $\varepsilon > 0$, there exists n with

$$\|n^2 \alpha - \beta_1\| < \varepsilon, \qquad \|n\alpha - \beta_2\| < \varepsilon.$$

Fix ε and inductively define $\{s_n\}$ so that $\|s_n^2 \alpha - 1/4\| < \varepsilon$ for all n, and $\|s_n s_m \alpha\| < \varepsilon$ whenever $m < n$. Then

$$(9.2) \qquad \left\| (s_n - s_m)^2 \alpha - \frac{1}{2} \right\| = \left\| \left(s_n^2 \alpha - \frac{1}{4} \right) + 2 s_n s_m \alpha + \left(s_m^2 \alpha - \frac{1}{4} \right) \right\| < 4\varepsilon$$

for $m < n$. We shall now show that if ε is small and δ is appropriately chosen, and

$$A_\delta = \{(\theta, \phi) : \|\theta\| < \delta, \|\phi\| < \delta\},$$

then

$$N_2(A_\delta) = \{n : A_\delta \cap T^{-n}A_\delta \cap T^{-2n}A_\delta \neq \varnothing\}$$

does not meet the difference set $\{s_n - s_m\}$. This will prove that $N_2(A_\delta)$ is not a Δ^*-set.

Suppose $(\theta,\phi) \in A_\delta \cap T^{-p}A_\delta \cap T^{-2p}A_\delta$. Then among the pertinent inequalities we have

$$\|\phi\| < \delta, \qquad \left\| \phi + p\theta + \frac{p(p-1)}{2}\alpha \right\| < \delta, \qquad \|\theta + 2p\alpha\| < \delta,$$

$$\|\phi + 2p\theta + p(2p-1)\alpha\| < \delta.$$

This implies

$$\left\| p\theta + \frac{p(p-1)}{2}\alpha \right\| < 2\delta, \qquad \|2p\theta + p(2p-1)\alpha\| < 2\delta$$

whence

$$\|2p\theta + p(p-1)\alpha\| < 4\delta$$

and

$$\|p(2p-1)\alpha - p(p-1)\alpha\| < 6\delta$$

or

(9.3) $\|p^2\alpha\| < 6\delta.$

Now if p has the form $p = s_n - s_m$, then by (9.2), $\|p^2\alpha - 1/2\| < 4\varepsilon$. This will contradict (9.3) if $6\delta + 4\varepsilon < 1/2$, and this proves that the sets $N_l(A)$ of (9.1) need not be Δ^*-sets.

It is significant that the collection $\mathscr{S}_T$ of thick sets is translation invariant, whereas the two other collections $\mathscr{S}_{IP}$ and $\mathscr{S}_\Delta$ are not. This property goes over to the dual; the translate of a syndetic set is syndetic, whereas the translate of an IP^*-set or a Δ^*-set need not be of the same kind. For example, the evens, $S = 2\mathbf{Z}$, form a Δ^*-set (and therefore an IP^*-set) since every infinite difference set contains the difference of two numbers having the same parity. On the other hand, the odd integers, $2\mathbf{Z} + 1$, do not meet an IP-set consisting only of even integers; hence they do not form an IP^*-set (and therefore not a Δ^*-set). Since $2\mathbf{Z} + 1$ is syndetic, we see that

$$\mathscr{S}_T^* \supsetneqq \mathscr{S}_{IP}^* \supsetneqq \mathscr{S}_\Delta^*.$$

Another distinction between $\mathscr{S}_T$ and our other two collections relates to the "Ramsey property."

DEFINITION 9.1. *We say that a family of sets $\mathscr{S}$ has the Ramsey property if whenever $S_1 \cup S_2 \in \mathscr{S}$, then either S_1 or S_2 contains a set of $\mathscr{S}$.*

Glasner in a forthcoming work [1] calls this property *divisibility*.

LEMMA 9.4. *If $\mathscr{S}$ has the Ramsey property and $\mathscr{S}^*$ is the dual family, then the intersection of a set of $\mathscr{S}$ and a set of $\mathscr{S}^*$ contains a set of $\mathscr{S}$.*

Proof: Let $S \in \mathscr{S}$ and $S^* \in \mathscr{S}^*$. We can write $S = (S \cap S^*) \cup (S-S^*)$ and one of the two sets $S \cap S^*$, $S-S^*$ contains a set of $\mathscr{S}$. Since $S-S^*$ does not meet S^* and $S^* \in \mathscr{S}^*$, the set $S-S^*$ cannot contain a set of $\mathscr{S}$. Hence $S \cap S^*$ contains a set of $\mathscr{S}$.

LEMMA 9.5. *If $\mathscr{S}$ has the Ramsey property, then the dual family $\mathscr{S}^*$ has the following intersection property: $S_1 S_2 \in \mathscr{S}^* \Rightarrow S_1 \cap S_2 \in \mathscr{S}^*$.*

Proof: Let $S \in \mathscr{S}$. Since $S \cap S_1$ contains a set of $\mathscr{S}$, $(S \cap S_1) \cap S_2 \neq \varnothing$. Hence $S_1 \cap S_2 \in \mathscr{S}^*$.

The collection $\mathscr{S}_T$ does not have the Ramsey property since we can decompose $\mathbf{Z}$ into two sets, neither of which is thick. On the other hand, $\mathscr{S}_{IP}$ and $\mathscr{S}_\Delta$ both have the Ramsey property.

PROPOSITION 9.6. *The collections $\mathscr{S}_{IP}$ and $\mathscr{S}_\Delta$ have the Ramsey property.*

Proof: To show that $\mathscr{S}_\Delta$ has the Ramsey property, we use the following theorem of Ramsey: If the set of pairs $P = \{(i,j), 1 \le i < j < \infty\}$ of natural numbers is divided into two sets, $P = P_1 \cup P_2$, then there is a sequence $\{i_n\}$ for which either all pairs (i_m, i_n), $m < n$, are in P_1, or all these pairs are in P_2. If now $S = \{s_j - s_i\}$ is a difference set and $S = S_1 \cup S_2$, then set

$$P_l = \{(i,j) : s_j - s_i \in S_l\}, \qquad l = 1,2.$$

By Ramsey's theorem there exists a subset $\{s_{i_n}\}$ with all differences $s_{i_n} - s_{i_m}$ in the same S_l. So S_l contains a difference set.

The fact that $\mathscr{S}_{IP}$ has the Ramsey property is a special case of Proposition 8.13.

2. $\mathscr{S}$-Convergence and $\mathscr{S}$-Recurrence

Let $\mathscr{S}$ be a family of nonempty sets with the finite intersection property of Lemma 9.5. If $\{x_n\}$ is a sequence of points in a topological space indexed by $n \in \mathbf{N}$ or $n \in \mathbf{Z}$, we shall say

$$\mathscr{S}\text{-lim } x_n = x$$

if for every neighborhood V of the point x the set $\{n : x_n \in V\} \in \mathscr{S}$. We obtain ordinary convergence for $|n| \to \infty$ if $\mathscr{S}$ is the family of cofinite sets. The notion of *convergence in density* introduced in Chapter 4 can be regarded as $\mathscr{S}$-convergence where $\mathscr{S}$ consists of all subsets of $\mathbf{N}$ of density 1.

The finite intersection property of $\mathscr{S}$ ensures that $\mathscr{S}$-convergence shares two basis properties with ordinary convergence. Assume the topological space in question in Hausdorff. Then if the $\mathscr{S}$-limit of a sequence $\{x_n\}$ exists, it is unique. For if $\mathscr{S}$-lim $x_n = x'$ and $\mathscr{S}$-lim $x_n = x''$ and $V_{x'}, V_{x''}$ are disjoint neighborhoods of the two points, then the $\mathscr{S}$-sets $\{n : x_n \in V_{x'}\}$, $\{n : x_n \in V_{x''}\}$ are disjoint in contradiction to the finite intersection property.

In addition we have that if

$$\mathscr{S}\text{-lim } x_n = x \qquad \text{and} \qquad \mathscr{S}\text{-lim } y_n = y,$$

then

$$\mathscr{S}\text{-lim } (x_n, y_n) = (x, y),$$

as is readily checked. In particular sums and products of $\mathscr{S}$-convergent sequences are $\mathscr{S}$-convergent and the $\mathscr{S}$-limits are respectively the sums and products of the $\mathscr{S}$-limits in question.

With $\mathscr{S}$ as above we also define $\mathscr{S}$-*recurrence* for points in a dynamical system.

DEFINITION 9.2.   *Let (X, T) be a dynamical system. We say that a point $x \in X$ is $\mathscr{S}$-recurrent if $\mathscr{S}$-lim $T^n x = x$.*

We shall specialize all these notions to the cases $\mathscr{S} = \mathscr{S}_{IP}^*$ and $\mathscr{S} = \mathscr{S}_{\Delta}^*$. In these cases we shall speak of IP^*-convergence and IP^*-recurrence, or Δ^*-convergence and Δ^*-recurrence. More explicitly, we write

$$IP^*\text{-lim } x_n = x$$

if for every neighborhood V of x, and every sequence $\{p_n\}$ of integers, we have

$$x_{p_{i_1} + p_{i_2} + \cdots + p_{i_k}} \in V$$

for some $i_1 < i_2 < \cdots < i_k$. Similarly we write

$$\Delta^*\text{-lim } x_n = x$$

if for any neighborhood V of x and any sequence $\{s_n\}$ of integers, we have

$$x_{s_n - s_m} \in V$$

for some $m < n$.

Note that since $\mathscr{S}_{\Delta}^* \subset \mathscr{S}_{IP}^*$, Δ^*-convergence implies IP^*-convergence. To appreciate how remote both these notions are from ordinary convergence, notice that Δ^*-lim $(-1)^n = 1$, since for any sequence $\{s_n\}$ some $s_n - s_m$ is even for $m < n$, and $(-1)^{s_n - s_m} = 1$.

We proceed to investigate IP^*-recurrence and Δ^*-recurrence. We shall assume throughout that (X,T) is a dynamical system with X compact metric.

PROPOSITION 9.7. *If $x \in X$ is Δ^*-recurrent or IP^*-recurrent, then it is uniformly recurrent and its orbit closure is a minimal set.*

Proof: If x is IP^*-recurrent and V is a neighborhood of x, then $\{n : T^n x \in V\}$ is an IP^*-set and so, by Lemma 9.2, is syndetic. The same is true if x is Δ^*-recurrent since this implies IP^*-recurrence. According to Theorem 1.17, if x is uniformly recurrent, then its orbit closure is minimal.

The following provides examples of both of these forms of recurrence.

PROPOSITION 9.8. *In a Kronecker system, every point is Δ^*-recurrent.*

Proof: If (X,T) is a Kronecker system, then $X = G$, a compact group, and $Tx = ax$ for a fixed $a \in G$. Let $\{s_n\}$ be an arbitrary sequence in $\mathbf{Z}$, and let $y \in G$ be some accumulation point of a^{s_n}. Let $x \in G$ and let V be a neighborhood of x. Let U be a neighborhood of y such that $UU^{-1}x \subset V$. Since $a^{s_n} \in U$ for infinitely many n, we have for some $m < n$, $a^{s_n - s_m}x \in V$, or $T^{s_n - s_m}x \in V$. Hence x is Δ^*-recurrent.

PROPOSITION 9.9. *Let (X,T) be a dynamical system and $\pi : (X,T) \rightarrow (Y,T)$ a homomorphism. If $x \in X$ is IP^*-recurrent or Δ^*-recurrent, then so is $\pi(x)$. If $y \in Y$ is such that $\pi^{-1}(y)$ consists of a single point x and if y is IP^*-recurrent or Δ^*-recurrent, then so is x.*

Proof: The first assertion is clear since $T^n y \in V_y$, a neighborhood of y, whenever $T^n x \in V_x$, a neighborhood of x satisfying $\pi(V_x) \subset V_y$. In the other direction, if $\pi^{-1}(y)$ is a singleton $\{x\}$, then for a neighborhood V_x of x, $\pi^{-1}(V_y) \subset V_x$ provided V_y is small. Then $\{n : T^n x \in V_x\} \supset \{n : T^n y \in V_y\}$, and if the latter is an IP^*-set or a Δ^*-set, so is the former.

LEMMA 9.10. *Suppose (X,T) is a minimal dynamical system and that $x,x' \in X$ are proximal. Let V be a neighborhood of x'. Then there exists an IP-sequence S such that $T^n x \in V$ for all $n \in S$.*

Proof: Indeed $\{n : T^n x \in V\}$ is a central set (Definition 8.3) and by Proposition 8.10 it contains an IP-set.

THEOREM 9.11. *Assume x is a point of a compact metric dynamical system. The following conditions on x are equivalent.*

(i)   *x is uniformly recurrent and x is not proximal to any point in its orbit closure.*

(ii) *x is IP^*-recurrent.*

(iii) *For any recurrent point y of a system (Y,T), (x,y) is a recurrent point of $(X \times Y, T \times T)$.*

(iv) *For any uniformly recurrent point y of a system (Y,T), (x,y) is a uniformly recurrent point of $(X \times Y, T \times T)$.*

Proof: Condition (i) $\Rightarrow$ (ii). Suppose x were not IP^*-recurrent and that $S = \{p_{i_1} + p_{i_2} + \cdots + p_{i_k}\}$ is an IP-set along which $T^n x$ fails to meet a neighborhood V of x. Consider the IP-system $\{T^\alpha\}$ with

$$T^{\{i_1, i_2, \cdots, i_k\}} = T^{p_{i_1} + p_{i_2} + \cdots + p_{i_k}}.$$

Apply Lemma 8.15 to find subsequences $T^{\phi(\alpha)} x \to y$, $T^{\phi(\alpha)} y \to y$ in the sense of $\mathscr{F}$-sequences. In particular x is proximal to y, and y is in the orbit closure of x. If condition (i) is valid, we must have $y = x$, and this cannot be if $T^{\phi(\alpha)} x \notin V$. This contradiction shows that condition (ii) holds.

Condition (ii) $\Rightarrow$ (i). Suppose x is IP^*-recurrent. x is then uniformly recurrent by Proposition 9.7 and its orbit closure is minimal. Suppose x were proximal to a point x' in its orbit closure. We can apply Lemma 9.10 to the orbit closure of x, so that if $V_{x'}$ is any neighborhood of x', $\{n : T^n x \in V_{x'}\}$ will contain an IP-set. Now let V_x be a neighborhood of x; $\{n : T^n x \in V_x\}$ is an IP^*-set and so meets every IP-set. Hence any two neighborhoods V_x and $V_{x'}$ of x and x' meet and so $x = x'$.

In dealing with conditions (iii) and (iv), we should bear in mind that the products of open sets containing x and y, respectively, form a basis of neighborhoods of (x, y). Moreover,

(9.4) $\{n : (T^n x, T^n y) \in V_x \times V_y\} = \{n : T^n x \in V_x\} \cap \{n : T^n y \in V_y\}$.

Condition (ii) $\Rightarrow$ (iii). Assume y is a recurrent point of (Y, T). Then by Theorem 2.17, $\{n : T^n y \in V_y\}$, V_y a neighborhood of y, contains an IP-set. If x is IP^*-recurrent, $\{n : T^n x \in V_x\}$ is an IP^*-set and meets every IP-set, and so by (9.4), (x, y) is recurrent.

Condition (iii) $\Rightarrow$ (ii). Again using Theorem 2.17, we can find for any IP-set S a system (Y, T) and a recurrent point $y \in Y$ with $S \supset \{n : T^n y \in V_y\}$. If condition (iii) holds, then by (9.4), $\{n : T^n x \in V_x\}$ must be an IP^*-set.

Condition (i) $\Rightarrow$ (iv). Let x be an IP^*-recurrent point of (X, T) and y a uniformly recurrent point of (Y, T). According to Theorem 8.7, (x, y) is proximal to some point (x', y') in its orbit closure which is uniformly recurrent. This, of course, implies that x is proximal to x', and by (i), $x = x'$. Thus (x, y') is uniformly recurrent. Let V_x and V_y be neighborhoods of x and y respectively. Apply Lemma 9.10 to the orbit closure of y in (Y, T). It follows that $\{n : T^n y' \in V_y\}$ contains an IP-set. Since x is IP^*-recurrent, $\{n : (T^n x, T^n y') \in V_x \times V_y\}$ is nonempty. Hence (x, y) is in the orbit closure of (x, y'), and since the latter is uniformly recurrent, it follows that (x, y) is uniformly recurrent.

Condition (iv) $\Rightarrow$ (i). First take Y to be a trivial one-point system. Property (iv) implies that x is itself uniformly recurrent. We shall show that (iv) implies that x is not proximal to any $x' \neq x$ in its orbit closure. If x' is in the orbit closure of x, then x' is a uniformly recurrent point of

(X,T). Apply (iv) to $(x,x') \in X \times X$ to conclude that this point has minimal orbit closure. But if x is proximal to x', the orbit closure of (x,x') meets the diagonal of $X \times X$. By minimality, it is contained in the diagonal, and so $x = x'$.

One nontrivial consequence of the foregoing is the following.

COROLLARY 1.   *If $x \in X$ is a distal point of a minimal system (X,T) and $\pi:(X,T) \rightarrow (Y,T)$ is a homomorphism onto, then $\pi(x)$ is a distal point of (Y,T).*
Proof: In the case of minimal systems a point is distal if and only if it is IP^*-recurrent and the result follows from Proposition 9.9.

We also have the following corollary.

COROLLARY 2.   *If $\pi:(X,T) \rightarrow (Y,T)$ is a homomorphism and y is an IP^*-recurrent point of (Y,T) such that not two points of $\pi^{-1}(y)$ are proximal, then each $x \in \pi^{-1}(y)$ is IP^*-recurrent.*
Proof: Suppose x is proximal to x', a point in the orbit closure of x. Then $\pi(x) = y$ is proximal to $\pi(x')$ which belongs to its orbit closure. Since y is IP^*-recurrent, $\pi(x') = \pi(x) = y$. But no two points in $\pi^{-1}(y)$ are proximal.

It follows that the property of IP^*-recurrence is preserved by isometric extension (Definition 1.5). It is also preserved by "almost 1-1 extensions", i.e., extensions $\pi:X \rightarrow Y$ where π is 1-1 at the IP^*-recurrent point $y \in Y$ in question. It is also preserved by passage to homomorphic images (Proposition 9.9). A deep result of Veech and Ellis (Veech [2], Ellis [3]) which is beyond the scope of this exposition asserts that by a (possibly) transfinite application of these three procedures, beginning with a one-point system, one can arrive at all systems and their IP^*-recurrent points. One consequence of their structure theorem of which we take note is that not every minimal dynamical system has IP^*-recurrent points. The following argument, due to B. Weiss, shows this in a more elementary fashion. We begin with a definition.

DEFINITION 9.3.   *A dynamical system (X,T) is said to be topologically weak mixing if for every two nonempty open sets $A,B \subset X$, the set $\{n: T^{-n}A \cap B \neq \varnothing\}$ is thick.*

There exist minimal systems that are topologically weak mixing. For example, if (X,T) is minimal, and X carries a T-invariant borel measure with $(X,\mathscr{B},\mu,T)$ a weak mixing measure preserving system, then $\{n: T^{-n}A \cap B \neq \varnothing\}$ has density 1 for A,B open and nonempty, and, as a result, is thick. Such systems exist, since by Jewett's theorem ([1], see also Krieger [1]) any ergodic *m.p.s.* can be realized as a minimal dynamical system. An explicit example is the horocycle flow.

THEOREM 9.12. *If (X,T) is minimal, topologically weak mixing system, then no point of X is IP*-recurrent.*

Proof: We show that every point $x \in X$ is proximal to some $y \in X$. Let A_1, B_1 be disjoint open sets with $x \in A_1$. Since x is uniformly recurrent, $\{n : T^n x \in A_1\}$ is syndetic, whereas $\{n : T^{-n}A_1 \cap B_1 \neq \varnothing\}$ is thick. The two sets intersect, so let n_1 be such that

$$T^{n_1}x \in A_1, \qquad T^{-n_1}A_1 \cap B_1 \neq \varnothing.$$

Now let $A_1 \supset A_2 \supset A_3 \supset \cdots$ be a sequence of open sets with diameters tending to 0, all containing x. We repeat the foregoing argument and define inductively $n_1, n_2, n_3, \ldots$ and open sets $B_1 \supset B_2 \supset \cdots$ as follows:

$$T^{n_1}x \in A_1, \qquad B_2 \subset \bar{B}_2 \subset T^{-n_1}A_1 \cap B_1$$
$$T^{n_2}x \in A_2, \qquad B_3 \subset \bar{B}_3 \subset T^{-n_2}A_2 \cap B_2$$
$$\cdots$$
$$T^{n_k}x \in A_k, \qquad B_{k+1} \subset \bar{B}_{k+1} \subset T^{-n_k}A_k \cap B_k.$$

We have $\bar{B}_{k+1} \subset \bar{B}_k$ so that $\bigcap \bar{B}_k$ is nonempty. Let $y \in \bigcap B_k \supset \bigcap \bar{B}_{k+1}$. Then $T^{n_k}y \in A_k$, and $T^{n_k}x \in A_k$ and it follows that x and y are proximal.

We turn now to Δ^*-recurrence.

DEFINITION 9.4. *A point $x \in X$ is an almost automorphic point of a dynamical system (X,T) if $T^{n_k}x \to x'$ implies that $T^{-n_k}x' \to x$ for any sequence $\{n_k\} \subset \mathbf{Z}$.*

Here for the first time in this discussion we require a *group* action so that T will be invertible.

THEOREM 9.13. *A point x is an almost automorphic point of a system (X,T) if and only if it is Δ^*-recurrent.*

Proof: Assume x is Δ^*-recurrent and suppose (passing to a subsequence, possibly) that $T^{n_k}x \to x'$, $T^{-n_k}x' \to x''$. If $x'' \neq x$, let $V_x, V_{x''}$ be disjoint neighborhoods of x and x'' respectively. The set $\{n : T^n x \in V_x\}$ is a Δ^*-set and so intersects the difference set $\{n_k - n_j, k > j\}$. Moreover, by Lemma 9.6 this intersection contains a difference set. Tracing back the proof of that lemma via the Ramsey property of difference sets, we see that, in fact, the difference set in question can be taken to have the form $\{n'_k - n'_j\}$ where $\{n'_k\}$ is a subsequence of the original sequence $\{n_k\}$. Passing to a further subsequence, we shall have $T^{-n'_j}x' \in V_{x''}$, and for each n'_j, if k is large, $T^{-n'_j}(T^{n'_k}x) \in V_{x''}$, since $T^{n'_k}x$ is arbitrarily close to x'. But $T^{n'_k - n'_j}x \in V_x$ by our choice of $\{n'_k\}$. This is a contradiction, and so $x'' = x$.

Conversely, suppose x is almost automorphic. To show that x is Δ^*-recurrent, we must show that for any neighborhood V_x of x, and for any sequence $\{n_k\} \subset \mathbf{Z}$, some $T^{n_k - n_j}x \in V$ with $k > j$. Pass to a subsequence and suppose that $T^{n_k}x \to x'$. Then $T^{-n_j}x' \to x$. From this it follows readily that $T^{-n_j}(T^{n_k}x) \in V_x$ for j appropriately chosen and k sufficiently large.

COROLLARY. *An almost-automorphic point of a dynamical system is distal from every point in its orbit closure.*

A more careful analysis yields the following more precise statement. We omit the proof.

PROPOSITION 9.14. *Let (X,T) be a dynamical system, and let $x \in X$. If x' is in the orbit closure of x and x' is proximal to x, then there exists a point $y \in X$ and a sequence $\{n_k\}$ with $T^{n_k}x \to y$, $T^{-n_k}y \to x'$.*

Veech has obtained a "structure" theorem for all systems having almost automorphic points, thereby showing how all Δ^*-recurrent points come about (Veech [1]). We have already noticed that all points of a Kronecker system are Δ^*-recurrent. Also by Proposition 9.9, an "almost 1-1 extension" of a Kronecker system leads to Δ^*-recurrent points. According to Veech [1], all Δ^*-recurrent points come about in this way.

It should be observed that both IP^*-recurrence and Δ^*-recurrence are preserved under passage to products, since the families of IP^*-sets and Δ^*-sets have the Ramsey property. This provides some information about multiple recurrence.

LEMMA 9.15. *If $T_1, T_2, \ldots, T_l$ are l maps of a compact metric space X to itself and x is an IP^*-recurrent point of each (X, T_i), then $(x, x, \ldots, x) \in X^l$ is an IP^* recurrent point of $(X^l, T_1 \times T_2 \times \cdots \times T_l)$.*

Now if x is IP^* recurrent for (X,T), it is clearly so also for (X, T^i). (We could define IP^*-recurrence for any group action. It is then a property that passes on to any subgroup.) So $(x, x, \ldots, x)$ is IP^*-recurrent for $T \times T^2 \times \cdots \times T^l$. In particular it is recurrent for $T \times T^2 \times \cdots \times T^l$. We thus have

PROPOSITION 9.16. *If x is an IP^*-recurrent point of (X,T), then we can find a sequence $\{n_k\}$, $n_k \to \infty$, with $T^{n_k}x \to x$, $T^{2n_k}x \to x, \ldots, T^{ln_k}x \to x$.*

This should be compared with the results of Chapter 2 on the multiple Birkhoff recurrence theorem. Since we have already seen that not every system possesses IP^*-recurrent points, we could not have proved the multiple recurrence theorem via Proposition 9.16.

We conclude this section with a brief discussion of $\mathscr{S}$-recurrence for $\mathscr{S}$ the dual to the family of central sets. Let us say that a set is a C^*-set if it intersects every central set. Central sets have the Ramsey property, as we pointed out at the end of Section 3 of the last chapter. Consequently

C^*-sets have the finite intersection property, and we may talk of C^*-convergence and C^*-recurrence. It is easy to verify that every thick set is a central set; hence C^*-sets are syndetic. In particular, C^*-recurrence implies uniform recurrence. In the other direction note that every central set contains an IP-set, so all IP^*-sets are C^*-sets. There are, however, C^*-sets that are not IP^*-sets. For by Proposition 8.9, every central set contains arbitrarily long arithmetic progressions, and the set of integers with only 0's and 1's in their decimal expansion is an IP-set with no arithmetic progressions of length 3. Hence the set of integers with some digit >1 in its decimal expansion is a C^*-set that is not IP^*.

Nonetheless we have the following result.

PROPOSITION 9.17.   *If a point of a dynamical system is C^*-recurrent, then it is IP^*-recurrent.*

Proof: Let x be a C^*-recurrent point of a dynamical system (X,T). We will show that x is not proximal to any point x' in its orbit closure. Since x is uniformly recurrent, the proposition will then follow by Theorem 9.11. Suppose x proximal to x' in its orbit closure. Then x' is uniformly recurrent and for any neighborhood $V_{x'}$ of x', $\{n: T^n x \in V_{x'}\}$ is a central set. But if x is C^*-recurrent, $\{n: T^n x \in V_x\}$ is a C^*-set for any neighborhood V_x of x. Then $V_x \cap V_{x'} \neq \varnothing$ for any neighborhoods of x and x' and so $x' = x$.

3. Algebras of Recurrent Functions on Z

The Bohr almost periodic functions on a group form an algebra that is closed under translation and under passage to uniform limits, and for which each function exhibits certain recurrence properties that approximate periodicity. Bochner has extended this class by defining "almost automorphic" functions (Bochner [1]). These can be characterized in the case of the group $\mathbf{Z}$ in terms of the notions presented here as the bounded functions $f(n)$ satisfying for each $k \in \mathbf{Z}$,

$$(9.5) \qquad f(k) = \Delta^*\text{-}\lim_n f(k + n).$$

It is obvious that this family is translation invariant, and it is easily shown that it is closed under passage to uniform limits. Moreover, on account of general properties of $\mathscr{S}$-limits for $\mathscr{S}$ a family with the finite intersection property, it follows that (9.5) defines an algebra of functions.

We can enlarge this class further by replacing Δ^*-limits with IP^*-limits. In the following we consider numerical functions on $\mathbf{Z}$.

DEFINITION 9.5.   *A function $f(n)$ is an IP*-function if it is bounded and if for each $k \in \mathbf{Z}$,*

(9.6) $f(k) = IP^*\text{-}\lim_n f(k + n)$.

By exactly the same considerations as before we deduce that the family of *IP**-functions forms an algebra that is closed under translation and under passage to uniform limits. Another property that can also be demonstrated for both classes of functions is that if $f(n)$ belongs to the class and $\psi(t)$ is a function defined on the closure of the range of f that is continuous at the values $f(n)$ (but not necessarily on the closure), then the function $g(n) = \psi(f(n))$ is again in the class.

The following proposition shows how to construct almost automorphic and *IP**-functions.

PROPOSITION 9.18.   *If x_0 is a Δ^*-recurrent point of a dynamical system (X,T) and $\psi(x)$ is a continuous function on X, then the function*

(9.7) $f(n) = \psi(T^n x_0)$

is almost automorphic. If x_0 is IP*-recurrent, then (9.7) defines an IP*-function. Conversely, every almost automorphic function and every IP*-function comes about in this way.

Proof: If x_0 is Δ^*-recurrent, then $\Delta^*\text{-}\lim T^n x_0 = x_0$ and for every continuous function ϕ on X, $\Delta^*\text{-}\lim \phi(T^n x_0) = \phi(x_0)$. If we take $\phi(x) = \psi(T^k x)$, then we get (9.5). The same argument shows that if x_0 is *IP**-recurrent, then $f(n)$ is an *IP**-function. For the converse, take the Bebutov system generated by $f \in \Lambda^{\mathbf{Z}}$, where Λ is a compact set containing the range of f. If f satisfies (9.5) or (9.6), it is easily seen that f is then a Δ^*-recurrent or *IP**-recurrent point of $(\Lambda^{\mathbf{Z}}, T)$, T the shift map. If we set $x_0 = f$, $\psi(x) = x(0)$, then $f(n) = T^n x_0(0) = \psi(T^n x_0)$. This proves the proposition.

To illustrate this, consider the dynamical systems on the d-dimensional torus $\mathbf{T}^d$ treated in Chapter 1, Section 7:

$$T_\alpha(\theta_1, \theta_2, \ldots, \theta_d) = (\theta_1 + \alpha, \theta_2 + \theta_1, \ldots, \theta_d + \theta_{d-1}).$$

This system consists of successive isometric extensions of systems on lower dimensional tori, beginning with a one-point system. According to Corollary 2 of Theorem 9.11 every point of the system is *IP**-recurrent and so functions on the torus $\mathbf{T}^d$ lead to *IP**-functions on the integers. We showed in Chapter 1 that for any polynomial $p(t)$ of degree d with real coefficients there exists an α and a point $(\theta_1, \theta_2, \ldots, \theta_d) \in \mathbf{T}^d$ such that

$$T_\alpha^n(\theta_1, \theta_2, \ldots, \theta_d) = (p_1(n), p_2(n), \ldots, p_d(n))$$

where $p_a(t) = p(t)$. It follows that for any polynomial with real coefficients $p(t)$ the function $e^{ip(n)}$ is an IP^*-function. Consequently the algebra of functions of the form

$$\sum a_r e^{ip_r(n)}$$

together with their uniform limits forms a subalgebra of the collection of IP^*-functions. In footnote 5 in Chapter 1, this subalgebra was denoted $\mathscr{W}$.

The following interpolation result shows to what extent the class of IP^*-functions is a restricted class.

THEOREM 9.19.   *Every IP^*-function is determined uniquely by its values on any thick subset of* $\mathbf{Z}$.

In particular, if $f(n)$ is an IP^*-function, its values for $n < 0$ are determined by its values for $n \geq 0$.

Proof: It suffices to show that if $f(n)$ is an IP^*-function and $f(n) = 0$ on a thick set $S \subset \mathbf{Z}$, then $f(n) \equiv 0$. We show that $f(0) = 0$, and then by translating f, we conclude that $f(n) \equiv 0$. Suppose $f(0) \neq 0$. Since

$$f(0) = IP^*\text{-}\lim f(n),$$

then for some ε, $|f(n)| > \varepsilon$ on an IP^*-set. But every IP^*-set is syndetic, and so the set in question meets S. This contradiction proves that $f(0) = 0$.

It should be noted that although in the proof of the foregoing theorem we seem to be using only the syndetic property of IP^*-sets, the finite intersection property is being used implicitly in the fact that the difference of IP^*-functions is an IP^*-function. If we weaken the definition and say that $f(n)$ is a *uniformly recurrent function* if, for each k, and $\varepsilon > 0$,

$$\{n : |f(k) - f(n + k)| < \varepsilon\}$$

is syndetic, then two uniformly recurrent functions can agree at all but one value of n and nonetheless be distinct. An example is the following. Let

$$sgn^{\pm}(t) = \begin{cases} 1 & \text{if} \quad t > 0 \\ \pm 1 & \text{if} \quad t = 0 \\ -1 & \text{if} \quad t < 0. \end{cases}$$

If α is an irrational multiple of π, the two sequences $sgn^+(\sin n\alpha)$, $sgn^-(\sin n\alpha)$ agree for all $n \neq 0$, but are not identical. It may be seen that both of these represent uniformly recurrent functions.

We refer the reader to Knapp [1] for a related treatment of certain algebras of functions.

4. Mild Mixing

In this section we shall apply the notion of IP^*-convergence to a discussion of mixing properties of measure preserving systems. In the ensuing discussion, we deal with IP-sets and the dual family of IP^*-sets as subsets of $\mathbf{N} = \{1,2,3,\ldots\}$, rather than as subsets of $\mathbf{Z}$. The distinction is partly technical and arises in part because we want to deal with sequences $\{a_n\}$ not necessarily defined for $n < 0$. A more fundamental aspect is that in $\mathbf{Z}$, IP^*-convergence represents a form of continuity at 0. If IP^*-lim $a_n = a$, then since $\{0\}$ is an IP-sequence, $a = a_0$. On the other hand, in $\mathbf{N}$ any IP-set goes to ∞ and so IP^*-convergence represents a form of continuity at ∞. To keep the distinction clear we shall write

$$IP^*\text{-}\lim_{n \geq 1} a_n = a$$

for IP^*-convergence in $\mathbf{N}$.

We begin with a lemma about dynamical systems.

LEMMA 9.20. *Let (X,T) be a dynamical system with X compact metric and assume that there exists a unique point $x_0 \in X$ that is recurrent for T in the sense of Definition 1.1. Then*

$$(9.8) \qquad IP^*\text{-}\lim_{n \geq 1} T^n x = x_0$$

for every $x \in X$. Conversely, if (9.8) holds for every $x \in X$, then x_0 is the unique recurrent point of X.

Proof: Assume (9.8) holds. According to Theorem 2.17, if x_1 is a recurrent point, then for any neighborhood V_{x_1} of x_1 the set $\{n : T^n x_1 \in V_{x_1}\}$ contains an IP-sequence. If, on the other hand,

$$IP^*\text{-}\lim_{n \geq 1} T^n x_1 = x_0,$$

then $\{n : T^n x_1 \in V_{x_0}\}$ is an IP^*-sequence for any neighborhood V_{x_0} of x_0. V_{x_1} and V_{x_0} must therefore intersect and $x_1 = x_0$.

Suppose now that x_0 is a unique recurrent point. If (9.8) were not valid, then for some neighborhood V_{x_0} of x_0 and for some IP-sequence $\{n_\alpha, \alpha \in \mathscr{F}\}$ ($\alpha = \{i_1, i_2, \ldots, i_k\}$, $n_\alpha = n_{i_1} + n_{i_2} + \cdots + n_{i_k}$), $T^{n_\alpha} x \notin V_{x_0}$ for some $x \in X$. We apply Lemma 8.15 to the IP-system of maps $T^\alpha = T^{n_\alpha}$. Passing to an $\mathscr{F}$-subsequence, we find

$$(9.9) \qquad T^{n_{\phi(\alpha)}} x \to y, \qquad T^{n_{\phi(\alpha)}} y \to y.$$

Now this shows that y is a recurrent point and so $y = x_0$. But this contradicts the first part of (9.9) since $T^{n_{\phi(\alpha)}} x \notin V_{x_0}$ and $y = x_0 \in V_{x_0}$. This proves the lemma.

We apply the foregoing lemma to the following situation. Let $\mathscr{H}$ be a separable Hilbert space, and let $T = U$ where $U: \mathscr{H} \to \mathscr{H}$ is a unitary operator. Take $X = $ unit ball of $\mathscr{H}$ endowed with the weak topology. Obviously 0 is a recurrent point of (X,T). In general $x \in \mathscr{H}$ will be a recurrent point if for some sequence $n_k \to \infty$, $U^{n_k}x \to x$ weakly. Now recall that if $x_k \to x$ weakly in $\mathscr{H}$ and if $\|x_k\| \to \|x\|$, then in fact $x_k \to x$ strongly. Thus for a recurrent point $x \in \mathscr{H}$, $U^{n_k}x \to x$ strongly. We now have

LEMMA 9.21.   *Let $\mathscr{H}$ be a separable Hilbert space and U a unitary operator on $\mathscr{H}$. If 0 is the only recurrent vector of $\mathscr{H}$ for U, then for every $u,v, \in \mathscr{H}$,*

$$(9.10) \quad IP^*\text{-}\lim_{n \geq 1} \langle U^n u, v \rangle = 0.$$

Conversely, if (9.10) holds for all $u,v \in \mathscr{H}$, 0 is the only recurrent vector.

Now let $\mathscr{H} \subset L^2(X,\mathscr{B},\mu)$ consist of all functions with integral 0, and let $U: \mathscr{H} \to \mathscr{H}$ be given by a measure preserving transformation $T: X \to X$ with $Uf = f \circ T = Tf$.

DEFINITION 9.6.   *We say a measurable function $f \in L^2(\mathbf{X})$ is rigid for T if for some sequence $n_k \geq 1$, $T^{n_k}f \to f$ in $L^2(\mathbf{X})$.*

For $f \in L^2(\mathbf{X})$ all the various notions of convergence of $T^{n_k}f \to f$—weak, strong, in measure, pointwise a.e.—lead to equivalent definitions of rigidity. If $T^{n_k}f \to f$ weakly in $L^2(\mathbf{X})$, then the convergence also takes place in the strong topology. This implies convergence in measure and this in term implies convergence a.e. of a subsequence. Conversely, convergence *a.e.* implies convergence in $L^2(\mathbf{X})$ since the functions $T^n f$ are uniformly integrable.

An example of a rigid function is an eigenfunction. If $Tf = e^{i\lambda}f$, then if $e^{in_k\lambda} \to 1$, we shall have $T^{n_k}f \to f$. It is possible to construct examples of operators U with continuous spectrum for which rigid functions exist. See Furstenberg and Weiss [2] for details.

DEFINITION 9.7.   *A measure preserving system $(X,\mathscr{B},\mu,T)$ is said to be mildly mixing if there are no nonconstant rigid functions in $L^2(\mathbf{X})$.*

If $(X,\mathscr{B},\mu,T)$ is mildly mixing, then in the space $\mathscr{H}$ which is orthogonal to the constants, only 0 is recurrent. In this space we have, by Lemma 9.21

$$IP^*\text{-}\lim_{n \geq 1} \int f(x)g(T^n x)\, d\mu(x) = 0.$$

This gives the following result for all functions $f,g \in L^2(\mathbf{X})$:

$$(9.11) \quad IP^*\text{-}\lim_{n \geq 1} \int f(x)g(T^n x)\, d\mu(x) = \left(\int f\, d\mu \right)\left(\int g\, d\mu \right).$$

We thus have the following proposition.

PROPOSITION 9.22. *A m.p.s. $(X,\mathscr{B},\mu,T)$ is mildly mixing if and only if for every $f,g \in L^2(X)$, (9.11) holds. Equivalently, $(X,\mathscr{B},\mu,T)$ is mildly mixing if for any two subsets $A,B \in \mathscr{B}$,*

$$(9.12) \quad IP^*\text{-}\lim_{n \geq 1} \mu(A \cap T^{-n}B) = \mu(A)\mu(B).$$

This is analogous to the results of Chapter 4, where we found that $(X,\mathscr{B},\mu,T)$ is weak mixing if and only if

$$(9.13) \quad D\text{-}\lim_{n} \mu(A \cap T^{-n}B) = \mu(A)\mu(B).$$

Mild mixing is stronger than weak mixing since an eigenfunction is also a rigid function. It is less obvious that the validity of (9.12) for every A,B implies that of (9.13) for every A,B, but this too can be shown. In the other direction it is easily seen that mild mixing is weaker than strong mixing. It is not hard to exhibit systems which are mildly mixing but not strongly mixing.

In Furstenberg and Weiss [2] it is shown that **X** mildly mixing can be characterized by the fact that $\mathbf{X} \times \mathbf{Y}$ is ergodic for every ergodic measure preserving system $\mathbf{Y} = (Y,\mathscr{D},v,T)$ with $v(T) \leq \infty$. One consequence of this is that the product of mildly mixing systems is mildly mixing. We can prove this directly using the characterization (9.12).

PROPOSITION 9.23. *The product of two mildly mixing measure preserving systems is mildly mixing.*
Proof: Let $(X,\mathscr{B},\mu,T)$ and $(Y,\mathscr{D},v,S)$ be mildly mixing. We shall show that for $f,g \in L^2(X \times Y,\mathscr{B} \times \mathscr{D},\mu \times v)$

$$IP^*\text{-}\lim_{n \geq 1} \iint f(x,y)g(T^n x,S^n y)\,d\mu(x)\,dv(y)$$

$$= \left(\iint f\,d\mu\,dv\right)\left(\iint g\,d\mu\,dv\right).$$

It suffices to show this for a dense subset of $L^2(\mathbf{X} \times \mathbf{Y})$; so we choose functions of the form $f = \sum \phi_i \otimes \psi_i$, $g = \sum \phi'_j \otimes \psi'_j$. Since we can add IP^*-limits, the matter reduces to showing that

$$IP^*\text{-}\lim_{n \geq 1} \iint \phi(x)\phi'(T^n x)\psi(y)\psi'(S^n y)\,d\mu(x)\,dv(y)$$

$$= \left(\int \phi\,d\mu\right)\left(\int \phi'\,d\mu\right)\left(\int \psi\,dv\right)\left(\int \psi'\,dv\right).$$

But this follows from

$$IP^*\text{-}\lim_{n \geq 1} \int \phi(x)\phi'(T^n x)\, d\mu(x) = \left(\int \phi\, d\mu\right)\left(\int \phi'\, d\mu\right),$$

$$IP^*\text{-}\lim_{n \geq 1} \int \psi(y)\psi'(S^n y)\, d\nu(y) = \left(\int \psi\, d\nu\right)\left(\int \psi'\, d\nu\right),$$

using the fact that IP^*-convergence carries over to products.

5. Mild Mixing of All Orders

We shall use IP^*-convergence to get information about expressions of the form $\mu(C_0 \cap T^{-n}A_1 \cap T^{-2n}A_2 \cap \cdots \cap T^{-ln}A_l)$ in the case of a mildly mixing system $(X,\mathcal{B},\mu,T)$. The results we present are far from being definitive, and are intended to illustrate the approach.

We begin with an IP^* analogue of Lemma 4.9.

LEMMA 9.24.   *Let $\{x_n\}$ be a bounded sequence of vectors in Hilbert space and suppose that*

(9.14) $IP^*\text{-}\lim_{m \geq 1} IP^*\text{-}\lim_{n \geq 1} \langle x_{n+m}, x_n \rangle = 0.$

Then with respect to the weak topology of the Hilbert space,

$$IP^*\text{-}\lim_{n \geq 1} x_n = 0.$$

To prove Lemma 9.24 we need the following.

LEMMA 9.25.   *Let $Q \subset \mathbf{N}$ be an IP^*-set and for each $q \in Q$ let R_q be an IP^*-set. Let $S \subset \mathbf{N}$ be an IP-set and let $k \geq 1$ be given. There exist k numbers $n_1 < n_2 < \cdots < n_k$ in S with each $n_j - n_i \subset Q$ for $i < j$ and $n_i \subset R_{n_j - n_i}$.*

Proof: We shall be using several basic facts regarding IP-sets. First of all, by Lemma 9.4, each IP^*-set intersects each IP-set in a set containing an IP-set. In particular each IP^*-set contains (many) IP-sets. Secondly, if S is an IP-set, $S = \{p_{i_1} + p_{i_2} + \cdots + p_{i_k}; i_1 < i_2 < \cdots < i_k\}$ and $q \in S$, there is a sub-IP-set of elements of S that can be added to q and such that the sum is in S. We denote such an IP-set (it is not well defined, but can be chosen as all the numbers of S constructed from p_i with indices i not occurring in some fixed representation of q) by $S(q)$.

We proceed to the proof of the lemma. Let S_1 be an IP-set in $Q \cap S$ and let $q_1 \in S_1$. Let S_2 be an IP-set in $R_{q_1} \cap Q \cap S_1(q_1)$ and choose $q_2 \in S_2$. Then $q_2 + q_1 \in S_1 \subset Q$, so that $R_{q_2 + q_1}$ is defined. Next take S_3

to be an *IP*-subset of $R_{q_2} \cap R_{q_2+q_1} \cap Q \cap S_2(q_2)$ and choose $q_3 \in S_3$. Continue to define $q_4, q_5, \ldots$ and $S_4, S_5, \ldots$ inductively by choosing S_{i+1} an *IP*-subset of

$$(9.15) \quad R_{q_i} \cap R_{q_i+q_{i-1}} \cap \cdots \cap R_{q_i+q_{i-1}+\cdots+q_1} \cap Q \cap S_i(q_i)$$

and choosing $q_{i+1} \in S_{i+1}$. At each stage we have for $l \leq i$

$$q_{i+1} + q_i + \cdots + q_l \in S_{i+1} + q_i + \cdots + q_l$$
$$\subset S_i + q_{i-1} + \cdots + q_l \subset \cdots \subset S_l \subset Q.$$

Hence $R_{q_{i+1}+q_i+\cdots+q_l}$ is defined and (9.15) is meaningful at each successive stage. Continue till $q_1, q_2, \ldots, q_k$, $S_1, S_2, \ldots, S_k$ have been defined and set

$$n_1 = q_k, \qquad n_2 = q_k + q_{k-1}, \ldots, \qquad n_k = q_k + q_{k-1} + \cdots + q_1.$$

Then

$$n_j - n_i = (q_k + q_{k-1} + \cdots + q_{k-j+1}) - (q_k + q_{k-1} + \cdots + q_{k-i+1})$$
$$= q_{k-i} + q_{k-i-1} + \cdots + q_{k-j+1} \in Q$$

for $i < j$. Moreover,

$$n_i = q_k + q_{k-1} + \cdots + q_{k-i+1} \in S_{k-i+1} \subset R_{q_{k-i}+q_{k-i-1}+\cdots+q_{k-j+1}}$$
$$= R_{n_j-n_i}.$$

This proves the lemma.

Having proved this lemma which is the analogue of Lemma 4.8, the proof of Lemma 9.24 proceeds exactly as that of Lemma 4.9.

We now pursue the analogy between *D*-convergence and *IP**-convergence further and define a *totally mildly mixing* group *G* of measure preserving transformations of a measure space $(X, \mathcal{B}, \mu)$, as one for which each $T \in G$, $T \neq$ identity, is mildly mixing on $(X, \mathcal{B}, \mu)$. We then have the analogue of Theorem 4.10 with an essentially identical proof:

THEOREM 9.26.   *Let G be a totally mildly mixing group of transformations of $(X, \mathcal{B}, \mu)$ and let $T_1, T_2, \ldots, T_l$ be distinct elements of G, $T_i \neq$ identity. Then for any $f_0, f_1, \ldots, f_l \in L^\infty(X, \mathcal{B}, \mu)$,*

$$IP^*\text{-}\lim_{n \geq 1} \int f_0(x) f_1(T_1^n x) \cdots f_l(T_l^n x) \, d\mu(x)$$
$$= \left(\int f_0 \, d\mu \right)\left(\int f_1 \, d\mu \right) \cdots \left(\int f_l \, d\mu \right).$$

Finally, we have as a corollary that a mildly mixing transformation is mildly mixing of all orders.

THEOREM 9.27.   *If T is a mildly mixing transformation of $(X,\mathscr{B},\mu)$, $A_0,A_1,\ldots,A_l$ are $l + 1$ sets in $\mathscr{B}$ and $0 < e_1 < e_2 < \cdots < e_l$ are distinct integers, then for any $\varepsilon > 0$, the set*

$$\{n: \left|\mu(A_0 \cap T^{-ne_1}A_1 \cap \cdots \cap T^{-ne_l}A_l) - \mu(A_0)\mu(A_1) \cdots \mu(A_l)\right| < \varepsilon\}$$

is an IP^*-set.

BIBLIOGRAPHY

Auslander, J.
[1] On the proximal relation in topological dynamics. *Proc. Amer. Math. Soc.* 11 (1960), 890–895.

Bebutov, M. V.
[1] On dynamical systems in the space of continuous functions *Bull. Mos, Gos, Univ. Mat.* 2 (1940).

Birkhoff, G. D.
[1] *Dynamical Systems.* Amer. Math. Soc. Colloq. Publ. vol. 9, Amer. Math. Soc., Providence, R.I., 1927.
[2] Quelques théorèmes sur le mouvement des systèmes dynamiques. *Bull. Soc. Math. France* 40 (1912), 305–323.
[3] Recent advances in dynamics. *Science* n.s. 51 (1920), 51–55.

Bochner, S.
[1] A new approach to almost periodicity. *Proc. Nat. Acad. Sci. U.S.A.* 48 (1962), 2039–2043.

Bogoliuboff, N. and N. Krylov.
[1] La théorie générale de la mesure dans son application à l'étude des systèmes dynamiques non-linéaires. *Ann. of Math.* (2) 38 (1937), 65–113.

Bohr, H.
[1] Zur Theorie der fastperiodischen Funktionen I, II, III. *Acta Math.* 45 (1925), 29–127; 46 (1925), 101–214; 47 (1926), 237–281.

Brauer, A.
[1] Uber Sequenzen von Potenzresten. *Sitzungsber. Preuss. Akad. Wiss.* (1928), 9–16.

Denker, M. and C. Grillenberger.
[1] *Ergodic Theory on Compact Spaces*, Lecture Notes in Math. 527, Springer-Verlag, Berlin, 1976.

Deuber, W.
[1] Partitionen and lineare Gleichungs-systeme. *Math. Z.* 133 (1973), 109–123.

Doob, J. L.
[1] *Stochastic Processes.* Wiley, New York, 1953.

Ellis, R.
[1] Locally compact transformation groups. *Duke Math. J.* 24 (1957), 119–125.

[2] A semigroup associated with a transformation group. *Trans. Amer. Math. Soc.* 94 (1960), 272–281.

[3] The Veech structure theorem. *Trans. Amer. Math. Soc.* 186 (1973), 203–218.

Ellis, R. and W. H. Gottschalk.

[1] Homomorphisms of transformation groups. *Trans. Amer. Math. Soc.* 94 (1960), 258–271.

Erdös, P. and P. Turán,

[1] On some sequences of integers. *J. London Math. Soc.* 11 (1936), 261–264.

Furstenberg, H.

[1] Ergodic behavior of diagonal measures and a theorem of Szemerédi on arithmetic progressions. *J. d' Analyse Math.* 31 (1977), 204–256.

[2] The structure of distal flows. *Amer. J. Math.* 85 (1963), 477–515.

Furstenberg, H. and Y. Katznelson.

[1] An ergodic Szemerédi theorem for commuting transformations. *J. d' Analyse Math.* 34 (1978), 275–291.

Furstenberg, H. and B. Weiss.

[1] Topological dynamics and combinatorial number theory. *J. d' Analyse Math.* 34 (1978), 61–85.

Glasner, S.

[1] Divisibility properties and the Stone-Čech compactification. To appear.

[2] *Proximal Flows.* Lecture Notes in Math. 517, Spring-Verlag, Berlin, 1976.

Glazer, S.

[1] Ultrafilters and semigroup combinatorics. *J. Combinatorial Theory* (A). To appear.

Gottschalk, W. H. and G. A. Hedlund.

[1] *Topological Dynamics.* Amer. Math. Soc. Coll. Publ., vol. 36, Amer. Math. Soc., Providence, R.I., 1955.

Graham, R. and B. Rothschild.

[1] A short proof of van der Waerden's theorem. *Proc. Amer. Math. Soc.* 42 (1974), 385–386.

Hadamard, J.

• [1] Les surfaces a courbures opposées et leur géodesiques, *J. Math. Pures Appl.* Ser. V, vol. 4 (1898), 27–73.

Halmos, P. R.

[1] *Lectures on Ergodic Theory.* Chelsea, New York, 1960.

[2] *Measure Theory.* Van Nostrand, New York, 1950.

Hardy, G. H. and J. E. Littlewood.

[1] The fractional part of $n^k\theta$. *Acta Math.* 37 (1914), 155–191.

Hilbert, D.
[1] Über die Irreduzibilität ganzer rationaler Funktionen mit ganz-zahligen Koeffizienten *J. Math.* 110 (1892), 104–129.

Hindman, N.
[1] Finite sums from sequences within cells of a partition of **IN**. *J. Combinatorial Theory* (A) 17 (1974), 1–11.
[2] Ultrafilters and combinatorial number theory. Notes of an address presented to Southern Illinois University Number Theory Conference, 1979.

Hopf, E.
[1] *Ergodentheorie*. Ergebnisse der Math. und ihrer Grenzgebiete, Springer-Verlag, Berlin, 1937.

Jewett, R. I.
[1] The prevalence of uniquely ergodic systems. *J. Math. Mech.* (1969–70), 717–729.

Jones, L. K. and M. Lin.
[1] Ergodic theorems of weak mixing type. *Proc. Amer. Math. Soc.* 57 (1976), 50–52.

Khinchin, A. Y.
[1] *Three Pearls of Number Theory*. Graylock Press, Rochester, N.Y., 1952.

Knapp, A. W.
[1] Distal functions on groups. *Trans. Amer. Math. Soc.* 128 (1967), 1–40.

Krieger, W.
[1] On unique ergodicity. In *Proceedings of the Sixth Berkeley Symposium on Math. Stat. and Probability*, vol. 11, University of California Press, Berkeley, Calif., 1972, 327–346.

Maak, W.
[1] *Fastperiodische Funktionen*. Grundlehren der Math. Wissenschaften in Einzeldarstellungen, Springer Verlag, Berlin, 1950.

Margoulis, G. A.
[1] Discrete groups of motions of manifolds of non-positive curvature, *Amer. Math. Soc. Transl.* 109 (1977), 33–45.

Martin, J. C.
[1] Substitution minimal sets. *Amer. J. Math.* 93 (1971), 503–526.

Morse, M.
• [1] Recurrent geodesics on a surface of negative curvature. *Trans. Amer. Math. Soc.* 22 (1921), 84–110.
• [2] *Symbolic Dynamics*. Institute for Advanced Studies Notes by Rufus Oldenburger.

Moser, J.
[1] *Stable and Random Motions in Dynamical Systems*. Ann. of Math.

Studies 77, Princeton University Press, Princeton, N.J., and University of Tokyo Press, Tokyo, 1973.

Mostow, G. D.

[1] *Strong rigidity of locally symmetric spaces.* Ann. of Math. Studies 78, Princeton University Press, Princeton, N.J., 1973.

Namakura, K.

[1] On bicompact semigroups. *Math. J. Okayama University* 1 (1952), 99–108.

Nemytskii, V. V. and V. V. Stepanov.

[1] *Qualitative Theory of Differential Equations.* Princeton University Press, Princeton, N. J., 1960.

Poincaré, H.

[1] *Methodes nouvelles de la mécanique céléste.* vols. I, II, III, Paris (1892, 1893, 1899).

Rado, R.

[1] Note on combinatorial analysis. *Proc. London Math. Soc.* 48 (1943), 122–160.

[2] Studien zur Kombinatorik. *Math. Z.* 36 (1933), 424–480.

Riesz, F., and B. Sz.-Nagy.

[1] *Functional Analysis.* F. Ungar Publ. Co., New York, 1955.

Robbins, A.

[1] On a class of recurrent sequences. *Bull. Amer. Math. Soc.* 43 (1937), 413–417.

Rokhlin, W. A.

[1] Selected topics from the metric theory of dynamical systems. *Amer. Math. Soc. Transl.* 2, 49 (1966), 171–209.

Roth, K. F.

[1] Sur quelques ensembles d'entiers. *C.R. Acad. Sci. Paris* 234 (1952), 388–390.

Schur, I.

[1] Über die Kongruenz $x^m + y^m \equiv z^m$ (mod p). *Jahresbericht der Deutschen Math.-Ver.* 25 (1916), 114–117.

Sinai, J. G.

[1] A weak isomorphism of transformations with invariant measure. *Mat. Sbornik* 63 (1964), 23–42; *Amer. Math. Soc. Transl.* 2, 57 (1966), 123–143.

Stewart, B. and R. Tijdeman.

[1] On difference sets of integers. Seminaire Delange-Pisot-Poitou. To appear.

Szemerédi, E.

[1] On sets of integers containing no four elements in arithmetic progression. *Acta Math. Acad. Sci. Hungar.* 20 (1969), 89–104.

[2] On sets of integers containing no k elements in arithmetic progression. *Acta Arith.* 27 (1975), 199–245.

Thouvenot, J.-P.

[1] La demonstration de Furstenberg du théorème de Szemerédi sur les progressions arithmetiques. *Séminaire Bourbaki* 30 (1977–78), 518.

van der Waerden, B. L.

[1] Beweis einer Baudetschen Vermutung. *Nieuw Arch. Wisk.* 15 (1927), 212–216.

[2] How the proof of Baudet's conjecture was found. In *Studies in Pure Mathematics*, ed. L. Mirsky. Academic Press, New York, 1971, pp. 251–260.

Veech, W. A.

[1] Almost automorphic functions on groups. *Amer. J. Math.* 87 (1965), 719–751.

[2] Point distal flows. *Amer. J. Math.* 92 (1970), 205–242.

[3] Topological dynamics. *Bull. Amer. Math. Soc.* 85 (1977), 775–830. Weyl, H.

Weyl, H.

[1] Über die Gleichverteilung von Zahlen mod Eins. *Math. Ann.* 77 (1916), 313–352.

Wiener, N.

[1] *The Fourier Integral and Certain of its Applications.* The University Press, Cambridge, 1933.

• Zimmer, R. J.

[1] Extensions of ergodic actions and generalized discrete spectrum. *Illinois J. Math,* 20 (1976), 555–588.

INDEX

Library of Congress Cataloging in Publication Data

Furstenberg, Harry.
    Recurrence in ergodic theory and combinatorial number theory.

    (M. B. Porter lectures; 1978)
    Bibliography: p.
    Includes index.
    1. Ergodic theory.   2. Topological dynamics.
3. Numbers, Theory of.   I. Title.   II. Series.
QA611.5.F87   515.4′2   80-7518
ISBN 0-691-08269-3

H. FURSTENBERG is Professor of Mathematics at the Hebrew University, Jerusalem.

WOMEN'S WRITING IN
LATIN AMERICA

WOMEN'S WRITING in LATIN AMERICA

An Anthology

edited by

Sara Castro-Klarén
Johns Hopkins University

Sylvia Molloy
New York University

Beatriz Sarlo
University of Buenos Aires

WESTVIEW PRESS
Boulder • San Francisco • Oxford

Copyright © 1991 by Westview Press, Inc.

Published in 1991 in the United States of America by Westview Press, Inc., 5500 Central Avenue, Boulder, Colorado 80301, and in the United Kingdom by Westview Press, 36 Lonsdale Road, Summertown, Oxford OX2 7EW

Library of Congress Cataloging-in-Publication Data
Women's writing in Latin America : an anthology / edited by Sara
  Castro-Klarén, Sylvia Molloy, Beatriz Sarlo.
    p. cm.
  Includes bibliographical references.
  ISBN 0-8133-0550-0 — ISBN 0-8133-0551-9 (pbk.)
  1. Latin America—Women authors. 2. Women—Literary collections.
3. Latin American literature—20th century. I. Castro-Klarén,
Sara. II. Molloy, Sylvia. III. Sarlo Sabajanes, Beatriz.
PQ7083.W66 1991
860.8′09287—dc20   91-17691
                                                        CIP

Printed and bound in the United States of America

The paper used in this publication meets the requirements
of the American National Standard for Permanence of Paper
for Printed Library Materials Z39.48-1984.

10   9   8   7   6   5   4   3   2   1

Contents

PART 3   Women, History, and Ideology

Contents

Preface & Acknowledgments

THE LIFE OF THE BOOK COLLECTOR, according to Walter Benjamin, is marked by a dialectical tension between disorder and order. The seeming randomness with which volumes are collected, the memories of personal encounters with books are slowly replaced, in the spirit of the collector, by an order that, in retrospect, appears self-evident, even necessary. A similar tension, one might argue, rules the task of the collector of texts. No different from book collectors, eager to unpack a library and share with the public memories attached to the acquisition of each book, compilers of anthologies would share texts, and memories of their readings of those texts, with their own readers. And, as those texts are set forth, we text collectors confront our choices, make sense out of our individual preferences, discover our ideologies of reading, and assume our patterns of theoretical conjecture. Our personal collections become communal projects; our gathering of texts, deliberate itineraries we would follow with others. Thus, although this anthology of Latin American women writers doubtlessly responds to a historical necessity, calling attention to texts too long deprived of canonical status in their countries of origin and too long denied adequate translation and dissemination abroad, it should also be seen as a personal dialogue: it is the conversation of three readers who enjoy, love, have been marked by the texts they have chosen; three readers eager to learn from the mutual repercussions of their respective selections, eager to establish relations among texts not always seen side by side, let alone read together.

This book strives to bring together Latin American women whose voices have often been heard in isolation or not heard at all, women who, in addition, have often considered themselves isolated, marginal, unique. In a culture that has been loath to acknowledge a *tradition* of female expression, that is, a continuum of female voices of recognized authority, it is not surprising that the women writers themselves find it at times hard to perceive bonds and often echo the very same isolation from which they suffer. There are lonely voices in this book, texts that reiterate their singularity, their difference, their uniqueness, texts that even assume that singu-

larity—tentatively or defiantly—as a mask. And, next to them, there are the texts that work against that imposed isolation, laboring to create bonds and promoting awareness, the better to rescue a tradition of Latin American women's writing that has been too long belittled or denied. Joining in that rescue ourselves, we have tried to present the reader not with bits and pieces but with a chain, a meaningful sequence that need not be read in order but that should, we hope, reveal fruitful contacts, unexpected links. Our selection includes both writers and women who write. Chiefly interested in a practice of discourse that transcended professional specialization, we did not attempt to separate them but, on the contrary, learned a great deal from placing them side by side.

Growing interest in Latin American women's writing has contributed in the past twenty years to the discovery and expansion of a corpus that had remained, until then, well hidden from view. There has been a fruitful effort to recover texts written in colonial times by women in cloisters or at the margins of society, so as to open new paths of inquiry into the position of women as colonial subjects. A similar effort has reevaluated the work of women educators during the nineteenth and twentieth centuries. Yet another line of research has concentrated on travelogues, first-person accounts of various types, *testimonios*, diaries, autobiographies, and other hybrid modes of self-presentation from the nineteenth century onward, in which reflection on individual and national identity is forcibly conditioned by gender. This expansion of the field to include texts that had hitherto been considered secondary and therefore negligible has had beneficial repercussions on works by women that had already found a modest place in the canon. The last twenty years have witnessed a fresh interest in authors who have survived the unexamined rigors of patriarchal canonization. The works of Delmira Agustini, Gabriela Mistral, Clarice Lispector, and Rosario Castellanos, to name but four, are now considered in a different light; the hasty assumptions about them are questioned, and new and challenging readings are being put forth.

A world of reading and interpretation lies in wait as criticism begins to trace relational lines capable of engaging the productivity of this corpus of writing. These texts—poems, novels, essays, speeches, tracts—make demands of their own, ask to be scrutinized in detail, prompt a reflection on contexts, set their own pace for the reader. These texts, in short, invite the reader to recompose Latin American women's very complex scene of writing. If general categories developed by Anglo-American and French feminism are highly useful tools with which to analyze this scene, they should be used with caution. Yet another purpose of this anthology is to invite the English reader to dwell on the implications of the double marginality of Latin American women and to consider that double marginality in its very

difference instead of dissolving that difference with a homogenizing (albeit generous) glance.

Some of the writers in this anthology are well known and their works have appeared in English before. In most cases, we have tried to avoid their more familiar compositions—so often cited in anthologies that they have become trite—and have focused on lesser-known, equally striking texts not previously published in English. In some cases, however, we have gone back to those familiar compositions merely because we felt that the existing English versions were but mockeries of the originals and deserved better translations. In yet other cases, we have included existing English versions whose very excellence precluded any further attempt at translation on our part. Finally, this anthology includes many new writers unfamiliar to the public. Some are newcomers. Others are not but have authored texts that, because they did not quite fit the conventional categories of "the literary" (for long, the only discourse grudgingly allowed women in Latin America), have been deemed unworthy of attention. In order to include these vigorous new voices and the forgotten, no less eloquent older ones, we have had to sacrifice others: there are noteworthy omissions in this book, well-known contemporary figures that we have not included, confident that their considerable visibility as published writers in English will compensate, to some extent, for their absence here.

Friends, colleagues, and students have helped us in this venture, and we thank them all. In particular, we wish to thank Marta Peixoto for her careful reading of part of the text and her very helpful comments. James Fernández was a most thoughtful editor of sections of our manuscript. The following people, who shared with us the most difficult task of rendering into English those pieces that had not been previously translated, deserve our deepest gratitude: Rosario Ferré, Amy Kaminsky, Oscar Montero, Laura Beard Milroy, Patricia E. Paige, Marta Peixoto, Marcy Schwartz Walsh, and Nicolás Wey. It is thanks to them that these texts have been given new life.

We also thank Rosario Ramos for her recastings of the bibliographies here included. Finally, we gratefully acknowledge the research funds made available by the Johns Hopkins University for the preparation of the manuscript.

Sara Castro-Klarén
Sylvia Molloy
Beatriz Sarlo

PART

I

Women, Self, and Writing

Introduction

SARA CASTRO-KLARÉN

GREAT DEAL HAS HAPPENED to the quest for women's liberation since Virginia Woolf's *A Room of One's Own* pushed ajar the doors of the male domain to women's writing and action in the world. The once appealing simplicity of a "literature of their own" has given way to the complexities of gender formation, gender and class, and identities forged in the confines of hegemonic or colonial nationalities. Above all, we have tested the quicksands of the profound differences between competing hypotheses on the question of the feminine.

No longer a new field, women's, or gender, studies has introduced a thorough rethinking of the once naturalized and sustaining categories in the social sciences and the humanities. Concepts such as the Cartesian subject, the Oedipal complex, and self-identity as a relation of economic, political, and kinship status have undergone serious questioning. In fact, Alice Jardine postulates in her introduction to *Gynesis, Configurations of Woman and Modernity*[1] that the current epistemological crisis of the West circles around the central emptiness of woman as the unknown, or rather, as nonknowledge.

Closer to home, in feminist literary studies, the opposition male/female has been taken as a fundamental category of analysis. Such a category, however, involves areas of study traditionally outside what has been strictly defined as the sphere of the "literary." Feminist criticism has integrated linguistic, narratological, and stylistic categories of analysis with post-Freudian psychoanalysis, Marxism, and poststructuralism.[2]

Simultaneously, new feminist studies and the recovery, as well as the appearance, of a vital corpus of texts written by women have caused a reexamination of the processes of canon formation. Literary histories therefore seem woefully inadequate and insufficient. As the search for woman

3

legitimizes studies on topics previously thought unimportant, the tenets of autobiography, testimonial writing, and other forms of "private" writing such as letters, diaries, and even recipes receive the scrutiny and thoughtfulness previously reserved only for "important" texts.[3]

Such a myriad—poems, novels, political speeches, testimonial autobiography, and occasional essays—makes demands of its own. These texts set a new context and pace for their newly alert readers. For instance, the differences in the ideology of writing presented by texts tightly connected to a specific historical context and texts more inclined to a metaphysical meditation pose a difficult and unyielding problem for the reader who engages a text primarily as "feminine" or "feminist" discourse.

Not all women's writing can best be understood within the generalizing categories developed by recent Anglo-American "feminist" social or literary studies. Such an attempt brings forth, among others, difficult problems of cultural translation. The implicit comparative approach between Latin American women's writing and Anglo-American women's writing—an approach unavoidable for the readers of this anthology of texts in English translation—must begin by reexamining such stereotypical cultural differences between the Protestant, or largely secular, North and the Catholic South; the nuclear family of capitalist society in contrast with the extended kinship ties of people living in "developing" economies.

The emergence of woman is the central concern of the writings selected for this anthology; the texts assembled here, then, provide a unique space for reading women's writing as a series marked by its own sign. For a reader of the texts of Sor Juana Inés de la Cruz, Clarice Lispector, Rigoberta Menchú, and Magda Portal, to take a variegated sample, one paramount question surfaces over and again: "Who speaks?" This question, seeking to clarify the situation of the speaker, immediately brings forth a number of correlatives: It questions the identity of the recipient of the message, the content and rhetoric of the message itself, and the relation between the subject of the enunciation and her presumed interlocutor. Are the women presumed to be situated in the writing/speaking subject addressing a feminine/feminist reader? Or does the enunciation deployed address an "abstract" ideal though not necessarily a feminine/feminist reader? Clarice Lispector, Cristina Peri-Rossi, and Julieta Campos, women who practice a self-reflective writing, have advanced the position that neither writing nor reading can be so blatantly or abstractly "marked" feminine or feminist. And yet, when we read their texts, feminist criticism invites us not to lose sight of the fact that the historical being authorizing each of these texts is a woman.

Partly because of the vexing nature of this conundrum, the selections included here address the question of an ideology of writing as articulated by several of the most prominent women writers of this century. Placed spatially and temporally next to each other, and in light of the contextual

relations suggested by the texts in the other two sections of this book (on women's ideology and on history and textual self-figurations), these selections should offer a fresh opportunity for an interfacing inquiry about the subject of the enunciation. In some cases, it has been possible to include a text in which the woman writing meditates on the nature of her praxis, and another text in which the same writer produces a fiction or an allegory of being (female). Such has been the felicitous coincidence of texts in the cases of Nélida Piñón, Clarice Lispector, Rosario Castellanos, Cristina Peri-Rossi, and Lygia Fagundes Telles. Anthologies are undergirded by the notion that something greater than the specificity of the collective emerges in the reading of many brief samples of texts by various authors. In this case, these specific texts were chosen because it was assumed that together they provide an exploration, in a fictional or essayistic mode, into the constitution of a map of the self in language, that is to say, in the consciousness of being in the world. Such broadly conceived self-consciousness emerges as the counterpart of silence.

The Writer Before the Mirror

In the following pages I try to engage the series of selections included, guided by the provisional notion that at least seven positions between writer and writing can be detected and outlined in the praxis of women's writing in Latin America. Many writers other than those included here have addressed the subject of writing as it pertains to their own singular craft, the writing of women, or the Latin American writer. Particularly incisive essays or interviews, and even poems, have made the views of Luisa Valenzuela,[4] Tamara Kamenszain, Alejandra Pizarnik, Olga Orozco, and Sylvina Bullrich, to name only a few, available to Spanish-speaking readers. Some of these writers appear more confident than others in their ability to use the word to breach their silence. Cristina Peri-Rossi, for example, chooses the dialogue situation—between Woman and Word—as a metaphor for the erotic possibility of a mutual opening into a configuration of signs and thus of meaning:

> *Woman:* My woman is a word
> Woman's word, she hears me
> *Word:* She listens to me
> I speak to her in words of love
> my woman stretches out, in leisurely meter,
> When she is stretched full out
> I open her, like a word
> and she, like a word
> moans, weeps, implores, stalls, disrobes,
> speaks names, makes sounds, cries, calls, creaks, brays.[5]

In contrast to the yielding compenetration of woman and word in Peri-Rossi's poem, the promise of meaning turns into a flashing though unintelligible drumming in Pizarnik's pursuit of the sign: "Everything makes love with silence. They had promised me a silence like fire, a house of silence. Suddenly the temple is a circus and the light, a drum."[6]

Experimental writers such as Julieta Campos and Nélida Piñón have repeatedly and extensively commented not only on their own work but also on the writing of many of their European, North American, and Latin American contemporaries.[7] The writer posing as a critic of her own, or other writers', work offers many singular opportunities for a specular relation.

Sometimes the specular relation reaches back in time in pursuit of a progenitor. Such is the case of Magda Portal. Almost a hundred years after Flora Tristán visited Peru, Portal wrote *Flora Tristán, precursora* (1945). She acknowledges a feminist parallel between her writing and political action and the cutting and bitter analysis of Peruvian society in Flora Tristán's *Peregrinaciones de una paria* (1847). Similarly, Rosario Castellanos's essay, "Once Again Sor Juana," establishes Sor Juana's feminist concerns as prefigurations of her own self-consciousness. In Brazil Lygia Fagundes discusses her mystical approach to religion, evoking Santa Teresa's texts as unsurpassable precedents. Portal, Castellanos, and Fagundes see themselves in their chosen predecessors.

Other similar contemplating gestures are, however, more often than not deflected by the undulating surface of a mirror in which epochal, national, or gender differences seem to obstruct the emergence of a clean and sharp self-reflecting image.

Some of these movements of discordance and concordance on the surface of the mirror can be appreciated in the case of Gabriela Mistral as reader of José Martí, Victoria Ocampo, and Teresa de la Parra.[8] In each instance Mistral's reading seeks to identify points of implicit affinity or resemblance between her own ideology of writing and the writing of her fellow artists, a high level of affinity warranting her praise. In the specular relation that Mistral sets up between the reader and the writer, interpretation becomes an instrument that measures degrees of resemblance. For instance, while the Chilean poet examines the elements of Martí's originality, the reader is subliminally directed to assemble a chart of similarities between Mistral's own textual practices and what she highlights as the elements of Martí's unsurpassed inauguration of an American (as opposed to European) language. Breaking with the reasons adduced by her hagiographers—who portray Gabriela as obedient Virgin and Mother[9]—Mistral cuts through the directives of the patriarchy. Contradicting predominant values of the literary establishment, she chides those who believe that writers who live in the tropics cannot write "strong" poetry because the tropical climate of

this habitat robs them of the intellectual energy necessary to think in a "masculine," clear manner. Mistral advocates the topic of writing within the received oral and written practice of the language. Defending the wisdom of rooting the poet's language and aesthetics in her immediate lived experience, she propounds Martí's break with the heavy Spanish use of rhetorical figures as well as his choice of a "regional" lexicon. She advances the proposition of a newly privileged tropicalism for Martí. In doing so she indirectly compares the goodness of Martí's *viril* style with her own "maternal" embrace of the American landscape as a metaphor for an American sensibility.

As a counterpoint to the portrait she draws of Martí—a rooting of the logic of the concrete and the existential—the Chilean poet's sketch of Victoria Ocampo delineates a face closer to Picasso's *Cubist Nude*. If Martí's face is photographed, Victoria's face is rendered as a cubist resistance to the "natural" order of a face. Ocampo's face has strong features, but their organization spells out a discordant mode. The reason of the "natural" is missing from the Argentine's face. Ocampo's face cannot be articulated into a coherent whole because, according to Mistral, Ocampo, unlike Martí, has not claimed her nationality as the place of her identity. Mistral's logic of the authentic predicates for her Argentine counterpart the pampa as the cultural metaphor for a concrete and imaginary discourse. Mistral's preference for mimesis of the American natural and social canvases leads her to believe that Ocampo's penchant for the world of European masters is a waste of the best resources available to an American writer.

Mistral surmises that Ocampo's early readings of masters of languages "inimical" to Spanish have deprived the Argentine woman of a refuge and abode similar to Martí's tropics and Mistral's own slim little Chilean valley. However regrettable Ocampo's love affair with Europe appears to Mistral, the Chilean poet finally recognizes that it is precisely by way of Europe—in the form of two English woman writers—that Ocampo's imagination finds a reason of its own. Mistral finds that Ocampo's portrait as a writer flows into a centered design when the Argentine feminist writes on the work and life of Virginia Woolf and Emily Brontë.

Reading Ocampo's work, Mistral discovers that it is in the life of another woman, in the specular relation with another passionate and turbulent life that a meaningful image of the intimidated Ocampo emerges. Mistral's gaze upon Ocampo's texts pierces beyond the early *Testimonios* (1938) to surprise Ocampo's own specular contemplation in the mirror of Brontë's rough, rebellious, and destructive Heathcliff. As the reader of Ocampo's biography of Brontë, Mistral does not accept Ocampo's elaboration of a possible coincidence by admiration between herself and the bio-graphical Brontë. What Mistral sees in the awed Argentine reader of fetishized male European masters and in her study of Brontë as the author of *Wuthering*

Heights (1847) is a plain though releasing alliance between the tortured Ocampo and the subversive Heathcliff. In Ocampo's claim that Brontë is most truthfully projected in Heathcliff rather than in Catherine, Mistral finds the raison d'être for Ocampo's own writing. Furthermore, Mistral shares in the subversion and stubborn fidelity to self and passion that Ocampo acknowledges in Heathcliff. In the same vein in which she praises Martí for his "naturaleza anti-imitativa o sea anti-femenina" (his anti-imitative or rather antifeminine nature),[10] Mistral now recovers Ocampo's sensitivity and protest over the historical oppression of their sex. So it is that in the subversive figure of Heathcliff, Mistral finds the road for the liberation of Ocampo's writing from the tragedy of French governesses and even her much-flaunted but crippling bilingualism: "The male-genius always had a hallucinatory effect on her. She was also intrigued by the female-genius. Nevertheless, Victoria was always aware of the honor and misery of our sex, its humbling and marvelous avatars, seems to have finally found her voice in the biography of Emily Brontë."[11]

The key to a writer's success, in Mistral's mind, is in finding the juncture where tradition and innovation meet without betrayal of a certain irreplaceable something that she calls authenticity, truth, autonomy, or in the case of Latin American writers, "lo americano." For her, neither tradition nor authenticity refers at all times to a single and static concept or experience. In Martí, tradition is represented by his reading of the Greek and Roman classics, whereas Teresa de la Parra establishes a link to tradition in encompassing in her work the folklore of the Venezuelan llanos. In both cases, Mistral finds these traditions—written and oral—indispensable. They are a place of learning for the writer, for she recognizes that no one can be an "Adán literario," a writer without a tradition. In her own case, she acknowledges the "release" that she found in the books borrowed from Don Bernardo Ossadón, the loving owner of the only library in her hometown. What Mistral detests is the writing of those who do not or cannot find in themselves the reason for their own writing. Thus, she prefers Martí to Rubén Darío because Martí "procede de sí mismo" (writes from within).

For Mistral, originality devolves on the ability to learn enough from the established written tradition so as to be able to pluck the correct chords and rhythms given to the individual at the outset of the biological and historical self. Writing is therefore posited as auto-bio-graphy, or the writing of one's self. Writing, for the author, registers an "hambreadura de lo propio" (a hungering for one's own) in *Tala* (Feeling of Trees, 1924). The axis that traverses the multiple specificity of one's own in a writer is what Mistral calls unanimity of tone: "Unanimity of tone is achieved when the writer has stopped wondering, stopped walking. Then, the writer sits down with her treasures in her gut. She now possesses language, speech. . . . A newly found solvency has put an end to borrowing."[12] And so she extends

a compliment to Teresa de la Parra that she would not easily offer to many other Latin American writers whose work she has also reviewed. Praising the seemingly "natural," artless, and "conversational" tone of the *Las memorias de Mamá Blanca,* Mistral emphasizes "an easy grace and wit unseen in a woman's writing since our dear Santa Teresa left us."[13] For *Las memorias* constitutes the joyful balance between classical and folkloric sources, the influences resolved within de la Parra's own historical contexts. Thus sketched, *Las memorias* gives back to Gabriela Mistral the image that she herself has of her writing project and the image of her self forged in memory.

The Mother Tongue

The question of authorship for Mistral, Piñón, Poniatowska, Fagundes Telles, and others is inextricably linked to the metaphor embedded in the idea of a mother tongue. The writer must have full, exhaustive command of the language learned at the earliest and darkest moments of consciousness, for the mother tongue and the writer are the seamless parts of a mystical union. This is the reason Mistral finds the story of Victoria Ocampo, who learned French first (and from a governess, no less), profoundly tragic. For Mistral, this event in Ocampo's emotional and intellectual universe is the trauma that keeps the Argentine writer from identifying the true object of her desire. It is as if, separated from the body of the mother tongue, Ocampo could not reach the contents of her own consciousness.

Inverting the relationship of God to Adam in the primal biblical scene in which God confers upon man his greatest gift—the power to name all the objects in his freshly minted universe Mistral posits the mother as the source of language. In this case, it is a female principle, with its reproductive and sustaining body, that is to say, with her own flesh, which passes on to her offspring the pleasure of the word. All of language (system and speech act) and, therefore, all sweet consciousness emerges in Mistral's view from the mother's language, from the cadences of her voice, from the tangible shapes of her body. The materiality of language thus becomes one with the mother's body. The beauty of the earth appears before the child's consciousness in the mother's act of naming. The mother bequeaths her child a "lengua forestal" (a forestlike language).

In "El modelo mariano," Helena Araújo, following some of the most recent studies on the mother and *marianismo,* advances the idea that "the idealization of maternity stimulates a primary narcissism in women . . . a semiotic that linguistic communication does not encompass."[14] Such a hypothesis could be descriptive of the mystical union between the evoked little girl and her mother in Mistral's text. Nevertheless, basing herself on

Julia Kristeva's *Hérétique de l'amour*,[15] Araujo also notes that the Virgin, as the center of *marianismo,* does not speak. "Silence is the very sign of the suffering mother."[16] If silence embodies the difference between Mary and Christ, then it follows that the "cult" of the mother in Mistral cannot be thought of as part of a generalized *marianismo* in Latin America.

In Nelly Campobello's *Las manos de mi madre* (1938) the mother is nature herself, "graceful as the mountain's flowers when they dance savaged by the wind";[17] in Mistral's scene, the child delights in the mother's body and in her language. It brings forth the world for her possession. Delight with the mother's body is not at all a sin or a transgression of God's rules, for the mother, unlike God, does not see or speak with an alienating gaze. Rather than naming things in order to produce a cold and resistant distance between the word and the object (as it happens in the world of Alain Robbe-Grillet), in the sweet and vegetable world of Elqui, the mother's word forms a beautiful, though hard, habitat for the baby daughter. The power to name, the mind's faculty to conceive, differentiate, organize, and create, are all taught to the poet by the mother's language as system and as speech act. Later, schools and books merely gloss the plenitude of the mother's tongue. "There is no creating word that I did not learn from you," writes Mistral in her evocation of the death mother.[18] But the mother's legacy is not just language. Her endowment includes the discovery of God as the hidden presence in nature's harmony and with it there appears in Mistral's work a general positive look upon the feminine in the universe.

Perhaps only partially aware of the point reached in her meditation upon her mother tongue, Mistral, having more than once been reprimanded by the church because of her alleged pantheism, stops before drawing the final logical conclusions implicit in her elegy. Her views on the mother tongue not only substitute a female principle as the source of the word in the scene of creation, but they also reinscribe the good news of St. John's gospel. In St. John's gospel we learn that "in the beginning was the Word, and the Word was with God, and the Word was God." Mistral's text on the mother tongue simultaneously implies and represses her reinscription of John, for her text should read: "In the beginning was the Word and the Word was my mother tongue."

Malinche, the Twisting of Betrayal

Some observers have seen in Latin American cultural patterns a pervasive cult of the mother. In trying to understand the observed phenomenon and its inherent contradictions within a society otherwise characterized as intensely patriarchal and *machista,* social analysts (and even historians) have posited Marianism—the idealization of female values, the Catholic

cult of the Virgin Mother Mary—as the causal explanation for the so-called cult of the mother.[19]

Recently, studies that document the continuity of the traditional and endemic subordinate role of woman in Latin American societies have puzzled over the persistent way in which woman is primarily and almost exclusively still cast in the role of mother. When the interests of women find advocates, social change and improvement are measured in terms of women's care and responsibility for their children's welfare and education. Even the discourse of recent feminists continues to posit women primordially as the seat of biological and cultural reproduction within the domestic sphere. The space assigned to her activity, imagination, and discourse continues to be the cultural economy of the household. Thus woman's field of action remains confined within the circle of the patriarchal. This is not the place to examine the validity and the complexities of the discourse of Marianism or the patriarchal cult of the mother. In addressing the practice (as opposed to the content) of the texts selected here and the question of the formation of a self in and for public life in contrast with a self for domestic life only, however, it is necessary to acknowledge the joyful but also tense and conflicted relations of each speaker to her own mother, to the idea of the mother as a historical force, and to the possibility of the speaker's own role as mother.[20]

Woman as mother appears more often and decidedly at the center of stories, poems, and essays in texts written before World War II. Thus, Gabriela Mistral, Rachel de Queiroz, Victoria Ocampo, Nelly Campobello, and Magda Portal posit the figure of an endlessly tender, innocent, but nevertheless all-knowing mother. This woman, often poor (and often silent), fights bravely against all odds to make life beautiful and plentiful for her children. In Campobello's *Las manos de mi madre* and Queiroz's *As Três Marías* (1935; *The Three Marias*, 1985), the mother's portrait is painted from the point of view of a child. This mother-child relation is enormously joyous, for the mother not only nurtures and gives of herself unendingly but also endows her daughters with the power to survive and believe in themselves. In these scenes, in which the child's perception is assimilated into the situation of the narrator of the story, the relation between father and mother appears foggy and distant.

Nevertheless, when the writer organizes the experience of her self as a young woman open to the possibilities of sexual and amorous encounters, the horizon of experience appears darkened by a storm of uncertainty and a deep anxiety of motherhood. In Queiroz's *Three Marias,* Guta's disillusion in the "free" love that she and Isaac thought they had found is only made deeper by his sweet but callous disavowal of *his* possible parenthood. Citing his problems with Brazilian immigration authorities, Isaac leaves Guta to confront alone the consequences of *her* pregnancy. Planned or

accidental, her reckless day at the amusement park ends in the utter desolation of broken trust and a necessary, although much-lamented, abortion. Neither Guta's nor Corina's (*Parque industrial,* 1935) tender anticipation of the "little baby boy" is enough to mitigate the sense of solitude and betrayal etched by chains of guilt wrapped around these future unwed, undesirable girl-mothers. Seldom, for this generation, is there the possibility of a motherhood free of a puzzling anguish created by the consciousness that a pregnant woman has somehow lost her value as a "free" lover. "Loved" or "used" girls lose their *virginal* value for the ambitious and self-centered artist in "Sauna," for the handsome white boy in *Parque industrial,* and even for Romelia's father and mother in Castellanos's "The Widower Román." Once sexuality is encircled by the social value of pregnancy and motherhood, it becomes the woman's burden. This vision of the self as possible (degraded) maternal being stands in sharp contrast to the memory of one's own pure and joyful mother.

Rosario Castellanos gives yet another disturbing twist to the maternal figure. Girls obsessively experience a sharp sense of betrayal in their relation with their mother. She is seen as an accomplice of a *machista* universe. In practically all of the stories of *Los convidados de agosto* (1964), the young heroine—Reineire, Ermelinda, or Romelia—is trapped by the strict codes of Comitán. She can neither find happiness within the given restraints of the social system nor escape them. The young women of Comitán struggle alone and agonize before the watchful and almost hostile eye of the older generation of women. Aunts, *beatas* (devout spinsters), older sisters, gossips, servants, and mothers expect their pound of flesh. Daily they attend their own local theater at mass or at the dinner table. Domestic life, conceived as rituals of feminine subservience, consists of charged moments when girls must acknowledge their oppressed and reduced condition. No wonder the little girl and writer-to-be in the auto-bio-graphical *Balún Canán* (1957) fantasizes the death of the weaker and younger brother who by sheer virtue of his being male becomes the sole beacon and hope of their insecure mother. "Betrayal!" shouts the little girl in *Balún Canán.* "Betrayal!" shouts Castellanos in the name of Malinche, the little girl sold into slavery by her own mother. The biography of Malinche, as refashioned by Castellanos, depicts a little girl not only sold into slavery but also deposed from her inheritance of a throne by her mother in conspiracy with her lover. Saved by a simulated death and burial, the little girl wanders on to the domain of other lords and learns to serve them as a means to her own survival. But she never forgives the mother who sacrificed her only because she had been designated, like her, female.

The complexities of betrayal, or *malinchismo,* seem especially poignant and contradictory when women's desire points outside the realm of the domestic. Both the desire and the assertion of power leave women in

ambivalent, irreconcilable situations. Castellanos's interest in the tragedies of La Malinche and Anna Karenina and the self-exiles of Santa Teresa and Sor Juana leads her to hope and conclude that there must be another way of being female, a way by which one can surmount the impasse posed by desire, betrayal, exile, or self-hatred as dictated by patriarchal law.

The idealization of the mother made possible by *marianismo* is also questioned in other texts in which the object of inquiry is not a historical mother but rather the Virgin Mary. In Luisa Valenzuela's short story "Proceso a la Virgen" (*Los Heréticos,* 1967), the Virgin is taken to trial in a small fishing village in Latin America for failing to perform the miracles that the community needs in order to preserve its female sexual economy of virginity, marriage, or prostitution. In the story, the Virgin must preserve the virginity of a beautiful, wild, and solitary girl. María has become the object of mad desire for all the village's (married) fishermen. The men beg the Virgin for a miracle. Their wives also beg the Virgin, but their request is that she keep their men away from María. In her analysis of Valenzuela's story, Helena Araújo points to the unsurpassable contradictions embedded in the Virgin Mary as the patriarchal idealization of the "feminine":

> Inaccessible, mysterious, demiurgical, the Virgin and the adolescent Mary represent a mystification of sexual power. They embody the mythological tradition of some primitive cultures (Babylon, Sumer) which venerated prostitutes. The priestesses' cult of the divinities included both rites of abstinence as well as orgies. In "Proceso a la Virgen" the sacred together with the profane archetype are represented by a virgin who refuses to perform miracles and a girl who remains a virgin by sheer miracle.[21]

Similar deconstructions of the Virgin in *marianismo* can also be read in Peri Rossi's "Anunciación," a short story in her *La rebelión de los niños* (1980), and in Armonía Sommers's "El derrumbamiento" in the anthology *Narradores uruguayos* (1969).

Beyond the conflict predicated in the Virgin Mother of *marianismo* and the anguishing regard of the mother as an accomplice of the patriarchy, women face the question of motherhood in conflict with the pursuit of a public life. The clashing verities, the contradictory choices between an all-absorbing, public, professional life as a writer, an activist, or a politician *and* a mother become ever more painful, frustrating, and relentless in the texts and auto-biographies of Rigoberta Menchú, Castellanos, and Pagú. In texts written when a woman sees her public self as the paramount subject in the text, a clear and direct line extends between the girl consumed by a passion for reading, driven to an early decision to become a writer, and her writing subject. This commitment to craft and self as writer pierces through the dilemma of motherhood, and establishes her self in her

texts as a writing persona. In Henriqueta Lisbõa's "My Profession of Faith," as well as in Lispector's reflections on her life as a philosopher and novelist, or in Julieta Campos's and Peri-Rossi's writings, the anguish of motherhood in conflict with a professional vocation seems to have vanished from the surface of the poet's discourse as she thinks of her craft.

Contamination

Writing after World War II entails a conflicted reappropriation of the literary system inherited from the male tradition. This is a more demanding and more discordant task than writing conceived as exploration of the mother tongue's endowment. The writer in the texts of Piñón, Margo Glantz, Poniatowska, and Peri-Rossi posits an ideology of writing that scrutinizes the received (male) tradition critically, suspiciously, and often ironically. The mystical union between Mistral's *parole* posited as the mother's word, would be impossible to imagine in the irritant countertexts of Castellanos's "Poesía no eres tú" or "Malinche," for they are written to rub against the grain, to state a difference. The juncture between classics and folklore Mistral envisioned in her praise of Teresa de la Parra does not appear as the possible space for the search on which Diamela Eltit, Angélica Gorodischer, or Lispector have embarked. A breach that both differentiates and links is perceived in the reappropriating relation between the female writer and her inherited written male tradition.

Yet the (male) literary system is not rejected outright in search of a totally new and pure origin. The very wording of "poetry is not you," "contamination of language," and "apprenticeship of delights" establishes the dynamics of a dialectic in which the yet unknown, but always emerging, continent of woman's being is in the making. And she is not exactly alone. Thus Castellanos writes against the romantic poet Gustavo Adolfo Bécquer. Piñón inserts her voice and, with it, the voice of other women conscious of their place in the legacy of the Portuguese language. Lispector's Lori "uses" Ulysses, the philosophy professor, to discover the myriad facets of her own desire or her search for self-understanding. The terms of the relationship with the "masters" have changed. They are neither revered nor rejected. The masters have become part of an unavoidable past not to be evoked by memory but rather used and questioned in the text. Tradition is plowed under to enrich the soil for the new seed. Contamination, silence, reappropriation, subversion, and experimentation mark the terms of the exploration of the "tone" of one's own that Mistral found to be the sine qua non of any writer.

In Mistral's scheme the danger incurred by the Latin American writer was the blinding cult of European culture; writers like Piñón and Poniatowska, however, feel well beyond this anxiety of fetishization. These

women writers, as they make clear in their own essays, appear conscious of their vulnerable female position in the face of the overwhelming maleness of the patriarchal legacy. Yet they are more immediately concerned by their place as writers within a narrower linguistic, national, or cultural problematic than by the sheer or primordial male/female opposition. And this does not seem to be a characteristic limited or limitable to a given group of writers. The intensity of the writer's concentration on communal social problems, at the expense of a more singularly focused consciousness on the female or feminine, appears to be rather widespread. In her introduction to *Open to the Sun: A Bilingual Anthology of Latin American Women Poets*, Nora Jacquez Wieser observes that "many times, the poets in this anthology are expressing the same themes as their male counterparts. Social protest poetry is very prevalent in Latin America. . . . The poem is a denouncement of cultural imperialism with an economic base; a denouncement frequently heard in Latin America and virtually ignored in the United States."[22]

Rather than inventing or elaborating a place of her own conceived as the place of femininity, Piñón questions the notion of a sweet and disheveled writing as a manifestation of the "feminine." For her, the disheveled, nonsequential writing is yet another symptom of the marginalization of women by the patriarchal system. Feeling free because she is conscious of her condition as a woman, she sees the writing of women as a contaminating force entering the flow of the Portuguese language as it is received. Yet Piñón holds neither preconceived ideas nor utopian notions about what such a rearticulation and recircuiting of elements and relations will produce. Piñón delicately suggests an inversion of values not unlike the one Mistral proposed. Although she believes that a mere "feminine sensibility"—a male-created formation—will fall short of the enormous task of creating new values favorable to the liberation of women, she credits woman's praxis of writing with the divine (formerly male) power to impregnate language for the first time. And this is so not because woman has a language of her own, for woman has thus far been silenced or has been given a "sugary language," but rather because women are finally finding the cracks in the social structures that had so far bound and blindfolded their lives. Silence will give way to expression, and as women act, their works will contaminate and impregnate the world in unforeseen ways. As women find their voice, that is, the capacity to participate in the activity of discourse, they will bring forth something whole and new. Contamination is neither a fact of biological sexual exchange nor a metaphysical gift of the word. It is the result of consciousness and of the experience of women first as oppressed social beings and later as autonomous selves. For Piñón, consciousness of the female's subordinate place, of her derivation from the historically constituted male principle, and of her lack of a world wrought

in her free experience of the given, delineates the possible text of a contemporary woman writer. Nonetheless, she cautions that

> to write a story is more than to contaminate language; it is to expropriate it, taking back what was stolen from us, part of our body, soul and imagination. To be a writer requires a great organization of mind and discipline to command thoughts, to antagonize thoughts, and put them together to an end without dropping a single word through eight hours or two years.[23]

In her two most recent novels, *A República dos Sonhos* (*The Republic of Dreams,* 1989) and *Doce Canção de Caetana* (1987), Piñón subversively engages the principles of the historical novel. With Caetana's and Breta's voice, Piñón dares the limits of reality and the representation of passion.

Humility

Above all, a writer must not be afraid of anything. One of the great fears that writers must master is the sense of inability before the richness and complexity of the craft. Mastery of the craft, domain over the realm of *technos,* opens the door to freedom from the former silence, from past clever tricks of submissiveness. Consciousness without experience is not enough, but consciousness without a thorough command of the technology of writing is also insufficient.

This emphasis on the acquisition of a thinking and a writing craft is particularly acute in Brazilian writers. Lispector despairs at the very words that prevent her from conjuring up the right words that make her thought viable. She seems forever aware of the failure of language to free her characters from solipsistic existences. Likewise, Lisbôa postulates poetics as the search for, and the miracle of, finding the craft, the well-wrought urn for the containment of the unthinkable.

Women who, like Lispector, Fagundes, Lisbôa, and Mistral, navigate in the sea of subjectivity, find humility rather than expropriation or contamination a more appropriate metaphor for their commitment and their experience of being-in-the-world. For them, gender unavoidably marks the struggle for being. And yet, the wonder before the inscribed page that is, and is not, the representation of an absence, the unraveling of a mystery, seems to move these poets beyond the persistent contents of consciousness of women in history. Rather, these women center their meditation on the baffling and uncertain relationship between words (the sign) and experience.

Almost studiously, Lisbôa maintains a calm, even stoic silence on the male/female opposition. In her poetics, the mirage of signification wells above the differential line between male and female identities. In a system in which poetry itself is undefinable, in which signification lies next to and

webbed within nonsense, identities (that is, stable identities) cannot be sustained for any length of time. If a poem is the mere momentary connection between the realm of nonbeing previous to the word and to being, that is to say, art as ritual, then cultural constructs such as the feminine and the masculine can, but do not necessarily, drive poetry's movement toward its object of desire. Poetry for Lisbõa postulated itself as the object of desire. When a poem becomes viable, the signifier "achieves that aura irradiated by being itself."[24]

For Lisbõa, poetry does not name the world outside or inside subjectivity, as it would for Mistral. The power over the word, so confidently assumed by Mistral, appears diminished in Lisbõa's poetics. The subject posited by Lisbõa, not unlike Lispector's fictional creatures, inhabits a restless sea of silence and darkness. Lisbõa holds that poetry resides somewhere between darkness and revelation, "between silence and the word."[25] And this is how Lori's fruitful search for being in "Luminescence" ends with a blank space instead of a final sentence. After illumination there can only be silence again, not the action some would expect.

Exploration of language, a feeling of an irretrievable awe before the movements of the chains of the signifiers, marks the stance of humility both Lisbõa and Lispector adopted. Some call this sense of awe God; others call it being. This advocacy for abstract thought, this exploration of language as an obstacle to thought, keeps both writers away from a meditation on social problems. For Lispector, it comes as a surprise that all people should not be instantly and immediately aware of the prevailing injustices of the world's social order and that words should be called upon to point out the details of social injustice. For the author of *Laços de Família* (1960; *Family Ties,* 1972), the problem does not reside in discovering the reasons or causes for social injustice; although she does not say what they are, she feels that they are plain for everyone to see. The problem is rather her own perceived inability to do something about it. Suffering, exploitation, and injustice do not, in Lispector's mind, require further meditation. They demand action. But she has not been gifted with a representational imagination. She cannot tell a story, link events in a causal, temporal, and sequential series. Her imagination sidesteps all plots of adventure and fortune. Lispector's fiction is honed to plots of thought, and although these, like all plots, rely on change, what changes in Lispector's passive characters is their capacity to perceive and understand their own subjectivity. For Lispector, as for Lori, the protagonist of "Luminescence," "a human being's most urgent necessity was to become a human being."[26] In her writing, Lispector sets out to unravel this tautology. The passage from the need to be human to the satisfaction afforded in finding one's own humanity is the tunnel of pain, because pain and anxiety represent the first instances of consciousness. When Lori sets out on her apprenticeship, she

is so fearful of rejection and failure that she totally denies her pain and thus casts herself into gripping paralysis. Lispector's narrator explains that, "without pain, she had been left without anything, lost in her own world and that of others without any means of contact."[27]

In contrast to the cries of pain and anger sustained by the voice of Castellanos, Lispector and Fagundes serenely accept pain as the first step to knowledge. The two Brazilian women understand knowledge as a Christian existential knowledge capable of leading a human being to the discovery of his or her own "divine limits." In Lispector's *Uma Apprendizagem ou o Livro dos Prazeres* (1969; *An Apprenticeship, or the Book of Delights,* 1986) the final reward for Lori will be the sexual but no less mystical union with Ulysses; in Fagundes's fiction, as she herself states, fear and anxiety are also a way to the beginning of knowledge. Knowledge for Fagundes is, however, the doorway to a mystical moment with the divine. Although clearly existential, Lispector's search has been strictly linked to the modes of the Catholic mystical tradition. The intricate system of metaphors deployed in the text of "Luminescence" constitutes in itself a search for a "reality that is the impossible," for the plot of the narrative entails nothing less than the shared birth of two new souls into a single, passionate, almost indivisible union.[28] Lori and Ulysses attain such a state of divine humaneness that the narrator is prompted to warn the reader of the wordless impossibility of their mystical union: "never had one human being been so close to another."[29] Yet such wordless happiness conceals an unforeseen danger, for as one becomes humble in love, "there was a danger of dying from love."[30] In "Luminescence" the circle closes upon itself as Lori realizes that "she did not have a gift with words and could not explain how she felt or thought, much less thoughts that were almost wordless."[31]

Baffled in the presence of the noumena that invade consciousness before the words can dance with its movement, shape, and rhythm, Lispector's characters continue to live, hoping for the breach when, in Lisbôa's experience, the flight of sense/nonsense can be arrested and "the first word comes, the first verse and then the torrential outpour. The poet grasps them in a sort of chaos. Later they are shaved, moulded, placed where they belong."[32] Arresting the flow on a ground prepared by merciless meditation finally enables Lori to call herself "I." At the end of her search, she has understood what the commonplace "I" given to her by preexisting language might actually mean. This "I" is, however, entirely modified, open to everything that it is not, for Lori's love is in love with "your I": "So we exist, Ulysses. 'We' is something, is something original."[33] Lori's discovery enables her to leave behind the preexisting cultural forms that labeled her, thoughtlessly, a "woman": "She used to be a woman looking for a function and a form. And now . . . it was the . . . freedom of having no functions or forms."[34] Language, encrusted in decaying identities, forms and functions

of meaning, has been stripped bare in Lori's experiment in search of a new, perhaps impossible, freedom, the freedom of a "potential God." Or, as Ulysses tells her, the freedom of a "super-woman" now inhabits her modified self.[35]

To Write Is to Witness

The slippage inscribed in the terms *woman, women, feminine,* and *feminist* is observable not only in the enunciation of texts written by Latin American women but also in the problematic web in which women appear in the world evoked and represented by their writing. Although the subjects posited in these texts are definitely marked as feminine, their authors reject the idea of a feminine writing as a category for understanding women or their texts. From different perspectives, and for diverse social or philosophical and analytical reasons, Campos, Peri-Rossi, Piñón, Poniatowska, Lisbôa, and perhaps even Mistral, Ocampo, and Eltit argue that there is only *writing,* even if historically it has been dominated by men. And this is so even though what we now have as a history of thought and writing is what, on the main, has been written, imagined, and created by men. This, of course, includes "woman" herself.

This refusal to accept the notion of a gendered writing, this dismissal of the idea of women's writing as yet another fad (especially because of its implication of a biological determinism) has often prompted certain "feminist" quarters to label writers who posit the hypothesis or the conviction of a genderless writing as antifeminists. And yet a careful reading of these texts shows that their authors see themselves as women struggling to achieve a greater freedom for their kind, exploring the complexity of their (received) writing craft in order to leave behind their absence, to find a sense of themselves as historically situated and gender-specific human beings.

Campos seriously doubts that there can be "something specifically feminine in the texts written by women." The author of *Tiene los cabellos rojizos y se llama Sabina* (1974) agrees with Peri-Rossi and Piñón in postulating that "the feminine," not only the "eternal feminine" but also the notion of a women's writing, responds to the historical conditioning of patriarchal systems over women. For Campos, "the figure incarnating the art work is androgynous, each text, each canvas . . . is the celebration of that alchemical mystery."[36]

When Fagundes says that she is an "involuntary feminist" and that she writes out of fear and anxiety, her statement creates confusion in the reader familiar with descriptions of feelings of inadequacy and fear of rejection and failure on the part of "women" or feminists exploring their situation as middle-class women in the United States. But when Fagundes adds that, in the process of overcoming her fear, she found that being a Brazilian,

"speaking a minority language and belonging to what is called the Third World," was her salvation, the logic of her remark seems to contradict the stated ambition of "feminist" movements: to enter the mainstream of the dominant male power. Somehow a disjunction is perceived in the series: minority, language, Third World, and salvation. So far in the First World, the Third World has not been credited with salvaging anyone or anything. In some capitalist quarters it is portrayed as pure hell. So how does Fagundes, the merciless feminist analyst in "Sauna," reconcile a fearful woman, a mystical believer, with an "involuntary feminist"? She does so because she departs from the conviction that "writing has no sex," that is to say, that male practices of writing are simply a historical occurrence but not a precondition of writing as a possibility of signification. Fagundes reverses a customary paralytic sense of fear and anxiety as a condition of the feminine and uses it as the power that charges the sources of her creative drive: "I am possessed by the demon of anxiety and it moves me to write. When I write, I achieve my fulfillment. When I write, I possess the key. It is a moment of complete love—because to write is to act out of love. I feel invincible and powerful. I believe in my text."[37] She sees herself as a reluctant feminist because her struggle against fear makes her believe in her self, and that self is female.

The "I" implicit in the one who writes, however, is an "I" who, not unlike Lori's "I" in Lispector's fictional text, is situated in the context of a woman writing as a Brazilian. She is thus witness; she bears testimony to her immediate social world. This last function of the artist as witness to a whole society is, for Fagundes, the essential task of a writer and, as such, it traverses and overpowers her first self-identification as "a woman who fights with words."

Fagundes's appropriation of a seemingly disorienting series of identifications for her person, her authorship, her writing, and her poetics brings up the question of what has been called the *politics of enunciation*. If the question "Who speaks?" is fraught with problems, as *The Poetics of Gender* (1986) and *The Female Author* (1984), to mention only two prominent collections, have shown, the question of representation becomes ever more treacherous. If women as writing subjects are in effect colonized subjects, what and how many more layers of mediation need to be drawn in order to speak of, or as, a Latin American writer? Poniatowska, who claims to understand or care little about literature and much less theory, does not seem preoccupied with inquiry either on the subject or theory of representation. She declares flatly that "the voice of the oppressed in Latin America includes literature by women."[38] For the pathbreaking author of Jesusa's autobiography in *Hasta no verte, Jesús mío* (1969), women's writing in Latin America is one with the voices of the human rights groups of the Comadres in El Salvador and Las Locas de Mayo in Argentina, as well as the

disappeared, the tortured, and the hungry who suffer under the gaze of indifference of those who "have" more than enough. These "haves" include all constituted and legitimated subjects who not only possess economic power but also are and have been empowered by the word.

If the crisis of representation in the West is a phenomenon ascribed chiefly to the general epistemological crisis of the master narratives of religion, philosophy, and history, the representational problematics for women's writing in Latin America are taken by many to be, at least initially and on the surface, more *social* than epistemological in character. It is easy to see the social outline (race, class, nationality, and language) of the distance and incompatibility between Domitila, the Bolivian miner, and the upper-class, Mexican, "feminist" politician of the Institutional Revolutionary party (PRI). The title of Domitila's autobiography, *Let Me Speak!*, cuts deeply into the question of representation. Not only is the matter of mimetic representation vitiated by unbreachable differences of class, economic status, occupation, and race, but it is rendered even more unstable by a perceived disparity between oral and written traditions. Neither the world of Domitila nor the experience of the Guatemalan Rigoberta Menchú nor the life of Jesusa Palancares comes close to a writerly tradition in which books and reading are the supreme enjoyment of childhood. In the texts produced by the testimonies of these three marginalized women, there is no movement or progression from the confessional, intimist diaries to a "modern" literature. Nor is there a continuous line leading from the love poetry that women practiced in the nineteenth century to a politically committed literature, in spite of Poniatowska's insistence of the existence of a link between herself and her writing predecessors. These women affirm solidly the veracity rather than the verisimilitude of their auto-biographies: "I didn't learn it from a book, I never went to school. . . . When I was a little girl, I was very shy. . . . My father taught me how to speak."[39]

If, as Poniatowska claims, Domitila, Jesusa, Rigoberta, and Carolina Maria de Jesus are myth destroyers, it is because their auto-biographies irrupt into the institution of writing with all the differing force of their inherited oral strategies of narration and the rhetorical devices for legitimating the "truth" of the speaking subject. The enunciations of both Rigoberta and Domitila (*I, Rigoberta Menchú* and *Let Me Speak!*) reenact the situation of the subject before the missionary's confessional or before the judiciary. First, the given legal name must be divulged in an act of courage, for taking up the word (the word in Spanish, the language of the authorities) can lead to unprecedented and regrettable consequences. "I," Rigoberta Menchú, in admitting to her existence in a language that she did not receive from her ancestors, is liable to get caught in the intricate web of allusion lurking below the seemingly tranquil surfaces of lexicon, syntax, and grammar. If Rigoberta were to reveal her name under mistaken circum-

stances, she could wind up like her mother: a lacerated body left to rot in the jungle. The limitless vexations and physical suffering endured by Rigoberta, her family, friends, and kin are enough to empower these testimonial texts with a new and overwhelming mimetic force. No matter how disturbing the crisis of representation in the West, the lived experience of these women—survivors of hunger, forced labor, rape, illiteracy, political persecution, unimaginable poverty, exile, incarceration, and torture—encapsulates and deploys an undeniable sense of the real and the immediate. Because these auto-biographies constitute a segment of a life that is not yet over, and because the telling of the self in time is posited as yet another act in the struggle for survival and justice, these testimonials ("I didn't learn it from a book") are invested with a sense of urgency that keeps them outside of the secure and closed vault of literature (the past) and returns them to the cruel and insufferable world from which they originated.

It is this perception of writing as a link between the exploration for meaning and knowledge, and the ultimate act of protest against unmitigated injustice, that Poniatowska claims as the space for women's writing in Latin America. Together with Fagundes, Poniatowska claims that "we write in order to bear testimony. . . . We write like the men who write their name in prison walls. . . . Perhaps this has nothing to do with literature."[40] Partial as she is to direct intervention in the world, Poniatowska considers writing to be a complete act of insurrection.

White Heat, or the Exhaustion of the Sign

There is, however, yet another strategy for addressing this sense of insurrection against an insufferable world. Some younger writers have chosen to delve into the nature of language itself, especially into the ways in which language has been deployed in order to constitute the "feminine," the sugary feminine that Castellanos, Piñón, Campos, and Peri-Rossi see as the male, yet dominant, sense of such a historical construct. If, as Peri-Rossi states, "women" have been absent from the world, that is if women writers as predecessors evoke the sense of an absence; if the feminine as the consciousness of an other that is always present but also always denied is a subversive notion, then the investigation of this "nonbeing" or "nonsense" lying right next to "being" (male) and "sense" (male) requires not only the expropriation or the appropriation of language but also the dismantling of a series of categories that in themselves generate the logic of the subject.

Perhaps taking up Simone de Beauvoir's contention in *The Second Sex* (1952) that woman has never in history been thought of as subject ("I") but rather as object, Eltit attempts in *Lumpérica* to write a "story" in which the "protagonist" is a subject that is yet to be constituted. *Lumpérica* is an attempt to denaturalize and radically question the three main categories

upon which all narrative devolves: character, story line or plot, and mimetic action. Above all, this novelistic text refuses to produce anything close to a character that approximates a mimetic identity. The "actant" in the sentence lacks a set of desires. "She" is not endowed with the mental, physical, and psychological qualities necessary for the fulfillment of anything constituted as the goal of "her" actions. Most strikingly, "she" has no name. Her name, or rather the noun by which "she" is referred to, is "L. Luminated." "She" has no face; the surface of what we call a "face" is so overinscribed with the scars left by the frequent embroidery done on it that nothing is discernible.

"Her" face as text is not a palimpsest. What is under the top layer of inscription cannot be differentiated from what is at the surface. The jumbled nature of the overinscription does not yield to light or to further efforts of inscription. The whole face is a huge scar. Although it cannot be said that "she" hopes to discover an identity as her face and body come under the light of a huge neon sign, the entire text of the novel allows us to believe that the hope for a new baptism is a possible reason why "she" and the rest of the naked lumpen congregate on the blank cement plaza in Santiago de Chile. "She" is much closer to a sign, to a Derridian linguistic sign, than to a "person." Denaturalized and dehumanized, "she" no longer shines as she did when she was contemplated under natural light. L. Luminated no longer reflects light like the women's faces set in Renaissance portraits. Rather, her face, her whole naked body seems to absorb the light, and in doing so, produces an effect of undifferentiation between the subject in the painting and the background shadows. A similar undifferentiation marks the "orgiastic" movement of L. Luminated's body. Because there is no object of desire, not even a self-desire posited outside the mind or the body of L. Luminated, her movements become either gratuitous or the product of the reader's arbitrary decoding of a body's wriggling, twisting, and convulsing. L. Luminated is the opposite of a disembodied spirit, it is a dispirited body. Contrary to the idea of the eternal feminine, to the angel of the house, to woman as pure spirit, or to woman as poetry, L. Luminated appears as *body*. In the act of reading, however, that body functions as a sign that has been exhausted, squeezed dry of signification.

Upon reading *Lumpérica,* one could postulate that the quest of the novel is to elaborate the female body into a subject with new or modified identities. Nevertheless, one of the first descriptions or interpretations of L. Luminated can be said to be equally appropriate for the end of the text: "Nightfall sustains the plaza's decoration so that she may adopt the fugitive poses which conjugate her to exhaustion."[41] If Lumpérica's body is to be the site of an emerging subject, it is so by virtue of its heraclitean postures rather than by offering the discovery of a new entity.

Conclusions

In the foregoing pages, I have briefly sketched seven models of women's writing in Latin America. These models have emerged out of an effort to organize conceptually what the praxis of the texts establish as dominant themes in them. Displayed next to each other, "the mother tongue," "betrayal," "humility," "testimony," and the "sign exhausted" suggest a field of tensions and conflicts yet to be resolved. As evidence of women's participation in the production and reproduction of cultural formations given or introduced into the public spheres of action and discourse, however, these models can be, almost without a doubt, considered to have charted a course for women as discursive subjects and as subjects of a historically new discursivity.

Provisional as they are, these models may be thought of as a beginning in the endeavor to conceptualize and conjugate the seemingly diverse and conflicting images, realities, myths, and self-portraits of women divided by class, race, occupation, and nationality. It is surprising to observe, for instance, that the *pedagogical model* outlined in Sarlo's essay roughly coincides with the positions and public images assumed by Mistral and Ocampo. Although correspondences cannot always be established on a clean mathematical grid, it is possible to point out that Sarlo's outline of three main styles in the ideological discourse of feminists in Latin America obtains for the models described above. The *politics of reason* could describe Lispector's, Lisbôa's, and even Eltit's emphasis on a searing search for the word. The *politics of passion* could encompass Castellanos's or Poniatowska's and Menchú's engagé feminism. And finally, the *politics of action* blends the passion and the critique of discourse itself found in Eltit's, Menchú's, and Domitila's affirmations of experience as given in their own historical specificity.

Each of these models or structures, in which the imaginary and the real cross paths, becomes a galvanizing zone in which the acts of women in the world pass from an absence or a negation into a sense of being more durable and intelligible than the derivative *otherness* about which de Beauvoir wrote some thirty years ago. The discursive practices represented in this anthology offer an entry into a partially charted but no longer ignored domain of human history.

Notes

1. Alice Jardine, *Gynesis, Configurations of Woman and Modernity* (Ithaca: Cornell University Press, 1985), 25.

2. Terry Eagleton's *The Rape of Clarissa: Writing, Sexuality and Class Struggle in Richardson* (Minneapolis: University of Minnesota Press, 1982), Jane Gallop's *The Daughter's Seduction: Feminism and Psychoanalysis* (Ithaca: Cornell University Press, 1982), and Tania

Modelski's *Loving with a Vengeance: Mass-Produced Fantasies for Women* (New York: Methuen, 1982) are but three examples of a growing body of book-length studies that revamp our received knowledge in light of the "feminine."

3. See Jean Franco, *Plotting Women: Gender and Representation in Mexico* (New York: Columbia University Press, 1989); Asunción Lavrin and Edith Couturier, "Dowries and Wills: A View of Women's Socio-economic Role in Colonial Guadalajara and Puebla, 1640–1790," *Hispanic American Historial Review* 59 (May 1979): 280–304; Fernando Benítez, *Los demonios en el convento: sexo y religión en la Nueva España* (Mexico: Era, 1985); Sylvia M. Arrom, *Women of Mexico City, 1790–1857* (Stanford: Stanford University Press, 1985); Josefina Muriel, *Cultura femenina novohispana* (Mexico: UNAM, 1982); Josefina Muriel, *Los recogimientos de mujeres, respuesta a una problemática social novohispana* (Mexico: UNAM, 1974); Josefina Muriel, *Familia y sexualidad en Nueva España* (Mexico: Sep-Ochentas, 1982); Asunción Lavrin, ed., *Sexuality and Marriage in Colonial Latin America* (Lincoln: University of Nebraska Press, 1989). The works of Josefina Muriel, Fernando Benítez, Electa Arenal, Jean Franco, and many others have opened new paths of inquiry into the position of women and their relationship to their own consciousness as colonial subjects.

4. See, for instance, in English, Luisa Valenzuela, "Change of Guard," in *The Web*, ed. H. Ernest Lewald (Washington, D.C.: Three Continents, 1985).

5. Poem without a title by Cristina Peri-Rossi, trans. Naomi Lindstrom, in *The Renewal of the Vision*, eds. Marjorie Agosin and Cola Franzen (Boston, 1988), 84.

6. Alejandra Pizarnik, "Signs," trans. Suzanne Jill Levine, in *Renewal of the Vision*, 85.

7. See Julieta Campos, *Oficio de leer* (Mexico: Joaquín Mortiz, 1971); *Imagen en el espejo* (Mexico: Joaquín Mortiz, 1965); and *Función de la novela* (Mexico: Joaquín Mortiz, 1973).

8. *Gabriela piensa en . . .*, ed. Roque Esteban Scarpa (Santiago de Chile: Editorial Andrés Bello, 1978), is a collection of a number of Mistral's reviews and critical essays on the work of her contemporaries. Her reading of Spanish American, Spanish, and European writers expresses clear and strong preferences. More important, however, this collection provides the basis for Mistral's poetics, which values what she calls "authentic voices" and "simple," natural language.

9. See Virgilio Figueroa, *La Divina Gabriela* (Santiago de Chile, 1933); and Benjamín Carrión, *Santa Gabriela Mistral* (Quito: Casa de la Cultura, 1956).

10. Esteban Scarpa, *Gabriela piensa en*, 164.

11. Ibid., 53.

12. Ibid., 43.

13. Ibid.

14. Helena Araújo, "El modelo mariano, tema y variaciones," *Eco* (Bogotá) 248 (June 1982): 120.

15. See Julia Kristeva, *Hérétique de l'amour* (Paris: Tel Quel, 1978).

16. Araújo, "El modelo mariano," 120.

17. Nelly Campobello, *Cartucho* and *My Mother's Hands*, trans. Doris Meyer and Irene Mathews (Austin: University of Texas Press, 1988), 96.

18. Esteban Scarpa, *Gabriela piensa en*, 23.

19. See Evelyn P. Stevens, "Marianismo, the Other Face of Machismo in Latin America," in *Female and Male in Latin America*, ed. Ann Pescatello (Pittsburgh: University of Pittsburgh Press, 1973).

20. See Nora Jacquez Wieser, ed., *Open to the Sun: A Bilingual Anthology of Latin American Women Poets* (Van Nuys, Calif.: Perivale Press, 1978).

21. See Helena Araújo, "El modelo mariano, tema y variaciones," *Eco* 248 (June 1982): 118–126.

22. See the introduction to Wieser, *Open to the Sun*.

23. Nélida Piñón, "The Contamination of Language," interview with Catherine Tinker in *The 13th Moon* (1982), 73.

24. Henriqueta Lisbõa, *Vivência Poética, Ensalos* (Belo Horizonte: Impresa Oficial, 1979), 3.

25. Ibid., 4.

26. Clarice Lispector, *An Apprenticeship, or the Book of Delights,* trans. Richard A. Mazzore and Lorri A. Pavris (Austin: University of Texas Press, 1986).

27. Lispector, *An Apprenticeship,* 22.

28. Ibid., 75.

29. Ibid., 108.

30. Ibid., 110.

31. Ibid.

32. Lisbõa, *Vivência Poética,* 5.

33. Lispector, *An Apprenticeship,* 111.

34. Ibid., 113.

35. The Nietzschean superwoman who emerges in the body and soul of the socially insignificant and marginalized Brazilian schoolteacher has very little in common with the generic image of the Superwoman produced in the texts of North American newsprint and electronic media.

36. Julieta Campos, "Mi vocación literaria," *Revista Iberoamericana* 51 (1985): 3.

37. Lygia Fagundes Telles, "Fear," *Revista de la Universidad de México* (July 1981): 31.

38. Elena Poniatowska, "La literatura y la mujer en Latinoamerica," *Eco* (Bogotá) (1983): 11.

39. See Elisabeth Burgos-Debray, ed., *I, Rigoberta Menchú,* trans. Ann Wright (London: Verso, 1984). For the "moral economy" of another peasant rebellion in Guatemala, see Robert Wasserstrom, "Indian Uprisings Under Spanish Colonialism: Southern Mexico in 1772," *Power and Protest in the Countryside, Studies in Rural Unrest in Asia, Europe and Latin America,* eds. Robert P. Weller and Scott E. Guggenheim (Durham: Duke University Press, 1982). For a wider sample of testimonial autobiographic accounts of peasants and other "common folk" in Spanish America, see William H. Beezley and Judith E. Well, eds., *The Human Tradition in Latin America* (Wilmington: Scholarly Resources, 1978).

40. Ibid., 15.

41. Diamela Eltit, *Lumpérica* (Santiago de Chile: Ornitorrinco, 1983), 1.

Gabriela Mistral
(Lucila Godoy Alcayaga)

LIKE HER FELLOW CHILEAN Pablo Neruda, Lucila Godoy Alcayaga (1889–1957) is better known by her pseudonym, Gabriela Mistral. Growing up in a poor household headed by her mother, Mistral was self-educated. She published her first poems in local newspapers and magazines. At the age of fifteen she graduated from the regional teachers' college and took her first job as an elementary school teacher. Mistral's career is continually marked by the image of the teacher, even after she won the Nobel Prize in 1945. In 1914, she won her first major national literary prize for her *Sonetos de la muerte* (Sonnets of death).

With the publication of *Desolación* (incorporating her prize-winning sonnets) in 1922 in New York, Mistral became the most widely read and recited of all Spanish American poets. Mistral's poems were praised and widely acclaimed for their themes of maternal love and God, as well as for the absence of sensuality. The invitation to participate in educational policy debates by the Mexican intellectual and minister of education José Vasconcelos further confirmed Mistral's image as an educator of children and women. In later years, she traveled widely in North America and Europe, often as diplomatic consul representing her country. In recent years large collections of her letters and newspaper articles have been published. *Cartas de amor de Gabriela Mistral* (1978) and other publications offer enough evidence to broaden the image of Mistral as the perennial teacher of children and mothers fostered by *Lecturas de mujeres* (1923) to complete the portrait of a woman far more complex and unconventional than many critics have been willing to see.

There is abundant secondary literature on Mistral. Emir Rodríquez Monegal is hesitant about her accomplishments as a poet, writing that in her best collection, *Tala* (1938), "she achieved a sort of stark and uncompromising beauty that came very close to justifying the Nobel Prize she received at a time when Reyes, Neruda and Borges were still very active."[1]

---

1. Emir Rodríguez Monegal, *The Borzoi Anthology of Latin American Literature*, vol. 2 (New York: Alfred A. Knopf, 1977), 485.

Some of the existing studies demonstrate the poet's mastery of her craft and the innovations of her untidy but powerful metaphorical world. Anderson Imbert asserts that Mistral's metaphors are pure symbolism, antiintellectual, and therefore resist the sluggishness of logical phrasing.[2] Mistral's rejection not only of modernism but of urban artifice in general, together with her incorporation of "provincial" and local oral traditions into her poetry, caught many a critic unprepared to appreciate the freshness and originality of her linguistic preferences. Her writing shares with Santa Teresa and Sor Juana Inés de la Cruz a certain ironic stance in relation to the self-ascribed importance of the self in the conventional (male) poetic tradition. Mistral's language is a return to well-scrubbed and healthy basics, to the things of this world.

The Absent Mother

Mother: my eyes, my mouth, my hands, all took silent form within your womb. You irrigated me with your richest blood as the rain wets the hyacinth bulbs when they still lie hidden under the loam. My senses are all yours, and with this loan of your flesh I walk upon the world.

There is no sweeter rhythm, amongst those poured by the First Musician upon the earth, than that of your rocking mother, and of the tranquil feelings that took shape in my soul as I lay on your swaying knees and in your arms.

As you rocked me, you sang, and those verses were just playful words, excuses for your caresses.

In your songs, you would name the things of the earth: its hills, its fruits, its towns, its tame fauna, as if to house your daughter in the world, as if to number for her the beings of that strange family in which she had been put out to live.

And thus I began to get to know your harsh and at the same time smooth universe: There was no little word that served to name our creatures that I didn't learn from you. Later, teachers only repeated the beautiful names that I had already learned from you.

I never loved dead toys very much, you remember: The one I preferred, the one I thought was the prettiest, was your body. I played with your hair as if it were made of rivulets of water that slid through my fingers; with

---

2. Enrique Anderson Imbert, *Historia de la literatura hispanoamericana*, vol. 2 (Mexico: Fondo de la Cultura Económica, 1954), 53.

"The Absent Mother," from Gabriela piensa en *(Santiago de Chile: Andrés Bello, 1978).* Translated by Rosario Ferré.

your round chin, with your fingers, which I loved to twine and untwine. Your face, bent over me, was the whole spectacle of the world. I looked intently at your moving eyelids and at the light that broke in your clear green eyes, and I was always in awe at the strange expression on your face when you were unhappy.

Yes, my whole world was your face, your cheeks, like honey-colored hills, and the furrows that sadness dug out toward the corners of your mouth, two small, tender valleys. I learned about the shape of things looking at your head: The trembling of the newly sprung grass was already in your lashes, and the bending of the plant stems was in your neck, which, as it bent toward me, gathered its folds of intimacy about me.

And in this way, she who spread the earth's mantle for the first time before our eyes, in all its varied color and form, also unveils the God that lies hidden before us.

I was a sad child, mother, shy and intractable like the night cricket in daytime or the green lizard that drinks the sunlight. And you suffered because your daughter didn't play with the others, and you would say I had a fever when you found me in the vineyards talking to the gnarled stalks and to the slender, delicate almond trees that looked like daydreaming children.

All the teachers that came after you, mother, have taught me only what you taught me in the first place, and they needed many words to say what you could say in just a few. They tired our ears and took away our pleasure in hearing a good tale. With you, one would learn things lightly and with ease, the way a little girl leans on your breast. You would wrap your lessons in that special golden wax of affection. You never talked because it was your duty to do so, and thus you were never in a hurry, except when you needed to pour yourself out toward your daughter. And you never made her sit quiet and straight on a hard bench, listening to your lessons. As I listened to you, mother, I would play with the fringes of your blouse or with the mother-of-pearl button on your sleeve. And this was the only time that learning has been a pleasurable experience for me, mother.

After this, I became an adolescent, and later a grown woman. I have walked alone, without your body to lean on, and I know that what they call freedom is something without beauty. I have seen my shadow fall, ugly and sad over the fields, without your small shadow to keep it company. I have also spoken without your help. I wish that, as it used to be in the past, your helping words could be present in each one of my phrases, so that what we would finally say would be a garland woven by both.

Rachel de Queiroz

RACHEL DE QUEIROZ was born in 1910 in the northeastern state of Ceará. In the early 1920s she joined the group of Brazilian artists and intellectuals whose main concern was to forge an alliance between art and social criticism. Gilberto Freire's study of the Brazilian family in *Casa Grande le Senzala* (1933), along with the early novels of Jorge Amado, Graciliano Ramos, and Queiroz's first novel, *O Quinze* (1930), constitute some of the hallmarks of this movement. Like other Brazilian poems and narratives, *O Quinze* belongs to the literature written about the devastations brought on nature and society by the periodic droughts affecting the Brazilian northeast. Queiroz tells the disillusioning love story of the schoolteacher Conceiçao and the survival story of Chico Bento's family. Queiroz's narrator, making use of narrative strategies borrowed from the local oral style of storytelling, introduces a new sense of mimesis of the authentic and the "real" life of the *nordestino*. Her second novel, *João Miguel* (1932), follows the framework of *O Quinze,* though the story is about a man in jail. The broad social setting of the novels offers a rich view of the *sertanejo,* or backlands life.

During the 1930s Queiroz was quite active in politics. From 1930 to 1933 she was a member of the Communist party; though she quit the party in 1937, she spent some months in jail when Getulio Vargas struck against more leftist intellectuals. After the publication of *Camino de Pedras* (1937) and her autobiographical novel *As Três Marias* (1939), Queiroz began to spend a good deal of her time as a journalist. For years she has divided her life between journalism and caring for her family ranch at Queixadá, a place often depicted in her novels. In 1977 she authored countless popular *Crônicas* published in *O Cruzeiro* and became the first woman to be admitted to the Brazilian Academy of Letters. She has continued to write on a regular basis. Her most recent novel *Dôra, Doralina* appeared in 1975 (and was translated into English in 1984). Many critics believe that this novel exhibits an unusual maturity of thought and style and that it marks Queiroz's highest achievement. Once again, the narrator is a woman from the northeast, María das Dores. With *Dôra, Doralina* Queiroz seems to be making an effort to encompass all of Brazil, her female characters investing their existence with a voice particular to their struggle for survival.

from The Three Marias

Papa, however, is healthy and strong, and has only a touch of stomach trouble, and certain bad headaches that affect him after heavy meals. He is fat, sanguine, and serene. Nothing justifies this tender concern that I have for him, and that I suppose was developed in me by the thought of Mama's death. How can he help feeling cheated and cut off from the best and most beautiful and purest side of his life! Having learned to love his first wife with such a great and blinding passion, what all has he not had to suppress in himself in order to be able to adjust to his new measured, regular, immutable life! He, who learned to love and enjoy the subtle charm of life with Mama, who never pointed out her inconsistency, lack of order, childishness. (How he adored her, how he spoiled her! At times Papa would be lying down reading. She would ask to sit down beside him, just on the edge of the hammock, and she would gradually stretch out until she was lying down. She would put her head on his shoulder; Papa would close his book and lie gently contemplating her fine, merry face, so close to his. Then I would seize the opportunity and also jump into the hammock and what fun, what laughter, what hugging among the three of us! Papa always ended up by making both of us leave, looking for his book underneath him, finding it always crumpled, with the pages torn loose.)

Now all this is past history and forgotten. Papa is stern, is a different man, works very hard, looks fat, fat like the whole family. Where are his books? Now he reads only the newspaper.

Where are the poems that you used to teach me at night, out on the porch, with me lying beside you in the rope hammock, we two watching the great red moon as it came up, we two repeating the verses—the verses of the shipwreck—don't you remember, Papa?—about the ship's propeller "which beat like an enormous heart"? The frogs were croaking in the distance, the scent of water lilies came to us on the fresh night breeze, you stroked my hair, and my tiny heart beat, beat with such emotion, Papa, alongside yours, and I was so happy, so sad, the night was so vast and tender, the verses moved me so, although I didn't understand them fully, that I often kept still and let you go on saying the words, because emotion had closed off my breathing and that enormous, slow-turning ship propeller in the story, I could feel beat—inside me.

. . . . .

It was evident that Isaac was in love with me, but he had never spoken to me of love. He made no plans, sought no promises, took no mortgages on the future. In moments of most intimate tenderness, or in others of no importance, his words were always appropriate only to the sentiment at hand, the sensation of the moment.

And I, who did my dreaming and planning all by myself, never dared ask for a thing, imitating his neglect; I saw the day of leaving draw close and still kept silent, afraid to break the charm, afraid of disappointing him, of making him think me capable of some sort of design or scheme with my heart. I took pride precisely in the gesture of giving myself without asking in return, or at least without showing that I expected anything.

There was nothing left of myself that I might not have given him, either of heart or body. And he had accepted all, gravely moved, but without any great display of gratitude, without meekness or remorse, without any thought of owing me anything.

It was as if I had been his wife for a great while, and my surrender, which meanwhile cost me in tears, in terrible, secret repentance, had for him no other meaning beyond its own immediate content of pleasure and tenderness.

When he took me, he asked for nothing, accompanying his own desire gradually, inducing me to share it, smiling at my fears and hesitations, obstinately, gently, inflexibly.

Rather than of physical pain, that first surrender left me with a sensation of fear and of secret humiliation: the pleasure that he got from me was so much his alone, so separate from me, diminished me so much, that I didn't feel any of the mysterious pleasure whose approach made him gasp as if suffering, and which afterward left him somnolent and quiet, sprawled on the sand, in a kind of happy unconsciousness, with his face nestled against my breast.

I was lucid, lucid and deeply hurt, and extraordinarily sad and apprehensive. I wished for him to console me, to embrace me, to make up to me for everything. But Isaac, in his torpor, left me to myself, and it looked as though my function had ended there—at least until his desire revived.

from The Girls and Other Chronicles

The girls are between eighteen and twenty years old. They wear Lee shoes and smoke desperately. They swear. At the university they are science

---

Excerpt from As Menininhas e Outras Crônicas *(Rio de Janeiro: José Olympio, 1976).*
Translated by Sara Castro-Klarén: copyright © 1991 by Sara Castro-Klarén.

majors. At home no one can control them, even if they try, or so the girls boast. And with the freedom to come and go as they please—day and night—with the liquidation of the taboo of virginity, with the easy acquisition of the pill, the girls, beyond the liberation of domestic authority, feel sexually liberated.

The irony is that those free souls, college students and all, do not think about economic freedom. Even the girls who already have their degrees seem lost in a variety of vague postgraduate courses in art and foreign languages, which only extend the irresponsibility of college days.

The girls act with great aplomb. They are extroverted conversationalists; some claim that they even smoke marijuana. I don't know. Alas, the girls who took that sinister path are no longer our girls, they are part of another story. Here we speak of the girls who are still with us at all times, colleagues and possible girlfriends of our boys.

It turns out that their pretended petulance is but a thin mask used to cover serious problems. Because, in fact, these girls are extremely vulnerable and insecure. And their boyfriends are even more immature. They are equally unprepared for life. Trapped as they are within the patterns of an old machismo, their incompatibility with the liberated girls is overwhelming.

They are all for the new permissiveness, but because of a different personal perspective, the girls' notion of the meaning of such permissiveness is not the same as the boys'. For the boys, permissiveness is permissiveness—I like you, you like me, catch as catch can, no chains. But the girls still have in their blood, in their bones and wombs, the conviction that such beginnings are the preview of a long-lasting, reciprocal relation. Almost none fully accepts the idea of simple fornication for play, without future commitment. Or at least, once the discovery phase is past, almost all the girls fall into the old rut of sex-child-marriage, which is the true path because the species demands it.

Only under false pretenses do the girls play the game of sex without commitment. After each experience, they feel enormous frustration and resentment. They are even frustrated with the very idea of sexual pleasure, which had always been presented to them as a sublime and gratifying experience. It is not surprising to see them disappointed and even afraid of the consequences of free sex, for they have come to know it in improvised encounters with inexperienced partners in dirty and uncomfortable places.

Some of the girls who talk to me seem always anxious, vulnerable, bitter. Their basic feeling is a constant insecurity—with themselves, with the world, and with their boyfriends. Once the irresistible moment has passed, the pitiful boys feel equally dissatisfied and afraid. These boys know that because of biological reasons the girls are much more mature. And they resent that. It is then that they start the premature and impractical marriage

promises, or they fall into cynicism and flight. The girls then look for an understanding person, a good strong shoulder on which to cry. Or, fed up with the boys, they look for better-experienced and more established older men.

Sometimes it works out: The man falls in love and turns the love affair into marriage. More often than not, these girls go from man to man, they grow disappointed, they feel empty, worn down. If they can pay, they wind up on the analyst's sofa. It is grotesque and moving to listen to their bragging about the years they have been under analysis. The girls with artistic or professional interest save themselves from the solitude and despair of the others. Others simply become members of the "oldest profession." Some become drunkards; others commit suicide. This must be said even at the risk of sentimentality: No matter how daring, the girls are in fact pathetically fragile, because of their female biology and the world that awaits them. Their boyfriends cannot afford to give them the security they need. The poor things, so insecure themselves!

Yes, life is difficult, it is indeed a risky proposition. In the old days, with all the old guarantees, the girls drowned. What can we expect now that they jump into a deep and furious ocean with neither ship nor life preserver, with only their bare arms and beautiful bodies?

Patricia Galvão (Mara Lobo)

BETTER KNOWN by her nom de guerre, Pagú, Patricia Galvão was pronounced by her "modernist" colleagues in São Paulo to be the best embodiment of their "anthropophagist" credo. As a highly visible and often scandalous practitioner of the iconoclasm of the Brazilian avant-garde of the 1920s and 1930s, Pagú constantly and directly consumed and transgressed social and artistic taboos. It has been said that Pagú as anthropophagist "abolished the grammar of life. . . . She was capable of devouring several venomous bishops."[1]

Though from a middle-class family, she advocated the total restructuring of Brazilian society. Pagú was a radical and uncompromising Marxist who believed in the triumph of the proletariat. At a time when the Brazilian suffragist doctors, lawyers, and professors adopted conventional strategies in their quest for the ballot box, Pagú denounced them as mere accomplices of the established powers. She espoused a radical feminism; women needed to conquer much more than just the vote. They needed to possess the economic means of production and *then* transform the entire society into a brand-new and truly just social order.

Celebrated for her transgressions in dress, makeup, language, and ideology, she enjoyed the support and close friendship of the group headed by the writer Oswald de Andrade and his wife, the painter Tarsila de Amaral. Pagú's own drawings began to appear in the modernists' magazine *Revista de Antrofagia* in 1929. A few years later, under the pseudonym Mara Lobo, she published her proletariat novel *Parque Industrial* (1933). This time, not only did her behavior shock the bourgeois values of Brazilian society, but the critics were unprepared for her novel's brazen feminism, vulgar language, and sexual explicitness. Above all, her novel was a merciless rejection of the hypocrisy of well-meaning, "educated" groups.

Galvão's irreverence and aggressive satire finally earned her the status of nonperson when she betrayed Tarsila de Amaral by having an affair with her husband. She continued to insist on her intransigent feminist views at the bosom of the Brazilian Communist party, from which she eventually resigned: "Profoundly disillusioned and embittered . . . her experience had

---

1. Augusto de Campos, *Pagú, Patricia Galvão: Vida e Obra* (São Paulo, 1982), 321–323.

taught her to distrust the dogmatic left wing as much as she despised the reactionary right wing."[2]

Though she never recovered from her existential and political crisis of 1940, when she left the party and parted with her modernist friends, she did write a second novel, *A Famosa Revista* (1945). In this novel there is no longer any faith in the party as the possible savior of the Corinas of this world. Instead, she denounces the party's monolithic control of its members and the ensuing intellectual, moral, and political corruption of its utopian vision.

from Industrial Park

Looms

In the streets of Braz, a long line of illegitimate children. Illegitimate to distinguish them from the others who have had abundant inheritances and comfort all their lives. The bourgeoisie always produce legitimate offspring. Even when their virtuous wives are actually ordinary adulteresses.

•   •   •

The girls spin romantic tales at night while they pack their bag lunches in grey and green paper.

"I will only marry a laborer."

"Fat chance! I've had enough of being poor. Spend my whole life in this shit?!"

"Do you think that rich people court seriously? To them it is only a game."

"I already told Bralio that if he is just playing, I'll kill him."

"Peter's here!"

"Is he waiting for you? Then let me leave."

•   •   •

Seamstresses

Corina is the last to return to the garret. A long belt burns, red in the same old dress.

---

2. For further details, see K. Susan Besse, "Pagú: Patricia Galvão—Rebel," in William H. Beezley and Judith Ewell, eds., *The Human Tradition in Latin America* (Wilmington, Del.: Scholarly Resources, 1984), 112–113.

Excerpts from Parque Industrial *(São Paulo: Editorial Alternativa, 1933). Translated by Laura Beard Milroy: copyright © 1991 by Sara Castro-Klarén.*

A mouth swollen with kisses. Her happiness flushes her already tanned face. With runs in her hose, she lifts her legs taking long strides.

Below her flushed face, she flaunts a new and futuristic scarf on her neck.

·   ·   ·

"What happened to the money, Corina?"

"I spent it."

"Florino's going to hit you."

Corina thinks about her mother's repulsive boyfriend.

"I'm tired of working for some prick who isn't my father. I need to eat and then get back to the office."

"I don't have lunch."

"And the money from the record?"

Florino, fat and drunk, shows up at the door. His stomach jiggles. He shakes his wooden cane. He wants to hit some street urchins that follow him. He misses. The infernal little boys would like him to fall.

"Drunkard! Lush!"

·   ·   ·

Public School

"Fedorzinho! Don't wear yourself out."

"Cut it out. You're José Mojica[1] himself. Very dashing."

"The other day I saw him in Santana with Dirce."

"Oh! Did you know that her father caught her in a sleazy hotel on Aurora Street? With a married man. . . . "

"Who doesn't? That's why she hasn't come. They say that he's going to put her in Bom Pastor School."

"That's why the teachers of college students have a bad name. They corrupt people."

"Get out of here! She was examined and she's a virgin. She didn't do any more than you did in Recreio Santana or I did in Santo Amaro."

"But I never went into a room. . . . "

"Look at Edith's boobs! She goes around just to show off her breasts in the sketching class."

The moustached young men wait on the corners. The boss doesn't want to ruin the name of the school with the daily scandal of lovers. No man can stop near the doors. But the blue skirts crowd the corners.

·   ·   ·

---

1. José Mojica: a Mexican priest who was also a singing movie star.

They all come in. One arrives late. She gets off the trolley and runs. Her breasts are thrusting out. Pretty as a flower. Very blond silky hair.
"So, Eleonora? Are you getting married?"
"The day I graduate."

.   .   .

In the pastry shop frequented by the schoolgirls, Alfredo Rocha, a rich boy, kisses the hands of his girlfriend. He makes a great impression on Eleonora's sweet-toothed friends. He pays and says good-bye. The car pulls away in a hurry. The wide-eyed girls watch it go.

.   .   .

She never meant to give in completely. She would give everything but her virginity. She would be a virgin when she married. She wouldn't be a fool like the others.

.   .   .

Eleonora's father earns 600,000 reis in Repartição. He moonlights on the side. Her mother was trained in the kitchen of a feudal house, where she was brought up morally, with the precepts of honor and recipes. They dreamt for their daughter a little home just like theirs. Where the wife is a saint and the husband rehashes middle-aged passions.
"Until Nora marries, I won't rest."
A black woman helps with the housework.

.   .   .

"Who is it?"
"Count Green. . . . King of the transformation industries."
The bourgeoisie invent mediocre novels. Deceptions slide out from under the pillows. Malicious words bubble out of their expensive champagne. Caviar explodes behind their clenched teeth.

.   .   .

The Opium of Color

Squinting, Corina mends.
Why did she have to be born a mulatto? She's so pretty. And when she puts on makeup, wow! Color is a damn thing. Why is she so different from the others? The baby boy was his son also. And if he had come out like him, with his pale rose skin. Why do blacks have children? Jesus! If only Florino knew that she was pregnant. She wants to tell her mother. She

adores the unborn child. How big must it be now? Does it already have eyes? And little hands?

The vomit chokes the laughter. She thinks about the next encounter. As always, the cheap brunette from Amhangabahu. The perfumed apartment. His love of impeccable slacks that she herself takes off with slavish attention. She likes to make a nest with her legs and enjoy the slight noise of his silk underpants.

She still has the twenty thousand reis. And she plans to hold on to them. She's going to buy herself a pair of stockings. The ones she has have holes. They can't be worn. Florino could find the money if he searched for it. . . . She hides the money in a water jug.

She checks out the size of her stomach in the mirror.

"It's big! Who wouldn't be exhausted, my God!"

Her mother surprises her. She reacted badly. Later she seemed to repent. The old woman sobs.

The whole town of Simione knows.

"I always said that shameless hussy would end up in a whorehouse."

Manuel's thirty-two teeth peek out from under his moustache.

"Man! If it weren't for that stomach!"

Others envy the romance with a rich man.

"What of it! Corina's child will have a car and a servant. He'll die in a mansion. Do you think she'll take her mother with her? And Florino?"

Florino shows up staggering. Like always. Fighting with the boys on the street.

"Did your daughter eat dirt? Look at her stomach swelling up! The wind is going to blow her away!"

Florino doesn't understand. But he huffs his way into the house, tapping with his cane.

"Pamarona! Whose stomach is this?" A shout.

"Get away from me, drunkard!"

Corina pushes him and cries, surrounded by some women who talk with her. But the maids shout, unshakeable in their bourgeoise morality.

"Whore!"

"Look at her stomach!"

She spends the night walking. They mess around with her. She doesn't know where he lives. He's not at the bar. Arnaldo. He never gave her any other name. She knew the number of his car.

The morning takes her to the office.

Madame Joaninha shows up in the afternoon.

The girls whisper and giggle.

"Otavia, did you see? Corina's pregnant! I swear she is!"

One of them goes to gossip with Madame. A seamstress calls the mulatto girl. They all get excited. It's a party for the girls. No one feels sorry for their colleague's misfortune. Their sewing is forgotten.

"Have an abortion? Kill my child?"

A violent shake of the head. Her nostrils flare.

"You shameless bitch! Leave immediately. In my shop, the girls! I cannot allow them to associate with a bum like you."

"Where must I go?"

"And your stud?"

She smiles between tears. Much later, at night she will meet her lover. She's almost dead from hunger. At least he will have brought money. She has forgotten the water jug. She didn't have time for anything.

Otavia drops her sewing.

"Corina, wait for me at the door."

She's the only one who still speaks to her. The one who had been the least friendly with her before. She always had stayed away. The fool!

When they meet, Otavia says:

"Come home with me. Stay there until you figure out where to have the child."

"Can I see Arnaldo when I want to?"

"Corina, don't you see what Arnaldo is? He is nothing but a horrible bourgeois. Eventually he will tire of you. They are always like that. . . ."

"But we are sweethearts. . . ."

"He'll never marry you. He'll never be brave enough to marry outside his class. What he does is seduce little girls like you who don't realize what a huge gap separates us from him."

Otavia, caught up in her sermon, continues speaking. Corina hears, but she doesn't believe it and she gets angry. She is the only one to offer her a refuge. They arrive together at the little house of João Boemer Street. Rosinha Lituona is at the door, in a huge red apron.

"Good news Otavia. I found a place for you at the Italo. You can leave the office. And earn five thousand reis a month more!" She squeezes Corina's hand.

"She's coming to live with me," says Otavia. "Come by tonight. We'll go to the party together."

.   .   .

The Viaduto do Cha trembles under the occasional trolley. Corina wants to die. Die like her son. The agonizing trembling of the school friend who killed herself last year flashes through her mind. The girl splattered on the parapets of Formosa Street, after her falling flight. The blood of the girl, the split head, the broken bones.

.   .   .

The next day, an elegant man takes her to a brothel in Braz.

.   .   .

In twenty-five identical houses, inside twenty-five identical doors, live twenty-five identical wretched women.

She remembers that with the other seamstresses, she joked around about the women of Ipiranga Street. She feels repugnance, but she remembers. She does it between bouts of crying, just like the others.

"Psst! Sweetie! Come here! I'll give you a rosebud. . . . "

She broadens her erotic vocabulary bit by bit.

.   .   .

Lady of the Night

In the red-light district, there is a restless movement. A lot of people. Broken men pass by, barefoot or in worn-out shoes. Dirty black men. Adolescents.

"I prefer a humpbacked woman because no one else wants her. At least she is clean!"

.   .   .

Corina sells herself in another room. The hands of a huge black man encircle her body deformed by her advanced pregnancy.

"Your pregnancy gets to me!"

A rough voice laughs boisterously in the room.

"You can come without money. I'll even pay you!"

A man with crooked eyes gives life to an old phonolite. Sagging breasts hang in a greasy bra. Corina opens the door, tired. But one more and she will have the money for the crib for her little son.

Soldiers argue in the street. In the bar they drink and gamble.

Cruz Branca Street continues to release the flow of the sap of Braz.

"A workman can't even get laid!"

An unemployed onanist jerks off on the street. A mulatto girl sucks on flavored sugar cubes. She sees the young man rubbing himself against the wall. She gives the other girls a hysterical imitation, with her hands between her legs. The tragic gesture of the man. She lets out peals of laughter.

"If I could get out of this life!"

"Come off it! The rich are worse off than we are! We do not hide. And we do it out of need."

"If I had a job, I wouldn't be here, sick of this whole business!"

"The pain of the poor is money!"

.   .   .

Birthing Houses

The ambulance wails around a curve of Frei Caneca Street. It stops in front of the rusty gate of Maternity. On a very white stretcher, a very brown arm stands out against the whiteness of the sheet. One more for the indigent ward. In the huge room there is a row of identical beds. Lots of breasts on display. Of every color. Full ones, suckled ones. A collection of little, round, shaved heads.

"I left my son here! Bring him to me!"

She doesn't understand that differences are made even in the birthing houses. The babies from the classes who can pay stay close to their mothers. The indigent mothers prepare their children for the future separation that will come when the women have to work. Bourgeois babies are prepared from the beginning, tied by an economic umbilical cord.

In the indigent ward, white nurses smilingly caress, while doing the hardest work, the women giving birth, who are occupying now the poor beds that the nurses will occupy later.

"Bed 10. A delivery."

A very tall nurse organizes pillows, receives a new patient.

"Your name?"

"Corina."

"Husband's name?"

"Don't have one."

"Gracious! Almost none of the poor women have married names. Ever had a child?"

"No. I'm so tired. . . ."

"That will pass. You're going to have a beautiful child."

"Without a father!"

"But with a mother who can love enough for two."

"Arnaldo."

"That will be his name?"

"It will."

Corina suffers horribly.

If only her mother had been there. She enjoys affectionate attention so much. She hasn't anyone to keep her spirits up. She calls the nurse.

"Don't leave me! Stay close to me. Smooth my head. Oh, that's nice!"

She shouts without realizing it. She uncovers herself.

At the base of her legs, a huge opening. The enormous hole grows bigger. She splits. Black. More pain. In order to vomit, suddenly, a living, bloody thing.

The nurse backs up. The midwife backs away. The doctor remains. A lifting of his eyebrows shows his surprise. He examines the bloody mess that shouts, messing the linens. Two skinny arms reach for the child.

"Don't let her see!"

"He's a monster. Without skin. And he's alive!"

"This woman is rotten. . . . "

Corina begs for the child constantly. Her eyes are bandaged; the cry of the little monster is near her.

"It's that indigent mulatto who killed her baby!"

"Idiot! Only to avoid the work of having to raise him. Bum! She should die in prison. . . ."

"Let it be, my dear! Look at our little treasure, wonderful! What a chubby dear! Look at his dimples. . . . So healthy."

"I'm going to give him all my toys. Now I have a real doll. And you have to buy that tall baby buggy. It's the latest style from New York. So he can promenade in the park with his nurse."

.   .   .

Proletarization

Otavia becomes aware that she likes Alfredo.

In the end, his infidelities seem natural and insignificant. He is for her more than a friend and comrade.

His involvement in the proletarian cause fills her with a childish delight. Why?

He arrives. It's seven p.m. There is no meeting at the union.

There is no point in pretending.

"Do you want to be my girlfriend, Otavia?"

"Yes."

.   .   .

"All my friends know that he is my boyfriend. But I know that he is a traitor. I'll leave him. And I propose that he be expelled from our union!"

.   .   .

Industrial Reserve

Corina wakes up in the rough, provincial atmosphere of the country place of Penha.

A cold sun fills the dark, neglected gardens with light. The long, gray tweed jacket shows the green elbow patches worn with use. Two red hearts light up the face pricked with thorns. The bright eyes of the old seamstress

look with suspicion and defiance. The old tenderness disappears behind the eyelashes and exhausted eyelids. She sees among the new eucalyptus trees new girls that soap with red hands bolts of cloth. A little girl with nice legs shows beneath her skirt underwear smeared with dirt. The proud teeth of an earlier time show up rough and yellowed in a smile. The girl flees. She sinks her hands in a tub of suds. The shivering mulatto girl bundles into her jacket, raising the high collar up to her nose. She stops to watch the laundresses kneeling at their work.

She will never work again. When she's hungry, she opens her legs to men. A caged bird. She would like to start life anew. She got a job as a maid at "Diario Popular." She was ready to do any job at any price. She was always rejected. She turns to prostitution again.

Little live flags. A rustic house full of life. A restaurant. Corina sits on a bench at a stone table. She has fluttering pains that go from her empty stomach to her aching head. She lights the last cigarette from the empty wrapper that she transforms into colored scraps.

"And what would you like, ma'am?"

It's the small, fat waiter.

"Nothing for now, Paco. I'm waiting for a young man."

Corina isn't waiting for a young man. She's hoping for a sandwich. She can already taste the mortadella stuffed in a warm roll.

The waiter passes by her a fourth time. He notices the scared face of the seated girl. He opens his little eyes set in a whirlwind of wrinkles.

"You can wait all day and he'll never come."

Corina gets up.

On the balcony, a barefooted little girl licks her green snot.

"I will return tomorrow. If he comes. . . . He has a moustache. He's Portuguese, but he looks Brazilian."

"If you would like a drink, I'll treat you to one."

"If I accept. . . . I'm dying of cold. Look. I can't even put my hands in my pockets."

She shows him her stiff hands.

She's in front of a counter full of golden rolls.

"But if you wish. . . . I'd prefer bread."

Comforted by the burning alcohol, she loses the childishness of her voice and gestures. Relaxing, she crosses her bare legs, showing her knee with its bluish tinge. The wrinkles of the Italian multiply. Of the old, pretty Corina now only the famous legs are left. They haven't changed. A little skinnier, more or less bronzed, curved, perfect.

·   ·   ·

Paco was like a pig suckling at Corina's breasts.

"Aren't you going to give me anything?"

"Jesus! What about the bread and the salami and the drink? You think you are worth more than that?"

Night finds Corina's stomach empty again. Sadly she bumps into a man on Rangel Pestana Avenue. For her there is only one crisis. The crisis of the sexes that invades the whole working district. An old, unemployed man has already told her:

"There is never anything to eat. Only if it comes from camaraderie."

If only she had a decent outfit, she could pick someone up on São João Avenue.

She goes to the church in Braz. She goes in to rest. A thousand candles light the gilded altar. She counts them all. She counts on her fingers the money spent there. How many days she could eat with those tongues of tallow dripping down on the silver candlesticks.

Many years ago, she sat on the same pew on Good Friday, dressed like Mary Magdalene with the same headdress from Carnival. Her mother at that time earned a lot of money in that house with a big garden on Chavantes Street.

The contradictory thoughts fill her congested head as it rests on the red velvet seat. She sounds out the letters on a column.

Ma-da-le-na. . . .

"I wonder if Saint Mary Magdalene was hungry when she was a whore?"

She laughs.

A young priest wrapped in a robe shows up in the rounded nave. He approaches.

"This pew is reserved. It is forbidden to sit in it."

At the door she meets a canon eating peanut brittle. The little daughter of a beggar, a chatterbox, very dark, in the middle of the drivers at the bus stand. Guessing by her breasts, she must be thirteen or more. She sells boxes of matches

She crosses the street. She sees on the other side, on the corner, a group of young men in a cafe. Maybe she can get a cup of coffee.

At the bar everyone argues heatedly. Corina only sees and feels the warming milk and coffee. A very lively young boy shouts, chewing on the dry stub of a cigarette.

Corina hears a familiar voice from underneath a big hat.

"Pepe?"

"Damn, Corina! You look awful!"

"What nonsense! And you? How is Otavia? And the others?"

"Don't even speak to me about that woman!"

"She left you?"

"That worm! I am the one who didn't want to know about a slut who gives to everyone."

The two, arms entwined, victims of the same unconsciousness, marginalized by the same capitalist realities, take salted popcorn to the same bed.

Rosario Castellanos

ROSARIO CASTELLANOS was born in Mexico City in 1925, daughter of a wealthy rancher from the region of Chiapas on the Mexican border with Guatemala. The region is still heavily populated by one of the groups descendant from the Mayas, the Tzotzil. The conflict between the Tzotzil and the ranchers is one of the central themes in Castellanos's chief novel *Oficio de tinieblas* (1962) and her autobiographical novel *Balún-Canán* (1957) (*The Nine Guardians*, 1959).

Castellanos left her hometown, Comitán, to enter the National University of Mexico. There she studied philosophy and literature. She joined a group of young writers who later became known as the Generation of 1950. Castellanos has written poetry, essays, novels, and plays. Like many Latin American writers she has also contributed extensively to newspapers and magazines. She has published hundreds of essays in *Novedades, ¡Siempre!*, and *Excelsior.*

During the 1960s Castellanos resided sporadically in the United States, where she was often invited as a visiting writer to the universities of Wisconsin, Indiana, and Colorado. *Album de familia* (1971), *Mujer que sabe latín* (1973), and *Poesía no eres tú* (1972) appeared after her two award-winning novels. These short stories, essays, and poems are marked by a distinct feminist tone and posture.

Though she was always part of the Mexican intelligentsia, Castellanos grew increasingly aware of the historical silence with which women were and are treated in Mexico. In the essays of *Juicios sumarios* (1966) Castellanos observes and repudiates the often not-so-subtle dominant ideology by which women appear as the nameless "other." The same exploration of the constitution of women as "other" and the struggle to find for herself a voice beyond that nonpresence informs the poems collected in *Poesía no eres tú.* Her experimental and brilliant feminist play *El eterno femenino* (1975), written at the very end of her life, has placed the work of Rosario Castellanos at the center of any discussion of feminism or women's writing.

An Attempt at Self-Criticism

To say that I have made the transition from poetry to prose (narrative and critical) is to suppose that readers, that species we writers want to believe are not extinct, are acquainted with my book. But experience tells me that such a supposition is very risky, false, to be exact; that readers, if they still exist, will have exercised their profession on other texts and that my works have not been distributed as widely as I would have liked. Thus, the shortest explanations are to the point.

My first publication dates back to 1948: an extremely long poem in which I wanted to embrace and give meaning to the entire universe, thanks to the only permanence possible, which was, to my way of thinking, that of aesthetic creation. I used free verse and abused imagery to such an extent that any train of thought was often lost. To sum it up, *Trayectoria del polvo* was as ambitious as it was frustrating.

It taught me neither brevity nor sobriety. All of its excesses and defects were repeated in the following poem: *Apuntes para una declaración de fe.* As an original touch, I deliberately added clichés and the commonplace in order to paint a black picture of the contemporary world, in order to end in a hopeful, but completely unfounded, apotheosis of a better future, envisioning its development (and why not?) on exceedingly fertile American soil.

I still blush with embarrassment at the critics' reception of such a monstrosity with all the insults that it deserved. From that time on I was finally able to end a poem when it was finished and not go on writing through inertia, dragged along by the force of adjectives. The second book—*De la vigilia estéril*—in spite of its unfortunate title, made a certain amount of progress. The poems were less rambling, more modest in subject matter, and tighter in form (I was beginning to discover my individuality and validity which, in poetry, has to express the moods of the soul).

Ah, but deadwood still covered the better part of my discoveries and only my closest friends continued to have confidence in me, even though they advised me, at the same time, to try to cultivate other genres. In theater, for example, I perpetrated a "high drama," which, protected by the title of *Tablero de damas,* did me the favor of demonstrating that I did not know how to handle dialogue and that my characters were as rigid and fragile as though they were made from cardboard. But if *Tablero de damas*

"An Attempt at Self-Criticism," from Juicios Sumarios *(Xalapa, Mexico: Universidad Veracruzana, 1966). Translated by Laura Carp Solomon.*

was a complete disaster from a literary point of view, from a social point of view it helped me to make enemies of my women-writer colleagues, who felt that I had alluded to them in the farce (and I must confess, some of them were right) and did not find their portraits flattering. The conflict grew way out of proportion until it reached the magazine's editors, who had taken my text under their protective wings. The outcome was that such generosity had as its reward the extinction of a magazine that until then had been a forum for beginners, a place of honor for the successful, and a refuge for fossils.

Studies, travel, and the systematic practice of criticism helped me to overcome the most obvious difficulties that my intellectual work had stumbled over until then. I am talking about 1952, the year I wrote two poems, for which I hold myself fully responsible: "Joyful Mysteries" ("Misterios gozosos") and "Splendor of Being" ("El resplandor del ser"). The series of influences to which I paid tribute fit into place and came together harmoniously for the first time: I was able to integrate and transmit a coherent vision of things, and an authentic relationship was established between them and myself.

If the poems mentioned above were basically constructed from fragments whose unity was thematic and formal, in "Lamentación de Dido" I tried to use the wide rhythmical verse line to retell a story already told by Virgil. I did not aspire to add any perfection or new beauty to it, but rather a deep personal experience. I believe that in no single moment have my goals and my achievements better coincided; for me poetic language has never been so flexible or so precise. I have never again reached such fulfillment even though I recognize the partial success of some of the poems in *Al pie de la letra* and *Lívida luz*.

It is not easy to let go of the vantage point one has chosen in order to contemplate life lyrically, or the linguistic habits that we have relied on for years to translate these contemplations. So when I tried to write a novel for the first time—*The Nine Guardians* (*Balún-Canán*)—I was not armed with rigor, only profuseness. Metaphors glittered everywhere, but I saved myself from condemnation, arguing that I had intended to retrieve a lost childhood and a world presided over by magic and not by logic. But the balance would be thrown off each time this world required an understanding and an explanation from me, and not simply a description.

I have gradually yielded to this demand in my subsequent narratives. *Oficio de tinieblas* is a novel that is based upon a historical period, which is reconstructed along the lines that the imagination traces according to its laws. The coincidences between both planes are not forced, but their discrepancies could well be classified as differences in levels, where the deepest would correspond to the novelist.

In *Ciudad Real,* the prose—which attempts to be a precision instrument placed at the service of intelligence—necessarily strips away a great deal of its adornment. Unfortunately, by the same token, it did not strip away certain rhetorical obsessions and certain quite obvious and simple techniques that detract from its quality. In spite of that, this book compiles an inventory of the elements that make up one of the components of Mexican national reality: the reality wherein descendants of the conquered Indians live side by side with the descendants of the conquering Europeans. If the former have lost the memory of their greatness, the latter have lost the attributes of their strength, and they all conflict in total decadence. The daily social behavior of beings so dissimilar produces phenomena and situations that began by interesting the anthropologists and have never stopped appealing to writers, who struggle to get to the very root of these extreme forms of human misery.

As far as I'm concerned, *Los convidados de agosto* exhausts that vein of archaic provincial life which served me so richly. After "Ritos de iniciación," a still-unfinished novel, I am venturing into other fields, tackling other problems and, consequently, trying out a style that I feel I have not as yet completely mastered.

Henriqueta Lisbõa

BORN IN MINAS GERAIS in 1903, Henriqueta Lisbõa is considered one of the major Brazilian poets of the twentieth century. Lisbõa has published more than ten volumes of poetry and essays. With each of her books, starting with *Velario* (1936), through *O Menino Poeta* (1943), and on to *O Alvo Humano* (1973), she has endeavored to master new thematic as well as stylistic horizons. A sharp, intensely personal camera asserts the passing phenomena and produces a still but vibrant version of the instant. Lisbõa's symbolism not only captures the visual, but her poems also attempt to create an acoustic representation of an experience. Because Lisbõa often uses the metaphor of God as timelessness and thus the absence of pain, her poetry has been called mystical. A sustained contemplation of death can similarly be traced through her entire work, but *Flor da Morte* (1949) condenses her refined meditations on this theme. She has also written perceptive essays on poetry and the modern poetic imagination.

My Profession of Faith: Poetry

The privilege conferred upon me today to speak of my own poetry causes me great violence, for I have made my home in the midst of silence and shadows.

In order to address the problem without failing to respond to the invitation made to me, I will speak of poetry in the context of convictions and ideas before I go on to speak of my own experience.

The games of the little girl who wrote poems on the tile, at the time when she went to school at the Grupo Escolar de Lamberi, was the initial point of an unforeseen line that was to prolong itself into the future and persists to this day. I cannot pinpoint the moment when play stopped and a serious sense of writing began. To this day I am surprised with the grace

"My Profession of Faith: Poetry," excerpted from a speech to the Cultural Foundation of the Federal District at the National Meeting of Brazilian Writers, Brasília, April 1978. In Vivência Poética, Ensalos *(Belo Horizonte: Impresa Oficial, 1979). Translated by Sara Castro-Klarén: copyright © 1991 by Sara Castro-Klarén.*

of the game, in the midst of cogitations about life and death, before personal and social conflict. This is so especially in view of the tests poetry has endured under the impact of this century.

In this field of contradictory forces, I have looked for a balanced position from which to approach poetry, knowing that it can be found at the bottom of the well or in the flight of a bird. The marking of the pulse that regulates poetry is not merely a bundle of nerves but rather the flux of life itself. It does not wear out; authentic poetry does not tremble at the edge of the cup; it does not thin out into foam. It suggests plenitude, vigor. It is crystalline. It is to be followed in the ups and downs of the road. Sometimes the poet stumbles on rocks and fallen trunks, always advancing as if surrounded by a mirage. It is a road conceived for torment and triumph.

I do not dare define poetry, even though I have advanced the idea that poetry might be like the trapping of the eternal within the ephemeral. I feel it like an aura irradiated by being itself. It is present in the smallest gestures. It becomes real in a poem, in a work of art, in music. That is why I consider poetry fundamental and cosubstantial with human existence.

.   .   .

According to Lévi-Strauss, every poem holds within itself all its possible variants. They are ordered upon a vertical axis that allows for the superposition of a series of levels: phonological, phonetic, syntactic, parodial, semantic, and so on. In this way we can observe elementary sounds, phonemes, grammatical structure, pronunciation, transposition of the signifying into the signified, and even rhythm. Nevertheless, the poem will only be grasped in its unity and perhaps in its totality if the analysis is foregrounded by ideatic intuition, that is to say, the relative intuition into the essence of things, not its mere existence or function.

This is because poetry resides somewhere between darkness and revelation, between silence and the word. A fecund expressive silence, not unlike the word itself, limits poetry while at the same time extends its psychological fluidity, lightens its density, protects it from naked clarity.

On Poetic Creation

There are no rules applicable to poetic creation. Nor are there any appropriate contexts. The poet knows, without great certainty, that she must be alert, that there is something about to unravel. And it happens: a shapeless memory, a care without a cause, a certain attraction, a sense of repugnance, a faster beating of the heart, a design in the air, an empty page. Then the first word comes, the first verse, and then a torrential outpour. The poet grasps them in a sort of chaos. Later they are shaved, molded, placed where they belong. At the end of the composition, the verses repose, filled with

quiet and calm. All of this happens in either hours or minutes of work. But the poem is never finished. It is constantly being polished in each syllable, in the light of each prism, in each resonance. Sometimes, suddenly, between the time I fall asleep and wake up, I find the right substitute for the word that was holding up the fluency or the authenticity of the poem.

The Word

The word is neither the end nor the limit of a poem; it is merely an option, a choice. The core of the poem, the diamond to be chiseled, resides in the meeting of the words, in the magnetic force with which each word attracts the other words and forms a feeling sequence. The word maintains the dignity of the poetic work. It is what is vertically human in the poem. It is at the base of the poem's verbal articulation, understanding, expression, desire to communicate, and grace to be intelligible.

The word is not used as a singular, isolated value, but rather as an instrument. It will be representative inasmuch as the user of words can confer upon them the power of suggestion and seduction. Each word is itself a jewelry box waiting to be opened, an inanimate body in expectation of the breath of life. It is capable of coming alive as many times as the miracle occurs and breaks the stored silence in order to vibrate, to speak.

In this "speaking," the inner voice is an important factor. An inflection of the voice makes exteriorization possible in subtlety, in force, in magnitude, or in simplicity.

On Representation

Artistic expression should be understood as a value in and of itself. It is not escapism. It does not conceal anything. The work of art is a living presence. It is audible, visible, palpable. It is continually tested as the objectification of an authentic determination.

This object will be defined in terms of a new category, even if it must be referred back to what escapes definition, since it exists by means of the form that it takes in the convergence of diverse elements.

The work of art bears witness to the real, that realm of the real that is beyond the human sphere, outside psychology, but never alien, always within the aesthetic order.

The creative imagination of a given individual can be major or minor, discordant or harmoniously modulated to the created object, but each— the creator and the work—remain autonomous identities "in the field of knowledge."

·   ·   ·

While the individual holds previous norms and concepts or experiences problems within certain conditions imposed by nature, art finds its basis in factors external to nature. Art demands a method all its own in accordance with its essence and transcendence.

Making, cutting, polishing, revealing the unknown, and organizing a chaos are the aesthetic values that give creation its positive qualities. This is beyond any merely realistic or portraiture value.

What matters is not a subjective truth, or the truth portrayed in the work, but rather the truth of the poem. It resides in the structure of the poem because it has been tested and solidified within its inherent canons of art and technique, which are not necessarily responsive to psychological or biological influences.

If, for instance, I write a poem entitled "This Is Fear" or "The Face of Fear" in which I try to convey the feeling of the phenomenon of fear, that means that the barrier of "fear" has been crossed into a "fearless" plain. It is not a question of catharsis, but rather a question of knowledge.

In her role as a transformer of language, the poet practices in several different ways: invents characters, stories, and metaphors; traces horizontal lines; uses masks. Her goal is not to cover up or diffuse what she has in mind but rather to achieve representative syntheses capable of signifying by themselves at a different level.

It is not a rejection of reality to transfer linear geometry into another dimension, nor is it the permutation of the univocal voice for a plural enunciation. There is a profound and natural inclination to manipulate feelings and ideas. This tendency is guided by a taste, a sensibility, a personality that needs or wants to challenge the obvious, the vulgar, the repetitive. The very same phrase found in an elliptical phrase loses its evasive quality and functions effectively, as a piece kept in the shadowy corners of an impressionist painting, in relation to the whole.

Even when the effort of the poet remains in darkness for the neophyte or the insensitive reader, the creative act is nevertheless an act of will and an act of faith.

Lygia Fagundes Telles

LYGIA FAGUNDES TELLES was born in São Paulo in 1923. She recalls her youth as encapsulated in a sense of fear and ignorance of the world. Like that of Clarice Lispector, Fagundes's writing centers upon ethical and metaphysical questions raised by the existentialism of Jean-Paul Sartre.

Fagundes brought to Brazilian prose innovative rhythms, syntax, and a superb ability to handle first-person narrative. She also suggested a plethora of new themes. Situating the style and context of her stories beyond psychological and social realism, Fagundes introduced a fresh voice and a new point of view with her female characters. Her narrative technique emphasizes the dramatic in simple, everyday routines. Her characters weave a net of frustration as they pursue the object of their fantasy. The salient marks of Fagundes's crisp and yet detailed prose are particularly suited to her narrative project. With infinite care and insight she follows the meandering yet intense thoughts of her characters.

Her work won immediate critical praise. Her first novel, *Ciranda de Pedra* (1956; *The Marble Dance,* 1986), has been reprinted more than ten times. In her second novel, *Verão no Aquário* (1963), as well as *As Meninas* (1973; *The Girl in the Photograph,* 1982) and the short stories of *Seminário dos Ratos* (1977; *Tigrela and Other Stories,* 1977), Fagundes employs characters caught in the drama of duplicity, ambivalence, and self-definition as an exercise toward the discovery of being.

The Sauna

Eucalyptus—this scent, especially, marked Rose and her world of water-steeped woodland plants, greenish filters, and glass receptacles stagnating on the shelves. This the damp-green perfume I smelled when she leaned out over the windowsill to pose. It had rained, and a warm vapor rose from the garden as the sun came out. This is the first portrait I've ever done, I need to get it right, I warned her, and she shrank back into the window. So

"The Sauna," from Tigrela and Other Stories *(New York: Avon, 1977). Translated by Margaret Neves.*

I kissed her forehead, come on, relax, don't think about what I told you but rather about this orange you're going to hold, like this, you can talk if you want but don't move, nice and quiet holding the orange. When I sketched the oval of her face, she was so serious that she seemed to be posing for a front-and-profile snapshot with a date. Mugshot Rose, I said, and as she smiled, she moistened her lips with the tip of her tongue. They remained lusterless, her anemic lips. My Anemic Rose, you need to put an end to all these vegetables and eat bloody steaks, you need meat! The best portrait I ever did. But what became of it? Marina asked. She must have it, I answered. What became of both of them is what I'd like to know, wasn't it more than thirty years ago?

"Really a long time."

The white-aproned attendant thought I was referring to the duration of the sauna and wanted to reassure me, I could leave earlier if I wished.

"It's nothing. I was just thinking."

"Is this the first time you've come here, sir?" he asked, taking a white robe from the closet. He put the plastic slippers on top of it. "We have a few artists on our list of clients—most of them have massages. Would you like a massage, sir?"

"Just the sauna."

"They say that in Tokyo, these sauna institutes are run by gorgeous girls who do everything to serve the client. One finds that kind of place here, too, but the Orient is something else. Do you know Tokyo, sir?"

The plush fabric of the robe is warm. Music. And the eucalyptus perfume stronger. I take out my handkerchief and wipe my forehead, loosen my collar. To be pleasant is to give him the Tokyo smile, it's easy to be pleasant. And hard, it's beginning to get quite hard, pleasantry saturates quickly: a pleasant fashionable painter. Not first-rate, but the aspiring middle classes think it's first-rate and buy what I sign. I got rich, didn't I? Shit, wasn't that what I wanted? So, don't complain, everything's fine, what's the problem? I follow the white apron submissively, in places like this I become absolutely submissive. The rubber soles of his shoes stick to the green linoleum of the corridor.

"Are you at your normal weight, sir?"

In hell there must be an extra circle, the circle of the questioners, asking their little questions, your name? your age? massage or shower? fire or noose?—without stopping. Without stopping. Marina used to ask a lot of questions too, but lately she's taken to staring at me. A time to ask and a time to stare, and this stare adds, subtracts, and adds again, she's excellent at arithmetic. The women's movement should take advantage of her to do their accounts. But the big thing there, it seems, is interviews. Questions. During one period she asked so much about Rose, had such a fixation, her curiosity was large enough for both of us. One Sunday she forced me to

show her the house, I want to see the house, the garden, the window where she posed for the portrait! Where the house had been, they had built a somber apartment building with narrow balconies and washing hanging from clothes racks. Fine, it used to be there, I said. And though I felt a certain relief (it passed, it passed), I had the half-agonized sensation that something had been taken away from me, what? As if the period of those first aspirations had been preserved in the house, so much energy, plans, as long as the house lived the past would be intact. Industrious Rose making perfumes and frames, all the games yet to be played, my fervor, my thirst for recognition—how old was I, twenty? So many possibilities there waiting, this way? That? The building in front of me was the dust-colored answer, furrowed by dripping water and soiled with spittle. I wanted to get the hell out of there, but Marina detained me (just see if I could be let off that easily!). If it weren't for this mania of yours to keep digging up what should stay buried, look there, dear, wasn't that what you wanted to see? There's no more house, or portrait, or Rose—are you satisfied? She lit a cigarette, a sign she was disposed to talk. I think what really lasts after a long marriage is one's knowing whether the other wants to talk or be quiet. I lit up one too and waited. You mean Rose sold the house so you could take a trip? A statement in the form of a question, Marina is expert at this type of expression. But I won the trip as a prize, have you forgotten the prize? She hadn't, but she remembered that during one of our first encounters in Paris, when I said I intended to stay longer than the prize would allow, I also said I was going to get some money from a house which was being sold, wasn't that the house? And wasn't that the money Rose was going to send me? A sharp girl, this Marina, when we were married I had no idea how sharp she was. Nor that she had such a memory. I think I talked too much, if I ever get married again I'll only open my mouth to ask for the salt shaker. On the second floor of the building there was a yellow towel drying on the clothesline. And diapers, an enormous quantity of diapers. Or dish towels? Do you think those are dish towels there? I asked pointing to the diapered balcony. She threw her cigarette out the window and looked: diapers. I turned on the car radio; the dashboard lit up for the voice singing faintly, *"Ne me quitte pas . . . ne me quitte pas . . ."* to hear that phrase sung in such a way made it tender, gave one the desire to stay. To hear it without music? Rose didn't say it, she thought it. I turned off the radio. With this computer memory of yours, I began slowly, with this ponderous archive, I hope that you haven't forgotten that Rose was pregnant when I left, I didn't tell you this detail in Paris, I told you later on, remember now? Yes, of course, and she remembers that we didn't have the money for the abortion, a small hitch, we didn't have money, my dear, I hadn't managed to sell a single painting, Rose had left her job at the pharmacy, there was just the house and garden left, but we can't eat a

garden, we can't do an abortion in a garden, can we? I repeated, holding her hard by the wrist, we liked this type of game which could end in a bloody nose. She pulled away, you're hurting me, you brute! I kissed her wrist. Lowered my voice: I wanted to sell my ticket and give her the money for the doctor, but she wouldn't accept it, I already told you this, didn't I? She believed in me, she loved me. She knew how important it was for my career, this chance to take a course in Europe, meet people, make contacts. She insisted on selling the house, which was too big now, her uncle in the sanatorium and me going off with no idea how long I might stay—what good was such a big house? And the pregnancy advanced, and the doctor demanding the money in advance, was there any other choice? I know, an abortion wouldn't have been any problem for you, a little industrialist flitting about the world, today a love so poor as this one must seem ridiculous to Madam Marina. But that's how it was. Nowadays everything is so simple, as a liberating leader you know that your sisters have access to the pill, childcare centers, psychologists, at least that's what they claim in the speeches. But back then, have you forgotten? There wasn't any pill, there wasn't anything. And if she insisted on sending the check after the damned abortion, it was because she wanted me to take advantage of the prize, she had faith in my work as no one else ever did. I emphasized the *no one* with such eloquence, but it was no good, Marina was no longer paying me the slightest attention. She took out her comb, combed her hair. She tried to see my reflection in the little car mirror; she used to work in a pharmacy, didn't she? A homeopathic pharmacy, you told me, that sort of thing. She made good money, she was independent, she even supported her mute uncle, didn't she? And so you appeared and started living with her. You committed the uncle to the asylum because you needed more space to set up your studio. Rose quit her job because you needed someone to frame your pictures, right? Wait, let me finish, naturally you began to be successful, prizes, exhibits, and right exactly at that point there appears the *damned* pregnancy, which would add itself to the expense of the journey. Very logical, sell the house. Which means, she was left without a house, without a job, without a baby, and without you, as you were already leaving. Oh, and I almost forgot, without the old uncle who, in spite of being mute, was good company, at least he could hear. All added up, as the poet says, one concludes that your appearance wasn't a very good deal. But one can't talk in terms of deals, you loved her, and when one loves—she added taking a chocolate bar from the glove compartment. She chewed thoughtfully: Another thing, knowing my sisters as I do, I'd go even further, my dear. Even further. I think that your Rose's dream was to have this baby, she loved you, and a woman in love thinks immediately about a child, it's the first thing one thinks about at twenty, a child. You never married her, I know, I saw your papers, my husband isn't a bigamist. But not being

married doesn't mean that she didn't want (she paused to examine a fingernail torn on the glove compartment) this marriage, a batch of children, everything on the up-and-up. However, dear, this wasn't your style, and she knew it perfectly; she only adapted herself so as not to lose you. So as not to lose you. I would be astonished to learn that the idea of the abortion was hers, was it really hers? she asked and looked rapturously up at the sky, what a beautiful afternoon! The bluest blue, shall we go to the country house? I turned my face away, because I felt it darkening with hatred. Now she was satisfied, comforted in the certainty that I would continue to be eaten away by thoughts, poisoned. Alone.

·   ·   ·

from To Write Is to Conquer Fear

Q: In the stories of *Seminario dos Ratos* (*Tigrela and Other Stories,* 1987) and in *A Disciplina do Amor* (The Discipline of Love), . . . a recurrent theme appears: the condition of man. In effect, there man sticks out like a "traveler lost in the forest" who wants to escape from an archaic, primitive fear, who is on a permanent search for happiness, who is characterized—psychoanalytically—as always making the same mistakes.

Fagundes: What you point out is true. The condition of man is my principal theme. It is the obsession that flows throughout my work from beginning to end. When I was not much more than a girl, man seemed a sentimental, romantic, essentially good figure, and I had a lot of hopes pinned on him—I still have them, because otherwise I wouldn't continue writing nor would I be here speaking. So, and as a consequence of that vision, man embodied all the virtues; but later, and as if it were a death, I came to think with Charles Peguy [French Catholic and socialist poet, 1873–1914] that man, in the way in which he lives, corrupts himself. To live is to degrade oneself. Nevertheless, little by little I have regained hope and now I believe that, in spite of it all, man has enormous potential that will allow him, always, to survive. It was William Faulkner who maintained that man would not only survive his many challenges but that—and this is what interests me—the day when he loses his fear he will be capable of doing great works. To overcome fear is the most difficult test that confronts man, and that is so because God has made us very fragile and delicate instruments to exercise what we could call our task as human beings. But

Excerpts from an interview with Danubio Torres Fierro in Revista de la Universidad de México *(July 1981), 16–18. Translated by Laura Beard Milroy: copyright © 1991 by Sara Castro-Klarén.*

here there is something that we should point out immediately: opposed to this fragility is a powerful force that comes from, I don't know where—although it seems to me that it comes from God. It is God who makes us get up as the boxer gets up to continue his fight as if we have never fallen. So, and in spite of living on a planet that is sick—and how!—man stands out, for he possesses a terrible force capable of tying him solidly to the earth or making him touch the sky. It is necessary to preserve, above all, this force.

Q: You spoke of fear. Couldn't we say that the greatest human works are born out of fear, of our fears?

Fagundes: It's true. But here we should specify that it is a controlled fear. Let's think of the work of Dostoyevski. All of his characters are anguished creatures, they suffer an anxiety that comes from fear, from that obscure and cloudy fountain that is fear, and they only emerge from this quagmire to the extent that they overcome this fear, that is, to the extent to which they control it, even if in order to do so they must die. There is a good colloquialism for this: "Man must overcome himself." Have you ever thought what it is that we must overcome? For me it is nothing but fear. What happens with the characters of Dostoyevski is also true in Kafka. What fear there is in Kafka! But Kafka was a fighter who managed to overcome his fear. There is a precise moment in which it is possible to locate his triumph: when he writes his letter to his father. It is a letter in which many of his fears disappear and he speaks face to face with his father. I, a woman who fights with words, and with the insecurity and the fragility of these instruments granted by God, try, then, in my own way to exercise the force within me. In my struggle with fear, I sometimes come out triumphant. My victory is a direct consequence of my best writing. That is why I continue to appeal to that force that gives life and breath. That force, I repeat, is given to me by God. In any case, let us not forget that the fear that hides within us, like a dark and dirty beast, is death. When we find ourselves confronted by it, we must look it straight in the eye. . . .

Q: On one occasion, Marguerite Yourcenar said that she wishes for a slow death in order to be able to follow its process step by step.

Fagundes: I confess that I have not arrived at that state of perfection. But there is something that, in this context, I want to point out. After having been cast out of paradise, man should fight anxiety and frustration. And I believe that anxiety is the source of great creations. I am possessed by the demon of anxiety, and it is this that moves me to write. Because of this, when I write I achieve my fulfillment. . . . When I write . . . it is a moment of complete love—because to write is an act of love. It is then that I believe in myself, in my text, and I feel invincible and powerful. Omnipotent, yes.

•   •   •

In the moment in which I write I do nothing but develop the workings of a passion, and I am happy. The key of which Drummond de Andrade speaks belongs to me. Words are that key. But, as everyone knows, literary creation is a permanent search. So, in the same instant that I find the key I lose it. I end my text and there is nothing left. I start again on my search until I bring forth the next creation and the key, mysteriously, reappears in my hands.

Q: Tell me about your Christianity.

Fagundes: My Christianity is mysticism. I don't go to church, nor do I practice the Catholic rites. No confession or Mass. I have a special way of communicating with God and that way is very poor because it does not include any heroism nor is it generous—in the Christian sense of the word. But it is the form I found for myself, to channel my faith. I am greatly jealous of a woman like Saint Teresa of Avila, whose eyes were like two balls of fire and who saw and talked with God. I can't reach those heights, nor those temperatures. My faith is poor and delicate but, even so, I try to keep its fire permanently burning. That Christianity is also a way of believing in the immortality of the soul. I firmly believe that we do not disappear after death. I am, if you accept this definition, a spiritualist. I believe in God and in the immortality of the soul. I know it: At times I believe in very naive things, one might say childish things. But there is an infantile and immature side to me. Would you like me to make a confession? I believe that immaturity is one of the principal sources of artistic creation. In every artist there is a side of her/his personality that is immature. What is immaturity? It is the child that we all carry within ourselves, and that suddenly spies us, tickles us, or looks at us with disapproval. It is a pure, naive, and innocent side that dwells within us. And it can be the most powerful side of a creative person. Think of the works of Posada: They are said by a child who is sardonic, full of irony, lively. I give the example of a naif painter, but there are others too, whose vision is not primitive, who are fed by the mischievous tricks of a child who comes to pull on one's sleeves and demands that her curiosity be satisfied.

What I have just told you should make it clear to you that my Christian faith is immature. From there also surges the little girl that lives within me. When she gives me her hand, she takes me to imaginary countries. This is the important thing: Only our infantile side is capable of leading us to a pure and whole imagination. This child is the manifestation of the fantastic that exists in us, and through which we arrive at the highest point of intuition and perception. Rimbaud said that one must be a seer. The seer is, always, a child. A child has the thirst of curiosity and that thirst enables her to transpose limits and broaden narrow horizons. I say, with Rimbaud, that one must be perceptive because perception includes seeing. Is the mystery of writing anything else?

Clarice Lispector

CLARICE LISPECTOR was born in the Ukraine in 1925. Her parents emigrated to Brazil the year of her birth. Until she was twelve she lived in the port city of Recife. Her family was almost always poor and often beset by illness. Like many future authors, Lispector became an avid reader early and soon decided to write her own stories. Originally she intended to publish in the *Diario de Pernambuco,* which occasionally published children's stories. But her stories were already beyond the usual series of adventures and were never published.

Though she studied to be a lawyer and finished her law degree in compliance with a bet made to a friend, she never practiced law. Her manner of taking down disparate notes on ideas and feelings that she apprehended in passing consciousness did not seem fitting for any legal endeavor requiring a high degree of discipline and regularity. She took a job at the newspaper *A Noite,* where she carried on with a variety of writing assignments and also made useful contacts in the publishing business. *A Noite* financed the publication of her first novel, *Perto do Coração Selvagem* (1944). Many of her books appeared posthumously but Lispector's place as one of the most delicate stylists and epoch-making novelists in Brazilian letters had already been established with her story-telling in *Laços de Família* (1960; *Family Ties,* 1972), *A Maçã no Escuro* (1961), *Uma Aprendizagem ou o Livro dos Prazeres* (1969; *An Apprenticeship, or the Book of Delights,* 1986), and *Do Corpo Inteiro* (1975). Her novels and stories have been adapted to the stage as well as to film.

The chief characteristic of Lispector's work is the inner, psychological and metaphysical search. Both her essays and her novels have always been experimental and innovative. In her later novels, the autobiographical plane is especially stressed in the exploration of search for self and the mystery of death.

She married a fellow student, Mauri Jurgel Valente, who eventually entered the foreign service. Together they spent many years in Italy, the United States, and England. In 1959 they returned to Brazil. Lispector continued to write her novels and to contribute to newspapers. In 1977 she died of cancer.

from Luminescence

From Ulysses she had learned to have the courage to have faith—a lot of courage, but faith in what? In faith itself, which can be very frightening, which can mean falling into the abyss. Lori was afraid of falling into the abyss and she would hold fast to one of Ulysses' hands while his other hand pushed her closer to it. Before long she would have to let go of the hand that was weaker than the one pushing her and fall. Life is no joking matter because people can die in broad daylight.

A human being's most urgent necessity was to become a human being.

·   ·   ·

Because reality is in the Impossible.

Lori endured the struggle because Ulysses, who struggled with her, was not her enemy. He struggled for her sake.

"Lori, pain is no reason to worry. It's something we share with animals."

She clenched her teeth and looked at the cold moon. She looked at the zenith of the celestial sphere.

She crushed a leaf that had fallen from the tree onto the table. And as if he were giving her a present, he said, "Do you know what the 'sarcophyll' is?"

"I've never heard that word," she replied.

"The sarcophyll is the fleshy part of a leaf. Take this and feel it."

He handed her the leaf. Lori touched it with sensitive fingers and crushed its sarcophyll. She smiled. It was pretty to say and to touch: sarcophyll.

·   ·   ·

It was on the following day, as she entered the house, that she saw the single apple on the table.

It was a red apple with a smooth, firm skin. She picked up the apple in both hands: it was cool and heavy. She put it back on the table to take another look at it. And it seemed as if she were looking at the photograph of an apple in empty space.

After inspecting it, turning it and noticing as she never had before its roundness and scarlet color, she slowly bit into it.

And, oh God, it seemed like the Apple in the Garden, but as if she already knew good, too, and not only evil as before. Unlike Eve, when she bit into the apple she entered paradise.

Excerpts from "Luminescence," reprinted from An Apprenticeship, or the Book of Delights *by Clarice Lispector, translated by Richard A. Mazzara and Lorri A. Parris: copyright © 1969. By permission of the University of Texas Press.*

She took only one bite, then put the apple back on the table because something strange was happening to her subtly. It was the beginning . . . of a state of grace.

Only someone who had already been in the state of grace could recognize what she was feeling. It was not the same as an inspiration, which was a special grace that often came to those who sought it skillfully.

.   .   .

She was not startled to feel his hand rest on her stomach. His hand was caressing her legs now. At that moment there was no passion between them, although she was filled with wonder as if star-struck. Then she extended her hand and touched his sex organ, which was quickly transformed, but he remained quiet. They both seemed calm and a little sad.

"Do you think love is making a mutual gift of one's solitude? After all, it's the greatest thing that one can give of oneself," said Ulysses.

"I don't know, darling, but I do know that my search has come to an end. What I mean is I've come to the edge of a new beginning."

"My woman," he said.

"Yes," Lori said, "I'm your woman."

The quavering light of dawn was breaking. For Lori the atmosphere was miraculous. She had reached the impossible in herself. Then, because she sensed that Ulysses was again imprisoned by the pain of existence, she said, "Darling, you don't believe in God because we've made the mistake of seeing Him in our image. We've pictured Him this way because we don't understand Him, and it hasn't worked. I'm positive that He's not human. But although He's not human, there are times when He makes us divine. You think that . . ."

"What I think," the man interrupted her and his voice was slow and muffled because he was suffering from life and from love, "What I think is this:

from Since One Has to Write

13. "Since one has to write . . ."

Since one has to write, let it be without restricting the space between the lines.

Excerpts from "Since One Has to Write," in Para não Esquecer *(Rio de Janeiro: Atica, 1978). Translated by Sara Castro-Klarén: copyright © 1991 by Sara Castro-Klarén. The original is arranged as a numbered series of aphorisms and short prose pieces.—Trans.*

.   .   .

18. "But, an error . . ."

An error made by intelligent people is a very serious matter because they
have the arguments to prove it correct.

.   .   .

21. "Once upon a time"

I said that someday I should like to write a story beginning with "Once
upon a time. . . ." 'For children?' I was asked. 'No, for grownups,' I said,
absentmindedly. I was then busy remembering the stories I wrote when I
was seven years old. They all began with "Once upon a time," and I sent
them to the children's page of the newspaper in Recife. None, not one, was
ever published. It was easy to see why; they did not include any of the
necessary elements required for a story. I read the stories that they published
and they all narrated an event. But if they were afraid, so was I.

Since then I have changed so much, perhaps I am now ready for the real
"Once upon a time." I ask myself: why don't I start right now? It would
be simple.

I begin. Right after I write the first sentence, I see that it is still
impossible. I have written: "Once upon a time a bird, God."

.   .   .

25. "To lie, to think"

The worst thing about lying is that it creates a false truth. (No, it is not as
obvious as it seems. It is not a truism; I know that I am saying something
and that I barely know how to say it accurately. What irritates me is that
everything must be "accurate," a very limiting restriction.) Oh, what is it
that I was trying to think? Perhaps this: if a lie were merely the negation of
the truth, then it would be one of two ways (negative ways) of telling the
truth. But the worst lie is the "creative" lie. (There is no doubt thinking
irritates me, because before I started to think I knew very well what I
meant.)

.   .   .

28. "Writing, humility, technique"

This incapacity to understand, to perceive is what causes me—like an
instinct?—to try to find a way of thinking that might lead me to a quicker

understanding. This manner, this "style" (!), has already been called a variety of things, but never what it really and barely is: a humble search. I have never had any problems expressing myself, my problem is more serious. It is a problem of conception. When I speak of humility, I mean humility in the Christian sense (as an ideal that may or may not be attained); I am referring to the humility found in the full consciousness of being utterly incapable. And I am referring to humility as technique. Holy Mother of Christ, even I was appalled at my lack of modesty, but it is not really that. Humility as technique means that only when we approach a thing humbly do we have a chance of grasping it. I discovered this kind of humility. Ironically, it is a sort of pride. Pride is not a sin, at least not a capital sin. Pride is a childish thing which we succumb to, like gluttony. Except that pride has the enormous disadvantage of being a serious error, with all the loss and delay that error causes in life. It makes for a great waste of time.

·   ·   ·

35. "Literature and justice"

Today, suddenly, like a great discovery, I found that my tolerance for others was also reserved for me (how long will it last?). I took advantage of the crest of the wave to bring myself up-to-date with forgiveness. For example, my tolerance for myself, as a person who writes, must forgive me for not knowing how to approach "social problems" in a "literary" vein (this means to transform within the vehemence of art). Ever since I have come to know myself, social problems have been for me more important than anything else: in Recife the black shanty towns were the first truth I encountered. Long before I felt "art," I felt the profound beauty of the social struggle. It is just that I have a straightforward way of approaching social conflict: I want to "do something," as if writing were not itself doing something. What I cannot manage is to use writing for this purpose. How humble and pained I feel by my inability to do so. The problem of justice is for me so basic that it never surprises me—and without surprise I am unable to write. Also, for me, to write is to search. The feeling of justice was never a quest for me, it was never a discovery, and what appalls me is that it is not equally obvious for everyone else. I am aware that I am simplifying the problem to its bare essentials. But, because today I am being tolerant with myself, I do not feel totally ashamed of not contributing toward a social or human end with my writing. It is not a matter of not wanting to, it is a question of not being able to do so. What shames me is not "doing," my failing to contribute with actions. (I know that the struggle for justice leads to politics, and that, in my ignorance, I would get

totally lost in the maze of politics.) For this reason I shall always feel ashamed. But I don't even try to do penance. I do not want, through dubious means and excuses, to give myself absolution. I wish to continue to feel ashamed. But I do not apologize for writing what I write, for if I were to be ashamed of my writing I would be committing the sin of pride.

· · ·

90. "To write, to extend time"

I am unable to write when I am anxious or when I expect solutions because when I feel anguish I do everything to accelerate the passage of time. To write is to extend time. It is to divide time into tiny particles, giving to each second irreplaceable life.

· · ·

107. "Writing"

I no longer remember where was the beginning. It was as if the whole thing were written at the same time. Everything was there, or should have been there, as in the temporal-space of an open piano, in the simultaneity of the keyboard. I wrote searching with keen attention for what was taking shape inside me. Only after the fifth, patient draft, did I perceive it. My fear was that because of my sluggishness in understanding I would be grasping a meaning before its time. I had the feeling that if I did not hurry it, the story would, by itself, tell what it needed to say. With time I find everything to be a question of patience, a matter of love creating patience, of patience creating love. It came up all at the same time, emerging a bit higher here, a bit lower over there. I interrupted a phrase, let us say in Chapter 10, only to write what belonged in Chapter 2, itself to languish for months while I wrote Chapter 18. With patience I learned to bear, without any hope whatsoever, the great discomfort of chaos. For it is true that order restricts. As always, waiting was the greatest problem. "I felt poorly," is what the woman said to the doctor. "That is because you are pregnant." "But I thought I was dying," said the woman. The soul, deformed as it grew, gained in volume, without even knowing that "it" was called waiting. Sometimes, in the case of a stillbirth, one recognizes the waiting. Past the difficult wait, writing requires recomposing the vision. And as if that were not enough, I unfortunately do not know how to "narrate" an idea, how to "dress" an idea with words. What is perceived already comes in words or it does not exist. At the time of writing, I feel paradoxically convinced that what hinders writing is having to use words. It is vexing. If I could write using designs on wood or smoothing out a child's head, or taking a

stroll in the open fields, I would have never taken the path of the words. I would do what everyone else who does not write does, with the exact same torment and happiness with which I write. And I would feel the same inconsolable disappointments, but I would not use words. That might be the solution. If it were, I would welcome it.

Julieta Campos

CAMPOS WAS BORN in Havana. She has lived in Mexico since 1955 and is married to the Mexican writer and journalist, Enrique González Pedrer. She has a doctorate in philosophy and letters from the University of Havana. Her entire career as academic, journalist, translator, and writer has taken place in Mexico, and rather than an obscure or struggling writer, she has been part of a large intellectual establishment. Campos's journalism and fiction have been recognized both in Mexico and abroad. It is therefore not surprising to see that she has been the editor in chief of the important *Revista de la Universidad Nacional Autónoma de México* and president of the Mexican chapter of the Pen Club.

Like that of Victoria Ocampo, Campos's work is marked by a deep knowledge and understanding of the forms and history of European literature and plastic arts. Her fiction is highly self-reflective. Campos often makes her readings the subject of her meditations. A close correspondence can thus be observed between her fiction and her essays on contemporary literature collected in *Imagen en el espejo* (1965) and *Oficio de leer* (1971).

She has received many prizes for her novels. *Tiene los cabellos rojizos y se llama Sabina* (1974) is an open novel in which what matters is not the adventure of the protagonist—a woman looking at the sea through a vast window—but the inner adventure of a search for glimpses of meaning and awareness of the self in memory. The prose that tells this "story" is poetic and slow, initiating a serious questioning of language in its relation to memory and reality. In *El miedo de perder a Eurídice* (1979), Campos delves into the themes of love and death, which preside over almost all of her fiction. Like Julio Cortázar in *El último round* (1969), in this novel Campos indulges in the art of collage. For Campos, desire, the movement of Eros, is always linked not only to death but also to writing. Thus writing momentarily abolishes silence and recovers love, just as Orpheus recovers Eurydice through his desire.

68

My Literary Vocation

The writing process and my compulsive self-analysis have led me to reflect somewhat often on the origins of my literary vocation. This exploration reminded me, for example, of the pleasure I found in books as a child. With each incursion into the mysterious and beckoning domain of the antiquarian's store, I seek, surely, that piece of furniture where the legendary objects of my early love were kept. I looked then for the places where clocks don't rule the magic of the few grape leaves, and for the cluster of grapes that framed the little door leading to the treasure at the time of the myth. Inside was the little forest, the word *currant,* Little Red Riding Hood's cape, snow that still radiates warmth, and that vague, blond, fantastic image that crosses the night sky pouring stars every time I open a book. But this pleasure associated with books and reading became insufficient to feed me after a certain point: then I began to write my own books. The receptive pleasure of reading was followed by the active pleasure of writing. I felt in each the pleasing sensation of reunion. With what? With whom?

Abusing autobiographical material, I could also assert that death is responsible for writing. If the urge to spill signs onto the white page originates in an adolescent's transgressions in the family library, I did not put this urge to rest until much later with my mother's death.

I intuited then that writing, in my case, was equivalent to hallucinating about that fading episode in my life. Much later, I learned that Melanie Klein had proposed looking for resolution—through the literary work— to very archaic, destructive fantasies related to the primal scene. The work would thus serve to re-order the fragments dispersed by hostile pulsations of oedipal origin.

Isn't the symbolic murder of one's father (as Freud thought) or the collection of the fragments of a parricidal phantom and their rearticulation into a body previously dispersed by fantasy what Vargas Llosa calls "cannibalization" or "abolition" of the real toward the reconstruction of a new reality, a process that in itself might constitute the essence of the ambiguous craft of storytelling? The enigmatic origin of the creative act seems to be this: a ritual abolition game and restitution of the real (or the paternal and/or maternal bodies). Hence the guilt that the creator, man or woman, associates with the act of creation. Writing is born, then, like dreams or

"Mi vocación literaria," from Revista Iberoamericana 51 *(1985): 132–133. Translated by* Nicolás Wey: copyright © 1991 by Sara Castro-Klarén.

myths, emanating out of an ever-present "nowhere" that we used to call "muse" or "inspiration," and that we now term "the unconscious." Writing is marked by Eros and death.

The duel between opposites that must be reconciled comes from the oldest cosmovisions. Toward the wedlock of the archetypal pair, alchemy fuses masculine and feminine (Sun-Moon, light-darkness) together in the philosophal flask. The figure incarnating the artwork is androgynous. Each text, each painted canvas, and each musical composition is the celebration of this alchemical mystery.

What, then, do we talk about when we say that there might be something specifically feminine to texts written by women? Just as the unconscious, ruled by desire, toys with immortality (which religious belief elaborates into the notion of soul transmigration, reincarnation, and eternal life bestowed through Eleusinian initiation or through baptism), it can, within the phantom's domain, assign to us a body that is not our own. This spectral body is what's reflected on the work's mirror.

There is, of course, a conditioning of feminine "identity" that can be traced back often to cultural conventions. But the feminine and the masculine, present in all psychic structures, stage between them alliances and combats that determine options, choices, and vocations. Within this plot woven from the inside (a plot that also defines each of us), identifications at the various levels of the psychic structure might or might not match corporeal or material signs.

On the other hand, we cannot reflect about woman and about the creative process without facing the relative scarcity of creative women, even when their access to culture might have occurred under conditions similar to those of men. We would have to think that the ability to engender flesh-and-bone creatures (children) satisfies in a great many of them the need for reconstruction that man can satisfy only by producing cultural objects. The "labor pains" that male artists usually associate with the creative process are equivalent to childbirth pain.

The most immediate and natural of creations, a child, is the privilege of woman, who consequently would remain closer to the world of nature than to that of culture. The world of culture emerges through the symbolic. Anaïs Nin, who since the age of eleven has wanted to restore the image of a lost father through the magic of the written word, believes that woman is closer to nature, to instinct and to intuition. In her own words, ". . . what neither Larry nor Henry understand is that creation for woman, far from that for man, must be exactly like the creation of children. It must come out of her own blood, wrapped in her womb, nourished by her own milk. It must be human creation, made of flesh, different from male abstraction. She must descend to the real womb and show her secrets and labyrinths. . . . She is the fish-tailed siren submerged in the unconscious. Her feat will

be to have articulated that dark world which dominates man (although he denies it) and which confirms its dominion through destructive demonstrations of its presence, of its madness." For Anaïs Nin, the feminine is the night, the ocean. Djuna Barnes vs. Simone de Beauvoir. Nevertheless, this bias becomes tempered when one recognizes that every artist has a "feminine" or "intuitive" side, and that talking about "women writers" is only a phase that must be left behind.

Up until this century, women's confinement to the hermetic space of the house, without outwardly directed activities, determined the tendency in the few women who materialize their latent vocation into a book, to write mostly about passion and about the web of human ties in the domestic setting. Virginia Woolf said, "How peaceful one feels down here, rooted in the center of the world." She was alluding to the sense of fullness that woman can receive from her house. With the limitation of her vital space, she is deprived of a wide range of experiences that were forbidden to her, but she will simultaneously deepen her experience of what is fundamental. She will appropriate for herself the pulsating heart of what is real. Her realm was not that of political ambition, economic competition, power, or heroic deeds. History is the struggle for power, and woman remained marginal to history. When women sat down to write, the result was *Wuthering Heights,* not *Le Rouge et le noir.* But it is also true that a man wrote *Madame Bovary* and that, before him, others had written *Werther* and *La Nouvelle Héloïse.* Might there be a "masculine" or a "feminine" mode of tackling the eternal themes, which themselves take us back to the duel between love and death? I don't know. In the act of creation, we always set out from something dark and chaotic that becomes articulated and objectified through form. There are writers, women and men, who descend more deeply than others into the tenebrous spaces of interior cities, or who come closer in their labyrinthine travels to the heart of this darkness. Then, subjective phantoms become something exterior, living, and objective: the work of art.

Virginia Woolf's notion of the artist as an androgynous being seems to be best suited for the enigmatic, unpredictable nature of creativity. For her, to perceive oneself as only a man or a woman is to be isolated: a purely masculine or feminine mind cannot create. One is closer to truth when both sides of human nature are integrated. To write is to obturate a vacuum and, in this sense, its sign is masculine. But the movement that generates this penetration of the spaces of the real through words preserves the double sign of an identity duel that took place in the archaic *illo tempore* of each internal process. Hence the ambiguity inherent to works of art (even those composed of words), and at the same time, the integrating energy that they irradiate, transmit, and perpetuate through time.

Nélida Piñón

NÉLIDA PIÑÓN was born in Rio de Janeiro in 1937. She has said that from the time she was an adolescent she knew that she could be nothing other than a writer. Starting with the publication of her first novel, *Guia-mapa de Gabriel Arcanjo* (1961), on to her first collection of short stories, *Tempo das Frutas* (1966; "Brief Flower," 1969), Piñón's narrative has been considered experimental. Concerned with exploration of woman's condition, Piñón's plots are always intricate and filled with anguish. Her prose and her character treatment rely on a subtle irony that dismantles many cultural traditions and narrative conventions. Mixing humor and a keen intelligence, Piñón creates worlds in which her characters appear constantly besieged by passion. Her fiction has steadily grown in complexity and scope. *Fundador* (1969) was her first major attempt to explore by way of parallel plots the movement of history. With her fourth novel, *A Casa da Paixão* (1972; "House of Passion," 1977), Piñón brings into Brazilian literature a new and unabashed "feminine" eroticism. Her later novels, *A doce Canção de Caetana* (1987) and *A República does Sonhos* (1984; *The Republic of Dreams,* 1989) have been received with great acclaim by Brazil's critics. In these two rich and supple stories, Piñón engages time and consciousness in order to portray the history of Brazil as a nation of dreams, a nation yet in the future, alive in dreams of generations of families who migrated to the land of the Tupi Guaraní.

The ten novels she has written to date have been translated into English, French, Polish, and Spanish.

I Love My Husband

I love my husband. From morning to night. Scarcely awake, I offer him coffee. He sighs, exhausted by his usual poor night's sleep, and begins to shave. I knock on his door three times, lest his coffee get cold. He grunts in anger and I clamor in distress. I don't want my effort confused with a

---

"I Love My Husband" translated by Claudia Van der Huevel.

cold liquid that he will consume just as he consumes me twice a week, especially on Saturdays.

Afterwards, I fix the knot of his tie and he protests because I have fixed merely the smallest part of his life. I laugh so he can go off more calmly, ready to face life outside and bring an always warm and bountiful loaf of bread back to our living room.

He says that I'm demanding, that I stay home washing dishes, go shopping, and on top of that, complain about life. Meanwhile, he builds his world with little bricks, and though some of these walls topple to the ground, his friends compliment him on his effort at creating brickyards, all solid and visible, from clay.

They salute me, too, for nourishing a man who dreams of mansions, shanties, and huts, and so contributes to the nation's progress. And that is why I am the shadow of the man that everyone says I love. I let the sun enter the house, to brighten up the objects bought with our joint effort. Even so, he never compliments me on the luminescent objects. To the contrary, because of the certainty of my love, he proclaims that I do nothing but consume the money that he brings together in the summer. Then I ask him to understand my nostalgia for the terrain formerly worked by woman; he furrows his brow as though I were proposing a theory that disgraces the family and the definitive deed to our apartment.

What more do you want, woman? Isn't it enough for you that we are married with community property? And while he was saying that I was part of his future, one that only he, however, had the right to build, I noticed that the man's generosity qualified me to be merely the mistress of a past whose rules were dictated in shared intimacy.

I began to think longingly about how wonderful it would be to live only in the past, before this preterit time was dictated for us by the man whom we say we love. He applauded my scheme. Within the house, in the oven that was the hearth, it would be easy to nourish the past with herbs and oatmeal so that he could calmly manage the future. He definitely couldn't preoccupy himself with my womb, which must belong to him in such a way that he wouldn't need to smell my sex to discover who else, besides him, had been there, had knocked at the door, had scratched inscriptions and dates onto its walls.

My son must be only mine, he confessed to his friends on the Saturday of the month that we entertained. And a woman must be only mine and not even her own. The idea that I couldn't belong to myself, touch my sex to purge it of excesses, provoked the first shock to the fantasy about the past in which I had been immersed until then. So the man, as well as having shipwrecked me in the past while he felt free to live the life to which only he had access, also needed to bind my hands, so my hands wouldn't feel the softness of my own skin. Would this softness perhaps tell me, in a

low voice, that there was other skin equally sweet and private, covered with velvet fuzz, and that one could lick its salt with the tongue's help?

I looked at my fingers, revolted by the long, magenta-painted nails. Nails of a tiger that strengthened my identity and grunted about the truth of my sexuality. I caressed my body; I thought, Am I a woman only through the long claws, and by clothing them with gold, silver, the sudden rush of blood of an animal slaughtered in the woods? Or because the man adorns me in such a way that when I remove this warpaint from my face he is surprised by a visage he doesn't recognize, which he covered with mystery so he wouldn't have me whole?

Suddenly, the mirror seemed to me the symbol of a defeat that the man brought home to make me become pretty. Isn't it true that I love you, husband? I asked him while he was reading the newspapers to keep himself informed, and I was sweeping up the letters of print spit out on the floor as soon as he assimilated the news. He said, Let me go on, woman. How can you expect me to talk about love when they're discussing the economic alternatives of a country in which the men need to work twice as hard as slaves to support the women?

Then I said to him, If you don't want to discuss love, which, after all, could perfectly well be far from here, or behind the furniture where I sometimes hide the dust after I sweep the house, what if after so many years I were to mention the future as though it were a kind of dessert?

He put the newspaper aside and insisted that I repeat what I had said. I spoke of the word *future* cautiously, not wanting to wound him, but I no longer refrained from an African adventure recently initiated at that very moment. Followed by a retinue oiled with sweat and anxiety, I was slaughtering the wild boars, immersing my canines in their warm jugulars, while Clark Gable, attracted by my scent and that of the animal in convulsions, was begging for my love on bended knee. Made voracious by the effort, I gulped river water, perhaps in search of the fever in my innards that I didn't know how to arouse. My burning skin, the delirium, and the words that sullied my lips for the first time, as I blushed with pleasure and modesty, while the witch doctor saved my life with his ritual and the abundant hair on his chest. Health in my fingers, the breath of life seemed to go out of my mouth, and then I left Clark Gable tied in a tree, eaten slowly by ants. Imitating Nayoka, I went down the river, which had almost attacked me by force; avoiding the waterfalls, I proclaimed liberty, the most ancient and myriad of inheritances, with shouts.

My husband, with the word *future* floating in his ears and the newspaper fallen to the floor, demanded to know: What does this repudiation of a love nest, security, tranquillity, in short, our marvelous conjugal peace, mean? And do you think, husband, that conjugal peace allows itself to be bound by threads woven through guile, simply because I mentioned that word which saddens you so much that you start to cry discreetly, for your pride

doesn't permit you the convulsive weeping that is set aside for me as a woman? Oh, husband, if such a word has the impact to blind you, I'll sacrifice myself once more so I won't see you suffer. Could it be that there is still time to save you, by blotting out the future now?

His glistening craters quickly absorbed the tears, he inhaled the cigarette smoke voluptuously, and resumed reading. It would be difficult to meet a man like him in our building with its eighteen floors and three entrances. At the condominium meetings at which I was present, he was the only one to overcome the obstacles and forgive those who had offended him. I blamed my egoism, for having thus disturbed the night of someone who deserved to recover for the following day's efforts.

To hide my shame, I brought him fresh coffee and chocolate cake. He allowed me to redeem myself. He spoke to me about the monthly expenses. With the company's balance sheet slightly in the red, he had to be careful about expenditures. If he could count on my cooperation, he would dismiss his partner in less than a year. I felt happy to participate in an event that would see us make progress in twelve months. Without my backing, he would never have dreamed so high. I took upon myself, from a distance, his capacity to dream. Each of my husband's dreams was maintained by me. And, for that right, I reimbursed life with a check that couldn't be entered into the books.

He didn't need to thank me. He had attained perfection in tender feelings in such a way that it was enough for him to remain in my company to indicate that he loved me. I was the most exquisite fruit of the earth, a tree in the center of our living room terrain, he climbed the tree, he reached the fruits, he stroked their rind, pruning the tree's excesses.

For a week I knocked on the bathroom door with just an early-morning touch. Ready to make him new coffee if the first became cold, if, forgetfully, he stood looking at himself in the mirror with the same vanity that was instilled in me from birth as soon as it was confirmed that they were dealing with another woman. To be a woman is to lose oneself in time, that was my mother's rule. She meant, who conquers time better than the feminine condition? Father applauded her, finishing up: Time is not the woman's aging, but rather her mystery never revealed to the world.

Just think, daughter, what is more beautiful, a life never revealed, which no one has gathered except your husband, the father of your children? Paternal teachings were always serious, he gave the luster of silver to the word *aging*. I became certain that in return for not fulfilling the story of woman, that biography of her own not permitted to her, she was assured youth.

Only those who live, age, my father said on my wedding day. And because you will live your husband's life, we guarantee you that through this act you will always be young. I didn't know how to get around the jubilation that enveloped me with the weight of a shield, and to go to his

heart, surprise his sincerity. Or to thank him for a state I had not desired before, perhaps through distraction. And that whole trophy on the very night I would be transformed into a woman. For until then, they whispered to me that I was a beautiful anticipation. Different from my brother, for already at the baptismal font they had affixed to him the glorious stigmata of man, before he had slept with a woman.

They always told me that a woman's soul emerged only in bed, her sex anointed by the man. Before him, my mother hinted, our sex most resembles an oyster nourished in saltwater, and therefore vague and slippery, far from the captive reality of earth. Mother liked poetry, her images were always fresh and warm.

My heart blazed on my wedding night. I yearned for the new body they had promised me, to abandon the shell that had covered me in everyday life. My husband's hands would mold me until my final days, and how could I thank him for such generosity? For this reason, perhaps, we are as happy as two creatures can be when only one of them brings to the hearth food, hope, the faith, a family history.

He is the only one to bring me life, even though sometimes I live it a week late. Which makes no difference. I even have an advantage, because he always brings it translated. I don't have to interpret the facts, fall into error, appeal to those disquieting words that end up silencing liberty. For my whole life I will need men's words. I don't have to assimilate a vocabulary incompatible with my destiny, capable of ruining my marriage.

Thus I proceeded to learn that my conscience, which is at the service of my happiness, is at the same time at the service of my husband. Its duty is to prune my excesses. Nature endowed me with the desire to be shipwrecked at times, to go to the bottom of the sea in search of sponges. It is well suited to that purpose. If not to absorb my dreams, to multiply them in the bubbling silence of their labyrinths of seawater? I want a dream that can be grasped with a strong glove and at times be transformed into a chocolate torte for him to eat with shining eyes. Then we will smile together.

Ah, when I feel like a warrior, ready to take up arms and acquire a face that isn't mine, I immerse myself in a golden elation, I walk on roads with no addresses, as though from me and through my effort I should conquer another country, a new language, a body that sucks up life without fear or modesty. And everything within me trembles, I look, with an appetite for which I won't be ashamed later, at those who pass. Fortunately, it's a fleeting sensation; soon I seek the help of familiar sidewalks; my life is stamped on them. The shop windows, the objects, the friendly people, my pride, finally, in everything about my house.

These birdlike actions of mine are quite unworthy, they would wound my husband's honor. Contrite, I ask him forgiveness in my thoughts, I promise him I'll avoid such temptations. He seems to pardon me from a

distance, he applauds my submission to a happy everyday life, which obligates us to prosper each year. I confess that this anxiety embarrasses me; I don't know how to quell it. I don't mention it, except to myself. Not even conjugal vows keep me in rare moments from shipwrecking in dreams. Those vows that make my body blush but have not marked my life in such a way that I can point out the wrinkles that have come to me through their impetus.

I've never mentioned these short, dangerous gallops to my husband. He couldn't bear the weight of that confession. Or if I told him that on those afternoons I think about working outside the home, to pay for odds and ends with my own money. Obviously this madness seizes me precisely because of my free time. I am a princess of the house, he sometimes tells me, and with reason. Therefore, nothing should distance me from the happiness in which I am forever submerged.

I can't complain. Every day my husband contradicts the version in the mirror. I look at myself in it and he insists that I perceive myself wrongly. I am not, in truth, the shadows, the wrinkles with which I see myself. Like my father, he, too, is responsible for my eternal youth. He is kind in his feelings. He has never celebrated my birthday boisterously, so I have been able to forget to keep track of the years. He thinks I don't notice. But the truth is that at the end of the day I no longer know how old I am.

And he also avoids talking about my body, which was widened with the years. I don't wear the same styles as before. I have the dresses stored in the wardrobe, to be appreciated discreetly. At seven o'clock in the evening, every day, he opens the door knowing that I'm waiting for him on the other side. And when the television shows some blooming bodies, he buries his face in the newspaper; only we exist in the world.

I am grateful for the effort he makes in loving me. I struggle to thank him, though at times without wanting to, or some strange face that isn't his but that of an unknown man whose image I never want to see again, upsets me. I feel my mouth dry then, dry from an everyday life that confirms the taste of the bread eaten the night before, and that will nourish me tomorrow as well. A bread that he and I have eaten for so many years without complaint, anointed by love, bound by the wedding ceremony that declared us husband and wife. Ah, yes, I love my husband.

The Contamination of Language

We don't know any more what a society would be like if the organization had been contaminated by women. Language then would be affected, it would be

---

"The Contamination of Language," from an interview with Catherine Tinker in The 13th Moon, *1982.*

impregnated by the feminine point of view; it would not marginalize women, not bind us to certain things. We must make of the breaches in language our weapons, our woman's text.

—Nélida Piñón, interview in *Brasileiras,* edited by Maryvonne
La Pouge and Clelia Pisa (Paris: des femmes, 1978)

Q: Do you still see woman's language in the same way as you did at the time of this interview for *Brasileiras,* or have there been changes?

Piñón: Well, I can't repeat myself because I've fought against memorization as a conscious destruction I operate on myself. We're just starting to impregnate the language, haven't done enough yet to have changed it. Just a "feminine sensibility" won't give us an identity. What is a sensibility which makes a woman different from a man except a sensibility of a sociology, the articulation of her reality? Women are taught to be subtle, to be sensitive, to have a fake *politesse*—not to be the one who takes the first step, but to give back what she has just received.

Q: Can women use men's language at all to communicate our lives?

Piñón: Look at women writers who kept diaries in small towns at the beginning of this century. Those diaries were defined as feminine literature, but for me this is a man's literature. It's a "contrafaccion" [counterfact] of men's language, not a woman's literature, because woman doesn't have a language of herself. It's not her own. When you borrow something, you receive half of it, not all but some of it. It's men's language. Men are responsible for the fake sweetness of the fake women's literature. It's the only way, as if a man were writing the book, signing the book. That woman wasn't given any other chance to write any other kind of literature, so men are responsible for this "sugar" literature. I'm not talking about writers like Virginia Woolf or Djuna Barnes. Good women's literature is always confused with this sweet literature. In Brazil, people used to call Virginia Woolf "feminine" taste. I don't know what this means. Back to the "contrafaccion": men are responsible, although even in sweet literature there is an intellectual aspiration from the writer. In Brazil in the last century there were almost no women writers, but there were in existence unpublished and almost destroyed diaries. . . . It's the same as the poetic vision found recorded in recipes. Not today, of course. . . . But in the past, in the Iberian world, recipes were poetry, very beautiful; recipes had "volutas" [volutes] like a chapel, and beautiful names. Food was a construction, it was like raising a cathedral. . . . "Sweets" and "jams"—incredible; a kind of literary product. Of course, she wasn't supposed to write or have ideas of her own which deserved to be recorded. So she was ashamed of anything she would write. She was the first one to censor herself. She was a permanent vigilante of her own work. She was terribly scared of opening her own mind, of knowing more than she was supposed to know, of saying more than she

could say. Instead of opening her eyes, she had to close them. It was a work of shrinking, like African shrunken heads. She had a remote, hidden aspiration, the same as any human being. No matter who you are or what you know—even so, as a human being you have an aspiration. Even if it is the aspiration of dying or of producing a child, which is a creation, too, for someone who is miserable—to postpone death, to guarantee survival for yourself by creating "a person you have written." We were the books of our ancestors; we're their copyright. There's a contradiction: I say sweet literature is somehow a man's language, that he's responsible. We women are not responsible for such bad literature. There has always existed among women an aspiration for changing, for expressing, however remotely, what she was.

Woman has always been a remote soul, distanced from herself, as if she were a continent without land, a geography, waiting to be named by herself through a few very special women. But this did not correspond with the universe. For me it is impossible to believe that there would be only three capable women and the rest worth nothing. So I put these three women under suspicion; something is wrong. We can't accept these three women when we know that the totality of women are marginalized. We can't take that. I myself, I only think that I have something to say as a woman because there are nowadays thousands of women who are thinking, not exactly thinking the same as I am, but on the same level. Then I must be counted; then I can be numbered; now we are being numbered, giving women a sum. The contamination is happening at the heart of culture, in the articulation of realities. Men can say that women always contaminate from the cradle on, but in this role woman is only the speaker, the spokeswoman of a very severe code. Women have been stricter than men, very reactionary and conservative, but they had to be in order to survive, in order to be dignified. The code required pride and dignity for women, so a woman had to be more faithful than anyone to the code to be respected, to be part of the tribe. Otherwise she would be collectively raped and humiliated.

.   .   .

Q: How can we contaminate the language faster?

Piñón: It should be through consciousness and experience. That's how a young woman can see what to do with her own life. She has to suffer domination in order to change. She has to be hurt—offended, somehow, to feel almost morally strangled. Then she will know what to do with her life, how to fight for her own interest and direction. I can't transmit all my experience to someone; she has to experience it herself in a dynamic process, a reevaluation of life, of reality.

Elena Poniatowska

BORN IN PARIS in 1933, the daughter of a Polish count and a Mexican aristocrat, Elena Poniatowska has lived in Mexico since 1942. Though her latest narrative, *Flor de lis* (1988), has the subtitle "novel," much of the story of this Mexican family might be considered autobiographical. In a vast retrospective that goes back to the Paris of World War II, the narrator, Mariana, sketches the family's return to Mexico and the childhood of two dissimilar sisters.

Poniatowska was educated in Mexico and the United States; her first writing and public image are shaped by her hard-hitting, committed journalism. In spite of her French roots and her aristocratic upbringing, Poniatowska cultivates the mimesis of the language of Mexican rural and lower urban classes. Her ability to grasp the freedom and creativity inherent in rural Mexican speech has given her work an imprint all its own. She is not the first Mexican writer, however, to recognize the experimental vocation of Mexican oral Spanish; Juan Rulfo mapped the sarcasm and lyricism of that world in his classic *Pedro Páramo*.

There are two main sources in the fictional world of Poniatowska. One is the world she inscribes by means of fieldwork: examining personal correspondence and archives, holding interviews, and visiting people like Jesusa Palancares or the servants in *Flor de lis*. The second source comes from a more "literary" linguistic realm that tells the stories of "purely" fictional characters, such as her collection of short stories, *Los cuentos de Lilus Kikus* (1967).

The strongest and most successful of her creations are those that emerge from her research and her ability to give flesh and bone to the testimony of other people's lives. Both her acclaimed *Hasta no verte, Jesús mío* (1969) and *Querido Diego* (1978) stem from this cross between fieldwork and fiction. In what some have called a collective autobiography, Jesusa Palancares tells the story of her life in the most uninhibited language. Poniatowska renders into fictional form and writerly conventions Jesusa's journey from the fields of Oaxaca to the spiritualist churches of Mexico City, while still preserving Jesusa's style and conception of intimate, testimonial narration. Poniatowska carries out a similar feat, but with a totally different historical basis, in *Querido Diego*. This epistolary novel reconstructs the

story of love and betrayal between the Russian painter Quiela and her lover, the Mexican painter Diego de Rivera.

A parallel reading of *La noche de Tlatelolco: Testimonios de historia oral* (1971) (*Massacre in Mexico*, 1975) and *Hasta no verte, Jesús mío* shows the close similarity in theme, style, and commitment between Poniatowska's fiction and her journalism. At present, Poniatowska is finishing a biography of the Mexican anarchist, photographer, and femme fatale Tina Modotti.

Literature and Women in Latin America

The voice of the oppressed in Latin America includes literature by women. I believe this so deeply that I'm willing to make it a leitmotif, a ritornello, an ideology. This literature is the continuation, too, of that long shriek that Rosario Castellanos wrote about in her *Jornada de la soltera:* "and when the kinsmen gather / around the fire and the telling, / the shriek of a woman on a vast highland can be heard." We must think of women's literature today as flowing with the rich wordstream of the oppressed. Marta Traba bases her *Homérica Latina* and, above all, *Conversación al Sur,* on a single reality: repression, torture, disappearances. Claribel Alegría opts for writing on Nicaragua because she knows that nothing is more important than freeing a nation. María Luisa Puga's discovery of her own country's misery grows out of her experience in Africa. . . . The cause for this literature is a painful and outrageous experience. We want to bear testimony of this here and now, beyond self-deceit, and then we want to send our scroll in a bottle that will float anxiously to the other end of the ocean, to the other end of time, so that those to come will know how loud was the shriek of their predecessors.

We, Latin American women writers, come from very poor, forsaken countries. Poverty here is not that of the European indigent, or of the *clochard* who wraps himself philosophically in his coat for all seasons. Latin America's malaise is indifference. There is no one to whom to appeal, no one to whom one can say "I've gone for days without food," because no one cares. No soup from the Salvation Army awaits at the stove. There is no shelter for a night's rest. Hunger becomes like dust and spreads wearily over the earth. There is no cure for desolation. Try talking to someone who's hungry and you'll then see his lost gaze. The hungry lose their gaze forever to a dark crevice on the ground. It is as though they don't perceive their own existence, as though they don't know that they live. They grow,

"Literature and Women in Latin America," from Eco *(Bogotá) 3 (1983): 462–472. Translated by Nicolás Wey: copyright © 1991 by Sara Castro-Klarén.*

yes, but in pain, with a calculated resignation which they might not even suspect. They are alone among other people because they live at the margin. Even noise evades them. The world has drained life from them. Their silence is that of stone. This is why Elena Garro's novel, *Los recuerdos del porvenir*, opens by telling of a stone that contains the town's memory. It is also why Juan Rulfo tells us that, at the hour of his death, Pedro Páramo falls flat on the ground and disintegrates like a heap of rock. The sun in Mexico causes people to harden.

Latin America doesn't claim for itself the entire patrimony of misery in the world. We know that India is deeply marked by it, and that it is pervasive in Africa and in the Arab countries. We do claim, however, a particular type of misery, not Third World misery, as the term was coined by de Gaulle, but rather the misery of solitude, displacement, and indifference. Maps of Latin America display many unexplored regions in red. This is how we feel: virgin, intact, lacking a name or someone who will set out to discover us. When García Márquez wrote *Cien Años de Soledad* [One Hundred Years of Solitude] in capital letters on the cover of his manuscript, he was also defining us, except he was wrong about the number. It's been many more than a hundred years. Like Alejo Carpentier said, it's not that we want our ceiba trees to form part of the sweet family of European chestnut trees, nor that we want our local cuisine to be as distinct as Breughel's kitchen, no. Simply, we know that so far the existence we've led extends our lives without granting us rights. We exist, numerically, but states and oligarchies don't let us be.

I don't mean to say that feminine literature starts with twentieth-century revolutions in Latin America. I know that there were other battles, that the main victory was won by Sor Juana Inés de la Cruz, born in 1651, and that Octavio Paz now regards her as the greatest poet of the seventeenth century. But she was never recognized as such, because she was a woman. On the contrary, she wrote in a ballad, "I don't know of such things / I only know that I came here, / such that if indeed I am a woman / no one will know." Sor Juana at once discards the feminine exterior that weighs on her and that brands her. Not in vain does she cut her hair: so that she can retreat to her studies while it grows back, because a head lacking in knowledge ought not to display such an appetizing ornament. Because her love of letters is so pressing and powerful, Sor Juana leaves the court of the Marquis of Mancera and enters the convent of the Hieronymites at the age of nineteen. Only seclusion can offer what she looks for and what she calls "*las impertinencias de su genio*" [the impertinences of her character]. These are: the desire to live alone, deliverance from obligations that would have compromised her freedom to pursue knowledge, and distance from society's gossip that would have kept her from the tranquil silence of her books.

Three hundred years later, Rosario Castellanos—who hoped for a beautiful and true order—feels that the world not only deceives her, but that it is hostile, too. And the shrieks throughout the continent grow louder until Sylvia Plath yells that woman has been the most forsaken being in history. Woman in Mexico has been doubly subdued: once by the Conquest, which introduced Christianity, and again by the fact of her gender. Her perspective will always be one and the same with the defeated. A woman who does not subject herself to her husband is doomed to fail. The titles of novels about women in Latin America are very telling: *"Santa"* [Saint] or *"Monja, casada, virgen y mártir"* [Nun, married, virgin, and martyr]. . . .

After the colonial period, many men and women poets flourish in Latin America. They write poems out of thin air. On almost any occasion, women write. It's interesting to see that the women who continue writing remain single or become suicidal. If they are not outspoken against a regime, they protest against their internal condition. But since guilt is the best instrument for self-torture, they never cease to feel guilty for lacking "feminine" virtues: dependency on men, sweetness, obedience, deference, submission, discretion, innocence, bewilderment at human evil, inconsistence, and culinary disposition. When they give up writing, they take up embroidering handkerchiefs with edges designed for crying in dark corners. Passive before events, the *"versificadoras"* [verse composers], as the encyclopaedia calls them, bow down, like the branches on weeping willows toward the waters. But these same creatures of dovelike flight and fainting spells have also moved from the literatures of confession, diaries, intimist descriptions, mood, and love poetry to politically committed literature.

If privileged writers like Marta Traba, Griselda Gambaro, and Claribel Alegría have allied themselves with the oppressed, others have chosen to give a voice to those without one. Such are the cases of Jesusa Palancares, the protagonist of *Hasta no verte Jesús mío;* Domitila, the Bolivian miner; and their predecessor, Carolina Maria de Jesus. They tell their stories with great control of the material. Their stories have been called literature. I would like to discuss these women because, more than anyone, they represent the great majority of Latin Americans and contradictions unknown in Europe. In Latin America, unlike Europe, women are not a homogeneous mass sharing identical tasks.

Domitila stated this clearly when, during the Women's International Year in Mexico, she confronted a Mexican leader who had told her to quit talking about massacres and the suffering of her nation, and to start talking about "us women . . . about you and me . . . about woman, o.k.?"

" 'All right,' I said, 'we will talk about you and me. But if you allow me, I will begin. Señora, I have known you for almost a week. Every morning you wear something different; but not me. Every day you arrive, made up like someone who has the time and money to spend in an elegant beauty

shop; but not me. And the way you look at me, you must live in an elegant house in an elegant neighborhood, right?

"'Now, tell me: is your situation my situation? Is my situation yours? How, then, can we talk about equality between us? If you and I are so unlike each other, if you and I are so different, we cannot be equal right now, even as women, don't you think?'

"But then another Mexican woman said:

"'Hey, you: what do you want? She's the leader of the Mexican delegation, and she's got priority over you here. Besides, we have been benevolent toward you: we've listened to you on radio, and T.V., and in newspapers. . . . I'm tired of applauding you.'

"I felt very angry when she said this, because I thought that the issues I was addressing had only served to turn me into a theatrical actor to be applauded. I felt I was being treated like a clown. . . .'"

The poor are always disposable, exchangeable matter; they are the masses, the people, those who serve as a backdrop, those who live the other life, the exiles. Their condition as inferior beings dooms them, and they doom themselves. Jesusa Palancares in *Hasta no verte, Jesús mío* erases herself in one stroke:

"After all, I have no motherland. I am like the Hungarians—from nowhere. I neither feel Mexican nor do I claim Mexicans as my countrymen. It's all mere convenience and opportunism. If I had money and things, I would be a Mexican, but since I'm worse than garbage, I am nothing. I'm the garbage a dog pisses on as it passes by. The wind blows me away, and it's over. I'm garbage because I can be nothing else. All my life I've been the same microbe that looks out. . . ."

Jesusa's only hope is that death will not fall upon her by surprise, "because the majority of people come to laugh at the dying. Such is life. One dies so that others might laugh. They laugh at your hallucinations; you lie sprawled, crooked, your mouth deformed, crooked, your mouth open and your eyes bulging." This is why Jesusa wants to know the hour of her death, so that she will walk away, fall under a tree, and become carrion meat. Warning: "One day you won't find me here: You will find only the wind. That day will come, and when it does, there won't be anyone to tell you what happened, and you'll think it was all in your mind. It's true, we are not here for real; what they say on the radio is a lie; what the neighbors say is a lie; and it's a lie that you're going to miss me. If I'm of no use to you, why should you miss anything? They won't miss me at work either. Who do you expect to miss me if I'm not even saying good-bye?"

Jesusa and Domitila are destroyers of myths. In America, we never get a grip on reality. What's more, we don't know it. In America, for example, we'll never know who killed Kennedy even though we were all watching him on T.V. when it happened. Neither will we know what happened in

Mexico in October 1968. So instead we replace reality with abstract and moralistic by-products: fear, guilt; but we have no idea of what really is happening. Thus we're taken by surprise and soon find ourselves in the midst of terrifying confrontations. This is why Jesusa Palancares destroys a myth when she says that the Mexican Revolution was a bloodbath: "Many people killed each other left and right. I think it was a misunderstood war. Father against son, brother against brother: Carrancistas, Villistas, Zapatistas; it was all idiotic because in the end we were all the same, penniless and hungry. But, as they say, those are things we keep to ourselves because we keep their secret."

There can be no omniscient narrators in Latin America (only Carlos Fuentes). Everything escapes us. There is no past other than the immediate past. In Marta Traba's *Conversación al Sur*, the romantic sense of our societies is gone. Rather, everyday things are cruel. The police director who asks Irene, the actress who's been arrested along with the militant youths who admire her, for her passport, recriminates her: "You're an old hag; you should be ashamed of showing off your thighs like that." This line alone wipes out every romantic perception of Latin America. There are no more "Santas," "Doña Bárbaras," "Perricholis," or "Güeras Rodríguez." We are struck by the immediacy of reality, as we are struck by the misery of Jesusa in her little room.

In *Las posibilidades del odio*, María Luisa Puga transforms herself into a beggar who has lost one leg. His only aid is a wooden crutch with a black rubber tip. He considers this crutch a miracle, and he caresses it gently and evenly every day. How could María Luisa reach under a beggar's skin? How is it that she was able to move among awkward, smelly, querulous shadows? Simply because María Luisa is a Latin American writer: "Hunger and the man were one and the same. He had never known anything but hunger, and so he had gotten used to it, to the point that he no longer thought about eating. Sometimes in his alley, at night, when he chewed slowly on bread or even boiled potatoes that were left for him in a bag, the food would get stuck in his throat even though he chewed for a long time. Often, he slept with the food in his mouth. Fruit was something different: The juice would pour from the sides of his mouth, reminding him of old, unreachable things. But he ate everything very slowly, in silent fear."

Torture is the other great fear. You might argue that I'm not impartial, and that there are many things other than torture, disappearances, and hunger in Latin America. This is true. Within Latin American literature, themes are varied, rich, and not always focused on misery. No one can be told not to look at the world from above, from cosmic space; and it turns out to be truly poetic to bear testimony, along with Yury Gagarin, that the earth is blue. But we are not floating in the stratosphere. We're down here on the globe, and we add to the murmuring sound of human beings. I

would give the world to echo Carlos Pellicer's feeling that, in Latin America, nothing happens more transcendent than the roses. When sons, husbands, and friends disappear; when hundreds of men rot in their cells; when peasants are killed and their blood muddies the waters, then we have to admit that this is our reality (at least the one we choose voluntarily), and that we want to bring it into our writing because nothing is more real, more bitter, or more a cause for reflection than this pain. Nothing more formative or de-formative, constructive or destructive can happen to us. What can one write about but this infamy? If the majority of Latin Americans don't master the official language, if they have no access to Spanish, and if their Spanish is poor and timid, and if it disappears as it has for Panamanians, whose Spanish has been devoured by North American occupation, then we have to talk about disappearances, about that physical absence, about that man, that woman, and that child who are suddenly erased, just like that. Once there, he's not here now. He doesn't exist, and we doubt as we do in a bad dream. Did he exist? Did we really see him? Did he walk among us? Have we come to live on earth? Or to die? Or to dream that we live?

Slowly, we learn about more kidnappings in Latin America, and about political opponents whose houses are raided at four in the morning. Tires screech, doors slam, people run, and then there is nothing, nothing except the frightened neighbors who in the morning murmur: "Last night . . . a raid." But we didn't know that a woman who was walking next to us, holding our arm, held by our smile, could disappear just like that. Alaide Foppa went to Guatemala, and she died from a heart attack the second time she was tortured. The other torture is hunger, the prison of hunger, the disappeared by hunger, children imploring a father who kills himself because he's unemployed.

Life goes on thus on our side of the continent, parceled out by "reglas de vida" [rules for living] and by Carlos Monsiváis's *Catecismo para indios remisos* [Catechism for infidel Indians]. Our jobs and routine structure our lives. We are socially conscious. We read *Uno más Uno* (like the French read *Le Monde*) in order to reaffirm our civilized consciousness. We don't allow anyone to spoil our threepenny opera. And then one day a sixty-seven-year-old woman, fundamentally good and innocent, disappears just like that, like someone who leaves without saying farewell. And we look for her in clandestine jails and military camps until we begin to search for her inside of us. Nothing.

Nothing. This is why we write. We write in order to understand the incomprehensible, in order to bear testimony of things, so that our children's children will know. We write in order to be. We write so as not to be wiped off from the map. In Latin America, we write because this is the only way we know not to disappear, and in order to bear testimony about

those who disappear because of politics or hunger. We write so that they will know that for a span of time and light—scientists would call it a "lapse"—we lived here on earth; we were a point in space, a point of reference, a sign, a particle that moved and generated energy and heat, and then rejoined other particles. This is why we write.

Rosario Ferré

ROSARIO FERRÉ, born in 1938 in Ponce, Puerto Rico, to a very wealthy family, has been educated to a large extent in the United States. She holds a B.A. from Marymount College in Manhattan, and an M.A. in Hispanic literature from the University of Puerto Rico. She also holds a Ph.D. in Hispanic languages and literatures from the University of Maryland. She was editor of *Zona de cargada y descargada* (1970–1974), a literary magazine which introduced feminist inquiry into the study of literature in the Spanish-speaking world. Ferré writes poetry, short stories, and longer fiction, as well as literary criticism, with equal ease. Her first books, *Papeles de Pandora* (1976), *Sitio a Eros* (1980), and *Fábulas de la garza desangrada* (1982), were met with unreserved acclaim by the critical establishment, which noted her work for its enormous capacity for symbolism in conjunction with a clear feminist stance. In her first novel, *Maldito amor* (1986), the unresolved historical status of her island nation is a central theme, with women occupying a recurring and dominant presence. Her own successful English translation of this novel established her as an accomplished bilingual writer.

Ferré's short stories have won several awards. In *La muñeca menor* (1980; "The Youngest Doll," 1980), she evokes the disappearing culture of the sugar cane plantation with sharp feminist criticism of a charming lifestyle founded on the oppression of women of all colors and classes. Her first collection of short stories, *Los cuentos de Juan Bobo* (1981), was considered by some to be "children's literature" because of the childlike and stark conceptualization of the stories' social scenarios. In one of them, she models Juan Bobo, the idiot, after a character from Puerto Rican oral tradition; like many other great fools in literary tradition, Juan understands only too well how society has conspired in order to benefit the few at the expense of the many.

It is perhaps with the publication in Mexico of the collection of poems *Fábulas de la garza desangrada* that Ferré's imagination connects with her strongest and best-developed theme: the ambiguity and the double value of women in society, from Desdemona and Mary Magdalene to Julia de Burgos.

Destiny, Language, and Translation,
or Ophelia Adrift in the C & O Canal

A few weeks ago, when I was in Puerto Rico, I had an unusual dream. I dreamt I was still in Washington but was about to leave it for good. I was traveling on the C & O Canal, where horse-towed barges full of tourists still journey picturesquely today, led by farmers dressed up in costumes of colonial times. I had crossed the canal many times before, entering the placid green water that came up to my waist without any trouble, and coming out on the other side, where the bright green, daffodil-covered turf suspiciously resembled the Puerto Rican countryside. This time, however, the canal crossing was to be definitive. I didn't want my five professionally productive years in Washington to become a false paradise, a panacea where life was a pleasant limbo, far removed from the social and political problems of my country. I felt that this situation could not continue, and that in order to write well about my world's conflicts, as war correspondents have experienced, one has to be able to live in the trenches of the war and not on the pleasant hillocks that overlook the battlefield. Thus, I had finally decided to return to Puerto Rico for good.

As I began to cross the canal, however, and waded into the middle of the trough, I heard a voice say loudly that all the precautions of language had to be taken, as the locks were soon to be opened and the water level was going to rise. Immediately after this someone opened the heavy wooden gates of the trough at my back and a swell of water began to travel down the canal, lifting me off my feet and sweeping me downstream, so that it became impossible to reach either of the two shores. At first I struggled this way and that, as panic welled up in me and I tried unsuccessfully to grab onto the vegetation that grew on the banks, but I soon realized the current was much too powerful and I had no alternative but to let it take hold of me. After a while, as I floated face up like Ophelia over the green surface of the water, I began to feel strangely at ease and tranquil. I looked at the world as it slid by, carried by the slowly moving swell of cool water, and wondered at the double exposure on both shores, the shore of Washington on my right and that of San Juan on my left, perfectly fitted to each other and reflected on the canal's surface like on a traveling mirror on which I was magically being sustained. In the water of the canal, in the water of words "where all the precautions of language had to be taken," everything melted into one, blue sky and green water, North and South, earth and vegetation ceased to be objects or places and became passing states, images in motion. The water of words was my true habitat as a writer; neither Washington nor San Juan, neither past nor present, but the crevice in between. Being a writer, the dream was telling me, one has to

learn to live by letting go, by renouncing ever to reach this or that shore, but to let oneself become the meeting place of both.

In a way all writing is a translation, a struggle to interpret the meaning of life, and in this sense the translator can be said to be a shaman, a man dedicated to deciphering conflicting human texts, to searching for the final unity of meaning in speech. A translator of a literary text acts like the writer's telescopic lens; he is dedicated to the pursuit of communication, of that universal understanding of original meaning that may perhaps make possible one day the harmony of the world. He struggles to bring together immensely differing cultures, striding over the barriers of those prejudices and misunderstandings that are the result of diverse ways of thinking and of cultural mores. He wrestles between two swinging axes, which have, since the beginning of mankind, caused innumerable wars to break out and civilizations to fail to understand each other: the utterance and the inter- pretation of meaning; the verbal sign (or form) and the essence (or spirit) of the word.

I believe that being both a Puerto Rican and a woman writer has given me the opportunity to experience translation (as well as writing itself) in a special way. Only a writer who has been able to experience two cultures deeply can be called a bilingual writer, and being a Puerto Rican has enabled me to acquire a profound knowledge both of Spanish and English, of the Latin American and of the North American way of life. Translation is not only a literary but also a historical task; it includes an interpretation of internal history, of the changing proceedings of consciousness in a civilization. A poem by Gongora, written in the seventeenth century, can be translated literally, but it cannot be read without taking into account the complex cultural connotations that the Renaissance had in Spain. Lan- guage, in the words of George Steiner, is like a living membrane; it provides a constantly changing model of reality. Every civilization is imprisoned in a linguistic contour, which it must match and regenerate according to the changing landscape of facts and of time.[1] Only a writer who has experienced both the historical fabric and the inventory of felt moral and cultural existence embedded in a given language can be said to be a bilingual writer.

When I write in English I feel that the landscape of experience, the fields of idiomatic, symbolic communal reference are not lost but are relatively well within my reach, in spite of the fact that Spanish is still the language of my dreams. Writing in English, however, remains for me a cultural translation, as I believe it must be for such writers as Nabokov and Vassily Aksyonov, who come from a country (the USSR) whose cultural matrix is also very different from that of the United States. Translating a literary work (even one's own) from one language to another, curiously implies the same type of historical interpretation that is necessary in translating a poem of the seventeenth century, for example, as contemporary cultures often

enclosed in different epochs of time coexist with each other. This is precisely what happens today with North American and Latin American literatures, where the description of technological, pragmatic, democratic modern states coexists with that of feudal, agrarian, and still basically totalitarian states. Translating literature from Spanish into English (and vice versa) in the twentieth century cannot but take into account very different views of the world, which are evident when one compares, for example, the type of novel produced today by Latin American writers such as Carlos Fuentes, Gabriel García Márquez, and Isabel Allende, who are all preoccupied by the processes of transformation and strife within totalitarian agrarian societies, and the novels of such North American writers as Saul Bellow, Philip Roth, and E. L. Doctorow, who are engrossed in the complicated unraveling of the human psyche within the dehumanized, modern city-state.

Translating has taught me that it is ultimately impossible to transcribe one cultural vision into another. I am inevitably translating a Latin American vision, still rooted in preindustrial traditions and mores, with very definite philosophical convictions and beliefs, into a North American context. As Richard Morse has so accurately pointed out in his book *Prospero's Mirror: A Study in the Dialectics of the New World*,[2] Latin American society is still rooted in Thomistic, Aristotelian beliefs, which essay to reconcile Christian thought with the truths of the natural universe and of faith. Spain (and Latin America) have never really undergone a scientific or an industrial revolution, and they have never produced the equivalent of a Hobbes or a Locke, so that theories such as those of pragmatism, of individual liberty, and of the social contract have been very difficult to implement.

In translating my own work, I came directly in contact with this type of problem. In the first place, I discovered that the Spanish literary tradition permitted a much greater leeway for what may be called "play on words," which habitually sounded frivolous and innocuous in English.[3] In Puerto Rico, as in all of Latin America, we are brought up on a constant juggling of words, which often has as its purpose the humorous defiance of apparent social meanings and established structures of power. In undermining the meaning of words, the Latin American child (as does the writer) puts into question the social order that he is obliged to accept without sharing in its processes. This defiance through humor has to do with a heroic stance (*el relajo, la bachata, la joda*), often of anarchic origins, which is a part of the Latin personality, but it also has to do with faith, with a Thomistic belief in supernatural values. It is this faith in the possibility of utopia, of the values asserted by a society ruled by Christian, absolute meanings rather than by pragmatic ends, which leads the Puerto Rican child to revel in puns such as "Tenemos mucho oro, del que cagó el loro" (We have a lot of gold,

of the kind the parrot pukes), or "Tenemos mucha plata, de la que cagó la gata" (We have a lot of silver, of the kind the cat shits), which permit him to face and at the same time defy his island's poverty; or in popular Puerto Rican sayings of the blackest humor and unforgiving social judgment, such as, "El día que la mierda valga algo, los pobres nacerán sin culo" (The day shit is worth any money, the poor will be born without assholes).

A second characteristic that helps to define Latin American vis-à-vis North American literature today often has to do with magical occurrences and the world of the marvelously real ("lo real maravilloso"), which imply a given faith in the supernatural world that is difficult to acquire when one is born in a country where technological knowledge and the pragmatics of reason reign supreme. We are here once again in the realm of how diverging cultural matrixes determine to a certain extent the themes that preoccupy literature. In technologically developed countries such as the United States and England, for example, the marvelous often finds its most adequate expression in the novels of writers like Ray Bradbury and Lord Dunsany, who prefer to place their fiction in extraterrestrial worlds where faith in magic can still operate and the skepticism inherent in inductive reasoning has not yet become dominant.

As I translated my novel *Maldito Amor,* for example, the issues I have just mentioned came to my attention in a curious way. The first serious obstacle I encountered was the title. "Maldito Amor" in Spanish is an idiomatic expression impossible to render accurately in English. It is a love that is halfway between doomed and damned and thus participates in both without being either. The fact that the adjective *maldito,* furthermore, is placed before the noun *amor,* gives it an exclamative nature that is very present to Spanish speakers, even though the exclamation point is missing. "Maldito Amor" is something very different from "Amor Maldito," which would clearly have the connotation of "devilish love." The title of the novel in Spanish is, in this sense, almost a benign form of swearing, or of complaining about the treacherous nature of love. In addition to all this, the title is also the title of a very famous *danza* written by Juan Morell Campos, Puerto Rico's most gifted composer in the nineteenth century, which describes in its verses the paradisiacal existence of the island's bourgeoisie at the time. As this complicated wordplay would have been totally lost in English, as well as the cultural reference to a musical composition that is well known only on the island, I decided to change the title altogether, replacing it with a much more specific one, "Sweet Diamond Dust." The new title refers to the sugar produced by the De Lavalle family, but it also touches on the dangers of a sugar that, like diamond dust, poisons those who sweeten their lives with it.

The inability to reproduce Spanish wordplay into anything but an inane juggling of words not only made me change the title; it soon made me

begin to mercilessly prune my own sentences like overgrown vines because, I found, the sap was not running through them as it should. How did I know this? What made me arrive at this conclusion? As I faced sentence after sentence of what I had written hardly two years before, I realized that writing in English I had acquired a different instinct in my approach to my themes. I felt almost like a hunting dog forced to smell out the same prey, which has drastically changed its spoor.

My faith in the power of the image, for example, was now untenable, and facts had become much more important. The dance of language had now to have a direction, a specific line of action, if I wanted to succeed at verisimilitude. The possibility of utopia, and the description of a world in which the marvelously real sustained the very fabric of existence, was still my goal, but it had to be reached by a different road. The language of technology and capitalism, I said to myself, must above all assure a dividend, and this dividend cannot be limited to philosophic contemplations or to a feast of the senses and of the ear. Thus, I delved into a series of books on the history and sociology of the sugarcane industry in Puerto Rico, which gave me the opportunity to widen the scope of the novel, situating its events in a much more precise environment.

Is translation of a literary text possible? I asked myself, seeing that, as I translated my own work, I was forced to substitute, cancel, and rewrite constantly, now pruning now widening the original text. In the philosophy of language, and in reference to translation in general (not necessarily of a literary text), two radically opposed points of view can and have been asserted. One declares that the underlying structure of language is universal and common to all men. "To translate," in Steiner's words, "is to descend beneath the exterior disparities of two languages in order to bring into vital play their analogous and . . . common principles of being."[4] The other one holds that "universal deep structures are either fathomless to logical and psychological investigation or of an order so abstract, so generalized as to be well-nigh trivial."[5] This extreme, monadistic position asserts that real translation is impossible and that what passes for translation is a convention of approximate analogies, "a rough-cast similitude," just tolerable when the two relevant languages or cultures are cognate."[6]

I lean rather more naturally to the second than to the first of these premises. Translating literature is a very different matter from translating everyday language, and I believe it could be evaluated on a changing spectrum. Poetry, where meaning can never be wholly separated from expressive form, is a mystery that can never be translated. It can only be transcribed, reproduced in a shape that will always be a sorry shadow of itself. This is why, according to Robert Frost, poetry is what gets lost in translation; and José Ortega y Gasset evolved his theory on the melancholy of translation, in his *Miseria y esplendor de la traducción*. A bit more to the

right one could place prose and poetic fiction, where symbolic expression may alternate with the language of analysis and communication. Here one could situate novels and prose poems that employ varying degrees of symbolic language and are directed toward both an intuitive *and* an explanatory exposition of meaning. In the far right one could place literary texts of a historical, sociological, and political nature, such as the essays of Euclides Da Cunha in Brazil, for example, or Fernando Ortiz in Cuba or Tomás Blanco in Puerto Rico. These texts, as well as those of literary critics who have been able to found their analytic theories on a powerfully poetic expression (such as Roland Barthes, for example), are perhaps less difficult to translate, but even so the lacunae that arise from the missing cultural connotations in these essays are usually of the greatest magnitude.

Notes

1. George Steiner, *After Babel: Aspects of Language and Translation* (Oxford: Oxford University Press, 1975), 21.
2. Richard M. Morse, *El espejo de Prospero, un estudio en la dialectica del nuevo mundo* (Mexico: Siglo XXI, 1982).
3. As every reader of Cabrera Infante in translation (*Holy Smoke,* for example) knows, musical wordplay in Spanish, no matter how brilliant, can sound puerile in English. Another example of the nemesis of translation of a novel based on wordplay and popular language is the English version of Luis Rafael Sanchez's *La guaracha del Macho Camacho* (Buenos Aires: Ediciones de la Flor, 1976). In spite of Gregory Rabassa's masterful rendition into Bronx English, *Macho Camacho's Beat* (New York: Pantheon, 1980) still sounds like salsa played on a synthesizer rather than a bongosero's drum.
4. Steiner, *After Babel,* 73.
5. Ibid., 74.
6. Ibid.

Domitila Barrios de Chungara

DOMITILA BARRIOS DE CHUNGARA was born in Bolivia in the mining area of Pulacayo in 1937. She became a voice for the Bolivian miners, traveling to Mexico City for the Conference on the International Year of the Woman organized by the United Nations in 1975. As a representative of the Comité de Amas de casa de Siglo XX, the miners' wives' committee she had headed for many years, Barrios was noticed for her challenge to the hegemonic control of "feminism" as defined by a woman politician from the PRI. After this encounter, Barrios de Chungara accepted the invitation by the Brazilian anthropologist Moema Viezzer to set down in a series of interviews her "testimony," that is, her experience as woman, wife, mother of seven, and social activist for the miners of Siglo XX and in the Bolivian political process.

With the Yungas, So That I May Not Speak

I didn't know we were on our way to becoming exiles. The truck hit the road and I fell asleep.

At dawn I woke up . . . It was very hot . . . and you could hear birds singing something like "chiu, chiu." I looked up and saw a multitude of trees.

"Where are we?" I yelled.

My husband said, "Calm down. You're all right . . . be calm."

He began to sweet-talk me.

I looked at him, and suddenly I recognized him. "Where are we?"

"We're going where you'll get better, where you'll be healthy again. Calm down."

"Where are you taking me?" I started screaming.

My husband told the driver to stop the truck. My father, who was riding in front, stepped down. He hugged me, cried, and said:

"With the Yungas, So That I May Not Speak," from "Si me permiten hablar" . . . Testimonio de Domitila, una mujer de las minas de Bolivia. Ed. Moema Viezzer (Mexico: Siglo XXI Editores, 1977). Translated by Nicolás Wey: copyright © 1991 by Sara Castro-Klarén.

"The important thing is that we save your life, child. That's what matters."

And since he is so religious and sees everything as God's work, he added:

"God is so great that he has allowed you to remain alive. He will allow you to save yourself. We're heading for a paradise where you won't suffer any longer. You'll go to Los Yungas [the jungle people]; you'll live there. When you're strong again we'll come back to Siglo XX together."

He cheered me up like this.

We arrived at Los Yungas. With the little money that was left, we bought a little house and a piece of land for planting crops. Then my husband traveled to Siglo XX to bring our children.

Then I learned that I would have to report myself daily at the local DIC [detention center] to sign a book where I stated that I wasn't moving from the area. In other words, I had been, as they say, deported, along with all my family. I didn't have the right to leave Los Yungas.

There was no medical care there. There was no one to give me shots of the antibiotics that the doctor had prescribed. And in that heat I had never known in the highlands, and those mosquitoes I'd never seen . . . with all that and the wounds on my body, I began to rot. The wounds became terribly infected. I knew that I was near death. I had violent chills. And I felt so, so sick that, in my despair, in the end I began to drink the injectable antibiotics with my tea. I would shower in cold water all the time. I placed damp cloths over my body. Barely, barely did I save myself. How long I have suffered this pain!

Aside from that, in my dreams, I couldn't let go of my little girl. And when my husband returned to the mine to look for our other children, I would leave the house screaming at night. I would see my little girl . . . It was horrible . . . It felt somewhat oppressive . . . I saw the faces of my executioners . . . I heard their laughter . . . I saw that they were eating my little girl . . .

I thought I would go mad. At times I felt like jumping off a tall rock and ending things right there. The mosquitoes would bite me . . . Everything bothered me. If the thought of seeing my children hadn't been so strong . . . I think I'd have killed myself that time, because I was devastated. I didn't want to go on suffering. The wounds hurt; I couldn't rest. And when I fell asleep, I would dream horrible things. It was something . . . ah!
. . .

Then my husband arrived, with the children, and I felt some relief. He brought me some medicine and bandages. And with these things I was able to cure myself and heal. But barely . . . barely! It was so hard!

Everything was different in Los Yungas. In the highlands we used to eat meat, bread, sugar. In Los Yungas, people ate only yucca, plantain, things we were not used to eating.

Due to all our problems, my husband felt very bad. He said that I was to blame for the whole situation. At least at the mine one could eat a good lunch, he said. And when we needed clothes for the children he would tell me to go ask the Committee of Housewives or the Union for the money. He suffered a lot; after all, he felt unhappy, right?

My children collaborated unknowingly with my husband. They cried because they wanted a piece of meat to eat, or because they wanted candy on Sundays, or because they wanted a can of milk on Sundays . . .

All this mortified me because I wasn't conscious then, as I am now, and at times I myself questioned what I had done. I got close to surrendering.

Then I would go out to the fields to find a job. I worked until my hands bled, in order to forget all my troubles and to stop thinking altogether, and also to earn some money. At the end of the day I would come home exhausted.

I felt like a criminal . . . In the DIC prison cells, they had persuaded me so successfully that I was a great criminal, that I felt an unbearable guilt. I regretted having gotten involved with the Committee. Why did I say anything? Why did I point my finger at people? Why did I get involved? I asked myself. And I would grow desperate; I regretted everything. And sometimes I wanted to have a pack of dynamite to kill myself and my children so that everything would come to an end. It was so painful . . . !

After six months, my father came to visit me. He was happy to see that I was healthy, working, and making new friends.

The people from Los Yungas were good to me. They were surprised to see me working in the fields as hard as they did. They knew that we people in the highlands worked differently from people in the jungle. And they were in awe that, being a woman, I would work so hard.

My visit with my father was great. I was able to talk. My father told me that when he had been a politician and saw that he had only daughters, he grew anxious because he did not have a son. He wanted to have a son that would follow him in his aspirations, continue his work, free their people, and bring power to the working class. And he said that when he had seen me follow that same path, when he saw himself in me, he felt happy and proud of me. How could I be saying these things now?

"No, my daughter . . . ! What you've done is great," he said. "Let's see, what have you done? . . . All you did was protest against the injustice of the government against its people. That's not a crime! It's a truth. And because of your courage, people love you, and they ask about you. I always go to Siglo XX, and they're all waiting for you."

"No, Dad, not again! . . . After what happened to me, if I am still alive, if this regime ends, if I have a chance to go back, I'll never get involved in anything again. Never again! How could I do it again after all I've suffered?"

My father went to the university at La Paz, and to the university at Oruro. He told the entire story to the directors and told them that I needed to prepare myself for the future. He said that, in order to find myself and to know that my views were fair, I needed moral support more than I needed money. He asked them to help me sort out the situation.

My father came back and he gave me some books to read. They were books about Bolivian history and about socialism. A professor at Oruro had written some notes in the margins for me. These comments guided me through my reading. For example, if the book treated the history of another nation, she would write something like this: "Domitila, don't you think that what happens in this nation, also happens in Bolivia? What happened to the agrarian reform? Don't you see that when a socialist revolution takes place, peasants have all the things described here, and that in Bolivia, the agrarian reform was betrayed?"

These readings were very useful. At the same time, I was confirming something I had dreamt as a little girl: some day there would be no more poor people, and everyone would have food and clothes. I saw that these hopes of mine were reflected in these books. And the exploitation of man by man would end. And everyone who worked would have the right to eat and dress well. The state would take care of the old, the handicapped, and everything. All this I thought was beautiful. It was as though someone had collected my thoughts as a little child and written them in a book. In other words, I agreed totally with what I read about Marxism.

This gave me strength to continue fighting, because, I thought, if I had hoped for these things as a little child, it was about time that I start to live by this doctrine in order to go on, right?

Cristina Peri-Rossi

CRISTINA PERI-ROSSI was born in Montevideo, Uruguay, in 1941. She holds the equivalent of a master's degree in linguistics from the Instituto de Profesores Artigas, and from 1961 to 1971 she taught at several Uruguayan institutes. Peri-Rossi was one of the main contributors to the distinguished leftist weekly *Marcha*. In 1972 she left Uruguay to seek exile in Spain; she now has Spanish citizenship.

Very early in life Peri-Rossi began publishing poetry, short fiction, journalism, and essays. *Viviendo* (1963) was her first book of short stories. She has received the Spanish poetry prize Palmas de Mallorca and the Benito Perez Galdos prize for her short stories. Her work embodies a radical questioning of social and gender mores, literary tradition, and political structures. Her prose and poetry, like Cortázar's, have been especially noted for their deep erotic sensitivity and acute imagery. The exploration of the erotic in her collection of poems, *Evohé* (1971), however, is joined by a drastic sense of a world coming to an end. *Diáspora* (1976), the novel *El libro de mis primos* (1969), and *Los museos abandonados* (1968) portray a world in a continuous process of disintegration. As she writes in *Los museos,* museums do not hold the past as a fixed point in a continuum of time and history but rather alienate objects from their human origin. This symbolizes the emptiness of a world marked by the sense of its own end. Peri-Rossi's crisp, often satirical and self reflective prose engages in the provocative and extremely original critique befitting the postmodernist mode.

The Nature of Love

A man loves a woman because he thinks that she is superior to him. In fact, his love is based on his awareness of her superiority, since he could not love an inferior being or an equal. But she loves him too, and although this feeling satisfies him and some of his aspirations, it also generates great uncertainty for him. Therefore, she either lies when she tells him that she

"*The Nature of Love," from* Una pasión prohíbida *(Barcelona: Seix Barral, 1986). Translated by Nicolás Wey: copyright © 1991 by Sara Castro-Klarén.*

loves him, or she is not superior to him, which is why his own love for her can be justified as no more than a misjudgment on his part.

This uncertainty torments him and makes him distrust her. He mistrusts his initial insights (about the woman's beauty, her morality, and her intelligence), and he then scorns his own imagination for having invented a nonexistent being. He wasn't wrong, though: she is beautiful, wise, tolerant, and superior to him. Therefore, she cannot love him: her love for him is a lie. Now: if she is in fact a liar, a deceiver, she can't be superior to him, the epitome of sincerity. Her inferiority thus demonstrated, she is not worthy of his love, yet he's in love with her.

Devastated, the man decides to part from her for an indefinite time: he must figure out how he feels. The woman accepts his decision with apparent calm, which once again makes him doubt: either she is a superior being who has tacitly understood his uncertainty, and thus his love for her is justified, and he must run to her feet and ask to be forgiven; or she didn't love him to begin with, which is why she accepted their separation with indifference, and he mustn't go back to her.

Every night, in the town to which he has retreated, the man plays chess alone, or he plays with a purchased life-size inflatable doll.

Diamela Eltit

BORN IN CHILE in 1949, Diamela Eltit teaches Spanish American literature. Together with Raúl Zurita, Lotty Rosenfeld, Eugenia Britto, and other young artists and writers, Eltit belongs to the large number of writers who chose to stay and struggle under Pinochet rather than to leave Chile. In 1985 she was awarded a Guggenheim fellowship in recognition of her work.

The experimental and extremely intelligent novels *Lúmperica* (1983; currently being translated into English by Ronald Christ), *Por la patria* (1986), and *El cuarto mundo* (1988; scheduled to appear in French in 1991) are considered the most important landmarks in Chilean narrative after José Donoso. Her novelistic strategies range from the mimetic mode and psychological realism to sharp epistemological inquiry. In *Por la patria*, Eltit recovers marginal zones of Latin American life and consciousness in Coya, her female agonist.

Her latest narrative text, *El cuarto mundo*, attempts to construct an androgynous self. Conscious of their inseparable togetherness from the time of conception to the moment of their final corporeal dissolution, the twin characters of the novel challenge our assumptions of person, personality, and desire as given in bourgeois society. As in the case of *Lúmperica*, a vague sense of self emerges in the process of consciousness that the novel creates. It is centered on the mind's awareness of the body and its constant transformation as it adjusts and responds to the world of stimuli within and outside it. It is this sustained and minute perspective of the body as a zone for the exploration of the oneiric and erotic imagination that marks Eltit's originality.

L. Luminated

What is left of this night will be a feast for L. Luminated, the one who devolves upon her own incessantly embroidered face. She no longer shines as she did when she was contemplated under natural light.

"L. Luminated," from Lumpérica *(Santiago de Chile: Ornitorrinco, 1983). Translated by Sara Castro-Klarén: copyright © 1991 by Sara Castro-Klarén.*

That is why the electrical light makes her face up by fractioning the angles, those lines which butt up to the cables that bring the light. Her whole body languishes under the light. The face is shattered. Anyone can detect her half-opened lips and her legs stretched out onto the grass, caught in a scissorlike rhythm perceived against the light.

Nightfall sustains the plaza's decoration so that she may adopt the fugitive poses that conjugate her to exhaustion. Lit up by the neon sign that illuminates the center of the plaza, amidst benches and trees, she lies down on her back.

Because the cold in this plaza is the time that she has taken up in order to imagine a name for herself. Her name will be a gift constructed by the intermittent lights of the letters of the sign. Its rhythm will eventually give them life: their urban identity.

Pale and fetid, the ragged of Santiago arrive in search of their area: the name and the pseudonym, which like an identity card will authorize them a route regulated by the expenditure of the body that must fit the light of the neon sign.

In that way they will be provisionally named pale, by gender. They arrive from very distant points to the plaza, which, lit up by an electrical mesh, guarantees a fiction in the city.

There they stand in their countless poses: the cables are their point of reference in the parallel lines of the pleasure of seeing. She waits anxiously for the neon light. That is why she feels moved when she feels touched—a quivering breast and tearful eyes. The neon light does not stop. It goes on clicking the sum total of all the names that will confirm their existence. The sheaf of light ejected onto the center of the square that is literature produces indices, in the coldness of dawn, while the other pale members protect themselves under the trees in an almost beautiful design drawn by their silhouettes.

At the center of the plaza L. Luminated starts to convulse again. The pale ones turn their heads in order to see from a better angle, and only then do they spread out over the lawn. Attentive, they set their sights over the baptismal scene. The neon sign attracts her. Frenetic, she moves her hips under the light: her thighs lift up from the ground and her hanging head hits the pavement repeatedly.

Her name is ratified in two parallel colors. Amplified over the body the sign writes L. Luminated. Rhythmically, it reviews all the possible names and pseudonyms. It writes: fugitive. The letters fall like a cinematic take. However, with such a complicated name, she points once more to the meaningless sounds. That is why the sign acquires a classic function, somewhat medieval in its perseverance, in the orthodoxy of its forms, in the freezing cold of its construction.

* * *

They moan for the light. In their orgiastic convulsions they become a solid mass. No one would imagine that in Santiago de Chile she could be baptized so that they could become distended like buds. They are under the branches of the trees licking their faces and she rubs against the trunks for the sheer pleasure of the spectacle. She is subsumed in the ecstasy of losing her personal scabs in order to be born hairless in the company of those who, like her, offer themselves as a commercial product in this desolate city.

* * *

Literature could compare them to sapphires, opals, or celestial aquamarines.

Said once again:

The light of the neon sign placed on top of the building falls on the plaza. It is a mad attempt to blanket the pale ones of Santiago gathered around L. Luminated rather than to give a visual compliment to their forms.

This neon sign that lights up at night is fashioning a message for them. Only when the sign moves over its foregrounded routes does it reach its plenitude. These messages are like assassination attempts on their menacing presences. That is why the sign, making use of its full autonomy, gives them literary names. For example, it says sandalwood, but the impulse died as it fell on her bristling skin in the metamorphosis of such high speculation. It says flying heron. It says crucible.

She destroyed the impulse to decorate the lips that writing had fabricated for her.

That is why in the plaza there is a conjugation of two types of electrical meshings: on the one hand the gear assigned to the square and on the other the meshing that sheds from the signs, that green light that is for sale. That is why her lips have shed their obsequiousness and in a sort of counterpoint her deceiving figure turns under the rays converging upon the center. But she is not alone. All her possible identities have surfaced—nailing down her anatomical points—surpassing all her zones. She is ruled only by the schedule assigned to the electric light that in the molding of the sign leaves stretch marks upon her skin.

* * *

Names upon names, with intertwined legs, come close to each other in translation, in fragments of words, in the mixing of terms, in sounds, and in film titles. The words write themselves upon the bodies. Convulsions of nails scratching the skin: Desire cuts down into the skin and opens furrows.

Female Textual Identities:
The Strategies
of Self-Figuration

Introduction

SYLVIA MOLLOY

I AM WHERE I AM NOT," writes Gabriela Mistral in a well-known poem. Years later, in a deceptively simple text, Alejandra Pizarnik echoes the same uneasiness: "to explain with words of this world / that a ship sailed from me carrying me away." This disquiet, frequent in texts written by Latin American women, bespeaks a dislocation of being—more specifically, a *dislocation in order to be*—that could well be the main impulse behind their writing. One is (and one writes) elsewhere, in a *different* place, a place where the female subject chooses to relocate in order to represent itself anew.

Muses, Authors, and "Women Who Write"

The texts that follow were written between the early years of the twentieth century and the present by Latin American women who were and are, by profession, writers. These texts come from several countries, belong to two different linguistic traditions, and reflect a variety of national cultures; they belong to different periods and relate to the literary movements of those periods; they are generically and stylistically diverse. They are brought together by my own critical preoccupation, by my dialogue with Latin American literature, and by a question I have often asked myself and that these texts, in some form, answer. The question is, in fact, twofold: What do women's texts do when they say *I*? And, as a necessary sequel: What representation of woman do these texts posit and what cultural forms govern that representation? For the most part composed in the first-person singular, the texts I have chosen are not necessarily autobiographical, nor am I proposing that they be read autobiographically. Yet whatever their

genre—lyric poetry, fiction, essay, and, yes, avowed autobiography—they all tackle, in one form or another, the problems of a gendered self-representation.

I do not argue here for an essentialist interpretation of feminine literary difference, in which I do not quite believe, nor do I contend that the desire for a different space in which to be, and from which to write, is peculiar to Latin American women writers. Rather, I consider the forms taken by that desire (universally common to women) in a specifically Latin American context, and try to see in what way dislocation and difference inscribe themselves in these texts as a necessary response to, and a correction of, concrete cultural stereotypes.

Before we consider the texts themselves, it is useful to take a look at the conditions that inform their production. Antedating the difficulties of textual self-representation (a difficulty that, in general terms, is not exclusive to women writers) lies a more basic problem, that of society's perception of woman—and woman's perception of herself—as a writer, that is, as a public figure. Any consideration of women writers in Latin America must take into account that the very term *woman writer* refers to an unstable reality, one that, even now, is not accepted without qualifications. One must not forget that the process legitimating writing as a profession, as a valid, socially acceptable form of production, differs from culture to culture. In the case of Latin America, the literary specificity claimed by the writer is not dissociated, as is frequently the case in the highly compartmentalized literary establishments of countries such as the United States, from a broad intellectual reflection and a critical practice that exceed reductionist views of "the literary." Since the time of the writer-statesman, that frequent nineteenth-century figure, the image of the Latin American writer as one invested with an authoritative voice and engaged in the affairs of the polis, although more subtly formulated now than it was one hundred years ago, is still very much in effect.[1] In this context, then, to speak of a woman writer is in a way to postulate an antimony: a subject, traditionally perceived as being "private" and devoid of authority, appears endowed with intellectual power within the public sphere. The ambivalence with which nineteenth-century writers greeted the emergence of female colleagues (that is of female *authors,* and not, to use the condescending circumlocution, of "women who write"[2]) was only matched by the swiftness with which those male writers sought to curb the authority that those colleagues might eventually wield. Indeed, the very use of the term *colleague* is questionable here, as women authors were very much relegated to secondary roles in the cultural field. Only two modes of public figuration seem to have been afforded them as professionals: lyric poets and teachers. "Lady poets" conforming to convention were ideologically innocuous. Teachers did not so much exert authority as they transmitted the knowledge with

which to acquire it; in that function, they were useful.[3] It is not surprising, then, that conventionally sentimental lyricism and didactic literature—both, in a way, forms of noncritical repetition, one aesthetic, the other ethical—remained the preferred vehicles for woman's expression. They were acceptable and they were accepted.

It is not without irony that the literary movement so often acclaimed as the first concerted reflection on Latin American cultural identity—I speak of turn-of-the-century *modernismo*—should have excluded women. *Modernismo* sees woman exclusively as subject matter: it focuses on her as the passive recipient of its multiple desires, as a commodity that is alternately (or at times simultaneously) worshiped in the spirit and coveted in the flesh. A movement that prizes the crafting and collecting of precious objects, *modernismo* makes woman the most valuable piece in its museum. Yet within this brotherhood, there are no women poets, which should, on closer look, not come as a surprise. The dearth of women writing within the movement (Juana Borrero at the very beginning, Delmira Agustini at the very end) and the often conflictive nature of the texts they write testify to an authorial dilemma: women cannot be, at the same time, inert textual objects and active authors. Within the ideological boundaries of turn-of-the-century literature, woman cannot write woman.

Latin American literature does not begin with *modernismo*. I turn to it here, however, as a point of origin because, cannily adapting images borrowed from English Pre-Raphaelites and French Symbolists to express local desire, it produced persuasive icons of femininity. Artfully elaborated and culturally convincing clichés—woman as virgin, as child, as toy; woman as demon, as temptress, as witch—they were to channel the perception of woman in Latin America and inform cultural attitudes toward her for years to come. These stereotypes legitimated by *modernismo* were not limited moreover to literature; they were applied to all aspects of life, became ways of viewing women inside and outside texts, and, quite specifically, became ways of viewing—and controlling—women writers themselves. Rubén Darío, the most influential poet and cultural ideologue of *modernismo*, writing on the death of the Cuban poet Juana Borrero, pays homage to her less as a literary *author* than as a literary *creature,* "a virgin ascending to Paradise, where she will be like Rossetti's beloved or Poe's Rowena."[4] (He also takes care to etherealize Borrero in opposition to "those more earthly companions of hers, mindless and uterine, instruments of the obscure forces of evil.") Darío's perception of women writers, paranoid as it may seem,[5] is symptomatic of a movement whose homosocial and homoerotic characteristics remain to be studied. *Modernismo* has very definite ideas about the (non) place of women; after a devastating review of Alfonsina Storni's first book of poetry, an Argentine critic concluded his piece with the following advice: "Yes, miss: for your own good, I advise you not to

continue to write poetry. Instead of writing verses, seek to inspire them. This is more feminine."[6]

One cannot exaggerate the complexity of the woman's scene of writing during the first quarter of the twentieth century in Latin America. The need to swerve away from the past is twofold: not only must the woman writer deviate from the limited and unsatisfactory authorial roles allowed her, she must also, in her writing, revise old figures or invent new ones that will represent her as *modernismo* did not. It is no coincidence, I believe, that Delmira Agustini, writing in early twentieth-century Uruguay, should explicitly and repeatedly resort to a critical revision of the swan, the *modernista* icon par excellence, to stage her presence in the text. She has to insert herself into a preexisting literary tradition and a preexisting system of representation in order to be seen and, more importantly perhaps, in order to see herself.[7] Yet the image she fashions in "Nocturne," for example, is less a dutiful replica than a correction, even a distortion, of the *modernista* model. Instead of gracefully gliding on a translucent lake, Agustini's swan is bloody, sexual: it soils the pure *modernista* waters as it takes flight. The poem successfully questions—through overstatement, disruption, and irony—previous representations of the feminine. In so doing, it constitutes a gesture, at once revisionary and original, that will be repeated in literary texts (and, I dare say, in nonliterary texts) written by Latin American women in the twentieth century. My purpose here is to inquire into the strategies of that gesture in different authors the better to determine the way in which woman represents woman and constructs herself as subject of her text.

A Place to Sign, a Text to Be

Not surprisingly, for authors whose writing has been inhibited if not denied outright by the hegemonic discourse of their day, many Latin American women's texts are preoccupied with institutional inscription. Names, epitaphs, signatures, résumés, genealogies, eulogies, even wills often fantasize the boundaries of a legal and historical persona too tenuous to be fixed. That some of these texts openly parody forms of thanatography—the unchangeable inscription of a gravestone or the no less lapidary curriculum vitae—shows to what extent the woman writer needs to ritualize her trace, courting, and at the same time questioning, forms of "official" stasis. Each of the six volumes of Victoria Ocampo's autobiography, for example, not only bears on its cover the word *autobiography*—a form of epitaph, as has been argued[8]—but has a photograph of the author and, in lieu of her name, a replica of her signature. In this way, each volume asserts itself as a credential, a document giving legal proof of identity. Alfonsina Storni, who untiringly staged her own death in her work, wrote an ironic "Epitaph for

My Grave" that mimics both the mendacious inscription on the tombstone and the voice that questions the epitaph's serene message. In a less playful vein, Storni also wrote "Erased," a poem haunted by the documentation of death and the obliteration of self: "Across my name, in an old book, / An unknown hand will draw a line." Wielding the same combination of irony and apprehension, Blanca Varela presents a disquieting "Identikit" and a "Curriculum Vitae" that ends, too, on a disembodied note: "your shadow / your own shadow / was your only / disloyal rival." Ana Cristina Cesar's attempt at biography ends in flippancy: "I am faithful to biographical events. / More than faithful, oh so captive! These mosquitoes / which don't let me go!" And Alejandra Pizarnik's quasi-liturgical repetition of her own name, in a striking poem, is yet another version of the fallacious epitaph, of the lapidary inscription undermined—literally and graphically in this case—by a duplicitous subjacent voice:

> alejandra alejandra
> I am underneath
> alejandra

Pizarnik's poem allows one to take the notion of epitaphs and curricula, these fixed forms of life or death, one step further. If one can see, from the abundance of these forms of naming and representing self, a need to fix the boundaries of an elusive textual persona, it is clear, from the examples given and the many more that could be added, that the way in which these forms are mimicked problematizes representation more than it resolves it. Epitaphs are false; curricula self-destroy; the name "alejandra" is at odds with the "I" beneath it—also calling itself "alejandra." The desire for fixity can result in alienation. The irony and biting humor to which many of these texts resort as a way of working against monumentality can turn on the feminine subject itself, thwarting attempts at an *other* coherence and circumscribing it to the fragmentary. This does not necessarily imply failure or loss; instability, however, is a heavy burden. In a sense, socially conventional forms of female inscription function for these women writers like mirrors: they provide them with specular others that are both familiar and alien, images that do, and do not, signify them, images whose fixity they alternately yearn for and deride. Impossible to breach, this representational gap—so often and so effectively described in the poetry of Cecília Meireles—constitutes the space of the woman writer; and the ambiguous trace she leaves there, as a counterinscription, is, indeed, her only possible signature.

Nowhere is that gap between inscription and woman more eloquently expressed than in woman's encounter with the book and, in more general terms, with the patriarchal canon. This encounter is necessarily complex, in Latin America as well as elsewhere. Reading-as-woman is a process

permanently oscillating between identification and verification of difference.[9] Two texts here—Victoria Ocampo's autobiographical writings and a short story by Elena Garro—highlight diverse aspects of woman's scene of reading. For Ocampo, a voracious reader who, by founding the literary review *Sur,* made reading her profession, books are not so much cultural objects as vehicles for the self: they serve as ways out of upper middle-class convention and as ways into a freer world. Not surprisingly, Ocampo often referred to herself as an autodidact:[10] she had to teach herself new ways of reading and of relating to a canon to which, because of her gender, she had limited access. The inordinate intensity with which Ocampo, in her autobiographical writings, refers to reading as an expression of self betrays a relation with books that goes well beyond—and even goes against—the tame and ideologically limited cultural landscape in which she was raised.

Referring to the modeling function of readings done early in life, Ocampo highlights empathy and identification. Yet empathy and identification can be perplexing for a female child reader, as her mixed emotions when reading *David Copperfield* show. No less perplexing and at the same time, literally revealing, is the experience of the two little girls in Garro's admirable "Before the Trojan War." The reading of the book (stolen from their mother's hiding place) signals the passage from a state of indifferentiation into one of personhood. The book—*The Iliad,* here—functions precisely like Jacques Lacan's mirror, marks that moment of identification in which the child assumes her image. What the book/mirror has to offer this symbiotic duo is, however, a splintered image, one that is doubly alienating: it proposes an identification with the masculine (Hector, Achilles) and, as it pits one sister against the other—Hector against Achilles—introduces discord and isolation.

To those lacking representation, mirror images are not only specular, they are often spectacular. A strong theatrical stance informs many self-figurations created by Latin American women: the image becomes a role, the text a performance. If, in literature written by men, women are often mythologized, either positively or negatively—think of the many Medeas, Medusas, Judiths, Helens, and Ophelias of turn-of-the-century literature[11]—and mythologized to the point of abstraction, in texts written by women the same mythologies are often assumed actively, read against the grain, recombined creatively, inverted drastically, to fit individual images. Conventional figures of frailty are endowed with vigor, victims reassert themselves over their oppressors, traditional stories are retold, gaps in legends are filled. In the last poem she wrote, read by many as a suicide note, Storni, revising Desdemona's dialogue with her nurse, significantly rewrites the end of the story as a choice, not a sacrifice. The subject chooses to die alone, leaving the inopportune lover behind: "If he should telephone again, / tell him not to persist. I have gone out." Resorting to the same story,

Rosario Ferré, in "Wedding Banquet," offers another vigorous rereading: she turns Desdemona into Othello's slayer. Ferré, in "Requiem," again subjects another mythical figure, Ariadne, to an unexpected, uncanny revision. Meireles proposes a strangely restless Diana, the huntress never satisfied with her prey.

Perhaps the most striking revision of myth is that presented by Mistral's "Electra in the Mist." An overt poem of matricide, little known by general readers and neglected by critics (I shall return to this neglect), it constitutes an exemplary rewriting as well as a reversal of previous mythologizations. In it Mistral, amending the classical story, focuses on Electra as protagonist. It is she who is the true slayer of Clytemnestra; Orestes is but the instrument of her death, a pawn in Electra's hands. An aimless wanderer on a lonely road (a pose frequently adopted by the female subject of Mistral's poems), Electra speaks to herself, to her slain mother, and to her absent brother whose presence she vainly dreams up at her side. Her monologue, a defiant affirmation of freedom brought on by her mother's death, slowly veers toward the realization that that mother lives on, dispersed in the landscape, contaminating the very air the daughter breathes, filling the world Electra sought to rid of her presence. As much as it rewrites classical myth, Mistral's poem rewrites the more general but no less enduring myth of motherhood, a myth to which her own poetry has contributed and that her essays, on more than one occasion, have upheld. The soothing maternal stereotypes of Mistral's lullabies and faux-naif ditties, not to mention the maternal stereotypes Mistral herself resorted to for her public persona, suffer severe correction in light of this poem of matricide. In addition, "Electra in the Mist" allows for a recontextualization and a perverse rereading of one of Mistral's best-known poems on the death of her own mother, "The Flight." In its original context, "The Flight" is a poem of loss, a mourning for the mother-daughter bond broken by death. Yet read in conjunction with the matricidal "Electra in the Mist," a poem that formally resembles, even plagiarizes its precursor text, the separation from the mother in "The Flight" takes on, retrospectively, a more complex meaning. The mother figure, once soothing, becomes problematic. Going yet one step further, one might argue that Mistral's rereading of Clytemnestra's murder gives an unforeseen twist to that memorable matricide. If, as Luce Irigaray argues, Athena saves Orestes from being punished for his crime by the Furies because "the matricidal son must be rescued from madness to institute the patriarchal order,"[12] the conflation of Orestes with Electra in Mistral's poem, and the eventual replacement of the matricidal son by the matricidal daughter, rewrites that foundational moment, erases the institution of that order. Electra, not Orestes, centers Mistral's text, not a figure of order so much as a figure of permanent divergence, wandering away from the foundational scene.

In a lighter, ironic vein, Adelia Prado submits her rereading of maternal myths to a double twist in "Great Desire." She first debunks the grandiose representation of motherhood by resorting to the commonplace quotidian—"I'm no matron, no mother of the Gracchi, no Cornelia, / but a common woman, mother of children, Adelia"—and then she debunks the icon of domesticity, that "common woman, mother of children," no less a stereotype than the venerable Roman matron.

Traditional icons of (female) evil, such as the femme fatale or the witch, are sometimes revised in unexpected ways, reclaimed for purposes that go beyond personal rectification or revenge. The black woman in Nancy Morejón's "I Love My Master" dreams of slaying her white lover—"I see myself, knife in hand, skinning him / like an innocent calf"—but her sacrifice is recontextualized; it is not the pleasure of a decadent seductress but the revenge of the oppressed. Morejón's poem purposefully incorporates into its representation elements of a popular tradition that has been devalued as a way of escaping oppression: the black woman's memory of her language and her music obstructs her acculturation and serves to nourish her rebellion. In a different way, Silvina Ocampo's "The Basement" deals conventional female iconography a radical blow by foiling all attempts to decode female representation along familiar (and acceptably prejudiced) lines. Ocampo's narrator is a recluse in a basement, living off alms and refuse, with only mice for companions, in a city that is being demolished. There is something impressive about this woman; but how or what is she? Allusion to "clients" may prompt the reader to identify her, conventionally, as a prostitute or a soothsayer; allusion to chaos, to identify her as a madwoman. But the story refuses to let the reader *recognize* her: "I hear cries and none of them are my name." All that this strange and emblematic story allows us to keep is an image of woman, at once magnificent and squalid, comical and solemn, staying on amidst the ruins; a female *presence,* unnamed, defiant, that lives on.

As inventive as the rewriting of preexisting mythical figures in order to, literally, put on a face, is the Latin American woman writer's active creation of personal masks. A common trait of some of the texts that follow, especially those written in the early part of the century, is their defiant, confrontational nature. Questioning, as they do, conventional views of woman and women writers, they often incorporate into their fabric the more or less fantasized perceptions others may have of them: assuming them defiantly, they turn those perceptions to their own profit. This seems especially true of women writers who have experienced, in one form or another, a marginalization that is not only general but specific, affecting them individually. It is hardly necessary to delve very deep into Storni's personal experience, for example, to understand her more virulent reactions against societal pigeonholing or her resentment against those censorious

others her poems so obsessively summon. It is not my intention here to explain a writer's poetry by unmediated reference to her life. Yet if one *reads* the writer's life as another text (which it is), that is, as a social narrative whose acts, conducts, attitudes are observed, interpreted, and judged by the community of social readers, then one may find, in the life, grounds for profitable reflection on the text. For example, if one bears in mind that, for the Argentine society of her day, Storni bore the multiple stigma of being not only a woman but the daughter of an Italian immigrant and, to make matters worse, an unwed mother, one may more easily understand her need for defiant self-representations. From the melodramatic early poem, "The She Wolf" (not included here), in which she identifies with the facile image of the loner rejected by, and rejecting, the pack, through the mock voyeuristic "What Would People Say," to the most interesting poems of all, the vigorously self-sufficient bodyscapes of the end, of which "Tropic" and "Lighthouse in the Night" are good examples, there is a constant summoning of those others (referred to, alternately, as "the people" or "they"), as judgmental presences, as more or less disapproving witnesses, or, in the last poems, as dispensable onlookers. A similar casting of woman as social rebel—a presence scandalous or intolerable by customary standards—is to be found in other writers besides Storni. With variants, the figure will find echo in Elvira Orphée's nonconformist grandmother and rebellious granddaughter in *Aire tan dulce,* the two women lovers in Peri-Rossi's "I Love You This and Other Nights," and the recluse in Ocampo's "The Basement."

It may be argued that women, long represented by exaggerated images forged by male others, even as they question those images or as they construct defiant alternatives, fall inevitably into the trap of the very system of representation they shun. While this is partly (and inevitably) true, I would venture that what the woman writer takes from an androcentric system is not so much the representation itself as certain representational strategies that she finds useful. This calculated reappropriation is not un-related to the way a predominantly male literary establishment has viewed— and continues to view—women writers. One should keep in mind that critics, in Latin America, have tended less to read the work of women authors than to dramatize the anomalies they attributed to them as persons. Following the process described by Joanna Russ as "Denial by False Categorizing,"[13] Agustini is often viewed by critics as the lustful virgin, Storni as the ridiculous virago, Victoria Ocampo as the blue-stockinged hostess, Mistral as the spiritual mother, Norah Lange as the extravagant Dadaist, Silvina Ocampo as the perverse eccentric. These critical fictions, abetted by whatever phantoms haunt the male social imagination, oscillate between the ridiculous and the tragic. They strive, in a sense, to magnify the figure of the woman writer, to draw attention to her exceptional quality, thus distancing her, in reality, from her readers. It is she whom the critic would

have us read, not her texts. What women writers achieve, instead, in adapting the mythologizing strategies of male writers and critics, is a sort of reverse totemization: the exalted self-figurations they adopt are not ways out of their texts but ways in, not icons that replace the writer but icons that govern her text.

On occasion, too, dramatic self-figuration is otherwise meaningful: it is used not only to discover but to mask. I think here of perhaps the most recognizably "public" writer in this selection, Mistral, whose professional career is impressively successful—distinguished pedagogue, cultural reformer, ambassador-at-large of Latin America in Europe, Nobel Prize recipient. Very early on, and in part because of Mistral's own manipulation, critics invested her with a powerful *mater et magistra* image. This image of spiritual mother and teacher of Latin America was not only to mark the reception of her work but the way in which the unspeakable, in her life, was glossed over. Mistral was often presented as one who turned to children through a frustrated longing for motherhood, one who in Latin American solidarity found a love she was otherwise denied. This pathetic compensatory image that Mistral, for whatever reasons, endorsed, has strongly influenced the way in which she is read, the preference accorded certain texts (lullabies and children's poems, for example) to the detriment of others. More importantly, this perception of Mistral has successfully written her lesbianism out of existence. With that knowledge in mind, and with more than the restricted corpus to which she has been reduced, the shrewd reader will discover a totally new image of Mistral. Just a look at the poems in *Locas mujeres,* portraits of "mad" women that may be read, at the same time, as self-projections, will reveal recurring themes—voluntary exile, rejection of family ties, revision of the past, celebration of freedom, need for concealment and, yes, bonding with other women—that propose a figuration by allusion hardly conforming to the cliché of the serene schoolteacher.

The Writing Body

Turn-of-the-century representation of woman, in Latin America and elsewhere, is haunted by dismemberment. In a frenzy of synecdoche, (male) poets will exalt woman's hair, her eyes, her feet, one foot, a glove, a stocking, as loci of desire. Only through the mediation of the fragment can the female body be apprehended and coveted in its plenitude. Without that mediation, plenitude—woman in her totality, woman complete—proves intolerable and, more to the point, strong and threatening: she is then seen as agent, not victim, of dismemberment. Not coincidentally, turn-of-the-century literature took up, simultaneously, the figure of woman mutilator—Salome,

that crucible of fin-de-siècle fear and desire—and, the better to control her, that of woman mutilated.[14]

Salome lives on in several of Agustini's poems, those staging a scene that Miguel de Unamuno found particularly disquieting, "that strange obsession of yours of holding in your hands at times the dead head of your lover, at others that of God."[15] What is striking is that these poems, while not devoid of eroticism, are, above all, contemplative; in a macabre reversal of *modernismo*'s basic situation, in which the male poet contemplates woman as precious bibelot, here it is the woman poet who contemplates a dead head as art object: "inset in my hands, like a strange jewel, / your head glowed." I shall not dwell on these poems, interesting though they may be. I prefer to call attention instead to signs of physical fragmentation of the *female* body in texts written by women and to observe what use they make of that physical fragmentation in their attempt to represent woman. In such cases of the dismemberment of the female body by a woman writer, the erotic component, and by extension, the fetishizing impulse, becomes much more complex in nature.[16] For, unless the case can be made for a narcissistic fetishization of fragments of the self or for a lesbian fetishization of fragments of the other, the fragmented female body, in poems written by women, is not primarily engaged in an exclusively erotic transaction. It is basically involved in a textual transaction where mutilation and fragmentation are cleverly used not to subdue the other but to portray the self.

Storni is the first poet to use body fragments—a head on the shore, a heart on a rock at night, ovaries rooted in a forest, the jagged marble of a tooth, the cavern of the ear—for the purpose of self-portraiture. These bare fragments are not synecdoches, in the sense that they do not allude to a whole that remains unnamed, as did the fetishized body fragments of *modernismo*. On the contrary, they resolutely remain fragments, stripped of any larger meaning: "Storni turns around poetic icons, strips them of an accustomed mystery, and delivers them back to wander in unfamiliar territory," writes Gwen Kirkpatrick.[17] As loose pieces, these body parts are described by the poet with passionate minuteness and function as realms unto themselves. This fragmentation doubtless responds to the break in aesthetic perception brought about by the avant-garde. Given the fortune of the female body before Storni, however, I would argue that it is also a reaction against the tendency to recompose fetishized fragments according to one, central interpretation, that of woman as vehicle of male desire. It is a way, finally, of wresting interpretive control from readers motivated by that desire. Thus the body fragments in Storni do not allude, do not evoke, do not recompose themselves into one icon, do not point to one, invariable referent: by defiantly maintaining their independent nature, they vindicate for woman's body a plurality of readings. Fragmentation allows for reflection on the body of the text itself. Wandering through a woman's body

(her own, that of another woman) the poems in Soledad Fariña's *Albricia* carefully explore a territory asking to be discovered, establish its grammar, its syntax. Experimenting with cuts and sutures, Vanessa Droz's "scarred" lines in *La cicatriz a medias* inquire into poetic form as a metaphor for the body.

But not only the body is creatively dismembered in women's texts. Voice, woman's voice—hard to find, agonizing to enunciate—speaks too in fragments, is composed of shards. Many of the authors in this book (as well as others that are not included) have made this anguish of expression the main concern of their poetry, resorting at times to images of childbirth but also, very often, to images of sickness, of malignant growth within, of death. Poems by Storni, by Mistral, by Agustini thematize this difficulty, stressing not only the hardships of the process but the strangeness of its products, as in Agustini's "The Ineffable." In Agustini particularly, distress is often accompanied by wonderment at the birthing of the near monstrous: her poems stress the strangeness, the excessive nature of the word she bears, whose foundational nature she intuits. In another register, Orphée's novel, *Aire tan dulce,* a text of woman's submission and woman's rebellion in a small town in the Argentine provinces, attributes a similar excessive quality to woman's voice. There, in the crisscrossing of voices that weaves the text, a haughty, desperate old woman vainly tries to give her adolescent granddaughter the elemental words she has buried in the landscape of her childhood—those "words [that] do not shatter against the days and the repetitions"—so that the girl may fight the battle she herself has lost to silence and oppression.

Fragmented voices may, at times, be joyful. Such is the image presented by Lange at the end of her childhood memoir, *Cuadernos de infancia:* the child standing on the roof, dressed like a tramp, shouting at her amazed neighbors in a private language concocted from bits and pieces of all the languages she knows. This ritualistic performance—the child offering herself up as a spectacle, passionately chanting her cacophonous words, hurling her dissonance and her difference at the public—has a wasteful elegance to it, a kind of devil-may-care dandyism that is infinitely seductive. Lange brings out into the open a refreshingly humorous persona, one that may also be found in texts by Storni, Varela, Pizarnik, Peri-Rossi, that of the writer as mountebank, breaking up her image as she cuts up her words, and throwing the pieces, gladly, passionately, to the winds. More often, however, mutilation—what Storni called her "beheaded words"—distresses voice to the point of near extinction. Among Latin American contemporary poets, Pizarnik is doubtless the one whose texts express most forcefully that wordless impotence whose final outcome is, as "Arbol de Diana" so often implies, an emptying out of being: "This repenting song, watchful behind my poems: / this song denies me, muzzles me." Pizarnik carries to

the extreme, and stages in its starkest terms, a splintering of voice not infrequent in women's texts.

Many of these poems address a second person: a mother, a sister, a muse, a lover. Frequently, however, we are faced with mock interlocutory acts, positing the illusion of an I/you relation that progressively disintegrates into a plural, mosaiclike subject of enunciation. The "I" summons a "you" that is a reflection of, a contamination of itself, like the voyeuristic first-person narrator in Silvina Ocampo's "The Mortal Sin," an adult looking back on her childhood. Or the "I" chides, or corrects, or makes fun of a "you" that is itself, as in Varela's "Curriculum Vitae." The whole text then turns into an act of textual mirroring, is enunciated from a second person that doubles the "I," as in so many poems by Pizarnik: "you have built your house / you have feathered your birds / you have struck the wind / with your very own bones / you have finished alone / what nobody began." These texts resolutely maintain their discursive ambivalence and are often susceptible to two readings, in which the you is and is not a doubling image. This doubling, of course, is not unusual in modern poetry in general. For women, however, it provides an ideal setting in which to stage the double voice that results from their marginalization. In addition, it is interesting to note that, from among the many possible postures of the poetic "I," women's texts rarely adopt the prophetic stance. Not infrequent in male poets in Latin America (Darío, Huidobro, Neruda), it is seldom sustained in texts written by women. When women resort to that form of magnification, as in the case of Mistral, for example, or that of Olga Orozco, they question it more readily than men, are more prone to see through the cracks of its authoritarian façade.

To rewrite woman's body, or fragments of that body, is also to rewrite its desire. Since Sor Juana Inés de la Cruz's erotic homage to her Vicereine, the countess of Paredes, in the seventeenth century, Latin American critics have been hard put to accept the expression of physical desire in texts written by women as not mimetic, as not "taken," from other texts. Thus Sor Juana's celebration of the countess's body is categorically explained away by aligning her poem with the conventional body heraldry of the *blason* poetry of her time. Agustini's unbridled eroticism, shocking to her contemporaries, is seen as "cerebral"; only as a figment of her imagination is it tolerable to her critics. Pizarnik's sexual and scatological obsessions are attributed to her fascination with the French poet Alfred Jarry and with surrealism in general. The underlying attitude in all three instances seems to be that the poet could not "really" have experienced such erotic excess or, as the case would have it, such perversity. Now it is obvious that, for a writer, *all* expression of feeling is necessarily mediated by texts. What is insidious here is that the mediation signaled by critics does not stem from a particularly enlightened view of literature but from a repressive ideology:

the concept of textual mediation or vicarious experience is not applied here to all feelings voiced in women's texts, *only* to the expression of sexual desire.

In Agustini, desire emerges with all the trappings of decadent poetry. Sex melds with religion, violence and pleasure are one, the occult is hinted at, the perverse courted. What changes, of course, is the speaker's stance in the poem. In the erotic fantasies of *modernismo,* gendered roles appeared to be fairly straightforward: man desired and woman was desired. Agustini's poetry, written toward the end of the movement, seizes to perfection the vacillation announcing the erosion of those roles. In all these poems, the female "I" desires actively, calls for the lover, powerfully constructs and controls the scene of her passion. Yet the embrace itself is often anticlimactic. It may be postponed indefinitely; or it may drastically reverse the initial situation, replacing the very strong desiring "I" with a surprisingly submissive, conventionally "feminine" presence. The desiring woman, in Agustini's texts, cannot live out her own passion but must become, at some point, woman desired. A similar conflict between two diverging representations of female desire is to be observed in the work of Storni. In Storni's case the conflict often resolves itself in humor, killing desire in the process. As in many other countries and other literary traditions, woman's erotic expression remains tentative and needs to be explored further. It appears to be scarce but one wonders to what extent this perceived scarcity might be attributed to a self-censorship that, for reasons one would suspect to be mainly social, even now muzzles woman in that domain. For example, few of Pizarnik's bawdy or fastidiously scatological texts were published in her lifetime; and, when they were published posthumously, not a few critics deplored their poor "literary" quality. Curiously (or perhaps not so curiously), the expression of sexual desire and the physical celebration accompanying it seems to be less problematical in lesbian poetry: having renounced social disguise, these texts appear to be less preoccupied with and less beleaguered by the strictures of sexual role playing. However, if expressions of erotic desire may seem scarce in women's texts, it may well be that—accustomed as we are to expressions of phallocentric eroticism—we are looking at those texts in the wrong way. Women's eroticism appears to express itself in forms more diverse (and, I would add, more textually productive) than the primarily sexual: if it celebrates and desires other bodies, those bodies are not only the bodies of lovers, they can be other things as well. What one often finds in women writers, in terms of erotic desire, is a slippage from sex to text: the text itself is an erotic encounter in which the poet makes love to her words. Fariña has been mentioned in this connection; to her name may be added that of Ana Cristina Cesar. Not limited to the physical body, and certainly not repressing it, desire in these cases extends to the body of writing.

Family Portraits

I have concentrated so far on the representation of woman through image and voice in itself, purposefully isolating the subject from, as it were, her setting. Yet along with the need to achieve a different presentation of self and of woman, the inclination is strong in these texts to turn that presentation into a family affair, a communal practice that restructures lineage and invents new kinships to replace conventional ties. Family, home, and childhood play an important role in these pieces, directly or indirectly, not because they are particularly suited to "feminine" sensibility but because they function as powerful figures of bonding. Contrary to the (predominantly male) view that woman, as the stereotypical "angel in the house," speaks and writes best about a domestic realm only she knows, Latin American women writers, with few exceptions,[18] revisit home, family, and childhood and approach the economy of desire dictated by patriarchal practice with a quizzical, reflexive attitude. Their texts are far from being conventional restorations; they are inevitably critical, not because they are negative about the past (they often are not) but because they forcibly engage a meditation on and, be it ever so lyrical, a reassessment of a period and a mode of life that no longer represent the woman who writes them. Women's family tales, both in poetry and prose, are hardly havens of security. They are often told by exiles, misfits, voyagers (Mistral's disburdened wanderer, Vilariño's stoic loner, Pizarnik's hallucinated traveler) who leaf through bric-a-brac (photographs, scrapbooks, letters, personal memories, the memories of others), recognizing a past into which they do not quite fit. I think, for example, of Tamara Kamenszain's "Big House," where the reminiscing self, evoking her immigrant ancestors, remains very much an outsider spying in. The poem is an act of textual voyeurism, an affiliation achieved through a mediated vision that finally stresses the alienation of the "I": "The eye widens, home in a lock / I want to write an ancient habitat, / clothe myself in the wardrobe of letters: / black box that someone will read / through the albino's darkened lens." In a humorous vein, Margo Glantz, too, seeks out her ancestors, conspicuously playing with genealogy, indeed devoting a whole volume—*Las genealogías*—to the fabulation of her lineage. The plural of the title is telling: poring over homey family archives; listening to humorous stories, at once legendary and prosaic, told by her Jewish elders; resorting to happy and not so happy memories of her Mexican gentile childhood, she weaves a "colorful and grotesque" Jewish past "to which I only belong from a sleepy part of myself" into an equally cluttered and carnivalesque Mexican present, thus achieving a plural life.

Childhood homes, when revisited, are not necessarily joyful places. Port Supe, in Blanca Varela's poem of the same name, takes on the ominous aura of a terminal vision, a father's house purposefully destroyed: yet it is

there, in that desolate seascape where natural bleakness combines with human violence, that the poet sets new roots and establishes her scene of writing. And although Lange's *Cuadernos de infancia* ostensibly presents the reassuring world of five sunny and loving little girls, the memoir is fraught with intimations of death and disruption and the familiar routinely visited by the uncanny. Prado strikes a delicate balance between her content celebration of the everyday domestic and intimations of disruption: "When I write a book with my name, / and the name I'm going to give it on the cover, / I'll take it to a church, to a gravestone, to an open field, / and cry, and cry, and cry."

Revisitation of the world of childhood and of the family home provides, then, an unsettling foundation for women's writing, yet one that proves infinitely fecund. Returning, too, to a childhood landscape, the valley of Elqui, and to the games she played with other little girls, Mistral, in "We Were All to Be Queens," pits the utopia she and her playmates dreamed up in their ditties against the disenchanted reality of their adult lives. Once more, the poet plays the role of the outsider, nostalgically looking in: writing has made her more fortunate than her sisters but has also made her different, separating her from the home group. The locus of self-portraiture seems to be, inevitably, a family place left behind forever, a "Land of Absence," to cite Mistral once more. In Meireles as in Pizarnik, separation and loss of family are at the very foundation of the poetic voice. Not merely exiles, they speak from irreparable orphanhood. In their dealing with the past, with family, with community, women's texts are often informed by a testimonial stance. "We Were All to Be Queens" may be read as a dirge for innocence lost and sisterhood dispersed; but it may also be read as a tribute, as an active effort, on the part of the writer, *so that* innocence will not be lost and sisterhood not be dispersed. To remember, in these texts, is often, also, to witness. Witnessing may be dictated by affective or ideological reasons, most often by a mixture of both: Kamenszain and Glantz, in the texts I have mentioned, pay homage to a mythologized Jewish minority to which they belonged; Morejón, in "Black Woman," to an oppressed race she politically vindicates. Yet in all three texts the sense of community prevails.

The communities that these poems repossess do not necessarily coincide, of course, with real families; the familiar is used rather as a trope to effect reunion. Beyond the merely biographical, many of these authors postulate literary, cultural, ethnic, and ideological groupings that they very clearly need to cast in the feminine. In their works one may observe references, seize allusions to other women, recognize the emergence of other women's voices and follow their dialogue. It is remarkable that many of Latin American women's attempts at self-representation call on other women who came before them, following an impulse that may be dated back to the

prodigious chain of women evoked by Sor Juana Inés de la Cruz in her autobiographical letter to the Bishop of Puebla. In a moving poem to Agustini in her grave, Storni writes: "But on your breast, forever dead / My knowing breast still watches." This sororal solicitude is echoed by Mistral in a poem to Victoria Ocampo, by Victoria Ocampo in an essay on Mistral, by Ferré in a poem to Julia de Burgos, by Peri-Rossi in a poem to Pizarnik, by Pizarnik's frequent dedications of poems to Orozco and to Silvina Ocampo; all of them homages to sisters, acknowledging either a legacy or a shared effort, and, most importantly, all of them forms of oblique self-portraiture.[19] But lest these poems be read, too, as forms of self-aggrandizement—a way of giving oneself being through the figure of a prestigious other—let it be remembered that these galleries of women's portraits are not restricted to the writerly happy few. The impassioned evocation of anonymous sisters in Storni, in Mistral, in Morejón, is proof of a *sympathy* that goes beyond literary boundaries. Throughout that evocation, one may trace the profile of a vast sisterhood that, because it has been denied meaningful existence, needs all the more to be named.

Notes

1. For a profitable reflection on the Latin American writer as "master," see Roberto González Echevarría, *The Voice of the Masters* (Austin: University of Texas Press, 1985).

2. The euphemism is not peculiar to Latin American critics. R. P. Blackmur describes Emily Dickinson as "neither a professional poet nor an amateur; she was a private poet who wrote indefatigably as other women cook or knit," cited by Joanna Russ in *How to Suppress Women's Writing* (Austin: University of Texas, 1983), p. 80.

3. For further remarks on the matter, see my "Dos proyectos de vida: *Cuadernos de infancia* de Norah Lange y *El archipiélago* de Victoria Ocampo," *Filología* (Buenos Aires) 20, 2 (1985), pp. 279–293.

4. Rubén Darío, "Juana Borrero," *Obras completas,* vol. 4 (Madrid: Afrodisio Aguado, 1950), p. 846.

5. Ibid. His anxiety is made all the more manifest by the shrillness of some of his texts. He ridicules, with but a few exceptions, Spanish women writers: "The great majority are sentimental Corinnes or sugary Sapphos, members of that abominable international sisterhood [in English in the text] to which Great Britain has so greatly contributed with its thousands of authoresses." Rubén Darío, "La mujer española," *Obras completas,* p. 362.

6. José Fernández Coria, *Glosas y escolios,* cited in Rachel Phillips, *Alfonsina Storni: From Poetess to Poet* (London: Támesis, 1975), p. 30.

7. For an insightful discussion of the difficulties of self-representation in painting, specifically in the case of Artemisia Gentileschi, see Mary Jacobus, *Reading Woman: Essays in Feminist Criticism* (New York: Columbia University Press, 1986), pp. 132–133.

8. See Paul de Man, "Autobiography as Defacement," *Modern Language Notes* 94 (1979), pp. 919–930; also Louis Marin, "Variations sur un portrait absent: les autoportraits de Poussin," *Corps écrit* 5 (1983), pp. 87–108.

9. For general considerations on reading-as-woman, see Jacobus, *Reading Woman;* Nancy K. Miller, "Rereading as a Woman: The Body in Practice," in Susan Rubin Suleiman, ed., *The Female Body in Western Society* (Cambridge and London: Harvard University Press, 1986),

pp. 354–362; Marcelle Thiébaux, "Foucault's Fantasia for Feminists: The Woman Reading," in Gabriela Mora and Karen S. Van Hooft, eds., *Theory and Practice of Feminist Literary Criticism* (Ypsilanti, Mich.: Bilingual Press/Editorial Bilingüe, 1982).

10. "Malandanzas de una autodidacta," in *Testimonios*, vol. 5 (Buenos Aires: Sur, 1957).

11. See Mario Praz, *The Romantic Agony* (Oxford: Oxford University Press, 1933); Frank Kermode, *Romantic Image* (New York: Vintage, 1957); Jerome J. McGann, "The Beauty of the Medusa: A Study in Romantic Literary Iconology," *Studies in Romanticism* 11 (1972), pp. 3–25. For feminist revisions of nineteenth-century feminine iconography, see Hélène Cixous, "The Laugh of the Medusa" (trans. Keith Cohen and Paula Cohen), *Signs* 1 (Summer 1976), pp. 875–893; Sandra M. Gilbert and Susan Gubar, *The Madwoman in the Attic: The Woman Writer and the Nineteenth-Century Literary Imagination* (New Haven and London: Yale University Press, 1979), especially chapter 1; Nina Auerbach, *Woman and the Demon: The Life of a Victorian Myth* (Cambridge and London: Harvard University Press, 1982); Linda Nochlin, "Eroticism and Female Imagery in Nineteenth Century Art," in Linda Nochlin, ed., *Woman as Sex Object: Studies in Erotic Art, 1730–1970* (London: Allen Lane, 1973).

12. Luce Irigaray, *Le Corps-à-corps avec la mère* (Ottawa: Pleine Lune, 1981), p. 17.

13. Russ, *How to Suppress Women's Writing*, pp. 49–61.

14. As Gwen Kirkpatrick writes: "*Modernista* poets insist on showing the physicality of the referent. . . . In the case of the feminine icon, the litany of these parts and the bodily dismemberment underscore the traditional fetishization of the erotic image of the woman. The body of the woman is used by poets like a Parnassian sunset, a canvas on which to cut, decorate, and engrave its images" (*The Dissonant Legacy of Modernismo: Lugones, Herrera y Reissig, and the Voices of Modern Spanish American Poetry* [Berkeley and Los Angeles: University of California Press, 1989], p. 234). Kirkpatrick's acute analysis of Storni has inspired many of my views in this section.

15. Miguel de Unamuno, letter to Delmira Agustini, cited in Clara Silva, *Genio y figura de Delmira Agustini* (Buenos Aires: Centro Editor de América Latina, 1968), p. 155.

16. See Naomi Schor, "Female Fetishism: The Case of George Sand," in Suleiman, *The Female Body*, pp. 363–371. Also Sarah Kofman, "Ça cloche," in Philippe Lacoue-Labarthe and Jean-Luc Nancy, eds., *Les fins de l'homme: A partir de Jacques Derrida* (Paris: Galilée, 1981), pp. 83–116.

17. Kirkpatrick, *The Dissonant Legacy*, p. 239.

18. The Uruguayan Juana de Irbarbourou, continuously celebrating domestic bliss in her poems, is perhaps the most notorious of those exceptions. She was hailed enthusiastically by male critics as "Juana de América."

19. Victoria Ocampo, "Gabriela Mistral y el Premio Nobel," in *Testimonios*, vol. 3; Alfonsina Storni, "Palabras a Delmira Agustini," in *Ocre;* Gabriela Mistral, "Recado a Victoria Ocampo, en la Argentina," in *Tala;* Rosario Ferré, "Carta a Julia de Burgos," in *Sitio a Eros;* Cristina Peri-Rossi, "Alejandra entre las lilas," in *Diáspora*.

Delmira Agustini

DELMIRA AGUSTINI was born in 1886 in Montevideo, Uruguay, where she studied music and languages at home and frequented *modernista* circles. Agustini began writing for literary reviews (under the pseudonym Joujou) in 1902 and published her first book of poetry, *El libro blanco,* in 1907. The book was widely acclaimed by Latin American critics and received the endorsement of philosopher and critic Carlos Vaz Ferreira. In her lifetime, Delmira Agustini published two more volumes of poetry: *Cantos de la mañana* (1910) and *Los cálices vacíos* (1913). The rest of her work was published posthumously in two volumes, *Los astros del abismo* and *El rosario de Eros,* in 1924. The prurient interest attracted by the circumstances of Agustini's death—she was murdered by her ex-husband, with whom she used to meet clandestinely after their separation—has conditioned the reading of her poetry, which is often erotic in nature. In turn, that eroticism has often been explained away by critics whom it either disconcerted or embarrassed. The interest of Agustini's poetry lies in its very excess: baroque and nonconformist, disdaining conventions, both literary and social, it is *traumatic* (in the sense that Benjamin gave the word when speaking of Baudelaire). Agustini's scene of writing is fraught with violence. Destruction and mutilation are constant themes in her poetry, not as decadent fin de siècle postures but as vital forms of self-engenderment through fragmentation. Agustini stands out as one of the most original voices of Latin American literature.

Nocturne

Set at night, the lake of your soul
seems a gossamer of crystal and calm
woven by great sleepless spiders.

"Nocturne," from Los cálices vacíos *(Montevideo: O. M. Bertani, 1913). Translated by Sylvia Molloy: copyright © 1991 by Sylvia Molloy.*

125

Film of lustral water in an alabaster vase;
Mirror of purity lighting up the stars;
you reflect the abyss of Life onto the sky . . .

I am the errant swan who leaves a trail of blood,
I soil lakes as I go and continue my flight.

The Ineffable

Strangely I die . . . I am not killed by Life,
Nor am I killed by Death, nor killed by Love,
I die of a thought, silent as a wound . . .
Have you not felt, ever before, the pain

of a prodigious thought rooted in life,
devouring soul and flesh and never blooming?
Have you not borne, ever before, a sleeping star
consuming you and never giving light?

Oh cruelest of tortures! Eternally to bear
this ache, this sterile, tragic seed
sunk in my womb like a ferocious fang!

But oh one day to pluck it out a budding flower,
inviolate and miraculous . . . 'Twould be no greater
if in my hands I held the head of God.

The Miraculous Ship

Prepare for me a ship like a prodigious thought,
Some will call it "The Shadow," others will say "The Star,"
Guided not by the whim of a hand or the wind;
I want it fully conscious, indomitable, fair!

"The Ineffable" and "The Miraculous Ship," from Cantos de la mañana *(Montevideo: O. M. Bertani, 1910). Translated by Sylvia Molloy: copyright © 1991 by Sylvia Molloy.*

It will move to the rhythm of a great bloody heart
of superhuman life; and in it I will feel
strong as in God's embrace. For all winds,
for all seas, craft it a bow of light.

I shall load it with all my sorrow, and adrift
I shall float like the torn corolla of a flower
beyond the liquid limit of the sea . . .

Ship, sister soul, to what lands never seen
of deep revelations and unforeseen things
shall we go? I die from living in a dream . . .

False Rosary

Black crows hunger for rosy flesh.
In the wily mirror my statue I reflect,
They break their beaks, hammering against the plate.
And as I stand back—ironic, glorious and untouched—
Black crows fly off, glutted with rosy flesh.

Love of mockery and ice,
marble that boredom painted up with fire,
or lily masked pink by blush;
that let me always give, oh Lord . . .
Oh fecund rosary,
living beads clasping
the world's throat.
Earth's chain,
fallen constellation.

O rosary, serpentlike, magnetic,
glide for all time between my knowing fingers,
for on your fifty smiling beads
my life is fastened, with fiery kiss,
a full-lipped rose.

"False Rosary" and "Serpentine," from El rosario de Eros *(Montevideo: Máximo García, 1924). Translated by Sylvia Molloy: copyright © 1991 by Sylvia Molloy.*

Serpentine

In dreams of love I am a serpent!
I glide and turn like the tide.
My eyes are two dots, insomniac and hypnotic;
my tongue, the tip of charm.
I attract, like tears.
    I am a vessel for the void.

My body is a ribbon of delight,
it glides and turns like a caress.

And in my dreams of hate I am a serpent!
My tongue is a poisoned spring,
my head is a demonic diadem;
the face of death, fatal, oblique,
looks through my pupils; my jeweled body
    is a sheath for lightning!

If such I dream my flesh, such is my mind:
    the long, long body of a serpent,
ever vibrating—so voluptuously!

Gabriela Mistral
(Lucila Godoy Alcayaga)

MISTRAL'S VERSE IS AMPLE, rich with biblical and mythical echoes. Secure in its intonation, it often assumes a prophetic stance unusual for women. Displacement is an important theme in her poetry, as much for the loss it entails as for the strength and freedom it paradoxically confers. Mistral's poems repeatedly stage exile and are peopled with emblematic wanderers as oblique representations of the poet. Critics have tended to read Mistral too cautiously, gravitating toward her "safer" texts, that is, poems dealing with motherhood, children, nature. Yet Mistral is not a conventional poet nor is her poetry soothing and bland. In the folds of her simplest texts— take, for example, many of her lullabies—lie the seeds of disquiet and defiance.

Land of Absence

Land of absence,
strange land,
lighter than angel
or subtle sign,
color of dead algae,
color of falcon,
with the age of all time,
with no age content.

It bears no pomegranate
nor grows jasmine,
and has no skies

For biographical information on Gabriela Mistral, see pp. 27–28.

"Land of Absence" and *"We Were All to Be Queens," from* Selected Poems of Gabriela Mistral *(Baltimore: Published for the Library of Congress by the Johns Hopkins University Press, 1971). Translated by Doris Dana. Reprinted by permission of Joan Daves Agency: copyright* © *1961, 1964, 1970, 1971 by Doris Dana.*

nor indigo seas.
Its name, a name
that has never been heard,
and in a land without a name
I shall die.

Neither bridge nor boat
brought me here.
No one told me
it was island or shore.
A land I did not search for
and did not discover.

Like a fable
that I learned
a dream of taking
and letting go,
and it is my land
where I live and I die.

It was born to me of things
that are not of land,
of kingdoms and kingdoms
that I had and I lost,
of all things living
that I have seen die,
of all that was mine
and went from me.

I lost ranges of mountains
wherein I could sleep.
I lost orchards of gold
that were sweet to live.

I lost islands of indigo
and sugar cane,
and the shadow of these
I saw circling me,
and together and loving
become a land.

I saw manes of fog
without back or nape,
saw sleeping breaths
pursue me,
and in years of wandering
become a land,
and in a land without a name
I shall die.

We Were All to Be Queens

We were all to be queens
of four kingdoms on the sea:
Efigenia with Soledad,
and Lucila with Rosalie.

In the Valley of Elqui, encircled
by a hundred mountains or more
that blaze red like burnished offerings
or tributes of saffron ore,

We said it, enraptured,
and believed it perfectly,
that we would all be queens
and would one day reach the sea.

With our braids of seven-year-olds
and bright aprons of percale,
chasing flights of thrushes
among the shadows of vine and grape.

And our four kingdoms, we said,
so vast and great would be,
that as certain as the Koran
they would all reach the sea.

We would wed four husbands
at the time when we should wed,
and they would all be kings and poets
like King David of Judea.

Our kingdoms would be so vast
they would have, without a doubt,
green seas, and seas of algae
and wild pheasant.

Our lands would be so fruitful,
trees of milk, trees of bread,
that we would never cut the guaiacum
or eat the earth's metal.

We were all to be queens
and we would truly reign,
but not one of us has been a queen
even in Arauco or Copán.

Rosalie kissed a sailor
already wedded to the sea,
and in Guaitecas the one who kissed her
was devoured by the storm.

Soledad reared seven brothers
and left her life-blood in the bread,
and her eyes have remained forever black
for never having looked at the sea.

In the vineyards of Montegrande,
on her pure and faithful breast
she rocks the sons of other queens,
and never never her own.

Efigenia met a stranger
on the road, and wordlessly
she followed him, nor knew his name,
for a man is like the sea.

And Lucila who talked to the river
and the mountain and fields of cane,
under moons of madness
received a kingdom of her own.

In the clouds she counted ten sons,
and over the salt marsh she reigned,
in the rivers she saw her husbands
and in the tempest, her royal train.

But in the valley of Elqui,
among a hundred mountains or more,
others have come and are singing
and there will sing many more:

"On the earth we will be queens
and we shall truly reign,
and our kingdoms will be so vast
we will all reach the sea."

Electra in the Mist

In the sea mist, walking blindly,
I, Electra, feel my garments
and the face that in hours became another.
Now, I am just the one who killed.
Because of the mist, perhaps,
I give myself that name to recognize myself.

I wished to see the killer dead
but when I did
I did not see the killer, I saw Death.
What mattered to me then no longer matters.
Now she no longer breathes the Aegean Sea.
Now she is more silent than a pebble.
Now she no longer does good or evil. She is deedless.
She does not call my name, she does not love or hate me.
She was my mother and I was her milk,
just her milk turned blood,
just her milk and her profile, when I walk or sleep.

"Electra in the Mist," from Mundo nuevo *1 (1966). Reprinted by permission of Joan Daves Agency; copyright © 1966 by Doris Dana. Translated by Sylvia Molloy: copyright © 1991 by Sylvia Molloy.*

I walk free not hearing her cry
come back to me, not hearing her voices,
for she no longer walks, she now lies still.
And her words flutter in vain around her,
and her gestures, her name, her laugh,
while I and Orestes walk
the Attic lands of Hellas, hers and ours.
And when Orestes naps at my side,
his cheek hollow, his eye dark,
I shall see her hands
run up and down his body that they swaddled,
as if on mine,
and he will call her name, the four syllables
that do not break and will not be undone.
Because we called it out at dawn
and again when night fell and the hard name
lives on without her, even if she is dead.
Every time we look at one another
her name will fall like a fruit falls
sliding along marks of silence.

Only Iphigenia and her lover did she love
in the narrowness of her cold breast.
I and Orestes she left without a kiss,
without her fingers intertwined in ours.
Orestes, I know not where you walk, in what direction.
Were you with me tonight, I'd hear your soul
as you would mine.

This salty mist blots out all things
that speak to travelers and comfort them.
Roads, bridges, villages, trees,
no face looks out in recognition,
only the mist that with insistent hand
feels out our faces and our sides.

Where would we go to flee,
when the long name rolls in the mouth
or falls and lingers on the chest

like her breath, when her features,
dissolving in the air, perhaps look for me.

Childtalk returns to us, gliding
through our bodies, Orestes, my brother,
and our childhood games, your voice.

Follow my tracks and come to me, Orestes.
The night is riddled with her,
open with her, alive with her,
as if she were night's only voice, her only voyager,
her only password.
But at daybreak she will leave us.
Why does she not sleep by Aegisthus' side?

Is it that the milk she fed us
still flows from her breast and will flow eternally?
Is it that the salt the wind brings
comes not from the sea air
but from her milk?

Make haste, Orestes, for we shall always
be two, two of us like hands holding on to each other
or like the trotting feet of a scared turtledove.
Don't let me walk alone this night
toward the desert, groping in the mist.

I no longer want to know, yet I would like
to hear it all from your very lips
how she fell, what she said as she cried out,
and if she cursed or blessed you.
Wait for me at the crossroads
where there are flat stones, comforting sprigs
of rosemary and mint.

Because she—yes, you can hear her—she
calls and she will always call. Best that we die
together and unseen, Orestes,
die by the knife, dying our own death.

May the god that moved you give us such a grace,
as the three graces that moved me.

Our breaths are rationed, when
they break off we shall stop.
The mist has shroudlike folds, soft
to the touch, and in its saline mouth
you will again go forth to my song.
You always lived what I lived, you
will go a different path yet at my side.
Maybe the mist is your own breath, my steps,
bleeding and bare, are yours.
But why do you walk so quietly,
so silently go by my side,
with sickly step and smoky profile,
if, being one, the same thing we desired,
the same thing we accomplished, and our name
is Electra-Orestes, I, you, Orestes-Electra.
I am mist that flows unseeing
or you are mist that flows unknowing.
Let me halt that you may halt,
or drop my body to the ground and stop your flight.
This may all be a dream we dreamt
within the livid mist,
a floating mockery of mist, without a meaning.
But walking tires me and I need
to tear the mist or to be torn by it.
If soul we had, then let that soul
continue on its march, leave us behind.

But it is she who passes, not the mist.
She who was one, and in one palace,
is now mist-albatross, mist-road,
mist-sea, mist-hamlet, and mist-ship.
And though she killed, was killed in turn, she walks
sprightlier and lighter than when in her body.
We are overcome, not having overcome her.
Orestes, brother, either you have fallen
asleep as you walk or you remember
nothing. You do not answer.

Victoria Ocampo

BORN IN BUENOS AIRES in 1890, Victoria Ocampo at a young age experienced the cultural limitations imposed on women by society and devoted a good part of her life to fighting those limitations. A voracious reader, she was thwarted in her theatrical aspirations by her family and spent much of her youth "in literature." She began to publish her highly personal essays, with their canny combination of textual references and everyday experience, in 1924. *De Francesca a Beatrice* (1924), a text on Dante's *Commedia* highly revealing of Ocampo's reading approach as a woman, was followed by ten volumes of *testimonios,* essays recording her conversations with important cultural figures of her time and her reflections on current events. Ocampo also wrote a six-volume autobiography, which was published posthumously.

A tireless advocate of women's rights, Ocampo was one of the three founding members of the Argentine Women's Union in 1936. Her best-known cultural venture was the founding in 1931 of *Sur,* one of the most influential literary journals of the Spanish-speaking world. The first woman to be received into the Argentine Academy of Letters, in 1977, she died in 1979.

from Emily Brontë (terra incognita)

All imaginative and highly sensitive children are fascinated by certain heroes and tell themselves stories in which they play an important role in relation to their hero—they are pursued, loved, betrayed, saved, humiliated, or glorified by him. Later, once that stage is left behind, they usually act out in life those scenes so often rehearsed in childhood. When the magnificent or terrible moment finally arrives, the reply comes naturally, on cue. Impossible to change it, impossible to get it wrong . . . there have been too many rehearsals. It is no longer possible to choose another, it was

---

Excerpts from "Emily Brontë (terra incognita)," in Testimonios, *vol. 2 (Buenos Aires: Sur, 1941). Translated by Sylvia Molloy: copyright © 1991 by Sylvia Molloy.*

never possible to choose another reply. Each being carries within himself the same scene, the same drama, from the moment he awakens to consciousness till the end of his days, and he plays out that scene, that drama, no matter what events or what characters come his way, until he finds his own plot and his own character. He may never find them. But that does not stop him from playing out his scene, bestowing on the events and characters least likely to fit his play the shape of the events and character that are his own. He was born to play but one scene and one drama, and cannot help repeating them as long as he lives.

from From Primer to Book

I have already told once, if not more often, how the greatest punishment I received from my mother was the time she took *The Hound of the Baskervilles* from me when I had only read it halfway. I remember the place and the armchair where I was ensconced at the time of the dreadful confiscation. That chair, in which I have traveled the world over, is still in my possession. . . . While in that travel chair I trotted all over the globe, through the window came the noises of calm but bustling San Isidro: roosters, *benteveos*, frogs, garden rakes, the screeching of water mills, the shouting of children, the whistle of the train along the river, the noise made by the horse-drawn carriage as it daily took the engineer [Ocampo's father] who had built the bridge in San Luis to the station for the Buenos Aires train and daily went back to pick him up. I thought that none of this reached my ears; indeed I didn't seem to hear it, so engrossed was I in the imaginary world of my books. And yet I did hear, and even more, I smelled the garden that came in through the window. While my avid eyes heeded nothing but the book, my ears and my nose recorded things on their own. Just the fact that I was breathing made everything come together, connected everything, without my knowing, so that Conan Doyle's Devonshire moors—later turned by the Brontë sisters into Yorkshire moors—or the Luxembourg gardens, where Cosette and Marius looked ecstatically at each other, would, for the rest of my days, smell of the Argentine summer and echo with the amorous duets of ovenbirds and the resonant presence of cicadas. Leonardo wrote (and Proust, that great explorer of labyrinths, calls attention to it) that painting is a *cosa mentale*. Reading is all the more so, as are our most

---

Excerpts from "De la cartilla al libro," in Testimonios, *vol. 6 (Buenos Aires: Sur, 1963). Translated by Sylvia Molloy: copyright © 1991 by Sylvia Molloy.*

delirious adolescent infatuations. To underestimate the importance of these mental things, however, is a lack of discernment. . . .

The feeling, at once poignant and vague, of déjà vu frequently invades me in France and England. . . . As an adult, I have never had but a sense of returning to these countries, even to places where I had not set foot in my childhood or adolescence.

I finally understood that it was not precisely what one calls the "artistic treasures" that put me in this state of inner turmoil, but, at times, when I turned a corner, the unexpected and at once familiar name that I read, to guide my footsteps, on the white and blue streetsign, on the wall of any building. This and other equally insignificant details were the pumpkin turning into a carriage for this South American Cinderella. Like her, I was caught unprepared. How could the blue and white streetsign become a whole strange world, vast and familiar? How could this trivia overwhelm me with joy? A joy with no passport, apparently with no face, but whose face we conjure up from memory in the precise moment our memory seems incapable of recognizing it.

One day I understood. This is what I had lived in an armchair in San Isidro or in a room whose windows gave onto the "Bon Marché" (today, the Pacífico Arcades). I am where I am not, as Gabriela Mistral used to say. In those moments when I read, I wasn't in San Isidro or on Viamonte Street, I was in France, in England. And now, when I really find myself in France or in England, I am ensconced instead in an armchair protected by summer dustcovers in San Isidro or I am sitting by a window from where I see the pigeons from the cornices of the Convent of Las Catalinas flying by. The relish of being where we are not is one of the melancholy and fleeting pleasures that life repeatedly offers as it moves toward death.

from The Archipelago

Had it not been for books, for my growing inclination to write letters and for the promise of more books, that month at "La Rabona" (Socorro Station) would have been mortally sad. To be far from the corner of Florida and Viamonte was exile. . . . Books, books, books were a new world where blessed freedom reigned. I lived the life of books and had to account to no one for that life. It was my thing.

Excerpts from El archipiélago. *Vol. 1 of* Autobiografía *(Buenos Aires: Ediciones Revista Sur, 1979). Translated by Sylvia Molloy: copyright © 1991 by Sylvia Molloy.*

And then there was letter-writing. Another comfort, even if I could not write to the person I wanted. Writing, just for the sake of writing, also calmed me. After my first composition on the crimes of the British empire, inspired by the treacherous Miss Ellis, I had become convinced that writing was a form of getting even. The written word helped one to escape injustice, solitude, sorrow, boredom.

. . . Another type of book interested me now: Jules Verne and Dickens (we read *David Copperfield* out loud during our English lessons) were favorites. And the favorite of favorites, in the works of Jules Verne, was *Captain Hatteras*. David ran abreast of him, but on another track. In the same way that I would have liked to turn into Captain Hatteras's dog, in order to follow him (I also thought of imitating him and becoming an explorer of the South Pole), I wished to be little Emily, that is, practically Steerforth's dog. Steerforth fascinated me, as he did David. Captain Hatteras, on the one hand, personified invincible courage, and I was thrilled by courage. Steerforth, on the other hand, with his good looks and his cold insolence, made me fall in love with him. I could not understand why. I did not approve of his conduct, nor of his way of treating poor Emily, yet in spite of everything, I felt attracted to him. I saw his defects more clearly than did David. But even then, like David, I was attracted to someone I in no way approved of, to a boy who misbehaved; as if his charm and his conduct were two different things that could be separated, and that, in some mysterious way, reinforced each other.

When I got to the part where David finds Steerforth's body on the beach after the shipwreck; when I read that Steerforth slept in death as he had in life, "lying with his head upon his arm as I had often seem him lie at school," as David describes him, I could not bear it any longer and nearly burst out crying. . . . How could I find interest in the book without Steerforth, without the beauty and the smiling cruelty that radiated from him? I had always hoped to discover that he was not really bad so that I could have loved him wholeheartedly.

. . . In mourning Steerforth's death, in mourning the death of his beauty, his insolence, his charm, his violence, "his desperate way of pursuing any fancy that he took . . . the fervent energy which, when roused, was so passionately roused in him"; in mourning everything that, because of our resemblance and the no less breathtaking difference between us, attracted me to him, I mourned too for myself. I mourned the childhood that was leaving me for I had begun to see it from a distance. And I mourned the childhood that would not completely let go of me, that in vain resisted the triumphant onslaught of adolescence, as Steerforth's habitual gesture, so apparently full of life, in vain resisted death.

from The Insular Empire

I was an easy reader, voracious, omnivorous. The bad thing was that I could not go to a bookstore to buy any book that interested me, as Ricardo [Guïraldes, the novelist] could. Many books were on the family index. Some for incomprehensible reasons, since they did not deal with amorous passion (a forbidden theme when the love in question was not of the style "mon oncle et mon curé" and ended up in marriage). An example of this apparently gratuitous censorship was the confiscation of my copy of Oscar Wilde's *De Profundis,* found by my mother under my mattress at the Majestic Hotel in Paris. I was nineteen years old. Of course there was a memorable scene in which I declared that I would not continue to live in this manner and that I was ready to throw myself out the window. Unmoved by my threats, my mother did not return the book to me and left the room saying that I was hopeless. I immediately proved her right by throwing my stockings out the window. It was a symbolic act, much applauded by the chauffeurs stationed on the Avenue Kléber, who were enjoying themselves immensely. Fani, our nanny and maid, did believe my threat. It was she who, in opening my bed, had unintentionally revealed my private library and she had witnessed the scene. She was muttering to herself, "Dear me, the young miss is going to throw herself out the window! She's quite capable of doing it!" When she saw the stockings floating down she was quite relieved. She understood that I had vicariously made good my word.

. . . The Rostand case, because of the fervor it awoke in me, was typical of my adolescent passions. Coquelin's company was playing *L'Aiglon,* with Marguerite Moreno in the role of the Duke of Reichstadt. . . . Why did I recognize myself in the hero? It may sound preposterous. The plight of Napoleon's son was not my own. However, this boy with sick lungs (his galloping consumption seemed to me then a prestigious illness) was as much a prisoner at Schoenbrunn as I was in my parents' house. He was a *not-quite-a-prisoner-but.* He could not go out riding without feeling "the sweet honor of an invisible escort." He could not receive letters for fear that they would be opened. He could only read the books he was given.

> *Puis quelqu'un me passa des livres clandestins . . .*
> *Le soir, dans ma chambre, je lisais, j'étais ivre!*

Excerpts from El imperio insular. *Vol. 2 of* Autobiografía *(Buenos Aires: Ediciones Revista Sur, 1980). Translated by Sylvia Molloy: copyright © 1991 by Sylvia Molloy.*

Et puis, quand j'avais lu, pour cacher le délit,
Je lançais le volume au haut du ciel de lit.

And then someone brought me the forbidden books . . .
At night, in my room, in a rapture, I read,
And then, having read, so as to hide the sin,
I would hurl the book high, above the canopy.

My bed had no canopy but it did have a mattress. The operation was the same. We both had the same urge to do great (although different) things, of achieving glory through who knows what courageous acts. Not all courageous acts, I thought, were accomplished on the battlefield.

Alfonsina Storni

BORN IN THE TICINO REGION of Switzerland in 1892, Storni was taken to Argentina as a very young child by her immigrant parents. In 1911 she started a teaching career in which she would acquire considerable distinction; she also began to write poetry. Her first book, *La inquietud del rosal,* was published in 1916 and received mixed reviews. Storni became a very visible figure in the Argentine literary establishment and, because of her nonconformist attitudes, was a frequent subject of controversy. From 1916 on, Storni published steadily till her suicide in 1938. Her books of poetry include *El dulce daño* (1918), *Irremediablement* (1919), *Languidez* (1920), *Ocre* (1925), *Mundo de siete pozos* (1934), and *Mascarilla y trébol* (1938). Storni also wrote drama and occasional criticism. A prolific journalist, she devoted considerable energy to defending women's rights. Her poetry, gradually evolving from late *modernismo* to the metric and semantic experimentation of the avant-garde, constantly rewrites the feminine, either in the combative, polemical tone of her early work or in the less openly emotional, more experimental mode of her later poems. Storni does not scorn the sentimental, but her lyric expressionism is tempered by the timely use of humor and the deliberate introduction of colloquialisms into the fabric of her poetry.

What Would People Say?

What would people say, so shortsighted and empty,
If one day, just like that, from sheer fantasy,
I dyed my hair silver and violet, and then
Put on a Greek tunic, changed combs in my hair
For flower wreaths of jasmine and forget-me-nots?
If I sang in the streets to the music of violins,

"What Would People Say?" and "Premonition," from El dulce daño *(Buenos Aires: Sociedad Cooperativa Editorial Limitada, 1918). Translated by Sylvia Molloy: copyright © 1991 by Sylvia Molloy.*

And spoke out my verse, going from square to square,
Liberating my taste and unmuzzling my voice?

Would they gather on sidewalks to see me go by?
Would they burn me the way they burned witches?
Would the bells ring out calling the people to mass?

When I think of it, really, I just want to laugh.

Premonition

I have the premonition
That I shall not live long,
This head of mine, a crucible,
Purifies and consumes.
Without complaints, without a hint of horror,
I want, to end, a cloudless afternoon
Under the bright, pure sun,
And rising from a great gardenia, a white snake,
To sweetly, very sweetly bite my heart.

Erased

The day I die, news of my death
Will follow usual practice.
In registry after registry
They'll look for me, I'm sure.

And in a village far away,
Dozing in the mountain sun,
Across my name, in an old book,
An unknown hand will draw a line.

"Erased," from Languidez *(Buenos Aires: Sociedad Cooperativa Editorial Limitada, 1920).*
Translated by Sylvia Molloy: copyright © 1991 by Sylvia Molloy.

Lighthouse in the Dark

The sky a black sphere
The sea a black disk.

On the coast a lighthouse
Opens its gleaming fan.

Who does it search for,
Spinning so all night?

If in my chest it looks
For my mortal heart,

Let it turn to the black rock
Where it is nailed down.

A crow pecks at it endlessly
But no longer draws blood.

Tropic

White tombstone,
The burning sky
Falls on
The parched earth.

Woods burn
In red rings,
Curtains of smoke
Devour landscapes,
Scorch towns.

"Lighthouse in the Dark," from Mundo de siete pozos *(Buenos Aires: Editorial Tor, 1934).*
Translated by Sylvia Molloy: copyright © 1991 by Sylvia Molloy. "Tropic," from Obras completas
(Buenos Aires: Sociedad Editora Latinoamericana, 1976). Translated by Sylvia Molloy: copyright
© 1991 by Sylvia Molloy.

Stopped in its bed,
The river water sets
In a dull surface.

Demons,
Burning wings
Race through the fields
In a crazy jig.

Along the chalky bank
The glowing coals of a train
Shriek past.

Dragged across
This white hell
My ovaric plant,
Restored,
Will now root itself
In forests,
Not in men.

And from my breast
Will flow
Not milky sweetness
But sharp mountain stones.

I Am Going to Sleep

Oh, kindly nurse, you with the teeth of flowers,
With hair of dew and open hands of grass,
Lend me, I beg, some earthy sheets and quilt
Of plucked-out moss, in place of eiderdown.
For I would sleep, dear nurse. Put me to bed
And at my head gently set down as lamp
A constellation. Any one will do.
They're all so lovely. Turn it down a bit.

"I Am Going to Sleep," from Mascarilla y trébol. Circulos imantados *(Buenos Aires: El Ateneo, 1938). Translated by Willis Knapp Jones.*

Leave me alone where I can hear things grow.—
"A heavenly foot will rock you to and fro.
A bird will sing a constant lullabye
So you'll forget."—Oh, thanks! One more request:
If he should telephone again,
Tell him not to persist. I have gone out.

Cecília Meireles

CECÍLIA MEIRELES was born in Rio de Janeiro in 1901. A poet writing within the Brazilian modernist movement of the 1920s, she was also by profession a schoolteacher, journalist, and librarian. She traveled extensively in Europe and Asia. A prolific poet, she published tirelessly till her death in 1964. Her books include *Nunca Mais* (1923), *Viagem* (1939), *Vaga Música* (1942), *Mar Absoluto* (1945), *Retrato Natural* (1949), *Doze Noturnos da Holanda e o Aeronauta* (1952), *Romanceiro da Inconfidencia* (1953), *Poemas Escritos na India* (1953), *Metal Rosicler* (1960), *Solombra* (1963), and *Poesias Completas* (1976). In 1964 she was awarded, posthumously, the Machado de Assis Prize by the Brazilian Academy of Letters. Meireles's tightly woven poems, with their haunting combination of the pictorial and the conceptual, can be read as skeptical reflections on the elusive nature of reality and as sensual experiences in themselves. Self-definition is a central obsession in Meireles: there is a constant attempt to detain, to fix, to frame—in a mirror, in a poem—and that effort is constantly thwarted, for the "I" is no less tenuous and elusive than the rest of reality.

Portrait

I used not to have today's face,
So calm, so sad, so thin,
Nor these eyes so empty,
Nor this bitter lip, not then.

I used not to have these frail hands,
So still and cold and dead;
I used not to have this heart
Which refuses to be read.

"Portrait," *from* Latin American Literature Today *(New York: New American Library, 1977). Translated by Alexis Levitin.*

I didn't note this change,
So simple, smooth, and plain;
Lost in what strange mirror
Did my face remain?

Diana

From forest to forest,
ah, always following
disfigured tracks.

Flowers, edged in steel
touch her fleeting flesh.
And rain fills her eyes.

She wields her bow
with fine precision.
What victim would not comply?

At night, she returns
empty-handed, thoughtful,
alone with her shadow.

No arrow, no move
was worth the cry,
echoed in the wind

Presentation

Here is my life: this clearest sand
with footprints dedicated to the wind.

Here is my voice: this empty shell,
shadow of sound bearing its own lament.

"Diana," from Mar Absoluto *(Porto Alegre: Livrario do Globo, 1945). Translated by Marta Peixoto. "Presentation," from* Retrato Natural *(Rio de Janeiro: Livros de Portugal, 1949). Translated by Sylvia Molloy: copyright © 1991 by Sylvia Molloy.*

Here is my sorrow: this broken coral,
surviving its own pathetic moment.

Here is my legacy: this lonely sea,
on one side, love, and on the other, oblivion.

Survival

A survivor
(not of wars, shipwrecks, cataclysms)
but my own survivor,
I contemplate the world
to which one day we thought we belonged
(some brief day, fallen from use, from memory).

Wings of peace and melancholy bear me up,
wings of empty happiness.
Many frail wings allow me this light, free life.

My own survivor,
I wander with my own pensive freedom.

Around us a realm of exile:
no one dreams of surviving.
The world merely learns death.

Urns and Breezes

Amongst these urns, so clear, so smooth,
I shall choose one for my ashes,

"Survival," from Poesias Completas (Rio de Janeiro: Civilizaçao Brasileira, 1976). Translated by Sylvia Molloy: copyright © 1991 by Sylvia Molloy. "Urns and Breezes," from Solombra (Rio de Janeiro: Libros de Portugal, 1963). Translated by Sylvia Molloy: copyright © 1991 by Sylvia Molloy.

and yet I find that breezes
make clearer, smoother, finer urns

to carry longer the light ashes
remaining from such fragile ruins

Norah Lange

BORN IN BUENOS AIRES in 1906, Norah Lange began her writing career as a poet. With Jorge Luis Borges, she contributed actively to the *ultraísta* movement in Argentina. Her poems, published in avant-garde reviews such as *Prisma, Martín Fierro,* and *Proa,* were collected in a first volume, *La calle de la tarde,* in 1925. Lange published two other books of poetry, *Los días y las noches* (1926) and *El rumbo de la rosa* (1930). After 1930 she turned increasingly to prose, writing *Cuadernos de infancia,* a childhood memoir for which she was awarded the First Municipal Prize in 1937; *Antes que mueran* (1944), a book of prose poems; and two novels, *Personas en la sala* (1950) and *Los dos retratos* (1956). She also published a volume of festive, Dadaist speeches, *Estimados congéneres* (1968), which she delivered in banquets before similarly unconventional fellow souls like her husband, the surrealist poet Olivero Girondo. She died in 1972.

Lange took into her prose the taste for experimentation displayed in her poetry. Far from being a conventional memoir, *Cuadernos de infancia* allows Lange to contemplate the most uncanny aspects of childhood, its fantasies, its dreams, and, especially, its fears, with voyeuristic delight. It also allows her to explore the many possibilities of the first-person narrative, to experiment with language and chronology, and to destabilize the autobiographical subject. This experimentation, which made Lange one of the most appealing writers of her generation, are carried one step further in her two novels, *Personas en la sala* and *Los dos retratos.* Declaring "war on solemnity," Lange continually submits her texts to surprisingly humorous and often eerie twists.

from Childhood Copybooks

From a very early age, I liked to look at people in great detail. At six, this had become a firmly established habit. I would then laugh and laugh so much that Mother had to warn visitors who came to the house that I was

---

Excerpts from Cuadernos de infancia *(Buenos Aires: Losada, 1937). Translated by Sylvia Molloy: copyright © 1991 by Sylvia Molloy.*

very *cheeky*. Even if the word in English meant *impudent,* I know it was neither impudence nor aggressiveness that made me do it, for this habit was to follow me till I was much older and able to analyze it.

When I fixed my eyes on the people who came to see us—the parish priest, the town doctor, the Bishop of X., all those who were guests in our house—I imagined their profile from inside. It was as if I entered the person, physically, but only entered the face. In front of a hunchback or a one-armed man I never felt the desire to reproduce the whole body with my own. But the profile! . . . At six, I laughed when I noticed an exaggeratedly curved nose on one of the important men that came through the house. I would slide into their profiles, placing my whole body inside their faces and imitating their outline.

Sometimes I would kneel, holding my arms open· I was then the face of the parish priest, with his straight, tight nose and his barely marked eyebrows. Other times, I would slide into the face of the head doctor. Then it was necessary for me to squat so as to form his very wide nostrils; the tips of my feet sufficed to draw his nearly inexistent mouth; my folded fists copied his little eyes. Bok, the engineer, with his square reddish beard, required greater sacrifices. I had to stand on my hands, letting my hair down so as to give him a beard. . . .

One afternoon—I was about eleven—I tried to enter the face of a certain person in order to form the features with my body. I had to fabricate many imaginary figures, many limp arms, many legs entwined. When I finally got it, the result was so terrible that I felt afraid.

Two months later, that person died. I imagined the casket, I imagined the person in the posture that I had fabricated and that had been like an omen.

Thereafter, when looking at a newcomer, I would imagine myself bending in two, slowly, in order to slip into his face. But the game no longer satisfied me, no longer delighted me, and I finally gave it up.

.   .   .

All five of us had made great newspaper hats and, standing still before a mirror, each one of us peered at her face, at the effect of shadow on our eyes, at the different reflections the light from the window made on our newspaper-covered hair.

Suddenly the door opened and a draft made the hats totter on our heads. One of my sisters said:

"The first one whose hat falls will die before the others . . . "

Motionless before the mirror, arms crossed so as not to cheat, we played at who would be the first to die.

Slowly, a terrible fear began to invade me. The open door let in a brisk and dangerous breeze that, at any moment, would rob me of my hat. I

thought of Irene, of Marta, of Georgina, of Susana, of myself, and as I looked at my sisters out of the corner of my eye, as we smiled at each other, a dead twenty-year-old took over each one of their faces, a young and perfect dead woman, with a single flower on her pillow.

The breeze rustled the big triangular hats without quite knocking them over.

Georgina, her wide open eyes looking into some terrible vision similar to mine, cried suddenly:

"I don't like these games," and, turning away from the mirror, she pulled off her hat, crumpled it into a ball, and threw it on the floor.

For a time, that row of heads before the mirror always brought up in me sad and probable images, faces that were veiled forever, and I thought that it would have been better to wait for the breeze to single out the nearest death so that we might be kinder, gentler, to the sister who would be the first to die.

.   .   .

From the time I was four I was followed by the refrain, "She isn't pretty but she has such nice hair! She looks like a boy!" I was not affected by those remarks at the time, and when someone fondled my hair declaring that I looked like a boy I thought, naively, that it was a compliment.

One afternoon I was called to the verandah. The traveling salesman who delivered our orders from Buenos Aires was there, loaded with boxes and packages. I guessed that they had bought me a new dress to wear on Sundays, when the governess took us to the Club or to walk along Nevado Avenue.

When I stepped out on the verandah, I felt that they were all hiding something. I noticed a little light in my father's eyes, the light we usually saw when he was relaxed. I then saw Mother cutting away ribbons, straightening out bows, trying on a wide-brimmed hat.

My father took me in his arms and stood me on the table. The others had already snooped through everything, and I felt a slight twinge because I sensed that something was expected from me and from whatever was in the box that my mother was unwrapping at that very moment. I was surprised that Susana wasn't on the table with me. She and I were always dressed alike, as were Marta and Georgina, while Irene got different dresses.

Once Mother unwrapped the box, I saw her pulling out a dark suit, made of heavy cloth. After taking my pinafore off, she had me lift one leg, then the other. Used as I was to having a dress pulled over my head, I was infuriated by this new procedure and by my sisters' excitement. I thought that I was being disguised, and I didn't understand why they had chosen me, the ugliest one.

Once they fastened the suspenders, they put the coat on me and left me, alone, on the table.

—Now she really looks like a boy!—Mother cried, while my sisters looked on the results of my new clothes, smiling their approval.

My eyes started to itch and I suddenly felt abandoned and ridiculous. I thought they wanted to exhibit me and, little by little, I felt the first tears well up, indignant and rebellious. I didn't want to cry. I found it absurd to cry in men's clothes and I let out a shout instead:

—I don't want to be a boy! I don't want to be a boy!

Mother took me in her arms to strip me of the new strange clothes. I wept so much that my despair was more like a nervous fit. Frightened, repentant, Mother stroked me while the others, as if ashamed, hid the suit and took their things away. The incident was never mentioned again.

.   .   .

"When you girls grow up and get married, you'll have to kill chickens for your husbands' dinners . . ."

The day the gardener told us this, we did not pay much attention. I went to my room (suspecting that he was on his way to the chicken coop to pick out a chicken to kill) while he went on his way whistling, as if the idea of swinging the chicken in the air, by the neck, till it suddenly went limp was not disgusting or frightening. But one afternoon, long before Christmas, when a turkey was brought home to be fattened, I began to feel uneasy.

What if I were to marry someone who did not have a cook and who liked to eat chicken from time to time?

I remembered the alarm with which a cook once burst into the dining room, shouting that the turkey, already in the oven, had let out a last cluck, perhaps the last one choked in its gullet when its neck was wrung. And I began to imagine all the details of a killing. I felt the soft, granulated neck—repulsive, covered with little red dots—of a chicken wriggling from side to side. I saw myself holding its body tightly between my knees, trying to pull its head backwards, but the neck stretched like rubber and, when I left off pulling, it came back, with a little thud, to its usual position. Convinced that I would forget the head and miss my mark, at other times I imagined the chicken writhing at my side while my hands and apron were covered by a thick, repugnant, ghastly stain.

The only thing I was unable to visualize was the first gesture. How to grab it. Where to press in order to suffocate. How to twist its neck. Even if I used an axe, I would have to hold it down with one hand while I delivered the blow with the other. At that point I felt so disturbed by the horror of it all that, as I brought down the axe, I split its head in two. Or, I only tore off its comb and with it one of my fingers that lay bleeding, distant and unfamiliar, on the table.

At that point I imagined the chicken half dead, its head hanging over the edge of the table, looking up at itself, feeling the same fear I would have felt if a strange hand had tried to open the window of my room, or the fear I did feel when my sister Irene, without telling me, put out her hand toward me in the dark.

For a long time, I was convinced that I would never marry.

.   .   .

One of my favorite pastimes, when I was fourteen, was to shout outrageously and then, when I could no longer keep it up, to laugh, laugh softly at the beginning and then let the laughter grow till it resounded all over the block. Susana and Eduardo joined me in spending whole afternoons sitting on neighbors' doorsteps, laughing until people would beg us to leave.

Other times I would put on a man's hat and, wrapping myself up in a poncho, would climb onto the roof of the kitchen from where I could see inside the neighboring houses. After throwing some bricks on the tin roofs to draw attention, I would begin my speech.

After yelling two or three words in different languages, I would immediately call out to all the neighbors by name with a thundering voice and, when a few distrustful heads leaned out over the fences, my voice and my gestures became so emphatic they bounded off doors, off window panes, off the tin on the roofs.

With a tone that was at times inquisitive, other times ironic, I followed my taunts with tirades in English, in French, with dislocated phrases, the name of some neighbor, the few Italian or Norwegian words I knew, collective insults, a strident peal of laughter, a corny verse. If some neighbor felt inclined to disapprove or to applaud, I showered him with insults, with my faulty polyglotism, with erratic gesticulation, with more banging on the tin roofs.

When I suspected that my cries were about to dry up in my mouth, I performed balancing exercises on the fence, allowing for a pause before beginning the second part of my program. Never altering my serious expression, I would let a barely audible laugh gradually become a peal of laughter that reverberated like a shot, followed so quickly by others I smile to think of them even today.

Wrapped up in my poncho, my face red, my hat well over my eyes, I continued my work unruffled. It lasted for over an hour, until the moment when, having lost my voice, I came down and shut myself up in my room.

Silvina Ocampo

BORN IN BUENOS AIRES in 1906, Silvina Ocampo in her youth studied painting under Giorgio de Chirico and Fernand Léger before turning to literature. Ocampo has published several books of poetry, among them *Espacios métricos* (1945), *Poemas de amor desesperado* (1949), *Los nombres* (1953), *Lo amargo por dulce*—for which she received the prestigious National Prize for Poetry in 1962—and *Amarillo celeste* (1972). A prolific writer of fiction, she is the author of several collections of short stories: *Viaje olvidado* (1937), *Autobiografía de Irene* (1948), *La furia* (1959), *Las invitadas* (1961), *Los días de la noche* (1970), and *Y así sucesivamente* (1987). She has also written two books for children, *El caballo alado* (1972) and *La naranja maravillosa* (1977), and was an assiduous contributor to *Sur,* the literary review founded by her sister, Victoria. Together with Jorge Luis Borges and her husband, the writer Adolfo Bioy Casares, Ocampo edited two anthologies, one of fantastic literature and the other of Argentine poetry.

The casual, even humorous way in which Ocampo's stories deal with the perverse makes for an unsettling reading experience. Like many Argentine writers of her generation, Ocampo often resorts to the fantastic, both for its philosophical import and its narrative techniques. Her use of the uncanny, however, does not necessarily rely on supernatural effects. Everyday reality, these stories tell us repeatedly, is in itself sufficiently disturbing. Silvina Ocampo has an eye for trivial detail, for what appears insignificant and mundane but, when illuminated with a different light, can be monstrous and ultimately funny.

The Basement

This basement, exceedingly cold in winter, is an Eden in the summertime. Some people sit by the door upstairs searching for some cool air on the hottest days in January, dirtying the floor. No window lets in the light or the horrible heat of the day. I have a large mirror and a couch or cot given

---

"The Basement," from La Furia *(Buenos Aires: Sur, 1959). Translated by Daniel Balderston.*

me by a client who was a millionaire and four mattresses I have acquired over the years from other rascals. In the morning I fill pails (lent me by the doorman of the next building) with water to wash my face and hands. I am very clean. I have a hanger to hang my clothes behind a drape, and a mantel for the candlesticks. There is no electricity or water. My night table is a chair, and my chair is a velvet pillow. One of my clients, the youngest one, brought bits of old curtains from his grandmother's house, and I use them to decorate the walls, along with pictures I cut from the magazines. The lady upstairs feeds me lunch; for breakfast I have candy or whatever I can put in my pockets. I have to live in the company of the mice, and at first it seemed to me that was the only defect of this basement, where I don't have to pay rent. Now I have noticed that these animals are not so terrible; they are quite discreet. When all is said and done they are preferable to flies, so abundant in the fanciest houses in Buenos Aires, those places where they used to give me leftovers when I was eleven years old. While the clients are here the mice keep out of sight: they know the difference between one kind of silence and another. As soon as I am alone they come out in an uproar. They go running by, stopping for a moment and looking at me out of a corner of their eyes, as if they guessed what I think of them. Sometimes they eat a bit of cheese or bread from the floor. They are not afraid of me, nor I of them. The worst of it is that I can't store any provisions because they eat them before I have a chance to try anything myself. There are evil-minded people who are pleased at this turn of events and call me Fermina, the Mouse Lady. I don't feel like humoring them and so refuse to ask them to lend me traps to kill the mice. One of them, the oldest one, is named Charlie Chaplin, another is Gregory Peck, another Marlon Brando, another Duilio Marzio; one that is very playful is named Daniel Gellin, another is Yul Brynner, one female is Gina Lollobrigida and another is Sophia Loren. It is strange how these little animals have taken possession of a basement where perhaps they lived before I arrived. Even the damp spots on the wall have taken the form of mice; they are dark and rather long, with two little ears and a long pointed tail. When nobody is watching I gather food for them in one of the saucers given me by the man who lives in the house across the way. I don't want them to leave me. If some neighbor comes and wants to exterminate them with traps or a cat, I will make a row he will never forget as long as he lives. They have announced that this house will be torn down, but I won't leave here until I die. Up above they are packing trunks and baskets and incessantly making packages. There are moving vans by the front door, but I walk by them as if I didn't see them. I never begged for a cent from those people. They spy on me all day long and believe I am with clients because I talk to myself to annoy them. Since they are angry with me, they shut me in with a key; since I am angry with them, I don't ask them to open the door. For the last two days the mice have been acting

very strangely: one brought me a ring, another a bracelet, and another one, the sharpest one, brought me a necklace. At first I could not believe it and nobody else will believe it. I am happy.

What does it matter if it is all a dream? I am thirsty: I drink my own sweat. I am hungry: I chew on my fingers and my hair. The police won't come looking for me. They won't ask me for a health certificate or a certificate of good conduct. The ceiling is falling in, bits of straw are floating down: it must be the beginning of the demolition. I hear cries and none of them are my name. The mice are afraid. Poor things! They don't know, don't understand the way the world is. They don't know the joy of revenge. I look at myself in a little mirror: in all the time I have looked at myself in the mirror I have never seen myself so beautiful.

The Mortal Sin

The symbols of purity and mysticism are at times more aphrodisiac than pornographic pictures or stories: That is why—o sacrilegious one!—the days before your first Communion, with the promise of a white dress decorated with lots of lace, of linen gloves, and of a rosary of little pearls, were perhaps the most impure days of your life. May God forgive me for it, because in a way I was your accomplice and your slave.

With a red-flowered mimosa, which you picked in the countryside on Sundays, and a missal bound in white (a chalice stamped on the center of the first page and lists of sins on another), it was then that you discovered the pleasure—as I choose to call it—of love, so as not to call it by its technical name. You would not have been able to give it its technical name either, since you didn't even know where to place it in the list of sins you studied so diligently. Not even in the catechism was everything anticipated or clarified.

When seeing your innocent, melancholy face, nobody suspected that perversity, or rather, vice already had you caught in its complicated, sticky net.

When some girlfriend arrived to play with you, you would first tell, then show, the secret link between the mimosa flower, the missal, and your inexplicable exhilaration. No friend understood it or tried to participate in it, but all of them pretended quite the opposite, to oblige you, and they sowed in you that panicky feeling of solitude (which was stronger than you were) of knowing that you were being deceived by your neighbor.

"*The Mortal Sin*," *from* Las Invitadas (*Buenos Aires: Sur, 1961*). *Translated by Daniel Balderston.*

In the enormous house where you lived (from the windows of which were visible more than one church, one store, the river full of ships, sometimes processions of streetcars or carriages in the square, and the English clock), the top floor was devoted to purity and slavery: to the children and the servants. (You were of the opinion that slavery also existed on the other floors and that purity was absent from all of them.)

You heard someone say in a sermon: "Lust is greater, corruption is greater." You wanted to walk barefoot, like the baby Jesus; to sleep in a bed surrounded by animals; to eat bread crumbs, which you would find on the ground, like the birds: but none of these pleasures were granted to you. To console you for not walking barefoot, they dressed you in an iridescent taffeta gown, in shoes of gilded leather; to console you for not sleeping in a bed of straw, they took you to the Colón Theater, the largest in the world; to console you for not eating crumbs off of the ground, they gave you a fancy box of silvery lace paper, full of candies that barely fit in your mouth.

Only rarely that winter did the ladies, with their headgear of feathers and furs, venture into the top floor of the house, whose uncontestable superiority (for you) would attract them in summertime, when they wore light clothes and carried binoculars, in search of a flat roof where they could watch airplanes, an eclipse, or perhaps just the rising of Venus. Then they would pat your head as they went by, exclaiming in falsetto: "What lovely hair! Oh, what lovely hair!"

Next to the room full of toys that also served as the study room the men's bathroom was located, a bathroom you never saw except from afar, through the half-open door. The chief servant, Chango, the one granted most responsibility in the house, who had given you the nickname of Doll, would linger there much longer than the others. You noticed it, because you would often cross the hall to go to the ironing room, a place you found pleasant. From there, you could not only see the shameful entrance: You could hear the sound of the plumbing descending to the countless bedrooms and living rooms in the house, rooms where there were glass cabinets, a small altar with images of the Virgin, and the glow of sunset on the ceiling.

In the elevator, when the nursemaid took you to the room full of toys, you often saw Chango entering the forbidden room with a sly expression on his face and a cigarette in his lips, but even more often you would see him alone, distracted, baffled, in different places around the house, standing up, leaning endlessly on the edge of any table, be it a fancy or a plain one (any table, that is, except the marble one in the kitchen or the wrought-iron one decorated with bronze irises in the courtyard). "What's the matter with Chango; why doesn't he come?" Shrill voices could be heard, calling him. He would linger before moving away from the table. Afterwards,

when he did come, no one of course could remember why they had been calling him.

You would spy on him, but he also ended up spying on you: you discovered that the day the mimosa flower disappeared from your desk, a flower that later adorned the buttonhole of his lustrine jacket.

The ladies of the house rarely left you alone, but when there were parties or deaths (and they were very much alike) they would have Chango take care of you. Parties and deaths served to strengthen this custom, apparently preferred by your parents. "Chango is serious. Chango is good. He is better than a governess," they would say in chorus. "Of course, he has fun with her," they would add. But I know of one person with the mouth of a viper, the kind that's never lacking, who said: "A man is a man, but these people don't care at all, so long as they can save a little money." "How unfair!" the raucous aunts would mutter. "The little girl's parents are generous, so generous they pay Chango as if he were a governess."

Someone died, I can't remember who. That intense smell of flowers rose through the elevator shaft, using up and degrading the air. Death, with its countless shows of ostentation, filled the lower floors, went up and down with the elevators, with crosses, coffins, wreaths, palm fronds, and music stands. In the upper floor, under Chango's vigilance, you ate chocolates that he gave you; you played with the blackboard, the store, the train, and the dollhouse. Swift as the dream of a lightning bolt, your mother visited you and asked Chango if there was any need to invite some little girl to play with you. Chango answered that it was better not to, because two of them would make a commotion. A purple color passed over his cheeks. Your mother gave you a kiss and left; she smiled, showing her beautiful teeth, momentarily happy to see you acting so sensibly in Chango's company.

That day Chango's face was even blurrier than usual: we would not have recognized him in the street, neither you nor I, although you described him to me so many times. You spied on him out of the corner of your eye; he, who usually stood up straight, was curving over like a parenthesis; then he would come up to the corner of the table and stare at you. From time to time he watched the movements of the elevator: inside the black metal cage he could see the cables going by like snakes. You were playing, feeling a submissive uneasiness. You could foresee that something unusual had happened or was going to happen in the house. Like a dog, you could smell the awful scent of flowers. The door was open: it was so tall that its opening was the size of three doors of a modern building, but that would not make your escape any easier; besides, you had no intention of escaping. A mouse or a frog does not flee the snake that desires them; larger animals do not flee either. Chango, dragging his feet, finally moved away from the table; he leaned over the railing on the stairs so as to look down. A woman's

voice, shrill and cold, echoed from the basement: "Is Doll behaving properly?"

The echo, so seductive when you spoke to it, repeated the phrase without any sort of enchantment.

"She's behaving wonderfully," Chango answered, hearing his words resound in the lower depths of the basement.

"At five o'clock I will bring her her milk."

Chango's reply, "Don't worry, I will prepare it for her myself," mixed with a feminine "thank you," lost in the tiles of the lower floors.

Chango came back into the room and ordered you: "Look through the keyhole while I'm in the little room next door. I'm going to show you something very beautiful."

He stooped down next to the door and put his eye to the keyhole, to show you how you had to do it. He left the room and you were alone. You kept on playing as if God was watching you, as if you had taken a vow, with that deceptive zeal that children sometimes have when they play. Then, without a moment's hesitation, you went up to the door. You did not need to stoop: the keyhole was right at the height of your eyes. What headless women would you discover? The opening of the keyhole acts as a lens on the image that is seen: the tiles sparkled, a corner of the white wall was brightly lit. Nothing else. A slight draught made your hair blow around and forced you to close your eyes. You moved away from the keyhole, but Chango's voice resounded with a commanding and sweet obscenity: "Doll, look, look." You looked again. Bestial breathing could be felt through the door, no longer just the air from a window open in the adjoining room. I feel such sorrow when I think how horror imitates beauty. Like you and Chango, through that door, Pyramis and Thisbe talked lovingly through a wall.

You drew away from the door again and went automatically back to your games. Chango came back to the room and asked you: "Did you see?" You shook your head, and your straight hair flew around madly. "Did you like it?" Chango insisted, knowing that you were lying. You did not answer. With a comb you pulled off your doll's wig, but once again Chango was leaning on the edge of the table where you were trying to play. With his troubled look he was staring at the inches that separated the two of you, and almost imperceptibly he slid over next to you. You threw yourself on the floor, holding the doll's ribbon in your hand. You did not move. A series of flushes of red covered your face, like those thin layers of gold that cover fake jewelry. You remembered how Chango had rummaged around in the white underwear in your mother's drawer when he replaced the women servants in the housework. The veins of his hands were all swelled up, as if full of blue ink. On his fingertips you saw he had bruises. Without meaning to you looked carefully at his lustrine jacket, which felt very rough

where it touched your knees. From that moment on you would always see the tragedies of your life adorned with tiny details. You missed the neat flower of the mimosa, your misunderstood sickness, but you felt that this arcane production, brought about by unforeseen circumstances, must needs accomplish its goal: the impossible violation of your solitude. Like two very similar criminals, you and Chango were united by different objects but were pointed toward identical goals.

During sleepless nights you composed dishonest reports that would serve as confessions of your guilt. Your first Communion arrived. You could not find a modest or clear or concise form of confession. You had to take Communion in a state of mortal sin. On the kneeling stools were not only the members of your family, which was large, but also Chango and Camila Figueira, Valeria Ramos, Celina Eyzaguirre, and Romagnoli, the priest from another parish. With the sorrow of a parricide, of someone condemned to death for treason, you entered the church as if frozen, biting on the corner of your missal. I see you pale, not blushing before the high altar, with your linen gloves on, holding a bouquet of artificial flowers, like a bride's bouquet, by your waist. I would wander on foot across the whole world to search for you like a missionary to save you if only you had the luck, which you do not have, of being my contemporary. I know that for a long time, in the darkness of your room, you heard, with that insistence that silence reveals on the cruel lips of the furies who devote themselves to tormenting children, the inhuman voices, linked to your own voice, saying: it is a mortal sin, my God, it is a mortal sin.

How were you able to survive? Only a miracle can explain it: the miracle of mercy.

Elena Garro

ELENA GARRO was born in Puebla, Mexico, in 1920. She studied at the National Autonomous University in Mexico, was active in the university theater, and also worked as a journalist. Beginning her literary career as a playwright with three one-act plays, *Un hogar sólido* (1958), she began publishing fiction in 1963 with her novel *Los recuerdos del porvenir*. She has written several volumes of fiction—*La semana de colores* (1964), *Andamos huyendo Lola* (1980), *Rencuentro de personajes* (1982), and *La casa junto al río* (1983)—and a disguised autobiography, *Testimonios sobre Mariana* (1981).

Garro masterfully blends the fantastic and the realistic to create haunting, highly seductive narratives. She is adept at bringing Indian elements into her stories, either for ironic effect or as a way of deepening the mystery her texts often create and never dissipate through explanation. She is particularly interested in rereading the past, be it a national past (*Recuerdos del porvenir* may be seen as a revision of the novel of the Mexican revolution) or a personal past (as in many of her stories). Childhood illuminations and the unexpected workings of memory appear frequently in this revisionary process, not merely on an anecdotal level but as narrative strategies. Endowed with an unfailing sense of the dramatic, Garro has created unforgettable female characters in *Recuerdos* and has dwelled equally unforgettably on female experience in her short stories.

Before the Trojan War

Before the Trojan War the days touched the tips of their fingers and I crossed them with ease. The sky was tangible. Nothing escaped my hand and I was a part of this world. Eva and I were one.

"I'm hungry," Eva would say.

And the two of us would eat the same porridge, sleep at the same time, and dream the very same dream. At night I could hear the wind coming

---

"Before the Trojan War," from La semana de colores *(Xalapa, Mexico: Universidad Veracruzana, 1964). Translated by Oscar Montero.*

down from Big Hand Gorge. It made its way through the rocky peaks in the sierra, blowing heat over the backs of the iguanas, coming down to the town, scaring the coyotes, going into the corrals, burning the jacarandas' red blossoms, and breaking the papayas in the garden.

"It's going over the roof."

Eva's voice was my own. We heard it shaking the tiles on the roof. Scorpions would drop from the rafters and transparent lizards would break their tiny pink legs crashing against the slate floor in my room. Protected by the mosquito net, I touched Eva's heart running in mine down the valleys, fleeing the vapor blowing down from Big Hand Gorge. The wind did not burn us.

"Were you afraid last night?"

"No. We like the wind."

Then, the house would be a mess. With her braids all disheveled, Candelaria served us our oatmeal.

"That cursed wind—you have to tie her mane to a rock so we can have some quiet!"

"It's the burning rage of wild women," Rutilio added.

"That's why I say you've got to nail her mane to the rocks and let her howl."

Candelaria's rage was vast. We moved untouched in her voice and in the garden, where we watched the torn flowers.

"It was before Leli was born . . . ," my mother would sometimes say.

Those words were the only terrible thing that happened to me before the Trojan War. Every time that "before Leli was born" was uttered, wind, heliotropes, and words left me. I entered a world without shapes, where there was only vapor and where I myself was a shapeless vapor. Eva's slightest gesture returned me to the center of things, set the disheveled house in order and my parents' blurred figures recovered their impenetrable enigma.

"Let's see what the lady does . . ."

The lady was called Elisa and she was my mother. Every afternoon Elisa would hide in her room, go up to the dresser, and shut the doors on the mirror. She did not open them until evening, the time when she powdered her face. She lay on the bed, her blond braid dividing her back.

"Who's there?"

"Nobody."

"What do you mean, nobody?"

"It's Leli," Eva would answer.

Elisa hid something and then looked up. Through the net her face and her body looked like a photograph.

"Get out of my room!"

We went back to the hall, walking up and down, back and forth, from tile to tile, without stepping on the cracks and repeating: fountain, fountain, or any other word, until by dint of repeating, it merely turned into a sound that did not mean fountain. At that moment, we switched words, astonished, looking for another word that would not crumble. When Elisa threw us out of her room, we would repeat her name over each tile and ask, "Why is her name Elisa?" and the secret logic of names left us amazed. And Antonio? It was very mysterious that her husband should be called Antonio. Elisa-Antonio, Antonio-Elisa, Elisa-Antonio, Antonio, Elisa, and the two names repeated became one and then, nothing. Bewildered, we sat in the midst of the afternoon. The orange sky ran over the tops of the trees, the clouds came down to the water in the fountain and to the sink where Estefania washed the sheets and his shirts. Antonio had tiny green and yellow sparks in his eyes. If we looked at him closely, it was like being inside the grove in the garden.

"Look, Antonio, I'm inside your eyes!"

"Yes, that's why I sketched you as I pleased," was his answer on Sundays, when he trimmed our bangs.

Antonio was my father, and he did not send us to the beauty shop because "on little girls the nape should be smooth and the hairdresser might take it upon himself to shave you with a razor." It was a shame not going to the beauty shop. Adrian twirled among his multicolored bottles, sharpening razors and brandishing scissors in the air. He chatted as if cutting the words, trailed by a sharp perfume.

"Ah! the little blond ones are after me, but their daddy won't pay for a hairdresser."

Sitting in the round afternoon, we recalled the visits to Adrian and the visits to Mendiola, the one who sold "kisses" wrapped in little yellow papers.

"Here come those two tiny canaries!"

And Mendiola put a "kiss" in each of our hands. We were both visitors. When we went to the movies, we saw the two friends from afar. We could not chat with them nor with Don Amparo, the one who sold candles, because we were between Elisa and Antonio, who only greeted people with a nod of the head. They liked silence, and when we talked, they would say:

"Read, be virtuous!"

Peering at the gods drawn in the books, we found virtue. The Greek gods were the most handsome. Apollo was golden and Aphrodite silver. In India the gods had many arms and hands.

"They must be good thieves."

"Let your right hand know not what the left one does." We stole fruit with our left hand. And the gods from India? They had left hand, right hand, hand above, hand below, nice hand, mean hand, and hand in be-

tween. Impossible to tell which hand was the one that did not know what the other hands did.

"Oh, if we were like them, we would steal everything: screws, candies, tiny flags, all at the same time!"

The other gods were like us. Even Our Lord Jesus Christ had only two hands nailed on the cross. Huitzilopochtli was a dark little bundle, with hands and no arms, but he scared us very much and we preferred not to look at him, motionless on one of the bookshelves.

"What would a cross to nail Kali look like?"

"Like a windmill."

"I mean a cross, not a windmill."

"A cross? . . . Just like a cross."

You would have to nail one hand on top of the other and the other, with a nail as long as a sword.

"And the hand in the middle?"

"We'd leave it loose like a tail, to shake off the flies."

"Can't be done. It has to be nailed too."

"On the left or on the right?"

"Let's ask Elisa."

"What do you want?" Elisa asked with her voice inside the photograph.

"Nothing."

"Then get out of my room!" And she again hid something.

We went out into the hall with the shame of knowing that Elisa hid something in her bed. We walked over the tiles repeating her name and when only the sound was left, we went back to her room.

"What do you want?"

"Your husband is calling you. . . . He's in the henhouse."

The henhouse was no place for Antonio, and Elisa looked at us with curiosity. But the henhouse was behind the corrals, and Elisa would take a good while to go there and return to her bed. She went out. Her bed was warm and a vapor of cologne rose from the pillows. We found what she was hiding.

"Look!"

A dry leaf marked the pages of the book that Elisa kept under her pillow.

"Let's go!"

We left in a hurry, without the candies and with the book. We looked for a safe place where we could leaf through it. Every place was dangerous. We looked at the tops of the trees and we chose the greenest and the tallest. Sitting in a fork of the branches, we read: "The Iliad." This is how the wretched Trojan War began.

"Sing, oh Muse, the anger of Achilles!"

The anger of Elisa lasted several weeks. Deafened by the clamor of the battles, we hardly had time to listen to her.

"Where do you hide all day?"

"Hum! . . . Who knows . . ."

Up above, among the leaves Nestor, Ulysses, Achilles, Agamemnon, Hector, Andromache, Paris, and Helen awaited us. Without our realizing it, the days began to separate one from the other. Then, the days separated from the nights; later the wind left Big Hand Gorge and blew like a stranger over the trees, the sky left the garden, and we found ourselves in a world divided and dangerous.

"Do not allow the dogs to devour my body," said the wounded Hector, raising his arm to emphasize his plea. Achilles, over him, his spear resting on the neck of the vanquished, looked at him with disdain.

"Poor Hector!"

"I'm on the side of Achilles," Eva answered, suddenly a complete stranger.

And she looked at me. Before, she had never looked at me. I looked at her. She was straddling a branch, like another person who was not myself. Her hair, her voice, and her eyes amazed me. She was another. I felt dizzy. The tree went away from me and the ground went way below. She also did not recognize my voice, my hair, and my eyes. And she also felt dizzy. We came down hanging on to the trunk, afraid that it might crumple.

"I'm on Hector's side," I repeated on the ground, no longer feeling the ground under me. I looked at the house and the twisted roofs did not know me. I went to the kitchen certain that it would be the same as before, the same as my own self, but the stubborn door let me by angrily. The maids had changed. Their eyes shone apart from their hair. The sound of the knife was separate from the smell of the onions.

"I'm on Achilles's side," Eva repeated, putting her arms around Estefania's pink skirt.

"I'm on Hector's side," I said firmly, putting my arms around Candelaria's lilac skirt.

And with Hector I began to know the world all alone. The world all alone was merely sensations. I separated from my steps and I heard them resound all alone in the hall. My chest hurt. The smell of vanilla was no longer vanilla, only vibrations. The wind from Big Hand Gorge went away from Candelaria's voice. I touched nothing; I was outside the world. I looked for my father and my mother because the thought of being left alone frightened me. The house was also alone, rumbling as did the rocks that we threw up in a lonely valley. My parents did not know it, and words were useless, because they, too, had been emptied of their content. At dusk, separated from the evening, I went into the kitchen.

"Candelaria, do you love me very much?"

"Who's going to love a naughty blond!"

Candelaria started to laugh. Her laughter sounded in another time. The evening came down like a black bell. Above there was Glory and I did not see it. Hector and Achilles walked through the Kingdom of Shades, and Eva and I followed them, stepping on black holes.

"Leli, do you love me?"

"Yes, I love you very much."

Now we loved each other. It was very strange loving some one, loving everyone: Elisa, Antonio, Candelaria, Rutilio. We loved them because we could not touch them.

Eva and I looked at our hands, our feet, our hair, so enclosed in themselves, so far from us. It was incredible that my hand was I; it moved as if it were itself. And we also loved our hands as we loved other people, as strange as we or as unreal as the trees, the yards, the kitchen. We were losing our body and the world had lost its body. That is why we loved each other, with the desperate love of ghosts. And there was no solution. Before the Trojan War we were two in one, we did not love, we merely were, without knowing full well where. Hector and Achilles did not keep us company. They just left us alone, circling, circling each other, without touching, without ever touching anything again. They also circled all alone in the Kingdom of Shadows, never getting used to their lot as condemned souls. At night I could hear Hector dragging his weapons. Eva would hear Achilles' steps and the metallic sound of his shield.

"I'm with Hector," she asserted in the morning in the midst of the fading walls of my room.

"I, with Achilles," said Eva's voice far away from her tongue. The two voices were far from the bodies, sitting on the same bed.

Olga Orozco

OLGA OROZCO was born in Argentina, in the province of La Pampa, in 1920. A member of the Generation of 1940, a group of exceptionally talented poets, she was strongly influenced by German romanticism and by surrealism. Her first book, *Desde lejos* (1946), inaugurates what becomes, in *Las muertes* (1952), *Los juegos peligrosos* (1962), and *Museo salvaje* (1974), Orozco's unmistakable voice, with its incantatory first-person enunciation, biblical overtones, and dazzling metaphors. Few women (Mistral would be another exception) have ventured into the realm of the prophetic with the power and visionary imagination that Orozco has. Yet, unlike her male counterparts, who reinforce the prophetic voice with an authoritarian gesture, Orozco (like Mistral) turns prophecy on itself, recognizes its flaws, denounces its impostures.

Orozco's poetry is not a poetry of doom, even if death, destruction, fragmentation, and madness are its frequent, obsessive themes. These are poems that never tire in their violent exploration of the world and of the self, that ask questions without answers; they are poems in which reason gives way to hallucination and paradox—like the dark light of *La oscuridad es otro sol* (1967)—shows the way to revelation. Among Orozco's most important books are *Cantos a Berenice* (1977), *Mutaciones de la realidad* (1979), *Obra poética* (1979), and *La noche a la deriva* (1983). A fine translator, she has done excellent Spanish versions of Pirandello, Ionesco, and Adamov.

Olga Orozco

I, Olga Orozco, from your heart proclaim to all that I am dying.
I loved solitude, the heroic perpetuation of all faith,
leisure where strange plants and animals grow,

"Olga Orozco" translated by M. Bowman, from Contemporary Women Authors of Latin America: New Translations, *edited by Doris Meyer and Margarite Fernández Olmos. Copyright © 1983 by Brooklyn College Press. Reprinted with permission of the publisher and the editors.*

the shadow of a time of grandeur that moved among mysteries and
    hallucinations,
and also the candles' trembling at nightfall.
My story is in my hands and in the hands which others used to tattoo
    them.
From my sojourn the magic and the rites remain,
some dates worn out by the instant of merciless love,
distant smoke from the house where we never were,
and a few gestures strewn among the gestures of those who did not know
    me.
The rest still plays itself out in oblivion,
still cultivates grief in the face of one who searched for herself in me
as in a mirror of smiling meadows,
the one you will see as oddly alien:
my own ghost condemned to my shape in this world.
She would have liked to hold me in scorn or in pride,
in a final, lightning-like moment,
not in the uncertain sepulchre where I still raise a hoarse and whimpering
    voice
among the whirlwinds of your heart.
No. For this death no rest nor grandeur.
I cannot be so long seeing it for the first time.
But I should go on dying until your death
because I am your witness before a law deeper and darker than changing
    dreams,
there where we inscribed the sentence:
"They are now dead.
They have been chosen for punishment and pardon, for heaven and for
    hell.
Now they are a damp stain on the walls of the first room."

Between Dog and Wolf

They lock me up within me.
They cut me in two.

"Between Dog and Wolf," from Los juegos peligrosos *(Buenos Aires: Losada, 1962). Translated by Sylvia Molloy: copyright © 1991 by Sylvia Molloy.*

They give birth to me each day, with patience,
in a black organism roaring like the sea.
They clip me out then, with nightmare's scissors,
and I fall into this world, my blood turning in opposite directions:
one face carved from the inside by the fangs of solitary rage,
the other dissolving in the mist that protects the packs.
I am unable to find out who is master here.
I change beneath my skin from dog to wolf.
I decree a plague and race with flaming flanks
across the planes of the future, of the past;
I lie down to gnaw on the tiny bones of dreams,
dead in heaven's pastures.
My kingdom is in my shadow and is with me, wherever I go,
or it collapses, in ruins, an open door for the enemy's invasion.
Each night my fangs tear all the bonds that hold my heart,
each morning finds me burdened by my prison of obedience.
If I devour my god, I use his face beneath my mask,
and yet all I drink in men's troughs is the velvety venom of piety that
    sears my insides.
I have woven this joust on two sides of the tapestry:
I have won the trophy for wild and wandering beast,
and I have also given shreds of meekness for a prize.
Who triumphs within me?
Who defends my solitary bastion in the desert, the sheet of my dreams?
And who gnaws my lips, very slowly in the dark, with my own teeth?

Personal Mark

These are my two feet, my mistake at birth,
visibly dooming me to fall once more under the implacable wheel of the
    zodiac,
if they cannot take off in flight.
Neither base of a temple nor flagstones of a hearth,
They are just two feet, amphibious, enigmatic,
as remote as two angels mutilated along the jagged road.

"Personal Mark," from Museo salvaje *(Buenos Aires: Losada, 1974). Translated by Sylvia Molloy: copyright © 1991 by Sylvia Molloy.*

They are my feet for treading,
step by step, on all the dead,
for walking toward death with heel and toe,
caught in the jaillike night I must go through as it flows around me.
Feet on hot coals, feet on knives,
feet marked by the irons of the Ten Commandments;
two anonymous martyrs bent on going forth,
prepared to kick at the closed doors of the universe
or to leave their sign of obedience as one more track,
barely decipherable, in the whirling dust on the doorstep.
Feet that master the earth,
feet that pursue a fleeting horizon,
polished like jewels by the sun's breath, by the gravel's touch,
two prodigies who radiantly gnaw at my future in the present of my
    bones,
dispersing as they go all traces of a promised land
that changes place and slides beneath the grass as I go past.
What inefficient tools to enter or to leave!
And no evidence, no sign of predestination under my feet,
after so many trips to the same limits.
Nothing but this space between the two,
this impending absence that pulls me always forward,
and that breath of harmony and divergence at each step.
Prodigious, miserable condition!
I have fallen into the trap of these two feet,
like a hostage from heaven or hell asking in vain after her species,
not understanding her bones nor her skin,
nor this lonely beetle's perseverance,
nor this tam-tam that summons her to an eternal return.
Where does this huge and legendary, this incredible being go,
deploying her living labyrinth like a nightmare,
ever sustained
by two delirious fugitives of foam, under the rain?

Idea Vilariño

IDEA VILARIÑO was born in Uruguay in 1920. Among her many books of poetry are *La suplicante* (1945), *Cielo cielo* (1947), *Paraíso perdido* (1949), *Por aire sucio* (1951), *Nocturnos* (1951), *Poemas de amor* (1957), *Treinta poemas* (1966), *Pobre mundo* (1966), and *No* (1980). A shrewd literary critic, Vilariño has published essays on Rubén Darío, Delmira Agustini, and Julio Herrera y Reissig. She has also translated T. S. Eliot, Paul Valéry, and Raymond Queneau into Spanish.

Vilariño's poetry is economical, even laconic. Her voice is muted, striving for a curiously bleak tone. Although not expressionless, it is willfully reticent and morose, spare in its use of metaphor and suspicious of rhetorical devices. A meditative poet, Vilariño masterfully puts into practice what one of her critics has so aptly called "the evocative power of omission." Courting the prosaic, even the colloquial—her poems at times resort to tango lyrics for ironic effect—Vilariño's texts achieve a tone of quiet challenge that is strangely effective.

The Other One

If I say
I am afraid
out of panic
if I say
I don't like it
out of terror
and I can't this afternoon
so as to let myself die
and I never can
no

"The Other One," "Going Back," "In the Middle of the World," and "What the Hand Feels," from Nocturnos *(Montevideo: Arca, 1986). Translated by Sylvia Molloy: copyright © 1991 by Sylvia Molloy.*

I don't want I can't bear
to say, in truth, to say
exactly
where to find myself and who
I am at night at home
with my eyes shut
or when the hour of death is about to ring
and I remain without voice
buried in my air
invulnerable and blind.

Going Back

I'd like to be at home
among my books
my air my walls my windows
my worn rugs
my ancient curtains
eat at my little brass table
listen to my radio
sleep in my sheets.
I'd like to be asleep inside the earth
no not to be asleep
to be dead and without words
no not to be dead
not to be
that I would like
more than going back home.
More than going back home
and seeing my lamp
and my bed and my chair and my chest
that smells of my clothes
more than sleeping under the familiar weight
of my old blankets.
More than coming home one of these days
and sleeping in my bed.

In the Middle of the World

In the middle of the world
in the middle of the night
under the infinite sky
spread over me
quiet in the air's thick lukewarm broth.
Rough smell of pines
a clear faraway voice
of a bird of a child
a bolt of lightning
white and silent

and I
stay on without a name
without myself without meaning
no one someone
in zero.
A moment
just a moment
in the middle of the world.

What the Hand Feels

What the hand feels
what it holds up
what it sustains
is not my forehead my skin my wits
but the gentle bones
of my skull
with its warm disguises
its sockets
full right now
its nimble jaw that one day
will mimic a grin
the day I toss away
my light skeleton
my well-proportioned skull

and I am just
my lips my feet
my hair my cheeks
my eyes my color
and everything I was
slowly
clearly
rotting
rotting
turning to ash.

Elvira Orphée

Born in Tucumán, Argentina, in 1924, Elvira Orphée studied literature at the University of Buenos Aires, then traveled extensively in Europe and Latin America. She has published four novels, *Dos veranos* (1956), *Uno* (1961), *Aire tan dulce* (1966), and *En el fondo,* for which she was awarded the First Municipal Prize in 1969. Several collections of her short stories, *Su demonio preferido* (1973), *La última conquista de El Angel* (1977; *El Angel's Last Conquest,* 1986), and *Las viejas fantasiosas* (1981) have also appeared.

A fragmented, chronologically complicated novel, *Aire tan dulce* deals with several of Orphée's favorite themes: resentment, rebelliousness, denunciation of the strictures of provincial society lead to violence in many forms. Orphée's characters are misfits, extravagant in their desires and conducts. Because of their excessive feelings and their equal capacity for lyricism and hatred, they appear, in some way or another, doomed. *Aire* is particularly rich in such characters, many of them women. Of particular interest is the dialogue, or rather, the crisscrossing of monologues between Mimaya, a tired yet still combative old woman, and Sarah, or Atala, her rebellious granddaughter. In spite of the little contact between them—thwarted communication being one of the great themes of this novel—the complicity between women that these fragments evoke is particularly moving.

Silences

I wanted to bring her to this newborn world to speak to her in the same way that in my childhood I spoke to the stones, solitude to solitude.

With her fine face like a death mask, she is the only one who resembles me.

My face is broad; my eyes, narrow. But she resembles me and does not love me. Perhaps she will never suspect that we might have understood one another.

"Silences," "Do Not Mistake Eternities," and "Voices That Grew Old," from Aire tan dulce *(Caracas: Monte Avila, 1977). Translated by Oscar Montero.*

What made you so special that you married twice? she asks with her vague smile. She has learned the way to strike me unerringly, learned that incomplete beauty is my wound.

The monstrous character in the family, they say, the one that with her extravagances disturbs our calm and brings on the virtuous rage of the city. Does she think that I talk like the rest? I do not know. Or it is just that she has never loved me. She never even understood that my admiration, my envy, and my pity were for her.

All the things that I kept to myself in my life I wanted to tell her when I brought her here. Here I spent my childhood inventing shape and scent for the flowers. Only weeds grew in the red dirt. But I would have wanted, like her, to see for the first time the place our soul was torn from, the ground so red, the wind so parched, the agony of that red sun. I saw her awe; I did not speak. She said, if God made this, he must have been crazy. And her bracelets clattered like golden eyes. Had I arrived for the first time, I too would have suppressed with a hollow sound of bracelets the urge to raise my hands and praise the madness of God. I waited in vain for her laughter and her explanation: ah, it isn't true, but in a world of plesiosaurs and pumice stone one gets the urge to say that God exists and that he is not good, that he can strike unconscious the one whose conscience suffers, one gets the urge to say that his voice is like a sea serpent's. I just waited for her to say that, as if I didn't know that she hardly speaks unless it is to mock.

I lived here, in the cool, white house, when the air breathed fire, when it came down from the mountains burning red and the dogs dug frantically outside to find dirt that was not so red, that would be cooler, that would hide them from the great wind. When I left, I did not find words nor anyone to speak to. I was prepared for magic and night, for the incomprehensible, not for the trivial. What mysterious events in the still city prepared Sarah for the same thing, made her a stranger in her own city? Before now she had not seen the senseless current of a red river bent on appearing like many rivers through a spiraling mountain nor discovered the fear provoked by the ancient stones and this silence, different from the one coming from words often repeated.

I will remember for a long time, all the time that I have left, this air reddened by the setting sun, the soft wind that raised the dust from the red mountain, her terror, the yellow moon already shining and enormous, facing the sun dying violently, her scream *the name!* nothing more. Right away a gesture of bracelets, yellow golden like moons, shaped a halo around her head. Her arms, feigning nonchalance, wanted to deny that she had spoken.

I brought her so that she might find the words that I left here. Here in my childhood I invented the world. When I return—I have returned only

twice—I recover the words. Here words do not shatter against the days and the repetitions. Silence has its rumors; supernatural voices pierce it. From the ground to the mountain, from the mountain to the sky, from the stone to time.

What will become of Sarah if she does not find the words? She has seen the Indian women who used to come—who no longer come—selling cheese on palm leaves, and in the afternoon they would sit on the floor, smoking for hours, without saying a word. Centuries of solitude silenced those flattened bodies, painfully Indian. The remote silence that pierces the wretched figures will also pierce her if she does not do something. Something that will not be wrong like what she is doing.

We are home now. There are people at the door. The moon, stripped of its recent yellow, refreshes the parched skin of the women who are waiting for us. It is getting dark. A truce from the sun for the red one, for the earth. The burning wind stops, and the insatiable heat.

I will remember the moment she tried to get out of the car to walk toward the house and *smell the scent of the flowers that grow elsewhere.* I began to speak. Silly me, thinking I could tell her everything, leaving out no detail, my two marriages, the first with a much older man, pretentious and reserved, the second with a young man, not unlike the other one. Both times without love, for the sake of appearances, and only the second time knowing why.

Yet, why do I remember anything? It was all useless, the story I lived, the story I told.

Nora's daughters are left. . . . No one is left.

Do Not Mistake Eternities

That man who at first didn't love me and who married me because I owned the house that I built myself and because people told him so many times that an industrious young woman should not be a widow, that man, Atalita, began to love me with a restless love. As if something told him that my absence would be the absence of the spring that made him move. In some obscure way, something told him that, because with him words came out vague and misshapen. You can imagine, a few were good enough for everything. With him words had no shape. He lived in a world that seemed shapeless, soft and vegetal, without words to outline a skeleton. And he mistrusted me.

I was already full of the eternity revealed by the mountains. I pursued it through him.

Because of him I set out to become a reserved, efficient lady. I rose precisely at the same time, forgot the chaos of mountains and stars, kept

the impeccable perfection of my hair and a persistent lavender fragrance. My dedication to falsehood knew no truce.

Falsehood smoothed the sharp edges; piercing eternity found a soft sheath. Day by day the years did my living for me, falsely.

I too must have forgotten words. Until I heard them again, so different, so coarse.

That morning, summer woke me up with its air alive with insects and the sun about to shine. It was barely light. From my room, separated from his by the grape arbor, I understood the haste of the nocturnal flowers to hold their perfume before the coming of light. "Evening primroses" they call them, and they profit from the dark to tell what they are: ladies of the night. In the darkness I should have gone out to tell my name, the only real one.

I made a mistake, you see, eternity is not clear, it does not resemble what is repeated, nor does it have the fragrance of lavender. I made a mistake.

I went out of my room. Looking at his, I felt the usual weight all along my veins, already narrowing as if garlands bound them. Love. Love that weighed like pregnancy, perfected by falsehood, error, and renunciation. I walked the courtyards of the darkened house. Not fear, but something like it took hold of me, a brutal cold, and my heart paralyzed.

In the house next door, the one with jasmines and spiritualist pictures, the one belonging to the very old woman and her niece, someone was talking. And the voice came over the low wall bidding good-bye—with such words!—to that vulgar woman next door. And at his age.

There was no need to spy. Eudoro did not omit anything from his story when he found me leaning on the wall, neither the laughter nor the mocking account of my virtues.

I was not prepared for the dizziness. I fall and I fall in a strange void, hacked away from eternity. But isn't there a spot to grasp? Yes, there is! Atala. In her there is something delirious, criminal, phantasmagoric, everything I falsely renounced.

I have told you, Sarah. I have told you everything that I wanted in this red, dry place, my craggy eternity. I am waiting for you to speak. I know your cruelty. Do not use it now.

Oh, no! You are not using it. Without cruelty, only with disgust, you are saying: an old woman. You should be ashamed talking such nonsense.

Voices That Grew Old

My feet are tired now, not much I can do. Words are awkward at convincing. There is love and nothing more. But even my love is useless. Who will give her the love she needs? She must have lost count of the times that she

believed and missed. She just imagined, neither looked with her eyes nor understood with her head. She imagined, and her images barely coincided with those outside. They all wasted her sweetness. They were always hostile, clawing. Until she learned to approach them on the sly, knowing that they were deceitful images. She would have been happy to have them deceive her. They did not even do that.

I would prefer her to be a wanderer of cities, not a wanderer of images. Somebody always has something to tell me. They go by, see me on the balcony or send for me with Nora's daughters, who are playing outside. They have something to tell me. Sometimes they are strangers, plain-looking, who do not give their name, tell only where they live, and I have seen her go by, I have seen her go in the house next door, a house, you know, I tell you because it's so sad and she's so young.

Perhaps it is true; perhaps they mean well.

The late season's soft rain trickles down in large drops from the grape leaves. Somewhere she snuggles under a blanket, thinks of her childhood as if it were the past, remembers the adobe oven, Eudoro's window, the fragrance of the lime trees in my house, the guava trees where we tied her hammocks. She thinks: I cannot bear the ugliness, help me. She does not know whom she is asking for help.

If you would come, if you came out from under that dirty blanket, you would see the rain with me and your childhood would come back. We would snuggle up and look at the clean rain falling, the smell of grapes and lime trees would float by like a thread in the breeze, I would cover you with a white blanket with colorful flowers woven by women who lost their words long ago. What have you lost, Sarah? What have you lost, my tiny peach-fuzz soul? A cold bit of down that leaves me chilled when I want to caress it.

We would both sit and watch a lazy rain. You do not need a dirty blanket to have life turn bright. Rain is enough. But not even rain is necessary if you are drenched in words. That is what stories are good for when life falls short.

Do not stay under that dirty blanket. They will not love you that way. Nobody wants what they find discarded.

Night is coming on. You've got to go home. It has turned dark quickly with this sky. If you don't go home before Arturo closes the door, he will lock you out. It might perhaps occur to you to come here, but he will not have you tomorrow. He would have found a pretext for getting rid of you. And I can't have you, I couldn't stand you next to me so wounded.

"Let's eat, Mimaya. Don't stay there alone, thinking, hours and hours. We are all at the table."

"Give me your hand so I can get up. All of a sudden I'm so old. . . . Will this rain get to my red lands?"

"They're too far away."

"A little rain would be nice. It's as if that land were the bones of the earth. No moisture anywhere."

"There was so much before, they say, that it tired of water."

"Will you take us one day, Grandma?"

"I think I will not go back. I went back with Atala for the last time."

"And that never-ending lawsuit, Mimaya. If some day they rule in your favor, no one will challenge your title to the land."

"If that were the case, I would go back there to die. There I would feel better dead than anywhere else. Nothing worse for the dead than to die where they don't like it."

"And am I going to die too?"

"Who knows?"

"Everybody dies."

"What's it like to die?"

"I think it's being born elsewhere."

"In a star?"

"Maybe."

"You have to be sick to die. Will Atala die soon?"

"Hush, you've talked enough."

"What's more painful, dying or being sick?"

"Make your girls shut up. Nora, they talk just to talk."

How the plants smell in the rain. Closing my eyes I would recognize the dark by the fragrance of the evening primroses. They blossom in darkness. Only at night they know who they are and they say it in fragrance. You didn't recognize me, Atala. Darkness was also the time for me to come out and tell my name, darkness and its dancing demons.

"Good night, Mimaya. Sleep and don't think."

Night makes all still except our sorrows. They start to bite, the new ones right next to the old ones.

I married twice, both times without love. I wanted someone to talk to me to find out what happens in other lives. That's why I married again. From time to time I listened to the infrequent conversation of that man, and he never said anything but very worn, shattered words. What he said was useless. Perhaps out of the determination to give him words, love trapped me. Then I no longer allowed either work or children to erase my face. Mischief, jealousy, and the glance of some early dweller from the high peaks who chose me as his home still linger there. With violent zeal I set out to love him, but I didn't succeed in filling him with eternity. Our conversations were a front. If he ever said anything that was not worn, he said it like a ventriloquist's dummy. And he had begun to love me with a restless love.

They say I was sick for two years after Oriental died. They are dreaming. I was never sick. They call sickness caring so little for what happens that time does what it wants with our thoughts.

In my childhood I spoke with the mountain and with the hot wind, one on one. The great wind would become a breeze at the time when the setting sun soothed so much red, so much burning. I faced the sky, ancient like everything around me, believing without knowing that I believed. The day I left I stopped believing.

The mountain and the wind are there for all time. For the time I will no longer be anywhere.

But then, between them and me there was no difference. I was they or they I, who knows, because thousands of years ago some cataclysm tore living beings from the rock and made them walk. I come from them. Atala comes from them. To look at the mountain was the same as looking at the face of my soul. I didn't look at it that way. The soul allows no truce for a glance. We were of the same stuff, that's all, stone, wind, and I.

The day I left I lost hope.

And I could no longer live without eternity. I looked for it in that man and I, too, loved him.

I set out to love him. I dislodged temptations. And there were many: being alone, fleeing, following anyone who might pass by for another try that might turn out all right.

The rock and the wind were beating. Their heart was eternity itself. That is what I searched. From the time I began to love that man I was also beating.

Who's knocking on the door? Only people bringing misfortune knock on doors at this hour.

My feet are tired now. Moving to try to open is useless. I won't arrive before the misfortune.

Misfortune knows no hour. It comes at dawn, in broad daylight, whenever it wishes.

Whispers, murmurs, the knocks come again. . . . Sarah! Something's happened to Sarah. Good. Better that misfortune level her at once than have her going from misfortune to misfortune as if pacing the four corners of a chessboard. Sarah. Poor thing. My life will end, I won't have any life left to live after misfortune brings her down.

Aurora's steps, Nora's cough, steps upon other steps, voices whispering that I am asleep, silently frightened.

And that smell of rain.

Voices. Old voices. The voices of my children grown old. I loved them because they were children.

"Who is there?"

They are quiet. Now no more. A man who was once a boy shouts. A drunken shout.

"You wouldn't come in if my father were here."

That's Nora's voice.

Nothing's happened to Sarah! How soft my body goes. Nothing's happened to her. It's just Alejo Lamadrid drunk.

"I have to tell her . . ."

"Not at this hour."

"Any hour to tell her. Her beloved Atala. Nobody feels like telling her that her beloved Atala is the biggest . . ."

My feet are so tired. How have they carried me so fast right to Alejo Lamadrid to shut his mouth with one blow?

The rain is falling on him.

Clarice Lispector

LISPECTOR'S INQUIRY into feminine consciousness and awareness, at once sympathetic and implacable, has achieved memorable results in the stories of *Family Ties* (1972). They focus mostly on female protagonists, women visited by experiences that, perhaps trivial on the surface—a chance encounter, a mundane decision, a routine gathering—afford them a disquieting vision of the other side of things that permanently marks them. "Preciousness," a story of loss of innocence as experienced by a poignantly narcissistic adolescent, is one of the richest examples of Lispector's obscure revelations.

Preciousness

Early in the morning it was always the same thing renewed: to awaken. A thing that was slow, extended, vast. Vastly, she opened her eyes.

She was fifteen years old and she was not pretty. But inside her thinness existed the almost majestic vastness in which she stirred, as in a meditation. And within the mist there was something precious. Which did not extend itself, did not compromise itself nor contaminate itself. Which was intense like a jewel. Herself.

She awakened before the others, since to go to school she would have to catch a bus and a train and this would take her an hour. This would also give her an hour. Of daydreams as acute as a crime. The morning breeze violating the window and her face until her lips became hard and icy cold. Then she was smiling. As if smiling in itself were an objective. All this would happen if she were fortunate enough to "avoid having anyone look at her."

When she got up in the morning—the moment of vastness having passed in which everything unfolded—she hastily dressed, persuaded herself that she had no time to take a bath, and her family, still asleep, would never

---

For biographical information on Clarice Lispector, see p. 61.

"Preciousness," reprinted from Family Ties *by Clarice Lispector (Austin: University of Texas Press, 1972). Translated by Giovanni Pontiero. By permission of the University of Texas Press.*

guess how few she took. Under the burning lamp in the dining room she
swallowed her coffee which the maid, scratching herself in the gloom of the
kitchen, had reheated. She scarcely touched the bread which the butter
failed to soften. With her mouth fresh from fasting, her books under her
arm, she finally opened the door and passed quickly from the stale warmth
of the house into the cold fruition of the morning. Where she no longer felt
any need to hurry. She has to cross a long deserted road before reaching
the avenue, from the end of which a bus would emerge swaying in the
morning haze, with its headlights still lit. In the June breeze, the mysterious
act, authoritarian and perfect, was to raise one's arm—and already from
afar the trembling bus began to become distorted, obeying the arrogance
of her body, representative of a supreme power; from afar the bus started
to become uncertain and slow, slow and advancing, every moment more
concrete—until it pulled up before her, belching heat and smoke, smoke
and heat. Then she got on, as serious as a missionary, because of the workers
on the bus who "might say something to her." Those men who were no
longer just boys. But she was also afraid of boys, and afraid of the youngest
ones too. Afraid they would "say something to her," would look her up
and down. In the seriousness of her closed lips there was a great plea: that
they should respect her. More than this. As if she had made some vow, she
was obliged to be venerated and while, deep inside, her heart beat with
fear, she too venerated herself, she, the custodian of a rhythm. If they
watched her, she became rigid and sad.

What spared her was that men did not notice her. Although something
inside her, as her sixteen years gradually approached in heat and smoke—
something might be intensely surprised—and this might surprise some
men. As if someone had touched her on the shoulder. A shadow perhaps.
On the ground the enormous shadow of a girl without a man, an uncertain
element capable of being crystallized which formed part of the monotonous
geometry of the great public ceremonies. As if they had touched her on the
shoulder. They watched her yet did not see her. She cast a greater shadow
than the reality that existed. In the bus the workmen were silent with their
lunch boxes on their laps, sleep still hovering on their faces. She felt
ashamed at not trusting them, tired as they were. But until she could forget
them, she felt uneasy. The fact is that they "knew." And since she knew
too, hence her disquiet. Her father also knew. An old man begging alms
knew. Wealth distributed, and silence.

Later, with the gait of a soldier, she crossed—unscathed—the Largo da
Lapa, where day had broken. At this point the battle was almost won. On
the tram she chose an empty seat, if at all possible, or, if she was lucky, she
sat down beside some reassuring woman with a bundle of clothes on her
lap, for example—and that was the first truce. Once at school, she would
still have to confront the long corridor where her fellow pupils would be

standing in conversation, and where the heels of her shoes made a noise that her tense legs were unable to suppress as if she were vainly trying to silence the beating of a heart—those shoes with their own dance rhythm. A vague silence emerged among the boys who perhaps sensed, beneath her pretense, that she was one of the prudes. She passed between the aisles of her fellow pupils growing in stature, and they did not know what to think or say. The noise made by her shoes was ugly. She gave away her own secret with her wooden heels. If the corridor should last a little longer, as if she had forgotten her destiny, she would run with her hands over her ears. She only possessed sturdy shoes. As if they were still the same ones they had solemnly put on her at birth. She crossed the corridor, which seemed as interminable as the silence in a trench and in her expression there was something so ferocious—and proud too because of her shadow—that no one said a word to her. Prohibitive, she forbade them to think.

Until at last she reached the classroom. Where suddenly everything became unimportant and more rapid and light, where her face revealed some freckles, her hair fell over her eyes, and where she was treated like a boy. Where she was intelligent. The astute profession. She appeared to have studied at home. Her curiosity instructed her more than the answers she was given. She divined—feeling in her mouth the bitter taste of heroic pains—she divined the fascinated repulsion her thinking head created in her companions who, once more, did not know what to say about her. Each time more, the great deceiver became more intelligent. She had learned to think. The necessary sacrifice: in this way "no one dared."

At times, while the teacher was speaking, she, intense, nebulous, drew symmetrical lines on her exercise book. If a line, which had to be at the same time both strong and delicate, went outside the imaginary circle where it belonged, everything would collapse: she became self-absorbed and remote, guided by the avidity of her ideal. Sometimes, instead of lines, she drew stars, stars, stars, so many and so high that she came out of this task of foretelling exhausted, lifting her drowsy head.

The return journey home was so full of hunger that impatience and hatred gnawed at her heart. Returning home it seemed another city: in the Largo da Lapa hundreds of people reflected by her hunger seemed to have forgotten, and if they remembered they would bare their teeth. The sun outlined each man with black charcoal. Her own shadow was a black post. At this house, in which greater caution had to be exercised, she was protected by the kind of ugliness which her hunger accentuated, her features darkened by the adrenaline that darkened the flesh of animals of prey. In the empty house, with the whole family out and about their business, she shouted at the maid who did not even answer. She ate like a centaur. Her face close to her plate, her hair almost in her food.

"Skinny, but you can eat all right," the quick-witted maid was saying.

"Go to blazes," she shouted at her sullenly.

In the empty house, alone with the maid, she no longer walked like a soldier, she no longer needed to exercise caution. But she missed the battle of the streets: the melancholy of freedom, with the horizon still so very remote. She had surrendered to the horizon. But the nostalgia of the present. The lesson of patience, the vow to wait. From which perhaps she might never know how to free herself. The afternoon transforming itself into something interminable and until they all might return home to dinner and she might become to her relief a daughter, there was this heat, her book opened and then closed, an intuition, this heat: she sat down with her head between her hands, feeling desperate. When she was ten, she remembered, a little boy who loved her had thrown a dead rat at her. "Dirty thing!" she had screamed, white with indignation. It had been an experience. She had never told anyone. With her head between her hands, seated. She said fifteen times, "I am well, I am well, I am well," then she realized that she had barely paid attention to the score. Adding to the total, she said once more: "I am well, sixteen." And now she was no longer at the mercy of anyone. Desperate because well and free, she was no longer at anyone's mercy. She had lost her faith. She went to converse with the maid, the ancient priestess. They recognized each other. The two of them barefooted in the kitchen, the smoke rising from the stove. She had lost her faith, but on the border of grace, she sought in the maid only what the latter had already lost, not what she had gained. She pretended to be distracted and, conversing, she avoided conversation. "She imagines that at my age I must know more than I do, in fact, and she is capable of teaching me something," she thought, her head between her hands, defending her ignorance with her body. There were elements missing, but she did not want them from someone who had already forgotten them. The great wait was part of it. And inside that vastness—scheming.

All this, certainly. Prolonged, exhausted, the exasperation. But on the following morning, as an ostrich slowly uncurls its head, she awoke. She awoke to the same intact mystery, and opening her eyes she was the princess of that intact mystery.

As if the factory horn had already whistled, she dressed hastily and downed her coffee in one gulp. She opened the front door. And then she no longer hurried. The great immolation of the streets. Sly, alert, the wife of an Apache. A part of the primitive rhythm of a ritual.

It was an even colder and darker morning than the previous ones, and she shivered in her sweater. The white mist left the end of the road invisible. Everything seemed to be enveloped in cotton wool, one could not even hear the noise of the buses passing along the avenue. She went on walking along the uncertain path of the road. The houses slept behind closed doors. The gardens were hard with frost. In the dark air, not in the sky, but in the

middle of the road, there was a star: a great star of ice which had not yet disappeared, hovering uncertainly in the air, humid and formless. Surprised in its delay, it grew round in its hesitation. She looked at the nearby star. She walked alone in the bombarded city.

No, she was not alone. Her eyes glowering with disbelief, at the far end of her street, within the mist, she spied two men. Two youths coming toward her. She looked around her as if she might have mistaken the road or the city. But she had mistaken the minutes; she had left the house before the star and the two men had time to disappear. Her heart contracted with fear.

Her first impulse, confronted with her error, was to retrace her steps and go back into the house until they had passed. "They are going to look at me, I know, there is no one else for them to stare at and they are going to stare at me!" But how could she turn back and escape, if she had been born for difficulties? If her entire slow preparation was to have the unknown outcome to which she, through her devotion, had to adhere, how could she retreat, and then never more forget the shame of having waited in misery behind a door?

And perhaps there might not even be danger. They would not have the courage to say anything because she would pass with a firm gait, her mouth set, moving in her Spanish rhythm.

On heroic legs, she went on walking. As she approached, they also approached—and then they all approached and the road became shorter and shorter. The shoes of the two youths mingled with the noise of her own shoes and it was awful to listen to. It was insistent to listen to. Either their shoes were hollow or the ground was hollow. The stones on the ground gave warning. Everything was hollow and she was listening, powerless to prevent it, the silence of the enclosure communicating with the other streets in the district, and she saw, powerless to prevent it, that the doors had become more securely locked. Even the star had disappeared. In the new pallor of darkness, the road surrendered to the three of them. She was walking and listening to the men, since she could not see them and since she had to know them. She could hear them and surprised herself with her own courage. It was the gift. And the great vocation for a destiny. She advanced, suffering as she obeyed. If she could succeed in thinking about something else, she would not hear their shoes. Nor what they might be saying. Nor the silence in which their paths would cross.

With brusque rigidity she looked at them. When she least expected it, carrying the vow of secrecy, she saw them rapidly. Were they smiling? No, they were serious.

She should not have seen. Because, by seeing, she for an instant was in danger of becoming an individual, and they also. That was what she seemed to have been warned about: so long as she could preserve a world of classical

harmony, so long as she remained impersonal, she would be the daughter of the gods, and assisted by that which must be accomplished. But, having seen that which eyes, upon seeing, diminish, she had put herself in danger of being "herself"—a thing tradition did not protect.

For an instant she hesitated completely, lost for a direction to take. But it was too late to retreat. It would not be too late only if she ran; but to run would mean going completely astray, and losing the rhythm that still sustained her, the rhythm that was her only talisman—given to her on the edge of the world where it was for her being alone—on the edge of the world where all memories had been obliterated, and as an incomprehensible reminder, the blind talisman had remained as the rhythm for her destiny to copy, executing it for the consummation of the whole world. Not her own. If she were to run, that order would be altered. And she would never be pardoned her greatest error: haste. And even when one escapes they run behind one, these are things one knows.

Rigid, like a catechist, without altering for a second the slowness with which she advanced, she continued to advance.

"They are going to look at me, I know!" But she tried, through the instinct of a previous life, not to betray her fear. She divined what fear was unleashing. It was to be rapid and painless. Only for a fraction of a second would their paths cross, rapid, instantaneous, because of the advantage in her favor of her being in movement and of them coming in the opposite direction, which would allow the instant to be reduced to the necessary essential—to the collapse of the first of the seven mysteries so secret that only one knowledge of them remained: the number seven.

"Don't let them say anything, only let them think, I don't mind them thinking."

It would be rapid, and a second after the encounter she would say, in astonishment, striding through other and yet other streets, "It almost didn't hurt." But what in fact followed had no explanation.

What followed were four awkward hands, four awkward hands that did not know what they wanted, four mistaken hands of someone with a vocation, four hands that touched her so unexpectedly that she did the best thing that she could have done in the world of movement: she became paralyzed. They, whose premeditated part was merely that of passing alongside the darkness of her fear, and then the first of the seven mysteries would collapse; they, who would represent but the horizon of a single approaching step, had failed to understand their function and, with the individuality of those who experience fear, they had attacked. It had lasted less than a fraction of a second in that tranquil street. Within a fraction of a second, they touched her as if all seven mysteries belonged to them. Which she preserved in their entirety and became the more a larva and felt seven more years behind.

She did not look at them because her face was turned with serenity toward the void.

But on account of the haste with which they wounded her, she realized that they were more frightened that she was. So terrified that they were no longer there. They were running.

"They were afraid that she might call out for help and that the doors of the houses might open one by one," she reasoned. They did not know that one does not call out for help.

She remained standing, listening in a tranquil frenzy to the sound of their shoes in flight. The pavement was hollow or their shoes were hollow or she herself was hollow. In the hollow sound of their shoes she listened attentively to the fear of both youths. The sound beat clearly on the paving stones as if they were beating incessantly on a door and she were waiting for them to stop. So clear on the bareness of the stone that the tapping of their steps did not seem to grow any more distant: it was there at her feet like a dance of victory. Standing, she had nowhere to sustain herself unless by her hearing.

The sonority did not diminish, their departure was transmitted to her by a scurry of heels ever more precise. Those heels no longer echoed on the pavement, they resounded in the air like castanets, becoming ever more delicate. Then she perceived that for some time now she had heard no further sound. And, carried back by the wind, the silence and an empty road.

Until this moment, she had kept quiet, standing in the middle of the pavement. Then, as if there were several phases of the same immobility, she remained still. A moment later she sighed. And in a new phase she kept still.

She then slowly retreated back toward a wall, hunched up, moving very slowly, as if she had a broken arm, until she was leaning against the wall, where she remained inscribed. And there she remained quite still.

"Not to move is what matters," she thought from afar, "not to move." After a time, she would probably have said to herself, "Now, move your legs a little, very slowly," after which, she sighed and remained quiet, watching. It was still dark.

Then the day broke. Slowly she retrieved her books scattered on the ground. Further ahead lay her open exercise book. When she bent over to pick it up, she saw the large round handwriting which until this morning had been hers.

Then she left. Without knowing how she had filled in the time, unless with steps and more steps, she arrived at the school more than two hours late. Since she had thought about nothing, she did not realize how the time had slipped by. From the presence of the Latin master she discovered with polite surprise that in class they had already started on the third hour.

"What happened to you?" whispered the girl with the satchel at her side. "Why?"

"Your face is white. Are you feeling unwell?"

"No," she said so clearly that several pupils looked at her. She got up and said in a loud voice, "Excuse me!"

She went to the lavatory. Where, before the great silence of the tiles, she cried out in a high shrill voice, "I am all alone in the world! No one will ever help me, no one will ever love me! I am all alone in the world!"

She was standing there, also missing the third class, on the long lavatory bench in front of several wash basins.

"It doesn't matter, I'll copy the notes later, I'll borrow someone's notes and copy them later at home—I am all alone in the world!"

She interrupted herself, beating her clenched fists several times on the bench.

The noise of four shoes suddenly began like a fine and rapid downpour of rain. A blind noise, nothing was reflected on the shiny bricks. Only the clearness of each shoe which never became entangled even once with another shoe. Like nuts falling. It was only a question of waiting as one waits for them to stop knocking on the door. Then they stopped.

When she went to set her hair in front of the mirror, she looked so ugly. She possessed so little, and they had touched her.

She was so ugly and precious.

Her face was pale, her features grown refined. Her hands, still stained with ink from the previous day, moistening her hair.

"I must take more care of myself," she thought. She did not know how to. The truth is that each time she knew even less how to. The expression of her nose was that of a snout peeping through a hedge.

She went back to the bench and sat down quietly, with her snout.

"A person is nothing. No," she retorted in weak protest, "don't say that," she thought with kindness and melancholy. "A person is something," she said in kindness.

But, during dinner, life assumed an urgent and hysterical meaning.

"I need some new shoes! Mine make a lot of noise, a woman can't walk on wooden heels, it attracts too much attention! No one gives me anything! No one gives me anything!" And she was so feverish and breathless that no one had the courage to tell her that she would not get them. They only said, "You are not a woman and all shoe heels are made of wood."

Until, just as a person grows fat, she ceased, without knowing through which process, to be precious. There is an obscure law which decrees that the egg be protected until the chicken is born, a bird of fire. And she got her new shoes.

Blanca Varela

Born in Lima in 1926, Blanca Varela is a poet and painter strongly influenced by surrealist aesthetics. Varela founded with Emilio Adolfo Westphalen the review *Amaru*, to which she contributes assiduously. Her most important works include *Ese puerto existe* (1959), *Luz de día* (1963), *Valses y otras falsas confesiones* (1972), and *Canto villano* (1978). In true surrealist fashion, Varela successfully defamiliarizes the quotidian and endows it with the obsessive quality of a dream. Her metaphors are stark, often violent, as in the poem "Port Supe," a return to origins that is both devastating and exhilarating. In other poems, Varela's irony undercuts lyricism for epigrammatic effect. Varela's poetry is an extremely effective blend of fantasy, disenchantment, and acrid humor, a persuasive statement of lyrical rebelliousness.

Port Supe

My childhood is in this coast,
beneath the sky, high-domed
as no other sky, swift shadow,
forbidding clouds, dark swirl of wings,
blue houses on the horizon,
next to the great windowless mansion.
Next to the blind cows,
next to the murky liquor, the flesh-eating bird.
Oh, everyday sea,
mountain sea, rainy gorge of the wintry coast!
There I destroy with an enormous stone,
my father's house
there I destroy the cage for the small birds,
uncork the bottles so that black smoke spouts

"Port Supe," from Ese puerto existe *(Xalapa: Universidad Veracruzana, 1959). Translated by Sylvia Molloy: copyright © 1991 by Sylvia Molloy.*

tenderly tainting air and gardens.
There are my hours spent by the parched river,
amidst the dust and its quivering leaves,
in the blazing eyes of this earth
on which the sea casts its white dart.
One only season, one only time
of streaming fingers and a smell of fish.
One night, unending, in the sand.

I love the coast, that dead mirror
where air whirls like a madman,
that wave of fire devastating corridors,
circle of shade and crystal of perfection.

Here on the coast I climb up a black well,
I go from one night to another, deeper night,
I ride with the wind that runs blindly,
pupils luminous and empty.
Or I dwell inside a dead fruit,
smothering silk, dense space
teeming with water, with pale larvae.
On this coast I am the one who awakens
amidst the foliage of dark wings,
the one who occupies the empty branch,
the one who does not want to see the night.

Here on the coast I have roots,
imperfect hands,
a burning bed in which I weep alone.

Identikit

yes
the dark matter
aroused by your hand
is me

Va Eva

animal of salt
if you look back
you will turn into
your body
    and you will have a name

and the word
slithering behind you
will be your trace

Curriculum Vitae

let's say you won the race
and that the prize
was another race
that you didn't drink the wine of victory
but your own salt
that you never heard cheers
but the barking of dogs
and that your shadow
your own shadow
was your only
disloyal rival

Margo Glantz

MARGO GLANTZ, born in Mexico City in 1930, is a prolific writer and journalist who has published essays, poems, and fiction. She is an accomplished practitioner of fragmentary forms and literary bric-a-brac that owe as much to postmodern aesthetics as they do to popular almanacs and cultural gossip. Among her best-known collections are *Las mil y una calorías* (*novela dietética*) (1978), *Intervención y pretexto* (1980), *No pronunciarás* (1980), *Doscientas ballenas azules* (1979), *Las genealogías* (1981), *La lengua en la mano* (1983), *Síndrome de naufragios* (1984), and *De la amorosa inclinación a enredarse en los cabellos* (1984).

As with all of Glantz's texts, *Las genealogías* should be read against the grain, in this case, against autobiography. Mocking the usual self-gratulation that goes into such genealogical exercises when they are written by men, at least in Latin America, Glantz whimsically splinters her Jewish-Mexican woman's past, putting together a patchwork of trivial anecdotes and momentous events, a calculatingly untidy evocation that is at once poignant and implacably ironic.

Genealogies

Noble or not, we all have our genealogies. I am descended from Genesis no arrogance in that, just facts. My parents were born in a Jewish Ukraine quite different from today's and more different still from the Mexico where I was born, this Mexico, Federal District, where I had the good fortune of seeing life among the shouts of the merchants from La Merced, those merchants that my mother, all dressed in white, contemplated with such awe.

Unlike Isaac Babel, I can't be accused of preciosity or biblical erudition, because unlike him (and unlike my father) I studied neither Hebrew nor the Bible nor the Talmud (because I was not born in Russia and because I am not a man), and yet many times I catch myself thinking like Jeremiah

"Genealogies," from Las genealogías *(Mexico: Martin Casillas, 1981). Translated by Oscar Montero.*

197

and, like Jonah, avoiding the cries of the whale. Like Joan of Arc I hear voices, but I am neither a maiden nor do I want to die at the stake, although I do feel an attraction for that garish (and beautiful) color that Shklovsky reproached Babel for, when they were not yet old and that he remembers nostalgically now that he is old (Shklovsky, that is, for Babel died in a concentration camp in Siberia on March 14, 1941).

Perhaps what attracts me the most about my Jewish past and present is the awareness of those colors, clashing, grotesque, that awareness that makes real Jews a small people with a great sense of humor, because of their simple cruelty, their hapless tenderness, and even their occasional shamelessness. I am attracted by those old photographs of a Lithuanian street vendor with his pointed beard (inviting persecutions) and his oversized coat, looking at the camera with a "plump, drunken" grin as he offers his trinkets; next to him, solemn yet slovenly, is the seller of dead people's clothes, a jackal in the henhouse, because he knows how to smell out the coming death of the one who will sell him his suit. I am also attracted by those children from the *cheder* (the Jewish school) who go along with their grandfather, the boy barefoot and the grandfather with his worn gaze and white beard. But I do not belong to them, or barely, from a sleeping part of my self, the one that touches me closest, along with my father, a country boy, the Benjamin in an immigrant family, whose sister, Rachel, disappeared from home while still a girl, perhaps in Bessarabia (although perhaps somewhere else; what difference does it make at this point!) and whose brothers began to emigrate to the United States after the pogroms of 1905.

If I see a shoemaker from Warsaw or a tailor from Volhynia, a water carrier or a boatman from the Dnieper River, it seems to me that they are my father's brothers, although his brothers became prosperous merchants in Philadelphia and changed their caps and beards for department-store clothes, probably Macy's. If I see a group of children from Lublin who barely reach the table, sitting in amazement, always with their yarmulkes, before some old books, while the *melamed* (professor) holds his pointer on the Hebrew letters, it also seems to me that I see my father finishing his work in the fields, with his shoes muddied (on the other side his brothers wear Andrew Geller shoes), not being able to play because he must learn the Commandments, Leviticus, and the Talmud and the order of those feasts and celebrations that are so often alien to me.

I did not have a religious childhood. My mother did not separate the plates and the pots; she did not make a sharp division between recipients that could hold meat and those that would be filled with dairy products. My mother never wore, as my grandmother did, that wig that would cover her hair because only the husband may see the hair of his legitimate wife, and all this even though my grandmother Sheine was my grandfather's second wife (the first one died, during childbirth? nobody knows, nobody

remembers) and her daughter Rachel, the one who emigrated to the immense heart of White Russia, was a daughter from the first marriage.

On the other hand, I knew of the beautiful challah loafs for sale in a bakery with signs in Hebrew, proud of the braided product added to our own breads because an uncle of mine introduced them to this city before his helper, Mr. Filler, commercialized them in the supermarkets. Yet I have never seen my mother arrive at one of those bakeries (any of the ones owned by my two uncles), carrying her pot of *tcholnt,* stew made of tripe, meat, potatoes, and beans, putting it in the oven on Friday before dark so that it would keep warm until Saturday afternoon, so as to eat the main meal still warm with no disrespect for the Sabbath, but I do remember my uncle Mendel praying next to the window with his talithim and his yarmulke but without long sideburns, swaying to the sound of his prayers as if shaken by laughter, or rather it was I who shook with laughter during that long hour before we came to the table, just as now my two daughters shake with laughter, while some member of the family sings the prayers before Passover or the ones that sanctify Fridays. . . .

I for one have been in the ovens. On Uruguay Street, always through those streets with shabby, South American names, premonition and nostalgia of the various possibilities we had to emigrate to unknown lands. My uncle Guidale let us go into the warm Saturday oven, which yielded those cookies, their souls filled with jam, that I nibbled on endlessly, because my uncle knew that my teeth were tender like those of the mouse that left money in exchange for the teeth of good boys and girls. Those cookies came alongside some well-braided chocolate twirls, contrasting their hardness with the soft consistency of the cookie's jam nestled in an unforgettable dough. I always dreamed about having a bakery and selling loaves and every time I would hand a customer his bag full of wonders, looking this way and that, I would eat some of the cookies displayed in the windows, carefully arranged for the delight of the customers, goyim and Jews alike. Next door there was and still is The Danube, but back then I did not like seafood.

As if to clinch this thread, my mother recounts the final scene in the death of my father's brother, Uncle Albert, who died of cancer in Philadelphia, leaving as his only testament a paper where he asserted that cancer is not hereditary.

I also see, from afar, his veins and innards all flared up, an image of my uncle Guidale crying for his wife, my aunt Jane, father's sister, lying on the floor, wrapped in a thin sheet, dead after a long bout with cancer, and his sobs and his words are beautiful; so were my escapades with a sweetheart I had just at the time when Aunt Mira was dying, sick with liver cancer, cadaverous and yellow like the Jews from any concentration camp, and

whom I hardly visited before her death because I preferred to run off with the goy.

I have some Jewish things at home, handed down, a shofar, a sheepskin trumpet, almost mythical, which loudly proclaims the fallen walls, a candleholder for nine candles used during the commemoration of other fallen walls during the rebellion of the Maccabees, which another goy (like me) has already sung about in Mexico (José Emilio Pacheco). I also have an antique candleholder, from Jerusalem, that my mother loaned me and here it has stayed, but the candleholder is alongside some popular saints, some replicas of pre-Hispanic idols (the salesman said they were authentic, but Luis Prieto looks at them, wets his fingers with spit, feels them and says they are not), some altarpieces, some votive offerings, monsters from Michoacan, among them a Passion of Christ with its devils. Because of them, and because I put up a Christmas tree, my brother-in-law Abel tells me that I don't seem Jewish, because the Jews, like our first cousins, the Arabs, have a horror of images.

And everything is mine and it is not and I look Jewish and I don't look it and that is why I write—these—my genealogies.

Adelia Prado

BORN IN 1936 in Divinopolis, Minas Gerais, Adelia Prado is one of Brazil's best-known poets. She has published *Bagagem* (1976), *O Coração Dispar-ado* (awarded the Jabuti Prize in 1978), *Solte os Cachorros* (1979), *Cacos para um Vitral* (1980), *Terra de Santa Cruz* (1981), *Os Componentes da Banda* (1984), and *O Pelição* (1987). She has also translated poetry and works for the Department of Culture in her native Divinopolis. Prado's poems are at once quizzical examinations of the ordinary and poignant revelations, whose potential for devastation is only averted through an uncannily effective use of humor.

Great Desire

I'm no matron, no mother of the Gracchi, no Cornelia,
but a common woman, mother of children, Adelia.
I cook and I eat.
On Sunday, I bang the plate with a bone to call the dog
and throw him the scraps.
When it hurts, I say ouch,
when it's good, I am stupefied,
sensibilities out of control.
But I have my weepings,
a clarity behind my humble stomach
and a very loud voice for hymns on holidays.
When I write a book with my name,
and the name I'm going to give it on the cover,
I'll take it to a church, to a gravestone, to an open field,
and cry, and cry, and cry,
refined and exquisite lady.

"Great Desire," from Bagagem *(Rio de Janeiro: Nova Fronteira, 1976). Translated by Marta Peixoto.*

Dolores

Today I was sad
and suffered three kinds of fear,
increased by an irreversible fact:
I'm not young any more.
I discussed politics, feminism,
and the relevance of the penal reform,
but after all these matters,
I would pull out my little bit of mirror
and my eyes would fill with tears:
I am not young any more.
Science hasn't been any help,
neither do I have as a guaranteed comfort
the respect of the young.
I went to the sacred book
in search of forgiveness for my arrogant flesh
and there it was written:
"It was through faith that Sarah, despite her advanced age,
was able to have a progeny."
If only someone would portray me, I insisted,
in a painting, in a poem,
and if only my slack muscles were an object of beauty . . .
But no. I demand the common fate of women at their washtubs,
the ones who'll never see their names in print and yet
sustain the pillars of the world, because even dignified widows
won't refuse marriage, and think sex is pleasant,
a condition for the normal joy of tying back their hair
and sweeping the house in the morning.
A hope such as theirs I beg of God.

Day

The chickens open their beaks in alarm
and stop, with that knack they have,

"Dolores," from O Coração Disparado *(Rio de Janeiro: Nova Fronteira, 1978). Translated by Marta Peixoto. "Day," from* The Alphabet in the Park *(Hanover, N.H.: Wesleyan University Press/University Press of New England, 1990). Translated by Ellen Watson.*

immobile—I was going to say immoral—
wattles and combs stark red,
only the arteries quivering in their necks.
A woman startled by sex,
but delighted.

Alejandra Pizarnik

BORN IN BUENOS AIRES in 1936, Alejandra Pizarnik studied literature at the Facultad de Filosofía y Letras in Buenos Aires and painting with the Uruguayan surrealist Juan Battle Planas. She sought the company of surrealist poets, especially Olga Orozco and Enrique Molina in Buenos Aires and André Pieyre de Mandiargues in Paris, where she lived between 1960 and 1964. Pizarnik began writing poetry in 1954 and made good the surrealist tenet of living (in) literature. Her more important books include *La tierra más ajena* (1955), *La última inocencia* (1956), *Arbol de Diana* (1962), *Los trabajos y las noches* (1965), *Extracción de la piedra de la locura* (1968), *Nombres y figuras* (1969), *El infierno musical* (1971), and a biography of Countess Erzbet Bathory, *La condesa sangrienta* (1971). She committed suicide in Buenos Aires in 1972; *Textos de sombra y últimos poemas* was published in 1982.

A permanent reflection on writing, a permanent inquisition into the nature of poetry guides Pizarnik's work. Her short, elliptic poems—her illuminations, one might say—are composed of silence as much as they are of words, seem permanently on the verge of discovery, like suspended revelations. They take one by surprise, leaving the reader equally divided between wonderment and impotence, with a sense that the unsaid—what the poem will never surrender—is, finally, the very substance of this poetry. Pizarnik posits a subject in constant quest for her identity amidst fragmented voices that never fully name her, a poetic "I" who makes love to words and is destroyed by them.

from Diana's Tree

I have leaped from myself to dawn.
I have left my body next to the light
and I have sung the sadness of what is born.

*

Excerpts from Arbol de Diana *(Buenos Aires: Sur, 1962). Translated by Sylvia Molloy: copyright © 1991 by Sylvia Molloy.*

These are the versions proposed to us:
a hole, a wall that trembles
                    *
Only thirst
silence
no encounter

beware of me my love
beware of the silent one in the desert
of the traveler with the empty glass
of the shadow of her shadow
                    *

                                                            NOW THEN:
Who would not search in his pockets for a tribute for the forgotten little
girl. The cold will pay. And the wind. The rain will pay. And thunder.
                    *
Illuminated memory, gallery where the shadow of what I wait for roams.
It is not true that it will come. It is not true that it will not come.
                    *
These bones shining in the night,
these words like precious stones
in the living throat of a petrified bird,
this beloved green,
this warm lilac,
this heart, merely mysterious
                    *
now
   in this innocent hour
I and the one I was sit down
in the threshold of my gaze
                    *
The poem I do not say,
the one I do not deserve.
Fear of being two
on the way to the mirror:
someone asleep within me
is eating me, drinking me.
                    *

you have built your house
you have feathered your birds
you have struck the wind
with your very own bones

you have finished alone
what nobody began

Vertigo or Contemplation
of Something About to End

This lilac sheds its petals.
It falls from itself
and hides its ancient shadow.
I shall die of things like that.

Figures and Silences

Rigid hands confine me to exile.
Help me not to ask for help.
They want to nightfell me, they want to die me.
Help me not to ask for help.

Continuity

Not to name things by their names. Things have jagged edges, luxuriant growths. But who speaks in the room full of eyes. Who gnashes with a mouth made of paper. Names that approach, masked shadows. Cure me from the void—I said. (The light made love to itself in my darkness. I knew there was nothing when I found myself saying: it's me.) Cure me—I said.

Cristina Peri-Rossi

CRISTINA PERI-ROSSI taught literature, wrote fiction and poetry, and participated actively in left-wing politics in Uruguay till 1972, when she was forced into exile by the political situation. Peri-Rossi took up residence in Barcelona, where she writes and works as a journalist.

Peri-Rossi conceives her fiction and poetry not as a substitute for but as an expansion of her political reflection, a form of knowledge. The allegorical resonance of some of her titles, with their intimation of chaos and imminent destruction—abandoned museums, shipwrecks, ships of fools—points toward a double reading. Indeed, Peri-Rossi's work, without losing any of its specifically literary qualities, is a form of political reflection, an ideological statement in the Latin American political debate. That statement, for Peri-Rossi, is necessarily gendered. The speaker in her poems increasingly feels the urge to mark herself as woman and, more precisely, as a lesbian, thus incorporating yet another form of resistance and difference into this poetry of protest.

[For each woman]

For each woman
regal
dignified
mauve
that dies within you
another woman
is born under the full moon
for the solitary pleasures
of the translating imagination

For additional biographical information on Cristina Peri-Rossi, see p. 99.

"[For each woman]" and "[You come assembled]," from Diáspora *(Barcelona: Lumen, 1976). Translated by Sylvia Molloy: copyright © 1991 by Sylvia Molloy.*

[You come assembled]

You come assembled
by twenty centuries
of predestination
in which preceding men
made you so
to love you according to their needs
and their empires
that tradition
if unjust
and intrusive
is not, in sum,
      the least of your charms.

[I love you this and other nights]

I love you this and other nights
with our signs of identity
exchanged
as joyfully as we trade clothes
and your dress is mine
and my sandals are yours
As my breast
is your breast
and your ancient mothers are my own

(and at night we might go out for a walk
you or I dressed as a man
and the other as a woman
as the custom of the species
and divine counsel
decree:
Be fruitful and divide
Multiply in vain).

"[I love you this and other nights]," from Lingüística general (Valencia: Prometeo, 1979). Translated by Amy Kaminsky.

Rosario Ferré

IN HER POEMS, Rosario Ferré actively revises traditional images of woman. Resorting to mythology, *Fábulas de la garza desangrada* (1982) rescues a series of female icons, a precursor sorority, as it were. However, Ferré's evocation of these women—Ariadne, Antigone, Salome, Francesca de Rimini, Catherine of Sienna—goes well beyond conventional celebration. Reading them in a new light, she recombines these female figures creatively, questions the symbolic value traditionally assigned them, and finally allows them their full potential as heroines and rebels.

Requiem

Theseus was finally convinced:
the Minotaur was his identity.
He rose and buckled up his shield
and, as he strapped on his sandals,
the bracelets on his arms rang out
like lightning.
Legs apart, Cyclopean, he stood over Ariadne
and resolutely fit the dagger to his waist.
He wondered to himself whether the Minotaur
was a monster,
or whether it embodied the ferocity of good,
its head bent by the weight of its forelock
as it sowed love.
He whispered in her ear that one day he'd return
and present her with a rich horn of ivory
borne on damask cushions from across the sea.
Theseus clasped her to his heart one last time

For biographical information on Rosario Ferré, see p. 88.

"Requiem," from Fábulas de la garza desangrada *(Mexico: Joaquín Mortiz, 1960). Translated by Rosario Ferré.*

and strode off with a swagger
lighting up the labyrinth with the gleam of his sword.
Ariadne leaned against the wall
and followed him with her eyes as though unraveling
a string of fire in the dark.
A reed of icy bones
splintered along her back.
She slid slowly against the wall until she sat
upon the trampled dust
of unending nothingness.
A sudden pain tore at her insides
and then she felt the warm wine
as it ran down between her legs:
she had begun to abort the Minotaur.

Soledad Fariña

BORN IN ANTOFAGASTA, CHILE, in 1943, Soledad Fariña studied political science at the University of Chile. She has traveled in Europe and lived in Sweden, where she studied literature at the University of Stockholm. She currently lives in Chile, conducting the feminist literary workshop "Lecturas de Mujeres." Fariña has written two books of poetry, *El primer libro* (1985) and *Albricia* (1988). Explorations that literally twist the female body out of shape, these poems enumerate and celebrate a physicality that is both familiar and alien, a body (become) other. In Fariña's work, erotic ritual doubles as textual reflection: the text is the space where the body is dismembered and the poem composed.

This Tongue Unravels

With its frayed threads dancing
in the air like vines
it abandons the scene within the lips
(past the sentry of the teeth)
It winds its way down to the chest listening
to heartbeats from the surface
(it dwells on those beats)
Perplexed it returns to its scene
rolling up its vines   The lips welcome
this hollow tongue back into its prison

we I remain speechless

(the muddy loam does not coalesce into words)

211

I Raise a Mute Eyelid

I close my eye turning its beacon inward

    my spirit   my soul I search

among black lichens   algae   floating by
I see my face eaten away by tongues
There goes that eye like a gaping mouth   What does it seek

                    Above   Below

No One Writes My Body

      nails moan
scattered hands weep
toes soft as petals
Moist lips sing fish gasping
for air

      my spirit   my essence

and once more the shreds merge
into memories
from a half-open mouth

[The blue vein throbs]

    The blue vein throbs
and I plunge into it   Swimming
    against the tide
seeking the cause for its dark beat

I enter your breath   put my ear close
    to that pool
    till the breathing becomes
    a faltering gasp

Nancy Morejón

NANCY MOREJÓN was born in Havana in 1944 and studied French literature at the University of Havana. Her books of poetry include *Mutismos* (1926), *Amor, ciudad atribuida* (1964), *Richard trajo su flauta* (1967), *Parajes de una época* (1979), *Octubre imprescindible* (1982), *Elogio de la danza* (1982), and *Cuaderno de Granada* (1984). An anthology of her poems was published in 1980 in Mexico. She has also written several books of essays, among them *Nación y mestizaje en Nicolás Guillén* (1980), and has translated Paul Eluard, Aimé Césaire, and Jacques Roumain into Spanish.

Morejón's poetry, committed to the Cuban revolution, is both combative and elegiac. Throughout her works is a sense of community, made manifest in the effort (common to many authors writing in the aftermath of the Cuban revolution) to "give voice" to marginal or oppressed groups. Her texts strive for political empowerment and social justice. In addition, they are attempts at an affective restoration: the poet, turned witness, evokes her family group, rescues her anonymous African ancestors from oblivion, and reflects on her extended, Caribbean family. Morejón's poems are direct, urgent; they maintain a sense of oral immediacy, like those of her compatriot Nicolás Guillén, yet are fully conscious of a poetic legacy to which they often refer. In her poetry living voices successfully merge with memories of the cultural archive.

I Love My Master

I love my master.
I gather brushwood to light his daily fire.
I love his blue eyes.
Gentle as a lamb
I pour honey in his ears.
I love his hands

"I Love My Master," from Octubre imprescindible *(Havana: Casas de las Américas, 1984). Translated by Sylvia Molloy: copyright © 1991 by Sylvia Molloy.*

that threw me down on a bed of grass:
My master bites and conquers.
He tells me secret stories while
I fan his entire body, riddled with sores,
with bullets,
from days of sun and plunder.
I love his feet that roamed afar and pillaged
foreign lands.
I rub them with the finest powders
I could find, one morning,
coming out of the tobacco field.
He strummed his guitar and from his throat came
vibrant verses, as from Manrique's lips.
I would have liked to hear a marimbula play.
I love his fine red mouth,
speaking words I cannot yet decipher.
My tongue is no longer his.
And the silk of time in shreds.

Overhearing the old overseers, I learned
that my beloved
uses his whip in the vatrooms of the sugar mill
as if he were in hell, that of the Lord God
he used to tell me about so often.
What will he say to me now?
Why do I live in a place fit for a bat?
Why do I serve him?
Where does he go in his wonderful carriage
pulled by horses happier than I?
My love is like the brambles covering my little plot,
the only thing I can call my own.

I curse

this muslin robe he has forced on my body;
this futile lace he pitilessly makes me wear;
these household chores for an afternoon without sunflowers;
this baroque and hostile tongue I cannot chew with;
these stony breasts that cannot even nurse him;

this belly gashed by his ageless whip;
this cursed heart.

I love my master, but every night,
when I walk along the leafy path to the cane field,
where we hid to make love,
I see myself, knife in hand, skinning him
like an innocent calf.

Drowned by the drums, I cannot hear
his agony, his cries.
I hear the bells calling.

Rebirth

Daughter of ocean waters,
asleep in that womb,
I am reborn
out of the gunpowder
sown over the mountain
by a guerrilla rifle
so the world in turn
might be reborn,
and the vast sea
and all the dust,
all the dust of Cuba,

In the Streetcars' Shade

for Eliseo Diego

In the streetcars' shade . . .
grandmother, did she wear her hair
braided, or was that just

Translations of "Rebirth," "In the Streetcars' Shade," and "Coffee" © Kathleen Weaver. Originally published in Where the Island Sleeps Like a Wing: Selected Poetry of Nancy Moréjon *(San Francisco: Black Scholar Press, 1985). Translated by Kathleen Weaver. Reprinted with permission.*

the family tale
My poor grandmother . . . whom I never saw.

Coffee

Mama brings coffee
from over far-flung seas
as if her life story
encircled each phrase of smoke
that swirled between us.
Surprised by dawn, she smiles
and over her sugary hair
gold bracelets leap.
The somber thread of her childhood
endures between us.

We would like a towering tree,
a mountain *flamboyán,*
in whose noble shade
the troubadour might sleep.

Tamara Kamenszain

TAMARA KAMENSZAIN, born in Buenos Aires in 1947, studied philosophy at the University of Buenos Aires. She has traveled extensively throughout Latin America and the United States, lived in exile in Mexico during the Argentine military dictatorship of the 1970s, and teaches literature in Buenos Aires. A poet, Kamenszain published *De este lado del Mediterráneo* (1973), *Los no* (1977), and *La casa grande* (1986). *El texto silencioso* (1983) is both a critical reflection on other poets and an exemplary exercise in rereading-as-woman.

Kamenszain's poems are highly textured, stylized constructs, deliberately weaving loose strands of "life" (everyday trivia, childhood recollections) with memories of literature, echoes of other poets (such as the Cuban Lezama Lima, for example, seen by Kamenszain as a "maternal" influence in her poetry). Kamenszain has deliberately chosen margins as the locus of her poetry, metaphoric resistance as her poetic ploy. There is a theatrical quality to these poems (underscored by the reference in one of her titles to the Japanese No theater), one that prompts us to see texts as spectacles, as ceremonies, dazzling yet not undecipherable.

The Big House

1

Shapeless place written in the past
children's graveyard, backyard
frozen in a playful circle
long-legged cousins like statues
playing at the make-believe bridge linking
the complicity of parents to the distance
of aunts and uncles.
Sailors without ships in the fifties

"The Big House," from La casa grande *(Buenos Aires: Sudamericana, 1986). Translated by Oscar Montero.*

an anchor for a boutonniere, a family
that puts on the golden breeches
of procreation. There is a celebration:
the return through the promised land
to the place where slave to promises
great-grandparents took their refuge.
They serve the wine in cups of joy.
If the neighbors complain they define
another family, a ghetto without walls.
In the arms of women kneaded,
mother's leaven, growth,
sweet rolls peering at the pleasure
of a bite. The bread shop
makes deliveries. As long as they produce,
more children gather at the table.
The men read the other repertory,
ritual painted in prints
as in an old book or a Bible, open
with its leaves served upon that table.
Inverted letters come loose from the voice
of the reciter. And the stew's aroma
says: lineal narrative, a taste
for the facts, what came to pass
brings fame to the patriarch's greedy name.
The grandmother nods with her fingers
pressed to the texture of the species:
her people are not chosen nor her grandson—
his cap tousled on his head.
(To cover it would be to blot the ignorance
that ties the knot of ancient texts
and unleashes indifference from its bonds.)

2

Rat hole gravid with immigrants.
Here the floors prolong with each dance
the agony of their boards.
Rugs that come from Persia but retain
the obsessive rehash of a design
with each dusty step bite scraps
of a steppe that has turned native,

fields of gooseberries where maté
sweetens like hot tea
a thirst from the pampas that steeped in
the vastness of the south by now is bitter.
Women, amazons for birthing
—riding sidesaddle like a foreigner—
bringing their camisoles on ships
and at the right hour unraveling
in the strange land they seeded
a fashion for winter and warm coats
while, their braids tight as bridles,
Indian women, standing, bore the others.
Yiddish and its circle of words
keeps us children safe inside.
Ring around words weaving in the yard
an outside silence with voices from the home.
(Yet fleeing during siesta
furtively through the street they napped
in the soft shade of everyday words,
savoring a cushioned tenderness: Spanish.)
The neighbors tell my grandfather:
"the days that you call holy
not even the devil's out of work;
the days that by law we rest
the morning's busy for the Israelite;
if writing for you is from the right,
your calendar is woven upside down."

3

Who tells the record of these dead
has chosen to climb as far as nothingness
from the doleful margin on the left.
Between tall slabs words walk
the path of busy ants.
Heaving on their shoulders whitish leaves,
they banish me from now and cull
a joy, family in the old
daguerreotype, static avalanche
revival in sepia and oval

the recurring frame of a style:
a way of speaking, said and done.

4

And in the shapeless place of the beginning
there sits fabling the one who did not say
I am first, person, I am turning
my books toward the harbor of my childhood
gray covers, faded colors
back to the editor where they were born
to the Rio de La Plata's faded silver
to that Linotype enshrined
in bookstalls. Reader types who
never change: bookstore browsers
on the sidelines drinking up a lingo
of white wine understood in passion
but darkened in the dregs of sense.
Back to girlfriends taking refuge
in the stooped frame of doors
connected by hinges and by rumors
laying the creaking of a rhyme
on that verse measured out by mothers
toward home, inside, toward the parlor
seeking the trivial stitching of what's already said.
I walk on with them on the quiet
guided by the peephole in the doors
I see so small what I describe
the relatives, the fenced-in plain
the children waiting in the hall
for the arrival of the husband's steps.
The eye widens, home in a lock
I want to write an ancient habitat,
clothe myself in the wardrobe of letters:
black box that someone will read
through the albino's darkened lens.

5

I am ready. As a signal of return
I will fasten the verses to the package.
I loosen the measure, knot of what's been said

and send it back just like a letter.
Even to those sailor fifties
a postcard may arrive, a bit of news
the wrinkled thread of memory
the eye to thread such nearsightedness. For
they do not read me, the dead, my
grandparents.

Vanessa Droz

VANESSA DROZ was born in 1952 in Vega Baja, Puerto Rico. She studied history of art and comparative literature at the University of Puerto Rico and in the 1970s contributed to innovative literary reviews such as *Penélope o el otro mundo* and *Zona de carga y descarga*. In 1980, together with other Puerto Rican women writers, Droz founded the review *Reintegro*. *La cicatriz a medias,* from which the following text is taken, is a collection of poems written between 1973 and 1979. Experimenting with texture, surface, cuts, and sutures, these poems incessantly inquire into poetic form as a metaphor for the body—a body sensual or a body of texts—yearning for fixity and at the same time yielding to a seductive flux that works against it.

The Unnameable Vase

EQUAL   GOLDEN
like an egg
your name   unnameable
irreconcilable   and premature flight
of blood   how it flows, how it mounts
your name
situational   engraved
on the great vase
of the sea   of the ground
habitual   wherever
habitual   grievous
in the meeting
of the foot   against the ground
dual   doubled
your body and your inner silence

"The Unnameable Vase," from La cicatriz a medias *(Río Piedras, Puerto Rico: Editorial Cultural, 1982). Translated by Sylvia Molloy: copyright © 1991 by Sylvia Molloy.*

that is not   a duality of two
earth with sea   foot with ground
but rhythm   that emerges
completing   the sphere (the perfect vase)
and you move   feeling yourself
endowed unequal   endowed
                   with sick death
and the good   death
with all   its dead
you move
sexual   graver
and you approach   without pity to inaugurate
sadness   of blood and mind
throughout my   desolate realm
sensual   tamed
                   with my hands
soft with   soft death
virtual   giver
                   of life to your doe
before   silence and the silent
ritual   grieving
                   and thus
ritual   engraved
                   you are
conventual   grieving
                   without faith, without faithful
(depths   of blue waters without god
—god   transparent and blue
                   without sky)
without equal   grieving

I call and your   name
                   falls
diluted without   possible memory
in the words   I invent for myself
textual   reduced
                   by force
thus incomplete   vase without total form
actual   docile

in the poem
in language   (a necessary vase to hold you)
and to hold you   here is useless
(unnecessary vase to lose you)
liquor drunk   without brew
residual   inert

noose
I invent for you
I invent you   always   unnameable vase
lost vase   irretrievable

you return

Ana Cristina Cesar

BORN IN RIO DE JANEIRO in 1952, Ana Cristina Cesar studied literature and communications at the liberal Catholic University in Rio and literary translation in England, where she lived between 1979 and 1982. While a student in Rio, she became associated with the poets of the "mimeographic generation" (so called because they distributed mimeographed copies of their work). She translated Emily Dickinson and Sylvia Plath and, in 1982, published *A Teus Pés*, a collection of poems that included two previous books published privately in Rio in 1979, *Cenas de Abril* and *Correspondencia Completa*, plus poems she wrote after her return from England. In 1983 she committed suicide. *Inéditos e Dispersos* was published posthumously (1985).

Cesar's poetry is rebellious, disquieting. It struggles in search of identity and is painfully aware of the complexity of self-representation as woman. Constantly asking questions and implacably taunting any attempt at safe answers, it is, for all its restlessness, oddly in control and ever conscious of its literary texture. Ironic, ruthless, quirky, it offers an unsettling mixture of humor and despair.

[The weather turns.]

The weather turns.
I am faithful to biographical events.
More than faithful, oh so captive! These mosquitoes
which don't let me go! My cravings deafened
by cicadas! What am I doing here in the country
spouting yards and yards of emotional verse?
Ah but I feel emotional and Portuguese, and now
I am no longer, look, I am no longer harsh and severe:
now I am professional.

"[The weather turns.]," "Nothing, This Foam," and "June 16," from A Teus Pés (São Paulo: Brasiliense, 1982). *Translated by Patricia E. Paige.*

Flowers of More

slowly inscribe
a first letter
a slave
in the surroundings
created
by the hurricanes;
slowly measure
the first shy
swallow
that crosses
the stage curtain,
opened
to the winds;
slowly force
the pulse
which knows
how to bleed best
onto the knife
of the tides;
slowly press
the first
look
onto the dampened gallop
of beasts; slowly
ask for more
and more and
more

Psychography

I too leave furtively
and I seek a synthesis in delays

"Flowers of More" and "Psychography," from Inéditos e dispersos *(São Paulo: Brasiliense, 1985). Translated by Patricia E. Paige.*

I collect obsessions with a cold hand and I say
of the heart: it didn't know and I say
of the word: I won't say (I can't yet believe
in life) and I renounce poetry as one who waves good-bye
and I live as one who banishes the anger of having seen

Nothing, This Foam

As an insult to desire
I insist upon the wickedness of writing
but I do not know if the goddess climbs to the surface
or only punishes me with her howls.
From the railing of this ship
I so want the breasts of the siren.

June 16

I can hear my feminine voice: I am tired
of being a man. Angela denies with her eyes: a
woman left lonely. The day perishes. Come
children, come to Jesus. Bible and hymnal on my
little lap. White socks. The organ papa played.
The final blessing amen. Tossing in my
single bed. Mother came to sniff and caught on.
Mother sees into the eyes of the heart but I am
tired of being a man. Angela gets my attention
with her eyes painted lilac or some other sinister
color in the little box. My breasts have
hardened. Disfunctions. Cold in the feet. I am the
way, the truth, the life. Your word
is a lamp for my feet. A light for my way.
I can hear the voice. Amen, mother.

PART

Women, History, and Ideology

Introduction

BEATRIZ SARLO

W HAT IS THE MEANING of the triad that gives its title to this essay? Is it the verbal presentation of a continuity, the enunciation of a conflict, or the way in which a field of problems can be conceptually drawn and discussed? Are we facing a good-willed formula, the discursive expression of a gesture by which women are granted a place in history not only as shadows of the Other, not only as passive and receptive subjects, but also as producers of speech and practices? The triad "women, history, and ideology" turns out to be a plot of different conflicts woven by voices that can be heard today because they are finally getting the attention they have long demanded.

Our current atmosphere is more benevolent toward these voices that reach us from the past hundred years: Women's movements now cross paths with academic perspectives accepted or at least recognized by most insti tutions. Political parties move in a cultural and ideological climate in which it seems improper to support open discrimination, and they are pressed to place their discourse within the boundaries of egalitarian perspectives. In the production and reproduction of the public sphere, women's rights (the feminist and feminine programs) cannot be overtly ignored. Hypocrisy is, in the least favorable situations, a strategy by which women receive public recognition of their rights, although the practice of this recognition is still open to conflict. Even in the conservative forms of Latin American culture and politics, women are increasingly playing nontraditional roles.

Yet the precise relation among the words women, history, and ideology is unresolved in Latin American culture. It is true that we live in an ideological climate that has undergone great changes; nevertheless, it would seem strange and somewhat redundant to speak of men, history, and ideology because the latter two still "naturally" map out a masculine

231

domain where women can enter if they meet definite qualifications and are prepared to acknowledge privileges and differences. But things have changed over the last decades to the point that it is possible today at least to conceive of the sequence in feminine terms. Woman might not represent humanity as man does, but she is no longer excluded from *civitas*.

Models

I will not attempt here to follow the history of these developments. They are the frame for all of our thinking and living today. A radical narrative of the events that led to the present situation proceeds somewhat carelessly to a symmetrical inversion of the past differences and fosters the ideological turning upside down of cultural forms and traditions. This turning upside down (a recognizable aspect of feminism) does not seem the best starting point to judge a century of efforts that, for the most part, could be defined as reformist rather than revolutionary. This trait may explain one of the strategies of the writing compiled in this anthology, a strategy that consists of the twofold movement of conquering spaces and reassuring men that their privileges and hegemony are not at stake in each movement. By this feminine strategy of duplicity, men's authority is taken for granted and, at the same time, considered something that ought to be changed in the best interest of man, the family, society—and women. In the past, women's discourse about politics and the public sphere frequently started by accepting the masculine hegemony over these spaces and its topics: This seemed to be the precondition for feminine discourse and practice that, once assumed as a right, would conflict with power and equality vis-à-vis men.[1] Revolutionary thought, inspired by Marxism, translated this conflict into theoretical and transsexual terms.[2] However, the more challenging perspectives arise not from a mechanical translation of gender politics into Marxist terms but, first, from considering the structural place of women in capitalist societies, their role in the work force and in the mode of production, according to which feminine and feminist ideologies have emerged;[3] and, in later developments, from a radical reconstruction of the public and private spheres, beginning with new perspectives on women as subjects and public actors.[4]

In the development of these perspectives, education has had an enormous weight. The demand for educational rights for women, to benefit not only women but also children and, through maternal influence on them, society as a whole, is one of the obvious topics of feminist discourse related to the public and private realms. As in the case of men in the nineteenth century (politicians, ideologues, writers), women at the beginning of the twentieth century produced a program of affirmative action on women's rights to education, based on reasons that usually pointed to the common

good. Moreover, educated women were the principal characters in the staging of the social drama of inequality and the battle against it. Thus lawyers, journalists, writers, doctors, and teachers (in formal or informal groups, loosely knit or organized as parties) led the first episodes of the women's rights movements in Latin America.[5] As in Europe and the United States, the will to break with social conventions and to create new spaces for women as well as to widen their participation in public decisionmaking arose from what can be named the intellectual categories of society. Even though the cases of Luisa Capetillo in Puerto Rico, Tina Modotti in Mexico, or Teresina in Brazil[6] prove that this emergent feminine social consciousness was not limited to the middle classes, on the whole it is fair to say that urban middle-class women, often daughters of men who practiced the liberal professions, made up the first ranks of the movement at the beginning of the century.

In a sense, when these women wrote or spoke about education, they were occupying a place that society no longer entirely denied them. Women had been the first teachers in South America after the process of national organization in the second half of the nineteenth century. According to Domingo F. Sarmiento, the presence of women in the educational institutions contributed to both the political and cultural development of the United States. In fact, Sarmiento was the first politician to import women teachers from the United States to Argentina: the educational movement called *normalismo* (a term that pointed to the so-called *escuelas normales,* the schools where teachers were trained) was soon largely composed of women.

At the beginning of the twentieth century, teachers were intellectuals in the extended sense suggested by Antonio Gramsci. In societies that were undergoing complex processes of modernization, urbanization, and incorporation of immigrants, the school as institution of reproduction and imposition of national ideology had an importance that cannot be overrated. Women, after their experience in the classroom and often in the intermediate levels of direction and administration of the educational system in countries like Chile and Argentina,[7] had obtained a certain degree of independence and self-confidence, and, above all, were convinced of their importance as social communicators and organizers.

It is not surprising, then, that the pedagogical model provided a form of conceptualizing women's participation in the public sphere and the strategies for the first attempts at empowering women.[8] On the one hand, the pedagogical model gave legitimacy to women's participation (albeit subordinate) in the public sphere: Teachers were links between the family and the state, between private or individual needs and collective enterprises, between formal and informal institutions, between the highly ritualized space of the school and the loosely knit texture of the neighborhood or the

rural village. On the other hand, teachers defined their role in an ideological, semisecularized mode, according to which women in that profession participated in the virtues, qualities, and respect of mothers: They not only educated the mind but responded to the ideal of forming the character and instilling moral principles. But teachers were typical organs of re-production (of national ideology, of the mood a culture perceived as desirable) and not of production of new alternatives. Teachers were taught a sequence of highly codified practices that they repeated in an ideal order to obtain ideal responses and attitudes. During the first decades of this century, positivistic pedagogy was at its height, although it had to battle against an idealistic tendency in scope and methods. In fact, in Latin America positivistic pedagogical principles were often jumbled up with spiritualistic discourse without major epistemological or theoretical debate, at least at the level of the schools.

Thus, the pedagogical model offered women different alternatives of self-identification that did not inevitably imply conflict with the masculine hegemony over state and culture. But although highly controlled, the model did encourage women to start thinking that it was possible to cross the border between the school as institution and other institutions of the public sphere. If the right for women to teach was not only granted but actually stood at the basis of the educational system (in fact, women made this system possible in its daily operation), the idea of transferring the pedagogical model to other activities should be judged according to the same standards of acceptance and legitimacy. Women could think of transmitting ideas, their ideas, to a larger public, something they had already been doing through magazines but that they could now do through service to society as a whole.

And there is more to this point. Teachers offered a success story that could be repeated.[9] Children of different backgrounds, creeds, nations, languages, cultures (Latin American, African, and European) had been introduced to the so-called treasure of national identity through school and learning. School had been efficient and economic in the incorporation of large portions of the population to the hegemonic national culture in many Latin American countries, though not in all of them. In these countries where school had proved its usefulness, the pedagogical model could be extended to the education of society in general and to the special case of the education of women. Education could be thought of as symbolic capital; once women had obtained it, the powerful and the state would have to accept the integration of women in all spheres, including, of course, the decisionmaking process in public policy. If women were educated, no argument of good faith could hinder their participation in all social settings. Moreover, educated women would make excellent mothers to the collective well-being of society. Thus the pedagogical model completed the

turn of its virtuous circle. This circle is not only the expression of ideology nor a mere reflection of men's design over women's destiny, and it seems unacceptable to judge it only on the basis of women complying with plots and scenes designed solely by men. By the strategies of education and the pedagogical model, the avant-garde of the women's movement adopted an accepted role and, at the same time, modified it in a process of resignification and refunctionalization.[10]

During the first decades of the twentieth century, women in the liberal professions had learned that obstacles could be hard to overcome. Nevertheless, Brazilian women working as lawyers or doctors and Chilean women working as public officers had occasion to prove their own collective strength against the social resistance of traditional forces.[11] At the same time, although they belonged to a very small minority, they offered other models for a successful integration into an almost homogeneous masculine society. Learning was, once again, a key word with a double meaning. Women in the professions proved both that they could manage skills that were thought to be masculine and that they could transfer the acquired knowledge to fields other than those closely limited by a discipline. In other words, they learned the craft of the intellectual, as it has been defined by Michel Foucault. The secrets of the trade could be used in other trades—namely, in politics.

In this sense, women discovered what socialists in Latin America thought adequate for the working classes: Politics had to be learned, and the way of doing it was through acquiring a culture that, to this point, had been treated as the private property of the wealthy and the powerful. The socialist strategy of a pedagogical incorporation of the working classes into historical consciousness overlapped, in some general (theoretical) aspects, with the lived experience of professional women that had proved, in the first place to themselves, that specialized knowledge and skills could be translated to general and public discourses and practices. This might help explain the history of good relations between socialist parties and women (at least in the Southern Cone of Latin America), and it also might provide a hypothesis for the misunderstandings between women and populist nationalistic, charismatic parties.[12]

Suffragism was the first formal and extended strategy for women's politics.[13] Suffragists in Chile, Argentina, Brazil, and Mexico defended their right to citizenship and the enlargement of the political sphere that should include women on an equal basis to men; their discourse had a formal bias because they were demanding an extension of the formal regulations of democracy. Thus, organized women participated in a trend common to some countries in Latin America of extending rights to previously excluded sectors of the population. Women asked and fought for citizenship according to models that also served to convey the claims of peasants, cultural

minorities, and racial communities. In this sense the suffrage movement belonged to the democratic (and formal) expression of the will to open the public and the political sphere (in theory) to all the inhabitants of a country. It belonged to a period in which republics born in the nineteenth century evolved into some version of modern democracy. Consequently, the program was liberal in its claims and strategies.

But in the past there were and, more importantly, there still are women working and speaking the language of other political cultures. Anarchists in the early 1900s were libertarian on moral, political, and sexual matters; the fate of the prostitute and of the women workers held a place in their discourse that spoke to the victims of capitalism, religion, the army, and the state. Later, grass-roots politics in peasant unions and poor women's organizations in cities and in the country allowed the emergence of new voices. Women in these organizations speak for themselves in words that can often be understood as a refusal of the forms through which politics and ideology were processed in the past, by men and by politicians. Their discourse shifts from reformism and moderation to various modes of radicalism. Women that participate in grass-roots organizations frequently enter the domain of public experience coming from social origins that mean cultural as well as economic deprivation. They do not correspond to the pedagogical or the professional models. In fact, they contribute to the design of a completely different model, where the feeling of injustice provides a stronger basis for political and symbolic action than the certitudes arising from traditional or intellectual conceptions of politics and society.

Peasant women, working-class women, and poor urban housewives, unemployed and living in *favelas* (slums), relate to politics through their lived experience. In this sense, women produce a model of public action that evokes the violent eruption of the poor, the small peasant, the unemployed urban migrant. Women from such backgrounds often turn out to be understanding union organizers and tough leaders with little inclination to give in when negotiating their claims. In some Latin American countries, women form part of the avant-garde of an accelerated process of radical participation, through which they become masters of their own destiny outside the family and organizers of collective responsibilities.[14]

The old prestige of knowledge represented by the image of books, pamphlets, and print journalism, as it was defined for and by the first feminists in the twentieth century, tends to be subordinated to the authority of direct experience and of the knowledge that can be acquired through practice and action. Although "education" still holds a place in their discourse, it is no longer entitled to the privileged status recognized by reformists, socialists, and suffragists. Education is a need and a right, seldom a precondition of political action. And this is not surprising, because these women leaders and organizers have grown out of the grass-roots

movements making use of a special "education" obtained from their direct experience of injustice and subordination. The faith that the liberal democratic socialist-suffragist reformers had in literacy has not disappeared altogether but coexists with beliefs that come from experience rather than from theory and formalized ideology. A new style of women's intervention in the public sphere bears the social marks of their popular origin, as well as of the radicalized forms of conflict that, in many Latin American countries, call for the use of armed struggle.[15] This new style does not claim any kind of ideological purity; in fact, the mixture of traditional beliefs and practices with revolutionary discourse is one of its semantic and formal qualities.

Women's organizations and women leaders of unions or social movements strongly emphasize issues rather than elaborate programs. They respond to immediate needs (land, work, salary, health, or basic human rights) and thereby create close bonds between the leaders and the rank and file, although it would be difficult to assume that their relationship is always democratic and antiauthoritarian. That seems to indicate that feminine participation is not the only condition of a deep democratization of practice and discourse, being nevertheless one of the decisive conditions of new forms of ideology and politics in Latin America. The influence that women have on peasant and urban movements, as well as their participation in the armed struggle, can be read as a complicated and often contradictory tissue of attitudes and discourses: The "new model" is unstable, as it should be, and conveys the double aspect of its limits and its huge potential.

Its limits lie precisely in its relation to politics as a process of decision-making and consensus-building. Bound to issues (topics of discourse and practices), women face the difficult question of access to power, especially to institutional power, which seldom corresponds to power as it is thought of and formalized in the movements in which women implant their leadership and acquire their strength. Practical knowledge many times reveals itself as insufficient when extremely complicated questions of social organization, ideological struggle, and political options arise. Nevertheless, the possibilities of such knowledge are linked to the new social paradigm these movements convey in their practice and sometimes (although not always) in their discourse: the new subject of the public sphere, as it has been called. The human rights movements, for instance, oscillate between a political radicalization (which does not always mean political empowerment) and the localized extension of its themes and goals.

Styles

The styles that women produce or modify when they participate in the ideological debate or enter the political arena could be organized into a number of general models of discourse and practice that respond to three

broad trends: politics as reason, politics as passion, and politics as action. Although women's intervention across this century combines traits and qualities of each of these three ideal types, it is useful to analyze their formal and discursive aspects separately. The ideal types correspond to deep attitudes and beliefs and, in a way, relate to more general paradigms of politics, culture, and the public sphere.

Politics as Reason   The first type, politics as reason,[16] relates closely to what has been here called the pedagogical model. According to its premises, politics should be seen as a combination of knowledge, strategies, tactics, and goals that seem perfectly defined from a scientific, positivistic, and optimistic perspective. Reason is the backbone of politics, and ideology is conceived as a treasure that can be learned, discussed, and communicated through "classical" means. The written word, the press, and the propaganda apparatuses would enable millions to take their destinies into their own hands; fate and necessity would be redefined as option and liberty. Social reformers were convinced that historical, economical, and cultural knowledge offered valuable information on the present situation and on the paths that should be taken to change it. The extended assurance that change was not only possible but necessary and ultimately unavoidable is one of the main traits of politics as reason, and it is the source of its optimistic bent: No enemy would be powerful enough to stop a historical course that society, by its very nature, could not help but follow.

Despite the positivistic charm of this image, women who entered the public sphere conforming to this style believed that in a social nature thus conceived, women had an ideal place that society did not yet recognize but that women could, through a long process, come to occupy. Women's equality, they reasoned, was a more "natural" arrangement that would result in the mutual benefit of both sexes and of society. Conflict, according to this ideological perspective, is a state that can and should be overcome by evolution and not by disruptive change. In this sense, social homeostasis is the desirable horizon against which future gender relations will inevitably be projected. Gender politics and social politics are considered two noncontradictory levels of thought and action: Progress in one of them will entail improvement in the other. Both levels are mutually functional and strategies should be devised to open the space of interaction and avoid concurrence between actors objectively and basically interested in the same goals. The strains and conflicts of everyday action should not serve as a proof of conflict between gender and general politics but as an expression of underdevelopment of a given society or backwardness of their political leaders and masculine intellectuals.

Usually these beliefs were shared in different forms and proportions: Socialists were more inclined to recognize women's participation and lead-

ership than their colleagues of radical, national, or democratic persuasion. But leaders in these main lines of the Latin American ideological map had to recognize (or at least pay lip service to) the group of convictions that mobilized their female partners. This did not mean that they were ready to accept the suffragist or equal civil rights programs but that they did not launch strong, organic opposition to the programs' theoretical premises.[17] (The opposition, by the way, was victoriously led by the Catholic church and the conservative forces that organized their own women's movements against feminism and suffragism.)

Women tried different strategies: Participation in political parties (socialist, communist, anti-imperialist) ran parallel to the construction of independent alternatives. Professional women associated according to corporative as well as gendered interests; educators considered their work in public institutions as part of a strategy proving their empowerment; writers held strong opinions on sexual politics through their work but at the same time often abstained from participating in political organizations.

Politics as reason opens a wide range of options based on certain common beliefs, essentially that equal rights for women and men will result in more equitable private and public communities. Opponents emerged from the ranks of traditional institutions such as the church and the conservative wings of most parties, but literacy and education, on the one hand, and public intervention, on the other, were considered adequate responses to moral traditionalism and sexism. Propaganda, particularly through mass media and conferences and meetings, would create a powerful weapon whereby ideological innovation would defeat prejudice. An optimistic reading of universal history, coined from an enlightened mold, sustained this common sense, and Latin America was considered no exception to a trend that seemed to lead to sexual equality throughout the Western world. This optimism might be seen as naive idealism today, although such a careless judgment may inappropriately read the profile of the period. Social reform of the gender system and its expression on the political, institutional, and cultural levels acted as a powerful impulse for deep transformations in some Latin American countries through the first half of the twentieth century.

As do most forms of action and perception, politics of reason focused on a range of questions while ignoring others: The distinctive status of women in the private sphere—their subjection to stifling moral codes and traditions—was not viewed as a central issue, the assurance that everything would happily change through education and political reform highlighting transformation in the public sphere before and as a condition of changes in the private realm. Institutions were thought to be more important than informal networks or alternative organizations because reason and knowledge were linked to the traditional sources (literacy,

schooling, debate, journalism, etc.) inherited from the liberal democratic reformers of the nineteenth century. The assumption that political action would enable women to obtain equality was deeply rooted in the feminist and suffragist movements, as the women who participated in them had conformed to political and pedagogical institutions, through which they had acquired public respect and recognition. Their experience (which, of course, they considered both useful and applicable in a larger arena) projected its image on the type of instruments these women created.

Politics as Passion   Politics as passion describes women's relation to the public sphere within the space drawn by certain distinctive traits of the "feminine image." Values that tradition recognizes as typically feminine can be overturned in their social function and transformed into instruments of women's participation in the ideological and political world. If history, knowledge, and objectivity were at the basis of politics as reason, then subjectivity, intuition, and feeling confer the impulse to this style of intervention, which recognizes and underscores a distinctive set of "feminine" qualities. Women were given a place and a role in private and public institutions according to values that were considered essentially feminine; moreover, cultural, moral, and religious discourse had often warned women of the danger they were exposed to when they erroneously departed from those values to contend with men on a more equal and common basis. As revenge for what was labeled the "purely feminine," women reorganized those ideological themes in new forms of practice and discourse that, very obviously, did not obey any previous design but rather arose from the basic need of working with what culture had established as women's fields and legitimate feminine preoccupations.

Women adopt what can be thought of as a bricolage strategy, producing new public subjects from old roles and traditional functions. If society has defined the private as the quintessential feminine sphere, women transform private issues into public debates and interventions. A mother whose child was kidnapped can enter the public scene as mother and transform this private role into a public token of her ideological, political, and moral definition.[18] A victim of sexual violence can transcend the limits of private shame and suffering (traditionally considered her only appropriate reactions) and can publicize the offense, emphasizing that her claim to justice does not belong to the private sphere (as a matter of dishonor for her brother or father) and is autonomous of traditional masculine institutions (the church, the family). A woman whose link to politics is at best tenuous can find in her love for a political leader, in her attachment to a brother or a father, the private basis of a public commitment. Feelings of pity and charity (which religion considers appropriate to women) may find a political translation in the face of social injustice.[19] Womanly passions and private

virtues become the basis of public action: The private turns public, as feminists claim all over the world.

This mode of intervention does not grant knowledge in its traditional and institutional forms a privileged position. On the contrary, the postulation of feeling as guide to action is implicit in the discourse that could come under the heading of politics as passion. Injustice, deprivation, and violence are the original impulses to organization and participation, and they are felt, in the first place, as individual situations, although they may later be generalized and integrated into a collective design. Women are struck by violence or indignity in the private sphere, in the sphere to which traditional beliefs have rooted their meaning and values; this violence is sensed as a personal offense, but at the same time as an offense that cannot be amended in the sphere where it was produced. The feeling of injustice constitutes women as public actors; the passion it liberates transforms a passive sufferer into an active subject.

It is not the course of history or the progress of humankind that, as abstract certitudes and values, sustains this model. It is rather the concrete experience and the feelings emerging from it that give consistency to practice and discourse. New values emerge as the process of turning public the private as lived by women, who were thought of as secondary characters yet suddenly begin to play leading roles. New subjects, these women often speak a language profoundly different from the language of politics: Placing values such as life, dignity, security, and basic human rights as the axis of their action and discourse, they do not always offer holistic programs for the reform of society. Moreover, these new subjects tend to have difficult relations with traditional political parties, including the parties of the Left. Their fluid organization, the tone and color of their discourse, echoes with new voices and, at the same time, recalls well-known forms of populist claims and populist politics. Politics as passion could be also defined through a formula that links politics to direct experience. The mediation of the culture of literacy, of political traditions constituted through many decades and highly ritualized conventions, gives place to the culture of the lived world and the forms that arise when "outsiders" appropriate and employ some of the tools and devices created by other actors for other scenes.

Politics as passion more often than not produces an effect of ideological and semantic inversion. Women who enter the public sphere in the roles of mothers, housewives, daughters, or neighbors assume these social labels in order to reinforce claims that contradict the function (conservative of tradition, of the family, of the gender system) that these roles imply. Mothers, in particular, commonly considered as the last resource of order in a masculine-oriented family pattern, become strong supporters of what, from a traditional point of view, could be named disorder, challenging state

power, police repression, and hierarchical organization of society. Redesigning the conservative content of their function, they enter the public sphere as a disruptive force.

Politics as Action   Passion very frequently blends into a style of politics as action. The transitions between knowledge, literacy, propaganda, and action take place through the mediation of long-discussed tactics and their synthesis in a design that includes instruments and goals according to the rules of politics of reason. But a style sustained by beliefs rooted in experience and feeling may dispense with the very detours that constitute politics as politics, that is, the series of complicated mediations and the establishment of satisfactory relations between ends and means. Public intervention is often centered on a single goal (the respect of some basic right, for instance), which is not negotiable because it is singular and cannot be replaced by other goals and values. Politics as action presents this kind of irreducible condition by which women make their roles public in order to ask for something that cannot be denied to them. Their claim organizes and determines the style of their participation: They cannot help doing what they are doing, claiming what they are claiming. Modern Antigones, they find no way or reason to stop the impulse and alter the direction of their action. Claiming and resisting, they neither offer nor accept global schemes that may contradict their convictions and their sense of justice. Their legitimacy springs from their strong belief that the values they are fighting for are legitimate.

Politics as action does not necessarily entail violence. Through the last decades many Latin American women's organizations have shown that they can manage to stand precisely on the limit of pacific but not passive presence in the public sphere. They tend to occupy physically the symbolic ground of their claims: Standing in the agora, they embody the values they are fighting for, but their action seems to be at the same time resolute and not aggressive. Violence can obviously arise when these movements integrate into wider political and military organizations: Women as mass organizers and demonstrators turn into women as soldiers, frequently adopting the symbols of war that have been imagined and designed by men. Pacific action, the first and last resources in a value-oriented activity, may converge with politics as war.[20]

Writings

History, women's history, history of a resistance, history of a transformation, history of an inversion of function and role, history of practices and of the invention of practices, history of the invisible turned visible, of the private turned public, of the traditional that results in disruption—how

can all this be written? How do women write this process? What do they write about?

It is generally recognized that recent women's organization around issues of human rights has rescued a part of Latin American history from oblivion. Women not only cried out for justice but also contributed to the preservation of a collective memory; the narrative told by feminine voices overlaps with narrative told by peasants, workers, and *favelados,* often pointing to the same system of allies and enemies, similar patterns of topics, and comparable strains between discourse and practices.

In the writing of this history, women had to find a form, a tone, and an audible voice; in the nineteenth century, literate and cultivated middle-class women had frequently chosen the aesthetic dimension of discourse and discovered that their literary writing was tolerated by a society that often adopted a somewhat patronizing attitude. Fiction and poetry conveyed much of what women were publicly authorized to state, and frequently the conventions of the belles lettres served as refuge for those to whom other (more "masculine") forms of discourse were denied. Tradition also allowed women the practice of private forms of discourse: letters, journals, diaries, and travelogues were genres through which experience could be organized and given a set of highly codified meanings. Above all, these forms of discourse had the advantage that, being practiced in the first place by men, a clear line was drawn between what could be said and what should be kept silent. Women battled against this hierarchical code and, into the margins of what would be "normally" accepted, managed to slip topics, preoccupations, and twists of discourse, forcing the limits that defined the feminine subject and the feminine space. In fact, writing their history through their private stories, writing in plural under the appearance of the singular, women occupied with their voices marginal and not so marginal places in the cultural sphere. Disguise was often their attitude and their resource. Not only did they sign under masculine pseudonyms, but they also pretended to speak from a "proper" and accepted position, at the same time striving to modify the laws that defined the limits of the appropriate.

From the beginning of this century, women began to explore masculine genres such as the political and the ideological essay. To do so, they had to learn a form of reasoning, the strategies of demonstration, the logic and rhetoric of argumentation.[21] They had to do this in order to be admitted into the masculine world of politics and the masculine space of ideological discussion. In fact, they had to learn the rules of the game, formal rules that enabled them to turn their discourse public and contend with the substantive rules that organized actors and forces on the field.

Some aspects of the game were recognized as more feminine than others: Women could teach and write about teaching and public schooling as they

had been allowed to write fiction and poetry before. Women could also write for the benefit of other women: show the right way for feminine progress, teaching the pilgrim about the real obstacles. They could write about health, children's education and nurturing, and modern devices for good housekeeping—topics we might view with some irony today but by which women's journalism built a bridge between the home and the outside world, criticizing the backwardness of tradition and presenting modern woman as the new angel of progress in the private sphere, not as the slave of a capricious master. Women taught women about matters that an orthodox critique of feminine mystique would label enslaving. Nevertheless, these topics brought forward the occasion to speak up, to register thoughts and desires (no matter how codified) in a legitimate and public form. Besides, women were training themselves in the difficult art of public intervention and trying to find discursive strategies to speak as different and independent subjects.

One of these strategies has been to turn the well-known autobiographical genre into an instrument of ideological debate or political propaganda.[22] In recent decades women have told their lives and woven desires, claims, interpretations into the texture of personal stories that capture public interest. Some professional women writers have also invested in the different forms of personal/public genres: narratives that have social meanings developed under the guise of the first person. Moreover, women writing literature or journalism (in fact, crossing the limits between both discourses) have heard and registered the voices of different actors and reconstructed historical events by means of the presentation of these voices in personal texts that are, at the same time, strangely "objective."[23]

Life stories and interviews have not only been literary devices. The social sciences have learned from anthropology that silent voices become audible if asked the appropriate questions. The experience of women organizers is being registered in texts that freely combine the individual dimension with the collective experience, private origins with public action.[24] Some of these texts may seem hypercodified: voices that keep on telling similar stories with subjects and situations that look very much the same. Poverty has left its marks on the circular, repetitious narrative of a struggle for survival, punctuated by the search for work, food, and housing and the defense of personal dignity; in some cases, the narrator's voice is successfully integrated into a collective and public pattern of resistance.[25] Popular voices, as popular genres, use repetition and codification as strategies through which what has to be said can be said; simple narrative structures may add a sense of collective experience to individual stories told in the first person. If there is truth in these texts, it relies on the tone of the voices that attempt to communicate an experience through writing that, according to the rules

of historical and ideological genres in their classical versions, was kept in the margins of public space and discourse.

Notes

1. Francisca García Ortíz finished her speech in front of the First Feminist Congress in Mexico (1916) by clearly stating this defensive position. Advocating an adequate feminine education that would enable women to be better mothers but not to feel themselves superior to men (and in this point men should collaborate by educating themselves in order not to be looked down upon by their wives), she concluded: "Let us for God's sake educate man, our moral and social support: and let us not dominate him as many women intend, and which will be the outcome if our own development proceeds apace while his falls behind. Please God, let things work out this way; otherwise we would weep bitterly over the ruins of our feminine charms, which we would ourselves have trampled into the dust. Let us never forget that women should always be the delight of the home, the gentle comrade of man; she may indeed overcome him through her love and sweetness. But let her not dominate him with her intellect nor with her learning." Before, García Ortíz had stated her belief that man would continue ruling society "always and forever." Quoted from Alaide Foppa, "The First Feminist Congress in Mexico, 1916," *Signs* 1 (1979). For generous information on the feminist movement in Mexico, see Anna Macías, *Against All Odds: The Feminist Movement in Mexico to 1940* (Westport: Greenwood Press, 1982).

2. The history of the difficult relations between communism and feminism in Latin America has still to be written. An illuminating example of how a communist party tried to subordinate women's claims and read them only along the lines of the typical class-struggle model can be found in Esperanza Tuñón, "El Frente Unico Pro Derechos de la Mujer, 1935–1938," *Fem* (Mexico) 30 (November 1983). Nevertheless, considering Latin America as a whole, Marxism and anarchism were ideological alignments in which the belief in and defense of women's equality clearly prevailed from the last decades of the nineteenth century.

3. Theoretical development has been important since the late 1960s. Perhaps the first and most ambitious work is Heleieth Saffiotti's *Women in Class and Society* (New York and London: Monthly Review Press, 1978 [1969]). The "Conclusions" of her book form part of this anthology. The work of Isabel Larguía, especially the book with John Dumoulin, *Hacia una concepción científica de la emancipación de la mujer* (Havana: Editorial de Ciencias Sociales, 1983), shows the scope and method of an orthodox Marxist reading of the feminine question. Social and economic analysis as proposed by Lourdes Benería and Martha Roldán seem to render a very convincing description of women's situation in the workplace in capitalist societies without stressing the theoretical Marxist perspective. See *The Crossroads of Class and Gender: Industrial Homework, Subcontracting and Household Dynamics in Mexico City* (Chicago and London: University of Chicago Press, 1987). Similar concrete analysis can be found in the work done by Elizabeth Jelin, María del Carmen Feijóo, Larissa Lomnitz, and Catalina Wainerman. For a review essay on socioeconomic literature, see Marysa Navarro, "Research on Latin American Women," *Signs* 1 (1979). The essay points out the main trends of research and is still relevant, though written and published more than ten years ago.

4. Social movements are the public actors of this reconstruction of the links between the public and the private spheres. For a theoretical perspective, see Tilman Evers, "Identidade: a face oculta dos novos movimentos sociais," *Novos Estudos* (São Paulo), 2, 4; and Elizabeth Jelin, "Los movimientos sociales en la Argentina contemporánea: una introducción a su estudio," in her *Los nuevos movimientos sociales* (Buenos Aires: CEAL, 1985).

5. Almost all researchers emphasize the importance of professional women in feminist and suffragist mobilization. See Branca Moreira Alves, *Ideologia e Feminismo: A luta da Mulher*

pelo voto no Brasil (Petropolis: Vozes, 1980). Helpful information and insights can be found in June Hahner, "The Beginnings of the Women's Suffrage Movement in Brazil," *Signs* 1 (1979); Macías, *Against All Odds;* Elsa M. Chaney, "Women in Latin American Politics: The Case of Peru and Chile," in Ann Pescatello (ed.), *Female and Male in Latin America* (Pittsburgh: University of Pittsburgh Press, 1973); "The U.S. Suffrage Movement and Latin America," in June Nash and Helen Icken Safa (eds.), *Sex and Class in Latin America: Women's Perspectives on Politics, Economics and the Family in the Third World* (New York: J. F. Bergin, 1980); María del Carmen Feijóo, *Las Feministas* (Buenos Aires: CEAL, 1982).

6. Luisa Capetillo is a most interesting case of anarchist ideology mixed with spiritualistic themes and feminist claims. Born in Puerto Rico in 1880 or 1882, she considered herself a self-made woman with regard to cultural matters:

> I speak and perfectly understand every word I say, guided by a deep intuition; nothing could I learn according to the rules of school or university. . . . I introduce myself as agitator, journalist and writer, with no other authority than my calling and initiative; nobody recommends me but myself; nothing helps me except my own effort and I care very little for the criticism produced by those that had the opportunity to follow general and complete studies in order to expose their written views, protests or literary fictions. (Quoted by Julio Ramos in his pioneering work, "Luisa Capetillo: notas para una literatura menor," Mid-Hudson MLA Conference, November 1985).

Antonio Candido has written a beautiful text on Teresina Carini Rocchi, an Italian-Brazilian woman (1863–1951) who represented the various traditions converging in Latin American progressive thought and politics in the first decades of the twentieth century: "She argued against fanaticism inspired in Voltaire's *Moamé;* she defended feminine emancipation with Mary Shelley; she denounced poverty with Zola; she clamored for fraternity with Victor Hugo and against war with Baroness Suttner; inspired by Proudhon she was for the worker's union; criticized capitalism with Marx; proposed cooperation with Kropotkin." Antonio Candido, *Teresina etc.* (São Paulo: Paz e Terra, 1980).

7. The case of the Chilean Amanda Labarca shows the influence professional women could acquire. She was responsible for passage of the law on obligatory education and was an authority in matters concerning the organization of teachers' formal education in her country. She wrote extensively on comparative perspectives in education, launched reforms of the secondary level of schooling, and participated in debates on educational politics. She published on these topics, almost without interruption, from 1922 to 1948. In 1932 she was appointed to the Office of Secondary Education and in 1946 was elected representative of the government of Chile to the General Assembly of the United Nations.

8. On this topic, see Cynthia Jeffress Little, "Educación, filantropía y feminismo: partes integrantes de la femineidad argentina, 1860–1926," in Asunción Lavrin, ed., *Las mujeres latinoamericanas: perspectivas históricas* (Mexico City: FCE, 1985). Sylvia Molloy also points out that to be a teacher was one of the few legitimate professions for women ("Dos proyectos de vida: *Cuadernos de infancia* de Norah Lange y *El archipiélago* de Victoria Ocampo," *Filología* [Buenos Aires] 20, 2 [1985], pp. 279–293).

9. Gabriela Mistral is perhaps the most outstanding example: First a schoolteacher in the southern provinces of Chile, she eventually won the Nobel Prize in Literature. The well-known Argentine poet Alfonsina Storni also worked as a teacher for most of her life.

10. The importance of education is a topic in almost every feminist text written in the early part of the twentieth century in Latin America. The Chileans Amanda Labarca and Gabriela Mistral, the Argentine Alicia Moreau de Justo, the Mexicans Sofía Villa de Buentello and Margarita Robles Mendoza, and the Uruguayan Paulina Luisi underscore the need for developing institutions of formal education as a basic condition for political consciousness, opening the way to public participation and persuading men of women's worth and dignity. Bertha Lutz, the leader of the suffrage movements in Brazil, wrote in 1918: "One of the

greatest forces for emancipation and progress lies within our power: the education of women and men. We must educate women so that they can be intellectually equal and be self-disciplined. We must educate men so that they become aware that women are not toys created for their amusement, and so that, when observing their wives and sisters or remembering their mothers, they understand and are completely convinced of the dignity of women" (text reproduced by June Hahner, "The Beginnings of the Women's Suffrage Movement in Brazil," *Signs* 1 [1979]). Sofía Villa de Buentello placed education among the basic civil rights women should be entitled to in order to achieve a state that would enable them to participate in public decisions: "Woman has the right to study, to inform and enlighten herself in order to progress from an economical point of view, in order to be more moral and help man to be so, in order to avoid the obstacle that an ignorant being faces in life. Finally, in order to obtain as much happiness as possible. Nobody can hinder a cultivated woman from participating in the public affairs of man" (*La mujer y la ley* [Mexico, 1921], pp. 49–50).

11. Descriptions of this resistance in Mexico appear in Macías, *Against All Odds*, pp. 11–12. For a slightly different perspective on the Argentine case, see Mirta Henault, *Alicia Moreau de Justo* (Buenos Aires: CEAL, 1983).

12. Alicia Moreau de Justo, the Argentine socialist leader, and Magda Portal, the Peruvian politican of the APRA (Popular American Revolutionary Alliance) prove through their life stories the different relations established by socialist and nationalist or populist organizations with professional women and feminists. See, in this anthology, Magda Portal's short autobiography.

13. On suffragism, see n. 5 above. Women's right to vote was granted in the different Latin American countries during a period that begins in 1929 with Ecuador and ends in 1961 with Paraguay. Brazil and Uruguay granted the right in 1932; Cuba, in 1934; El Salvador, in 1939; the Dominican Republic, in 1942; Guatemala and Panama in 1945; Argentina and Venezuela, in 1947; Chile and Costa Rica, in 1949; Haiti, in 1950; Bolivia, in 1952; Mexico, in 1953; Honduras, Nicaragua, and Peru, in 1955; Colombia, in 1957. See Jane Jaquette, "Female Political Participation in Latin America," in Nash and Safa, *Sex and Class in Latin America*.

14. Outstanding narratives of this process of social and political participation and leadership by women can be found in Domitila Barrios de Chungara and Moema Viezzer, *Let Me Speak! Testimony of Domitila, a Woman of the Bolivian Mines*, trans. Victoria Ortíz (New York and London: Monthly Review Press, 1978); Elisabeth Burgos-Debary (ed.), *I, Rigoberta Menchú, an Indian Woman in Guatemala*, trans. Ann Wright (London: Verso, 1984); Audrey Bronstein, *The Triple Struggle. Latin American Peasant Women* (Boston: South End Press, 1982).

15. See Margaret Randall, *Cuban Women Now: Interviews with Cuban Women* (The Women's Press, Dumon Press Graphix, 1974); Adriana Angel and Fiona Macintosh, (eds.), *The Tiger's Milk: Women of Nicaragua* (New York: Henry Holt, 1987).

16. See, in this anthology, the text by Alicia Moreau de Justo, reproduced from her *La mujer en la democracia;* and "Feminismo" by Paulina Luisi.

17. See, for instance, the editorial essay published by the Mexican journal *El Universal* in December 1929, backing civil rights for women. The essay was probably written by Margarita Robles de Mendoza, who reproduces it as her own in *La evolución de la mujer en México* (Mexico City, 1931).

18. See, in this anthology, excerpts from an interview of Hebe de Bonafini, edited by Matilde Sánchez and published under the title *Historias de vida* (Buenos Aires: Fraterna/ Del Nuevo Extremo, 1985).

19. See, in this anthology, the chapters from Eva Perón's *My Mission in Life*.

20. For the various forms of open action through which women construct their public identities and become social or political militants, see, in this anthology, "Peasant Women and Silence," by Lourdes Arizpe.

21. Examples of early and successful management of the rhetoric of argumentation can be found, in this anthology, in the texts by Paulina Luisi and Alicia Moreau de Justo.

22. Eva Perón's text, included in this anthology, can be read from this standpoint; the same can be said of the *Historias* of Hebe de Bonafini.

23. See, in this anthology, the excerpts from Elena Poniatowska's *Massacre in Mexico.* Poniatowska's literature can obviously be read in the movement between fiction and document, produced through the rewriting of different social and cultural voices.

24. See, in this anthology, the interviews of Borjas and Antúnez. These form part of the growing corpus of women's life stories, to which belong the well-known texts of Domitila Barrios de Chungara and Rigoberta Menchú.

25. The terrible monotony of deprivation can be traced in Carolina Maria de Jesus's diaries, included in this anthology.

Paulina Luisi

IN 1921, MIGUEL J. FONT published in Buenos Aires a book based on no less than fifty answers to a general question about what feminism should be in its program, methods, and instruments. Font was planning the publication of a magazine, *La Mujer,* to deal with the issues that would arise from the answers to his inquiry. More than half of those who responded to his initiative defended rather traditional positions on the subject. Nevertheless, almost everybody in Argentina and Uruguay participated in the project that resulted not in the magazine but in an important book: *La mujer: encuesta feminista argentina.* The Uruguayan Paulina Luisi, president of the National Women's Council she had founded and organized, gave Font, as her contribution to the book, the speech she had delivered in front of the assembly of that council in 1918. The tone of Luisi's intervention is an accurate example of the type of argument that feminists presented in order to gain at least a mild acceptance of their program from the masculine intellectuals and politicians. However, Luisi advances some bold proposals on the rights that women acquire through motherhood, which society should recognize as a public service in exchange for which women should be granted civil and economic equality. The style of her intervention is an early example of what I have defined in my introduction as *politics as reason.* The moderate tone of her speech combines with the strategy of persuasion that feminism employed in the first decades of this century when embodied in women of the middle class.

Feminism

(Speech delivered at the annual assembly of the National Congress of Uruguayan Women on September 30, 1918.)

DEAR COLLEAGUES,

Two years ago today, September 30, 1916, we laid the initial foundations of our organization. A small but dedicated group of courageous women

---

"Feminism," from La mujer: encuesta feminista argentina. *Ed. Miguel A. Font (Buenos Aires, 1921). Translated by Marcy Schwartz Walsh: copyright © 1991 by Sara Castro-Klarén.*

gave of their intelligence, talent, and lucidity to support the cause of feminism. We embarked upon this journey with enthusiasm and undying faith in the social future of our sex, no longer willing to accept being reduced by our male counterparts to their sex objects or household decorations.

These first two years of struggle have been tough and seemingly thankless. We have fought energetically and tenaciously, even though our efforts have not yet crystallized into any tangible results to which we can point and say, "Look, here's what we've done." But the seeds spread by the winds of our work have begun to sprout.

Two years ago, we would scarcely dare pronounce the word *feminism;* it was synonymous with machismo, with revolution, with the family's dissolution. It evoked rebellion and rupture, a threat to the family and to the center of our social structure. Though we began this campaign envisioning its vast panorama, we barely ventured to whisper the most rudimentary ideas.

By now, we have convinced all attentive ears that what we want is not a destructive revolution at all, but rather a healthy and constructive evolution that will rebuild the crumbling structure of the family with the stone foundation of women's emotional and spiritual equality with men. This will make women the companions that men have been anxiously seeking in women other than their wives. Their legal spouses, the mothers of their children, will also be their friends, sisters, and spiritual partners. Husbands will find in their wives what they have looked for in educated lovers; and education of women will do more than train passive and unintelligent housewives, or lustful prostitutes with perverse morals.

Women's education until now has not sought to make women integrated beings who complement and comprehend the opposite sex. With mutual understanding and support, both mother and father will feel united by their children's delicate yet strong continuity; they will guide them along life's tortuous path, into the radiant horizon of the future. This, more than anything else, is the aim of our feminism.

Education has neglected to teach women how to grow into complete human beings. Education has forgotten to develop their intellectual abilities and talents, has not given women a clear sense of their own personalities. Women need to be offered the freedom to face and recognize all of their responsibilities, to be given an awareness of their worth in society and the consequences that exercising that freedom entails. Women must be given economic freedom and the means to earn a living. They must be freed from the sexual slavery that our society has connected to motherhood. Rather than economic slaves, mothers should rightly occupy a queen's throne since the survival of the species depends on them.

But instead, our society doubly enslaves women who are mothers. They must be constantly devoted to their child's every breath and move. Mothers are then economically limited in a society that refuses to recognize, with laws and policies, what the state owes them for providing it with new citizens, for risking their lives and enduring pain and anguish to supply the national capital with an abundant population.

It will be one of feminism's most illustrious triumphs when the state generously recognizes maternity as a social function, with all of its (unalienable) indisputable rights, including the community's obligatory responsibility to share and support women as mothers. Women's contribution to society, so dangerous and painful, is certainly more efficient, unavoidable, and permanent than men's military service. But men expect their service to justify centuries of abuse. Then they claim national rights and privileges all for themselves. In the name of this military service, which they only have to fulfill once and which some only go through by correspondence, men have appropriated the right to govern, to dictate the city's and state's laws and policies.

Based on the pretext of military service—though it is, granted, an unpleasant and thankless endeavor—women have been excluded from public administration. However, our society consciously forgets that women provide another service more than equivalent to that of the military, sometimes at the cost of their lives and always after long months of painful waiting and brutal suffering; women bring into the world those same young lives who will one day serve their country in the army.

We want to reclaim women's rights as social entities. As human beings who belong to the same community and contribute to the same public treasury, it is only right that women be called upon to help make decisions about their own investment in that community, to discuss the use of funds to which they are contributors. If women are subject to laws and ordinances, then it is only right that they take part, either on their own or through their representatives, in making the laws and policies that they, like everyone else, must obey.

And if we expect to ensure these rights for women, there is even greater reason and urgency to ensure them for mothers. In their hands, in their wombs, lies the essence of our future; the human beings that they forge and mold will determine the next generation and the destiny of our nation.

These are our goals, this is the vast project that will unfold year after year, bit by bit, like the stones that construct the largest monuments, carefully placed one by one.

The current war has shown us that women's skills are useful and needed in every facet of the work force, even in those areas where women were previously denied participation. Therefore, women deserve equal education, and equal pay for equal work; for neither hard work nor its production

has gender. Maternity needs to be elevated and dignified, legally and socially, as the greatest of all endeavors. We ask for legal protection for children, to forge the storage house of our future, too often neglected. We must fight with all our strength against immorality, gambling, pornography, and prostitution, social ills that reduce women to the slavery of an uncivilized way of life. These are the ills that destroy our young people's healthy principles of morality and honesty, which we untiringly plant and cultivate in them. We must wage war on alcoholism, that destruction of our future generations' health. Above all, we must wage war on war itself, the most frightening of all catastrophes. It robs us of our well-being and destroys what we have given all of our love and energy for, against the madness of bloody extermination: our children! Let us fight against the violence that, like an ancestral curse, continues to assault our hearts and souls.

This vast program of worldwide feminism advances in each country according to women's education and action. National feminist organizations reveal our concrete efforts to elevate and dignify women to the same social and legal status as men, with equal rights and responsibilities. Our sisters in the United States have been struggling for this for over a century, and so have the courageous women of France and England. Women in other countries continue the difficult struggle toward triumph, and in these last four years of war, they have gained ground in many important policies.

We, colleagues, a few determined women from a small and struggling country, must follow in their footsteps, with the same unbeatable faith, with the same persistence, carrying on the feminist cause. With our group's truly inspiring morale, before long, perhaps at the next international convention, after the war, Uruguayan women will have earned a position of respect among the women in the world who are marching toward the vanguard of progress.

Our congress has had to overturn controversial claims against feminism; however, with our continuing work we will encounter fewer and fewer adversaries and opponents. Our efforts are part of a struggle, and like any battle, it has its friends and foes. These ideas find enemies because they break down prejudices, they attack vested interests, because they engender new ideals that seek to transform old and established behavior. Serious struggle is a reward in itself; it shows that opposing forces are finally giving way. Recently we have even tasted the victory of our fiercest enemies making concessions.

Some of the most respected sociologists and politicians have been paying attention to our struggle. Recently President Brum shared his opinion of our work with me: "You all are not even aware of your feminist movement's

historical significance for our country. Your congress marks an important step in our social evolution."

His words, publicly expressed on several occasions, are a tribute to his serious commitment and his open spirit. May they be a hopeful sign of our slow and painful road to redemption.

Alicia Moreau de Justo

PUBLISHED IN 1945, *La mujer en la democracia* presents all the arguments that feminism had developed since the beginning of the century in order to advance the cause of equal civil and political rights for women. The text corresponds very clearly to the model I have described as *politics as reason*. In her book Alicia Moreau de Justo follows the history of feminine emancipation, introduces several national cases in a comparative perspective, describes the state of women's rights in the world, and sums up the arguments that lie at the basis of the feminist program in its political dimension. A socialist, Moreau nevertheless adopts a wide and open liberal democratic perspective.

Moreau was born in 1885; she belonged to a middle-class family of French socialists that had fled their country after the repression following the Commune of Paris in 1871. She studied medicine at the University of Buenos Aires and became one of the first women doctors in Argentina. In 1910 Moreau participated in the First Feminine International Congress, organized by professional women in Buenos Aires. From that moment on she lectured extensively on education, public health, and women's civil rights, visiting socialist centers, public libraries, and schools all over the country. In 1919 she became editor of *Nuestra Causa*, the organ of the Argentine Feminist Union. She was a member of the Argentine Socialist party and married Juan B. Justo, its most important leader. She held important positions in the party and directed its official journal, *La Vanguardia*. Moreau published several books (*Evolución y educación*, 1915; *La mujer en la democracia*, 1945; *Juan B. Justo y el socialismo*, 1984) and hundreds of articles and pamphlets. Moreau died in 1985 in Buenos Aires.

Women in Democracy

Argentinian women cannot affect coldness, indifference, or carelessness at this serious time. They must rise to fulfill their difficult duties, human duties, terribly important patriotic duties that cannot be avoided.

"Women in Democracy," from La mujer en la democracia *(Buenos Aires: El Ateneo, 1945). Translated by Laura Beard Milroy: copyright © 1991 by Sara Castro-Klarén.*

To be able to do so, and avoid limiting their efforts once again to the level of merely spoken aspirations, they must attain political independence and reach that path of action unique to a democracy, through which their idealistic goals can be realized.

The young women entering social activity today believe that joining political life, which was so difficult for their mothers and grandmothers, is quite easy.

Without their noticing it, the lead weight of the natural inferiority of women—which was almost a gravestone for their ancestors—has been removed.

Today no one wants to rehash the old debate about the superiority of men and the consequent stupidity of women.

The controversy that once filled pages and pages is not today as relevant as it is anachronistic and useless. So much paper wasted on the length of bones and the weight of brains. So many statistics, so many quotations cited from sacred and secular texts!

.   .   .

One of the weightiest arguments, cast like a stone to obstruct the new path, which women have more difficulty contesting because of the sentimentality that inhibits them, is that her involvement in politics will take her away from her natural mission of maternity and the formation of the family.

These arguments, wielded for so long, do not stand up to close scrutiny. Today more than three hundred million women vote; in some countries women are actively involved in social causes without any disintegration of their families, without refraining from having children.

It is true that there has been a decrease in the number of marriages and births, which has alarmed politicians and sociologists in some countries. France is one such country where, for several generations, the matter has been the topic of surveys, campaigns, and legislative measures. And in France women still haven't gained the vote. We could say the same of our own country, where not even the slow growth of the vegetation can be compared to the growth of women's suffrage, as the latter doesn't even exist.

Those that brandish the hackneyed argument have never questioned whether the act of voting has deterred man from his natural mission of paternity and the formation of a family.

If someone were to argue that upon fathering a child, a man takes upon himself such obligations that no other activity can be allowed him save that of feeding and protecting his offspring, an indignant chorus would cry out that such a belief would make a man deny his many talents and throw him back into a primitive state, close to that of the animals.

If it is true that many people, in their daily activities, do nothing other than those tasks that are only designed to provide for their and their children's lives, it is also true that others have other, superior, concerns in artistic, scientific, and social orders. These concerns are at times the most powerful impetuses for their behavior, and these are the people, quite justly, who contribute to society's progress. . . . These talents can be found in either sex and it is just as unfair to smother them in the name of maternity as it is to do so in the name of paternity.[1]

As human life becomes more complicated, each person encounters greater possibilities in a path that is open to all persons, regardless of race, class, or sex, where they can develop according to their various individual circumstances, circumstances that vary under the changing factors encountered throughout their lives. One of the greatest injustices, one of the strongest sources of irritation, pain, and rebellion, is the aim of enclosing a person in a rigid, traditional mold. A new life is weighted down with the beliefs of the dead as if biological heritage and the unavoidable need to adapt to one's circumstances were not enough to make young people follow in the narrow ruts humanity has trodden. This weight lies heavier on an individual in direct proportion with how much that person has been deprived of social rights; therefore, women and members of the proletarian class are the ones who have been molded with the greatest force, the former by the weight of masculine authority—whatever its range—the latter by the imposition of the possessors and administrators of wealth.

To narrow the limits of the spiritual development of women in the name of maternity is not only unjust but also absurd. The average cultural level of a woman from one of the civilized countries is infinitely superior to that of an Indian or a black woman. Which culture is best for motherhood? If we say one of the latter because, after all, the only form of birth control they know how to practice is, upon occasion, infanticide, should we lower the educational level of women in order to reserve her for procreative processes?

Even when the reproductive functions continue to be basic, the long evolution of humanity has constructed upon them a psychosocial structure so important that we cannot be sacrificed solely for the sake of those reproductive functions. This is as true for the woman as it is for the man. On the other hand, neither maternity nor paternity is what it used to be. If the physiology is the same as ever, profound and inevitable changes have taken place in the realm of psychological conditions. A child that could be an ignored consequence for a primitive man has become, at times, for modern man the most important goal of his life and his strongest source of happiness.

Early societies didn't recognize the rights of children; modern ones have reached a sufficient level of consciousness, a clear enough concept of their

continuity, to see in their children the incarnation of their own future and to prepare them for that future through education.

Besides, no matter how great the responsibility the adult bears to the newborn child, it cannot be such that the adult gives up her/his own life. The human species doesn't need to multiply to infinity in order to assure its continuation; it isn't necessary to make of each woman a queen bee or a termite. Peaceful coexistence in society yields the protection of life and extraordinary progress has assured that today more than ever, a blindly abundant production of children is pointless. A child today is so important that every effort is made not to lose or waste one. In the past an indifference or resignation to infant mortality was compensated for by repeated pregnancies that wore out and enslaved women. Today such a high number of pregnancies is a sure sign of low class, of poverty. This new consciousness of the value of life together with other psychological conditions results in a higher average marrying age and a lower birthrate. There are other contributing factors: a greater availability of jobs, a greater concentration of jobs in the heavily populated areas, and an increasing diffusion of a multifaceted culture that creates new worries in its members. These factors are observed in all countries of whites and, in fact, constitute a threat to the future of that race. Whites are faced with the fecundity of other races, a fecundity checked only by sterility or death. These processes are unstoppable because they are a part of the progress of the world, and, without a doubt, as soon as the whole world is accessible to the whites, the ways of the whites will influence the other races. For example, we believe the size of the Chinese family will decrease after the tremendous crisis it has suffered because even the poorest women will have caught a glimpse of a world where they won't be slaves appreciated for the rate at which they work and produce sons.

It would be wrong to see in women's emancipation the sole cause of the arrest of that slow growth. This stoppage was caused by a variety of changes, within a fatal historical process. Opposing women's emancipation in order to maintain a high population would be as illogical as stopping any increase in salaries, any improvement in the quality of life, any increase in cultural diversity. It would be as illogical as denying culture to the poor masses, urban and rural, that have been until now the great reserve of human life.

Whoever argues that the growing culture and the political activities of women are a cause of a decreasing population should, logically, recognize the same relation in the men upon whom falls with equal weight as upon the women, the responsibility of limiting or suppressing possible offspring.

Nevertheless, what legislator would dare to suggest revoking man's right to vote because the number of children is diminishing? The idea seems illogical and even offensive.

Masculine sterility (just as possible as feminine sterility) has never been considered a sign of social inferiority, and no one would find a relation between reproductive and political functions to be acceptable.

A counterargument is that maternity imposes a mental and physiological dedication so great that it changes the whole life of the woman. We have to concede this, because it is quite evident, but no one could argue that voting every two or three years takes up so much of a woman's time that she would be distracted from her important duties. There would be no distraction, using this term in its exact sense, but rather the attention, the intelligence of the woman would be applied to something new, elevating her, although momentarily, from her daily routine and linking her to the rest of humanity in a new way. She would achieve a new consciousness of herself and it is this, actually, that opponents of women's suffrage fear.

Proof of this assertion is that the majority of those who make maternity an insurmountable obstacle [in the path of women's suffrage] and present maternity as the only desirable activity for women, do not employ that same dogged determination in the fight against prostitution, against the exploitation of women workers, against the employment of mothers as laborers—a true form of slavery. Much less do we see them rising up against the religious vocation that encloisters so many women. Yet all of these activities hurt population growth much more than exercising the right to vote.

The deep-seated motivation of the systematic opposition of those that claim to want to save the home by denying women their political rights is the instinctive sentimental complex that resists anything that could destroy the age-old subjugation of women to which men are accustomed. . . .[2]

.   .   .

In the majority of civilized nations, women are no longer the stupid females who bear children without understanding their responsibility, passively, obeying now the blind call of nature, now the fulfillment of the marital duty. Women are more and more thinking mothers who want children, who know their obligations upon giving birth, who find in their children not only the satisfaction of their most noble instinct but also the natural complement to their personalities and the concrete objective that defines their existence.

The diffusion of scientific knowledge has brought to women the consciousness that they should prepare their bodies and spirits to be able to pass on to each child a clean heritage and to be able to form the soul as well as the body of each child.

Motherhood is not only pregnancy, birth, and upbringing, it is the sum total of the infinite attentions and cares that are more delicate in a greater intellectual and aesthetic environment. The dedication of almost all the activities of a woman to the little child, whose progress must be followed step by step, whose intellectual and moral formation must be guarded carefully, means so much more if one considers the great importance that

early childhood has for the future psychological balance of the youth and the adult.

This little life that begins, congregates around it all the great social problems. In order to assure the child's physical well-being, the best of medical care ought to be available. This presents, from the earliest moment, an economic problem. No matter how strong a mother's love, that love cannot, in response to the imperative needs of the child, make up for an insufficient diet or poor living conditions brought about by a lack of resources. Nor can that emotion or the will behind it make up for the intellectual poverty of the environment, erase its backwardness and give the child the preparation needed before her first studies.

The inability to fulfill these duties that is inherent in her condition as mother, should logically bring the mother to an understanding of the great deficiencies that the social organization has not been able to amend and that translate, in the end, to loss of lives or the production of defective lives.

Conscious, chosen motherhood, which is coming to replace accidental maternity, leads naturally, through a process of mental evolution, not only to a growing appreciation of itself, but also to its association with other expressions of collective life. . . .

In the moral and intellectual formation of the child, the influence of factors outside the home becomes stronger and stronger. From the nursery and the kindergarten to the school and later to the university, factors that are not in accord with the feelings and the ideas of the parents surround the personality in the process of formation. The home itself is infiltrated by the media, books, radio, and the telephone, which increase the intimacy of the union of the internal with the external.

These actions, previously the almost exclusive function of the church—which imposed its practices and beliefs   today belong to the group and are manifested by means of the state. In this situation a woman has two possibilities: She can be part of a government organization that is based exclusively on authority and force . . . or she can be part of a democratic regime in which the state is all the more the expression of the collective will when the conscience and the force of will of the masses is the strongest.

In the first case, maternity would be, as a social function, a passive condition; imposed upon the mother would be the practices and beliefs that the state needs to survive, in as dogmatic a form as in the periods of greatest religious dominance; and even when the formation of new lives would have a fundamental value for the survival of the nation, it would not mean a great appreciation for women as they would be divested of their lively influence.

In the second case, the mother would acquire an ever greater position, as she herself reaches the conscience of her social significances, as she fulfills,

in the most complete way, her maternal duties, as she understands the worth and the meaning of the ties that unite her home to the rest of society. Then, in defense of the life she has formed, she could intervene in the political arena to improve the general well-being by the improvement of economic relations. Then she could make politics work for the broadening and perfecting of education methods and, above all, for the protection and the prolongation of life. It would be then and only then that she could progress toward the realization of the pacifist ideal that, with more or less clarity, lives in the hearts of all mothers.

But this intervention, truly effective in politics, is only possible for women in a democratic organization when that democracy recognizes her political rights, when she is an active member in that democracy.

Far from being an obstacle to political activity, fulfilling maternity is really an impetus toward such activity. The more aware she is of her maternal responsibility, the more a woman will want to possess the means of collective action that will allow her to bear that responsibility more successfully.

Democracy would give to motherhood its complete meaning, because only in democracy is all human existence respected; only in a democracy is there the possibility that every life that begins will be able to live the full and satisfying life that each mother wishes for her child.

Nothing in life equals the blossoming out of tenderness and self-sacrifice that fills a mother's heart when she holds in her arms her newborn infant. In some women, however, that calling soon pales before the realities of a hard road before her. While in the animal kingdom motherhood turns a cowardly animal into an aggressive, violent one to protect her young, a woman almost always loses her fighting spirit because she is so convinced of her inferiority that she seems to forget that this child is hers more than anything.

In others the calling lights a fire that warms an entire life of work and struggle. All the ambitions, all the prides and worries, all the aspirations for success or glory center on the child and try to be realized through the child.

The day that a woman understands that through her effort alone it is difficult to provide the best home, the best diet, and the best education without damaging the equal rights of other mothers, the day that she understands that she alone is incapable of assuring peace and justice for her offspring, and that she should, toward that end, unite all the mothers in her country and beyond, that day an uncontainable force will bring together those that, in humanity, have worked until today for good and truth.

The perfection of democracy is closely linked to the perfection of each individual, not in the sense in which totalitarian regimes might wish—that is, for the purpose of producing strong men capable of the work that

corresponds to them, faithful to the regime in which they live. It is another form of perfection, one that consists in the internal liberation of the person, making that person feel equal to all others, without economic, racial, national, or religious differences placing some persons in superior levels. Above all, this perfection allows people to be free of fear and uncertainty and gives them the consciousness, in their heart of hearts, of their solidarity with society. In other words, the democratic perfection is in reaching that balance between the increasing union of the ties that bind the individual to the social environment that formed her/him and her/his capacity of self-analysis, of self-determination that, in any moment, might separate that individual from society. That interior freedom, psychological substratum of the social freedom, without which the sovereignty of the people would be an empty phrase, requires an early formation and is equally necessary for men and women. And here we find ourselves with another form of psychic equilibrium indispensable to democratic progress, as such progress cannot come from the union of emancipated men with women enslaved by ignorance, superstition, and mistakes. There is nothing more disastrous for a nation than to be formed of homes where the fathers feel that their ideas are alien, that civic life has not penetrated the home and hearth, that the women—and their children when they are under the maternal influence—are the fathers' enemies.

Notes

1. This conflict between the individual vocation and the duties imposed upon people by society is a topic that has often been explored by novelists, keen observers of reality. The life of all those creators, reformers, and adventurers always presents a period of struggle to decide their fate: Either their affections and family obligations win out, or an irresistible tendency, sometimes not even understood by those who live with and love them, takes over. This struggle can take place in the masculine personality as well as in the feminine, but it is in the case of the latter that moral considerations are so strong that the rebellious tendency is almost always snuffed out.

The strongest inhibitions are of a sexual nature and become a part of the female spirit during its very formation, which makes them consubstantial.

In order to understand it clearly, analyse and compare the meaning of the same terms when applied to a male and a female. A man's honor. A woman's honor. A public man. A public woman.

2. It is odd that man, who for some reason created the myth of Circe, still doesn't completely understand that the more limitations placed on a woman's activities, the more that woman will be ruled by her instincts, the most powerful of which—during the best years of her life—is the sexual instinct. That instinct-ruled woman will subjugate a man, in that she converts him into the sole objective of her life and, as wife or mother, bonds herself to him with a tremendous egotism. Christianity, upon exalting chastity in its reaction against paganism, could only create an artificial state. The cloister does not produce a creative solution, because the instincts are not killed. It is necessary to channel the impulses into various artistic

and social activities to give an individual life an objective that overcomes its enforced limitations. Religion has tried, but preparation for another world has—except for the mystics—little inducement compared to the enjoyment of this world. That is why encloisterment has seemed so artificial. The solution must be found in the development of the personality as a social being for the good of humanity.

Magda Portal

JOSÉ CARLOS MARIÁTEGUI, the great Peruvian Marxist writer, considered Magda Portal the first among the women poets of his country. Literary magazines throughout Latin America published her work in the 1930s and 1940s. But, above all, Magda Portal defined a new profile for woman as politician. She was ready to commit herself to a public cause under the dangerous and adverse circumstances that led to her exile and imprisonment. Born in Lima in 1903, Portal joined APRA, a nationalist-popular movement founded by Haya de la Torre, when she was twenty years old; she traveled widely throughout Peru and other countries in Latin America, organizing meetings, writing political tracts and pamphlets, discussing strategies, and articulating women's claims in front of a male party leadership that was reluctant to recognize equal political rights for women. Portal resigned from APRA to become a sort of legendary character in the Peruvian intellectual scene. She published poetry (*Una esperanza y el mar,* 1927; *Costa sur,* 1945), fiction (*El derecho de matar,* 1926; *La trampa,* 1956), biography (*Flora Tristán, precursora,* 1944), and political and ideological essays ("América Latina frente al imperialismo," 1929; "Hacia la mujer nueva," 1933; "Quiénes traicionaron al pueblo," 1950). Perhaps her most vivid text is the autobiography (edited by Esther Andradi and Ana María Portal) excerpted here. Portal summarizes the principal events of her political life as active organizer as well as the topics of her debate with APRA's leadership on feminist issues. The text displays revealing episodes of the difficult relation established between an anti-imperialist political party and an outstanding feminine militant.

from My Name Is Magda Portal

The Beginnings: APRA in 1923

I remember that during Leguía's presidency, when the dictatorship was trying to get in good with the church, a friend asked me if I would like to

"My Name Is Magda Portal," *from* Ser mujer en el Perú. *Ed. Esther Andradi and Ana María Portal (Lima: Tokapu Editores, 1979). Translated by Marcy Schwartz Walsh: copyright © 1991 by Sara Castro-Klarén.*

meet Victor Raúl Haya de la Torre.[1] I was totally ignorant then, politically speaking, but I had friends at the university who were involved in the protest against the church enthronement. Haya was really a strong leader, he spoke so impressively. I remember the police chasing him, and he threw himself into the river. It all seemed so heroic to me. Then they deported him; he went to Europe, he already had tuberculosis. I stayed in contact with those people. Then I met José Carlos Mariátegui. There were very few women in these groups, the Uruguayan poet Blanca Luz Brun, the Peruvian writer Angela Ramos, Mariátegui's wife, myself, and a few others coming and going sporadically. Women really didn't get together with Mariátegui much.

In 1927 I was going to Vitarte to give a reading of revolutionary poetry, but then I had to flee from the police and I wound up in Bolivia for a year. A group of students and I started a controversial newspaper called *Bandera Roja* [Red flag]. We attacked everyone. I also published my first book in Bolivia, *El derecho de matar* [The right to kill]. Before that I had been writing lots of poetry. *El derecho* was a book of revolutionary stories that were photostatted and distributed a lot. But soon enough the police pounced on us and the government sent us back to Peru, where we were thrown in jail. Soon after, we got together with Mariátegui again, and we founded the Laborers' Federation Press.

While we were involved in that, the police caught on again and sent us to Cuba. They imprisoned Mariátegui. Blanca Luz Brun was in the group, too. I remember the headline in *El Comercio* that day: "Two Women Are Implicated in the Communist Plot." They sent her to Buenos Aires. From Cuba I went to Mexico, where I immersed myself in socialist ideas and my awareness was sharpened there. At that time I didn't know anything about anything. I was only a poet.

There was a big group of us there, and we kept in touch with Haya and Mariátegui. They were invited to speak a lot at conferences. It was right around that time that we formed APRA, the Popular American Revolutionary Alliance. There were about eight or ten of us: Cox, Vásquez Díaz, Serafín del Mar, Haya, myself. It was in 1928, and I was the first secretary. I was sort of an acquisition for them, since I had won some literary awards and was well known as a revolutionary poet.

In 1923 I won the Juegos Florales Literary Prize at San Marcos together with Alberto Guillén. But Haya told me, "You can't write poetry anymore. Now you need to study political economy." So I started studying. I remember standing and talking with some friends near the river. I took out my book, *Anima absorta* [Absorbed soul], and ripped it up, throwing the torn pages into the river; I watched the river carry away the scattered pieces. . . . I felt heartbroken, but I had made a decision: "I have to study. Haya told us that we have to study." All of the early members of the group were

very young, and none of us had a university degree. However, Haya told us he didn't care about degrees, that he didn't have one either.

After studying diligently for two years, I decided to tour all of Latin America, giving lectures on anti-imperialism. How I ever had the nerve! I began in Central America, then went to the Antilles, Colombia, Venezuela, Panama, and then I ran out of money. I was in Puerto Rico to give a talk, and they deported me to Colombia. From there I went to Costa Rica, where I got my first false passport.

Mariátegui

I stayed in touch with Mariátegui. He hadn't joined APRA, and he urged me to become a member of the socialist party that he was forming. I told him that Peru needed APRA more than a socialist party. He thought that all of the leftist intellectuals should get together to meet about this and wanted me to pick a time and a place. I was in Colombia then—and don't get the idea that it was all fun and games; we suffered a lot; sometimes we couldn't even pay for our room. I answered him, "You're the one who is sick. Just decide on a place and a time, and we'll be there." I will never forget his response, "I know you—wherever they tell you, whatever they ask you to do, you'll be there." We decided to meet in Chile. We took a tiny boat, but we made it. Seoane, another APRA leader, was with us, too. I remember that Haya never attended the meeting in Chile. He was in Germany, and he was no longer working with Mariátegui, nor with Ravínez who had joined the Communist party. Ravínez was an evil man.[2]

Unfortunately, Mariátegui died April 30 of that year. Then Leguía fell from power, and we all returned to Peru. The police got hold of all my correspondence then, which is why I don't have any of those letters. It would have been interesting to have Mariátegui's letters. He was a sincere, loyal, and talented man.

"The Feminine Vote Was Premature"

The possibility of women voting was certainly being talked about, and that's what we were all hoping for. And the time came when these guys started to give in. But then they decided it was "premature," not yet feasible, that we needed to wait a bit longer. They even talked about ambassador positions for me, but it never came through. They suggested to Bustamante and Rivero the possibility of my going to Mexico, but Bustamante refused. He must have remembered the time when I was deported to Bolivia. They said women weren't "prepared," that they "had to be educated first." They said the same things about the peasants, that they weren't ready to take on managing their own affairs.

Women in Peru have hardly counted politically. We were at the embryonic stage. When I joined the party I was the only woman, and I had an executive position. Then they formed the Feminine Command and gave me that position, although I never was for men and women being separated. I always said, "It's one party, and there shouldn't be any divisions." But they never understood. My job was to distribute political information to women. I organized a women's convention for all of Peru in Lima. I have all the documentation. For the first time ever, women came to spend a week at a convention. Some married, some with children, housewives, Indian women, teachers . . .

I was terribly disappointed in Haya when we asked him to come speak at the convention. We weren't on very good terms with him by that time anyway. He started speaking to all these women about the home, about taking good care of their husbands, about marital bliss, that a marriage could only be happy "when the woman understands her husband's situation." Standing next to him, I let him know that they weren't interested in that, and I told him to talk to them about other things. They had come to hear about politics, and then I asked them, "What would you like him to speak about?" And they all responded, "About Marxism!" Women from small towns and rural areas had heard the word Marxism, and they wanted to know what it was all about! And Haya gets up there to talk to them about how to be dutiful housewives . . .

Within the Feminine Command was a special committee for complaints against husbands who were also party members. Even though they were activists, still the man was always the one in power, automatically discriminating against women. Lenin himself said, "In the home, the man is the exploiter, the woman the exploited."

The command was there for women to come and express their personal problems. But I never reacted solely as a feminist, only defending women's rights; I felt I was there to defend any worker's rights. I always said, "Women have to fight alongside men and learn to struggle with them without differences." You see, feminism started as a rift between men and women, which goes against all the ideals of social struggle. We have to educate men, to help them understand that women are soldiers in the fight just like them. Think of all the heroic women who have been fellow members of the same party, women who have hidden weapons inside their bodies so that they could hand them over to men in the middle of the battle! That's what women have been able to do, so how can they consider us cowardly, inferior, prone to teary outbursts? Think of all the women who have been in prison, kicked and beaten all the way to their cells. Women have suffered a lot for the party, and thus they have earned the right to be considered equals, but no one ever wanted to admit it.

APRA vaguely espoused a social doctrine without carefully defining its position regarding capitalism, and it eventually identified with the upper class, becoming more closely aligned with those on top rather than with those on the lower end of the social scale. When I saw that APRA was no longer a revolutionary party, I left.

"Travel Companions"

By the party's second convention in 1948, I was already openly against the party heads. This all revolved around a motion, or maybe just a declaration from the floor, specifying the following: "Women are not active members of the party, they are simply companions, because they are not citizens." Why did we fight for twenty long years, if only to hear men taking all the credit for our accomplishments? When I opposed it, they told me that it was simply a declaration, that there was nothing to discuss. But this was going to limit us in a way I could not accept. So, I got up and I left.

Maybe they did it just to alienate me, or maybe to alienate all women. They didn't take us into account. When I was out of the country or away from Lima, Haya wouldn't even meet with women. Of course some leaders have women who adore them. I remember that it was his birthday once when I was out of the country, and they gave him a silver vase full of flowers. He received it, but he never even thanked them. Basically, he had no use for women. But they respected me, at least, because I was strong and I had some influence. Someone once told me that I was the only person they couldn't treat like that, because if they did, "all the women would come out protesting." But even when they did, no one protested. Most of them just went along with them, or stayed home and that was the end of it.

Most of the women at that second convention stayed with the party. Five or six women, and a few men, did leave with me, when I said, "This I cannot accept. This is fascism." I left, and I never went back. However, the party didn't even notice; when they chose the new party leaders, they named me undersecretary general. They put another woman in charge of the Feminine Command, since I was getting too influential. Think of it, though: an undersecretary is like a vice-president. Your only role is to replace someone else. And that's how it was. My big office in the Feminine Command, where all the women met, was taken away from me.

1948: Resignation

They wanted to avoid fighting with me, and that's why they gave me the undersecretary position. It was easier to neutralize me. But once I left, I never went back. Soon after the October 3 revolt, when APRA leaders attempted revolution, a lot of people left, and others were indecisive because

in the interim we went underground. We had to hide for a year. Then the Potao[3] trial began, and during the proceedings, I said, "I can't keep on like this, because I'm against them. I've already told them that they betrayed the movement!" Haya had put a stop to three other movements before. The only one that actually happened was the one led by Colonel Pardo and other military leaders, like Victor Villanueva who was deported to Venezuela. But then someone turned us in, and the troops occupied the fort Real Felipe, killing lots of people. They had hidden all the weapons, taking them from their usual storage places before the people could get to them.

This could have been the road to revolution in Peru, if we had had capable leaders who had the courage that it takes. Their strategy was to push, and when something was about to explode, they gave the order to retreat. The whole Trujillo thing also happened without Haya. APRA's history is really dramatic. But in spite of it all, it was a positive step to awaken all those apathetic people. That year I wrote the pamphlet "Quiénes traicionaron al pueblo" [Those Who Betrayed the People], and a bit later the novel *La trampa* [The Trap], the book they almost killed me for writing.

They always thought I was armed. Some of the other women who were ex-members warned me so that I wouldn't walk around alone. When I published "Quiénes traicionaron al pueblo" they requisitioned the edition. Now I am revising, correcting, and adding to it. I don't know when I'll publish it.

When I left, Haya was hiding in various different places. Finally the Colombian embassy took him in, and he met with people there. One of the women told me, "Haya says that you don't have any political sense, that you don't know how to measure the consequences of violent political action, and that you've always been against the party." But they said to him, "We know that Magda Portal is for the revolution, because what she wants is for us to take power." But, as usual, their response was, "We can't, not now, it's not the right time." It was never the right time, and we're still waiting.

"Women Don't Understand Politics"

I knew what was going to happen on October 3, 1948, although I didn't actively participate. None of the women were involved, just the leaders and the rebellious military men. Do you remember when the Guadalupe strike left one man dead? APRA provoked that strike and had the young man killed. León de Vivero, one of the APRA leaders, kept saying, "We need a death, we need for someone to be killed." And as we were burying the poor fellow, Vivero said, "Now we have a martyr." Incredible: They wanted to march through the streets of Lima to create a public scandal, but Odría was prime minister then and he wouldn't allow it.[4] So a young man died,

for nothing. I was furious, and at the executive committee meetings I let them know it, I told them they had betrayed the party.

They were keeping an eye on me, and that's why there was a small committee that I was excluded from. Vivero used to tell me, "Women don't understand politics, they just don't understand." But we have better political sense than men. Look how together their political sense was on the October 3 uprising. I remember telling Vásquez Díaz, an old friend for years, the night before the revolt, "Now everything is ready and the revolt will proceed, but I'm sure the right will be back in power." And he answered, "Sure, because they're the only ones who know how to govern." "If that happens," I answered, "I'll leave Peru forever. I won't renounce the party, because I don't want to damage APRA, but I'll leave and I won't come back."

Machismo is a fact, of course; and this is a *machista* country. But they're a bunch of *machistas* who can't stand on their own two feet. Colonel Pardo always told me, "You're more of a man than all the men in the party," but always added, "even though you're a woman through and through." And now that I think about it, maybe I would have liked to have been born a man. I look at these men who go from place to place, free, they go out at night alone, no problem. And us, they watch over us everywhere. Because of that I might have wanted to be a man.

If there is such a thing as reincarnation, I'd like to come back as a revolutionary leader. And as a woman, just to avoid limiting ourselves to masculine models. If I've failed, then, it's because I'm a woman; if I had been a man, I would have succeeded. I never had the opportunity to go beyond my position in the party, and being a woman made me more vulnerable to attack by the opposition. They invented love affairs between me and one or another of the leaders. Once they printed a false photograph of me with I don't know how many faces kissing all the leaders. I never saw it, because I was in prison then. I never paid any attention to all their gossip; if I had, it would have driven me crazy and kept me from carrying on.

The first thing they attack in women is their sexuality. Even APRA spread these kinds of rumors about upper-class women who were in power then. I wouldn't stand for it, and I told them, "Why do we have to go and attack a woman's reputation?" Whether or not it's true, it's in bad taste. "The public likes it, it grabs their attention," they would tell me. Well, then that means whatever they say against APRA women grabs their attention, too, especially if it's about me. But I never worried about what people were saying about me.

They used to say that I was in love with Haya, and with all the others, too. When I was in Mexico, Diego Rivera and his wife were getting divorced, and she was in love with Haya. But he always said, "I would only

get married as a party obligation. It's like putting a saddle on a horse; I would do it, but only if I needed to politically. Otherwise, I have no reason to get married."

"My Name Is Magda Portal"

I traveled all over Peru spreading campaign propaganda for senators and representatives. And look, this was before women had the vote. I was the star. Once I went to Piura to speak at a rally with Arturo Sabroso, an influential labor leader, and others. It was a primarily socialist region, supporting Sanchez Cerro and Luciano Castillo's party, both opponents of APRA. The rally was in Sullana, and the theater was packed. The workers were sitting in the balconies, and the upper class in the orchestra. We walked out and were met with dead silence, no one applauded. They were our enemies. Sabroso spoke first, followed by the others, and I had to close the show.

I started with, "And now I am going to address the people up above, even though I always speak to those down below." They loved it. I continued, criticizing false revolutionaries who wear their leftist ideas like carnations on their lapels; when they wither, they simply replace them. It was interesting—they didn't clap for anyone else. But they gave me a standing ovation and cleared a path for me in the crowd. I knew it was risky, that any one of them could have shot me. Nothing happened, but when we got to our car to go back, we found that someone had let the air out of the tires.

Later, we took a trip up north in a van, and the driver seemed hostile to us. Just outside of town, he suddenly stopped and said he had to repair something in the van. It was around six o'clock in the evening, already getting dark. As we were getting out of the van to see what was going on, I saw a group of about six men approaching us. I heard, "We want to meet that Magda Portal." I looked down as I got out of the van, thinking that they were party supporters who just didn't know how to express themselves any better. It was a little cool, and I stood with my hands in my pockets. "My name is Magda Portal," I said to them. Then they all withdrew, pulled back and scattered, without saying a thing to me, as if they were afriad. Maybe they thought I was armed. I was puzzled for a bit, but then I realized that the others were really scared, too. Actually one of us did have a gun, but only one of us. The driver returned, and the van was fine.

I have lots of stories like that. In Arequipa, the priests wouldn't let any women come hear me speak; but all of them sat piously in the front row.

Love and Marriage

I fell in love a few times, and there's love in all of my poetry. But love has its time, and it's brief. A Chilean poet who is a friend of mine says that "eternal love lasts three months." Me, I'm not very romantic, and I always thought that love should last forever, but it doesn't, it fades. Even friendship fades. My love stories, my marriage, everyone knows about them; but I wish I could forget they ever existed because they just weighed me down.

That man was a drag, he was wimpy and had no personality, and he was so jealous. I never traveled with him, only with my daughter. I remember being thin then, and I wore sexy dresses. In Santo Domingo, I was getting off the boat and they took me for one of the dancers who had come to perform there. The reporters came up to me, and so I told them that I was Magda Portal. "Ah, Magda Portal . . ."; they expected me to be walking around with a bomb in each hand. It didn't occur to them that I might look like any other normal girl.

I was twenty-three when I had my daughter. Ah, what sad stories . . . Once he stole her from me and I had to go after him with a gun in my hand. I was ready to kill if he didn't give her to me.

It's funny—a lot of things that have happened in my life remind me of Flora Tristán, one of the most incredible female figures in history.[5] Her life is fascinating, and so controversial. I discovered her rather late, when Felicia Vergara, the director of Chilean Socialist Women, asked me to give a lecture about her. And then I had to get out of there, so that I wouldn't be persecuted.

Someone once told me that I was made for marriage, because I'm very feminine. And it's true, but I can't stand being dominated by a man or a family. Men usually don't want to be on equal terms with women. But giving up privileges is not the answer either, because that only demeans us. I especially did not want to be looked at as a weak, poor little thing, or even worse, put on a pedestal. Although at times I've felt the need to be protected, I never thought less of myself for not having a man. And basically I didn't need a man to protect me. I've lived for years without missing it at all.

Marriage has nothing to do with love; monogamous relationships are a thing of the past. I have my religious ideas, but I wouldn't call myself a Catholic. I freed myself from all that when I was about fifteen, because religion diminishes women more than anything. It has always put the brakes on anything that would improve women's lives, impeding us, reducing us, especially through marriage. Wedding vows still say that women must "obey their husbands." And to think that I once got married in the church!

Legal marriage is worthwhile for children, but women should remain completely free, and not be dominated in any way by their husbands, economically or otherwise. Flora Tristán's greatest accomplishment was to help mothers regain their rights, for themselves and their children. She won the right for a mother to give her children her name if she wants to, or if the marriage ends, and a woman's right to not take her husband's name. There are children who are not allowed to start high school because their father never recognized them by giving them his name. And that is unfair. The important thing is for women to fight for new laws that are in our best interest, laws that will free us from our husbands' power.

Abortion is one of women's most serious problems now, because it is being used as a method of contraception. In my era, of course, most of them were underground. After I was married, I had my tubes tied. I was very young and the doctor asked me, "But aren't you going to want to have more children later on?" But I said no, pretty defiantly. Of course that operation was illegal, too, but he was a friend of mine.

I feel that a woman who gets married and wants to have children should have them, but three at the most, because that's the most that you can manage to support. But what women really need is a better quality of life so that they aren't just converted into baby machines. Just giving birth over and over again, no way. Being a housewife, taking care of husbands day after day, we've had enough of that! That's how our society has negated women and made them anonymous. That's what was running through my head when I decided not to have any more children, and I never did. For me, having a child was a sacred act, and I knew that I would never abandon a child of mine. But if I had had more children, I could never have done all the rest, what sprang forth from my being, my true path, my vocation.

Notes

1. Victor Raúl Haya de la Torre (1895–1979), founder of the APRA, was one of Peru's long-standing politicians.—Ed.

2. A Peruvian politician initially connected to Mariátegui, Eudocio Ravínez was one of the founders of the Communist party. He later renounced his Marxist ideas and became an extreme conservative.—Ed.

3. Potao is a prison in Lima.—Ed.

4. General Manuel A. Odría deposed the democratically elected Bustamante y Rivero and governed Peru for eleven years. He strongly repressed the APRA and all other leftist parties.—Ed.

5. Flora Célestine Thérèse Tristán (1803–1844) was one of the first French feminists; she fought for divorce rights and international socialism. Her mother French and her father Peruvian, Tristán wrote *Peregrinaciones de una paria* (1847; translated from the French) after a visit to her family in Arequipa, Peru, in 1833. *Peregrinaciones* was an epoch-making book because of its naked criticism of Peruvian colonial society.—Ed.

Eva Perón

BORN IN 1919 in a small town in the province of Buenos Aires, Eva Duarte, the illegitimate child of a prosperous landowner, began her public career as an actress. In 1944 she met Colonel Juan Domingo Perón, the ascending leader of a nationalist military group and friend of trade unionists. From that moment on, Eva's life underwent dramatic changes: She participated in the organization of the political basis of Perón's government, had a prominent place in social planning, and founded and directed the Fundación Eva Perón, which invested huge public resources in a paternalistic, popular, and original set of welfare policies. Beautiful, strong-willed, loved, and feared, Eva Perón was a centerpiece of the Peronist strategy with regard to the popular sectors. Her book, *La razón de mi vida* (published in 1951, when she was very ill with cancer), develops in a simple way the topics that conferred symbolic dignity to the strong and vertical link between Perón and the popular sectors. Evita (as she was called by the working class and the poor) is the effective mediator between the leader and the masses; she entrusts herself to Perón's political guidance and pledges her absolute loyalty to his program. Her perspective is at the same time deeply political and extremely personal: Wife and disciple, representative and communicator, concrete organizer and spiritual messenger, Evita turned public, humanizing state and government with feminine qualities and values.

from My Mission in Life

Anyone who has read the preceding chapter can readily understand that it is Perón's greatness, mixed with the simplicity of his genial nature, which makes me what I am, fervently and fanatically "Peronista."

Sometimes the leader himself is wont to tell me fondly that I am "too Peronista."

I remember one afternoon when, after I had been talking to him for a long time of . . . of what would I be speaking to him except of himself, of

Excerpts from My Mission in Life *(New York: Vantage Press, 1953). Translated by Ethel Cherry.*

his dreams, of his achievements, of his doctrine, of his triumphs? . . . he interrupted me to say: "You talk to me so much about Perón that I shall end by hating him." So anyone who searches in these pages for my portrait should not be surprised instead to find the figure of Perón.

It is—I admit it—that I have stopped being myself and it is he who exists in my soul, owner of all my speech and my feelings, absolute owner of my heart and of my life.

On the other hand, this is an old miracle, an ancient miracle of love, which by dint of repetition in the world no longer seems to us a miracle.

One day I was told that I was too Peronista to head a movement by the women of my country. I thought it over a good deal, and although I felt immediately that it was not true, I tried for some time to learn why it was neither logical nor reasonable.

Yes, I am Peronista, fanatically Peronista.

Not too much. It would be too much if Peronism were not, as it is, the cause of a man who, by identifying himself with the cause of an entire people, is himself of infinite worth. And when confronted by infinite things, it is impossible to praise them too highly.

Perón says I am "too Peronista" because he cannot measure his own greatness by the yardstick of his humility.

Others—those who think, without telling me so, that I am too Peronista—come under the category of the "ordinary man." And these do not deserve an answer.

Because I am a Peronista, I cannot lead a woman's movement in my country? That indeed deserves an explanation!

"How," they asked me, "are you going to direct a woman's movement if you are fanatically in love with the cause of a man? Is not that a total acknowledgment of the superiority of man over woman? Isn't that a contradiction?"

No. It isn't. I felt it. Now I know it.

The truth, logical and reasonable, is that feminism cannot be separated from the very nature of woman.

And it is natural for woman to give herself, to surrender herself, for love; for in that surrender is her glory, her salvation, her eternity.

Then may not the best woman's movement in the world be, perhaps, that which gives itself for love of the cause and of the doctrine of a man who has proved to be one in every sense of the word?

In the same manner that a woman attains eternity and her glory, and saves herself from loneliness and from death by giving herself for love of a man, I think that perhaps no woman's movement will be glorious and lasting in the world if it does not give itself to the cause of a man.

What is important is that the cause of the man be worthy of such a total surrender.

I believe that Perón and his cause are sufficiently great and worthy to receive the total offering of the woman's movement of my country. And, further, all the women of the world may support his justicialism;[1] for with it, surrendering themselves for love of a cause which is that of humanity, they will increase in womanliness.

And if it is true that the cause itself will grow in glory by receiving them, it is not less true that they will be exalted by the surrender.

That is why I am and shall be Peronista until my dying day: because Perón's cause exalts me, and because its productiveness will continue forever in the works I perform for him, and live in posterity after I am gone.

But not only am I Peronista because of Perón's cause. I am a Peronista because of him personally, and I would not be able to say which of the two reasons is the strongest.

I have already said why and in what measure I am a Peronista because of his cause. May I say how and in what measure I am a Peronista because of him, because of him personally?

At this juncture, perhaps it might be convenient for those who think a "political marriage" took place between me and Perón to turn over the page.

Those who believe this happened will find here only propaganda.

We got married because we loved one another, and we loved one another because we both loved the same thing. In different ways we had both wanted to do the same thing: he with intelligence, I with the heart; he, prepared for the fray; I, ready for everything without knowing anything; he cultured and I simple; he great and I small; he master and I pupil.

He the figure and I the shadow.

He sure of himself, and I sure only of him!

That is why we married, even before the decisive battle for the liberty of our people, with the absolute certainty that neither triumph nor defeat, neither glory nor failure, could destroy the unity of our hearts.

Yes, I was sure of him!

I knew that power would not dazzle him nor change him.

That he would continue to be as he was: sober, smooth, an early riser, insatiable in his thirst for justice, simple and humble; that he would never be otherwise than as I knew him—giving his large, warm hand generously and frankly to the men of my people.

I knew that drawing rooms would be superfluous for him, because there are too many lies told in them for a man of his caliber to endure.

Neither did I ignore what my behavior would have to be like to harmonize with his.

I knew that to harmonize with him I needed to scale very high peaks. But I was also aware of his marvelous humility in coming down to my level.

I am not ashamed to admit that I seriously intended that he should see one fault less in me each day till none remained.

How could I wish and do anything else, knowing, as I knew, his designs and plans?

For he did not win me with fair and elegant words nor with formal and high-sounding promises. He did not promise me glory or greatness or honors. Nothing marvelous.

Indeed, I believe he never promised me anything! Speaking of the future, he always talked to me only of his people, and I ended by convincing myself that his promise of love lay there, among his people, among my people. Among our people!

It is the path which all we women take when we love a man with a cause.

First, the cause is "his cause." Then we begin to call it "my cause." And when love reaches its greatest perfection, the feeling of admiration that made us say "his cause," and the selfish feeling that made us say "my cause," are superseded by a feeling of complete unity, and we say "our cause."

When that moment comes, it is impossible to say whether love for the cause is greater or less than love for the man of that cause. I think the two things are in reality one.

That is why I say now: "Yes, I am Peronista, fanatically Peronista!" But I would not be able to say which I love most, Perón or his cause, for to me it is all one and the same thing, it is all one love; and when I say in my speeches and in my conversation that Perón is the nation and is the people, I do no more than prove that everything in my life is sealed by one single love.

·   ·   ·

"Evita"

When I chose to be "Evita," I chose the path of my people.

Now, four years after that choice, it is easy for me to prove that this was certainly so.

Only the people call me "Evita." Only the *descamisados*[2] learned to call me so. Men of the government, political leaders, ambassadors, men of business, professional men, intellectuals, etc., who call on me usually address me as "Señora"; and some of these address me publicly as "Most Excellent Señora" or "Most Worthy Señora," and even, at times, as "Señora Presidenta."

They see in me only Eva Perón.

The *descamisados,* on the other hand, know me only as "Evita."

I appeared to them thus the day I went to meet the humble of my land, telling them that I preferred being "Evita" to being the wife of the president, if that "Evita" could help to mitigate some grief, or dry a tear.

If a man of the government, a leader, a politician, an ambassador, who normally calls me "Señora," should call me "Evita," it would sound as strange and out of place to me as if a street-urchin, a workingman, or a humble person of the people should call me "Señora."

But I think it would seem still stranger to him!

Now, if you ask me which I prefer, my reply would be immediately that I prefer the name by which I am known to the people.

When a street-urchin calls me "Evita," I feel as though I were the mother of all the urchins, and of all the weak and the humble of my land.

When a workingman calls me "Evita," I feel glad to be the companion of all the workingmen of my country and even of the whole world.

When a woman of my country calls me "Evita," I imagine myself her sister, and that of all the women of humanity.

And so, almost without noticing it, I have classified in these three examples the principal activities of "Evita" relating to the humble, the workers, and women.

The truth is that, without any artificial effort, at no personal cost, as though I had been born for all this, I feel myself responsible for the humble as though I *were* the mother of all of them; I fight shoulder to shoulder with the workers as though I *were* another of their companions from the workshop or factory; in front of the women who trust in me, I consider myself something like an elder sister, responsible to a certain degree for the destiny of all of them who have placed their hopes in me.

And certainly I do not deem this an honor but a responsibility.

I believe that every one of the men and the women forming the mass of humanity should feel at least a bit responsible for all the rest. Perhaps then we would all be a little happier!

I attend to the trade-union problems of the workingmen.

I receive from the humble those complaints and needs which are not related to the state, even though at times in these cases I also act as the government's diligent collaborator. In the end it is all grist for the mill of our common leader.

I attend to women's problems in their multiple social, cultural, and political aspects.

If anyone should ask me which of my activities I prefer, I would not be able to answer with precision.

If the question were put to me when engaged in trade-union matters, my choice would be for these. If I were attending to my *descamisados* or to

women, perhaps I would vote for the matter on which I was engaged at that very moment. And I would not do so through "diplomacy" or "politics." No! But because, when I am working, what I am doing at the moment seems to me best, the most to my taste, to my vocation, and to my liking.

I realize, though, that at the bottom what I like most is to be with the people in their most genuine forms: the workers, the humble, the women . . .

With them I do not need to pose at all, as I sometimes have to when I am Eva Perón. I speak and feel as they do, simply and frankly, sometimes smoothly and sometimes roughly, but always loyally.

We never fail to understand one another. On the other hand, sometimes Eva Perón is not wont to understand those attending functions at which she must be present.

Do not think by this that "Evita's" work comes easily to me. Rather, it always turns out to be difficult, and I have never felt quite satisfied in that role. On the other hand, the part of Eva Perón seems easy. And it is not strange. For is it not always easier to act a stage part than to live it in person?

And in my case, it is certainly as Eva Perón that I interpret an ancient role which other women in all ages have already lived; but as "Evita" I live a reality which perhaps no woman has lived in the history of humanity.

I have said that I am guided by no personal ambition. And perhaps that isn't quite true.

Yes. I confess that I have an ambition, one single, great personal ambition: I would like the name of "Evita" to figure somewhere in the history of my country.

I would like it to be said of her, even if only in a small footnote to the marvelous chapter which history will certainly devote to Perón, something more or less like this:

"There was, at Perón's side, a woman who dedicated herself to conveying to the president the hopes of the people which later Perón converted into realities."

And I would feel duly compensated—and more—if the note ended like this:

"All we know about that woman is that the people called her, fondly, *Evita.*"

•   •   •

Women and Action

I firmly believe that woman—contrary to the common opinion held by men—lives better in action than in inactivity.

I see them every day in my work of political service and social welfare.

The reason is very simple. Man can live exclusively for himself. Woman cannot.

If a woman lives for herself, I think she is not a woman, or else she cannot be said to live. That is why I am afraid of the "masculinization" of women.

When that occurs, women become even more egoistic than men, because we women carry things to greater extremes than men.

A man of action is one who triumphs over all the rest. A woman of action is one who triumphs *for* the rest. Isn't this a great difference?

Woman's happiness is not her own happiness, but that of others.

That is why, when I thought of my feminist movement, I did not want to take woman out of what is so much her own sphere. In politics men seek their own triumph.

Women, if they did that, would cease to be women.

I have not wanted women to look to themselves in the woman's party . . . but rather that right there they should serve others in some fraternal and generous form.

Woman's problem everywhere is always the deep and fundamental problem of the home.

It is her great destiny—her irremediable destiny.

She needs to have a home; when she cannot make one with her own flesh, she does so with her soul, or she is not a woman!

Well, for this very reason I have wanted my party to be a home . . . that each basic unit should be something like a family . . . with its great loves and its small disagreements, with its sublime fruitfulness and its interminable laboriousness.

I know that in many places I have already attained this.

Above all, where the women I have appointed are most womanly!

More than political action, the feminist movement has to develop social service. Precisely because social service is something that we women have in the blood!

Our destiny and our vocation is to serve others, and that is social service. Not that other "social life" . . . which is contrary to *all* service . . . !

Notes

1. "Justicialism" is Perón's own neologism.—Ed.

2. Perón's followers from the working classes were called *descamisados* (the unshirted) because many of them, during meetings or mass demonstrations, used to take off their shirts, sometimes because of the warm weather, sometimes so they could wave them as flags.—Ed.

Hebe de Bonafini

IN APRIL 1977, a year after the military coup had implanted an authoritarian regime in Argentina, a group of no more than twenty women, whose children were among the thousands of *desaparecidos* (disappeared ones), organized in order to obtain information and justice for those victims of state terrorism. By July, more than 150 women occupied the Plaza de Mayo every Thursday afternoon, wearing white scarves over their heads and turning silently around the pyramid that commemorates the independence of the country. In October, they published their first text ("We Only Ask for the Truth") in *La Prensa,* the only newspaper that accepted it. In December, one of the first members of the group was kidnapped and killed. From that moment on, Hebe de Bonafini was the visible leader of the Madres de Plaza de Mayo, who continued facing repression and a campaign of discredit launched by the military dictatorship. Their struggle was, during 1978 and 1979, solitary, except for the voices coming from the Permanent Assembly for the **Human Rights**, the Association of Relatives of Prisoners-Desaparecidos, and the **Service** of Peace and Justice. But during 1981 and especially in 1982, the Madres de Plaza de Mayo began to find the collective response for which they had been fighting. By the end of the military dictatorship, in December 1983, they were respected and well-known figures in Argentina and throughout the Western world. Hebe de Bonafini, a strong-willed woman of popular origin, is the present leader of one of the two organizations into which Madres de Plaza de Mayo have split. Her narrative, recorded and edited by Matilde Sánchez, echoes with the tones of private feeling and experience that have turned public and communicated strength and persistence to the human rights movement in Argentina.

from Life Stories

My love for my children had been transformed into another kind of love, almost without my realizing it. As time passed, the change took place bit

Excerpt from Historias de vida. *Ed. Matilde Sánchez (Buenos Aires: Fraterna/ Del Nuevo Extremo, 1985). Translated by Marcy Schwartz Walsh: copyright © 1991 by Sara Castro-Klarén.*

by bit. By the time I noticed it, there was already an established feeling of respect between Raúl, Jorge, and me. Perhaps for me it was a little less powerful and magical, but nevertheless mature. After so many years, my sons were actually nurturing me, they were teaching me.

Their younger rebellious days were not wasted but illuminated some real insights, or at least some questions. We must have left them a world with some dark regions, some obscure feelings of shame. In our own way, we, their parents, were their history, and they our mirror that magnified our mistakes and virtues. They became our best judges.

And as they demanded that we take stock of ourselves, they challenged us to look at things with a fresh perspective. They reflected, they thought so much more than we ever did in our youth or even as adults. They weren't conformists. They didn't automatically accept givens; those givens were only allowed to exist after being tested. I could learn from this approach to the world: at times it was like seeing things for the first time, like being born all over again, almost as though my sons had brought me into the world. Seeing them reach such maturity—Jorge's marriage, so different from mine; Raúl's commitment to his studies and his work—was a rebirth for me, but this time not innocent.

But I challenged them, too. I wanted to be less of a mother to them and more of a friend, to allow them to nourish me with their thoughts, their tremendous energy, their incessant youth. And they inspired every part of my being, every attempt to shake the laziness out of our stagnant world. They wanted me to study, to learn to read, to give of my time—even if it were only through reading the newspaper. They chased away all of my ridiculous fears.

Now, so many years after that terrible episode in my life and in history, I find my mind returning again and again to the years when we were a family, our years together in Dique, in La Plata. My sons' words, their acts, loomed larger, embellished, in all of those nights that they were missing. There are things that one never gets used to. I have had a lot of time, and will have the entire future ahead of me, to get to know my sons better. After all the slander, after all the indifference and persecution, no one can take away my pride in these two men. If they cannot be here, then I have had to take their place, to shout for them. I have had to recapture them with honesty and return to them, if nothing else, a piece of their lives. I feel them present in my banners, in my unending fatigue, in my mind and body, in everything I do. I think that their absence has left me pregnant forever.

·   ·   ·

I started to notice the repetition of faces in the police stations and court offices, but I never said hello to any of the other women. And it was obvious

that what was happening to me was also happening to these people, because eventually I exhausted all of the possibilities in search of my son in La Plata and had to travel to Buenos Aires. I began to meet the same faces in the Buenos Aires court office, in the Department of the Interior, which at that time was also housed at the Casa Rosada, the presidential palace. But there were only looks of contempt. At first, I was only interested in getting a little information. Since we had to stand in the same lines, there was no need to ask questions. Each one of us felt obligated to repeat her incriminating story dozens of times to the changing faces behind the counters and across the desks.

Around April I think I started to talk to a few of the other women. Our conversation revolved around the legal system we had to face, and the best way to compile a case for the initial lineup of judges that would reject it. But I did not give them my name or telephone number. We remained anonymous, suspicious people, united by the procedures we had to participate in, and by what we were each trying to recover. We were not just diligently tracking down any old relative, but showed the faces of mothers without their children, wives without husbands, or sisters without brothers. There for each other, we formed a precarious companionship based on the solitude of our individual procedures.

When I walked through the doors of the Casa Rosada, I was never thinking about Videla, or the junta, or the doctrine of national security. I only wanted access to that hopeful window, where I would be waited on by a nauseatingly friendly female police officer, with the promise of "interceding and personally taking care of your problem." "Leave me your telephone number and your current address and I'll let you know. I'll try to do everything possible, I promise. Poor thing! Your case is so dramatic . . ." And we all gave our addresses and kept the promise a secret, believing that it was real. First of all, no one would take care of any of this personally: It was simply a ruse for them to accumulate more information in their sinister files. There they held on to our facts and statistics, our homes, x-rays of our lives.

One of the women's faces on the train from La Plata to Buenos Aires was already anticipating her fate; later on I would recognize her in the capital offices. One afternoon our steps and our itinerary coincided. I recognized her gray hair, her pronounced sunken eyes, from a few days before on the train, when I was on my way to Army Corps 1. We both got off the metro at Constitution and continued to the Plaza de Mayo, to the Interior Minister Albano Harguindeguy's office. I said hello; we had not seen each other for two weeks.

"We could go back together on the express train," she said to me one afternoon when we saw each other again, "It's a little more expensive but much more comfortable."

I agreed. We sat in a double seat at the back, in silence. I started to wonder why I had agreed to travel with her if I was going to keep silent. Actually, I was thinking about whether or not she was missing a son; maybe her son shared a cell or was eating with mine while we shared our seat, quiet, not knowing how to get into a deep conversation under these circumstances. She was the one to speak first. She never said her name, she was looking for her twenty-four-year-old daughter. They had taken her five months ago. She was pregnant.

I was astonished. It seemed inconceivable. How could this woman keep on living, while her daughter was imprisoned and soon to give birth? I had denied the truth that these things were really so serious, so grotesque. I did not know this woman's name, yet I began to feel tremendous solidarity with her pain. I began to think that my case was less serious, but it was mine. However, while the train sped forward, probably late because it was leaving behind lines of passengers at all the stops, I felt a sense of sisterhood with this woman. I felt understood. She kept on speaking about her daughter, about the steps she had already taken, but although it sounded sad, she seemed a bit detached from her own grief, as if she had already overcome certain moments of confusion that I was then going through. "There are many more of us than you think," she said at the end, "and we are starting to get to work."

"Work?" I asked. I couldn't imagine what that word could mean in this context. She explained that a group of mothers had been working together to get interviews with influential people who could help them. ("I feel so alone; every day it takes more energy to keep up with the process. I'm beginning to get the impression that they're pulling my leg, that we're doing all of this for nothing.")

The woman smiled; our initial distance had vanished. She said that the following Thursday a group of mothers would be meeting at 2:00 in the Plaza de Mayo. They were going to sign a petition and have an interview with a priest. It wouldn't be a bad idea for me to come, "the more mothers that come, the better." I said that I would think it over (something in me still wondered if we weren't making a big scandal over a small issue, over some minor confusion that might be resolved at any moment). The woman stood up and shook my hand and started to walk down the aisle of the train.

"Don't forget. Thursday at 2:00 sharp."

.   .   .

It wasn't 2:00 yet, so I decided to walk around the block. I hadn't seen the sunken-eyed woman on the train, and I was nervous; I was hoping that she would be in the plaza before I got there. And then I asked myself why I was so worried. What harm was there, after all, in meeting with these

women? The billboards on Bolivar Street displayed the official slogan: "The Nation Advances," sky-blue letters on a white background, like the flag. I went back down that street again toward the plaza.

A few women were already there, near the pyramid, at the right, standing next to a bank. There couldn't have been more than ten. The woman from the train still hadn't arrived, but another woman came over to me. Brown-haired, strong and pleasant features; she was short but had muscular arms and the hands of a hard worker. Her body gave the impression of a great fortress. For April, the weather was rather cool, but she was wearing a loose cotton t-shirt. I was about to explain who I was, the woman from La Plata, I had heard about the meeting on the train. All vague information, but as I approached the woman began to smile. "Ah!" she said, "She told me you would be coming. Join us."

They all seemed terribly rushed. They spoke fast, in low voices that ran into one another, passing a paper around.

"Fortunately we got hold of a typewriter," the woman whispered to me, "We wrote a letter to the president, to Videla, asking for our children, and now we're signing it, before we give it to the secretary. If you would like to sign . . . by the way, call me Azucena."

While she was talking with the others, one woman was passing out pens from her purse and another approached me.

"You must have gone to the Human Rights League, right? Oh, then you haven't been to the Permanent Assembly, or to Emilio's group either . . ."

Wait, I said, I think I know that man, I saw him at the base. Then she told me that the patient, bald Emilio Mignone was a lawyer and was trying to form a parents' group. But she and the other women wanted to keep their group separate. "As women, it seems to me that we feel different and the other groups just don't accommodate us. We think it's best to take steps on our own." Then she told me that she and Azucena had gone to speak with Emilio Gracelli, captain general of the armed forces, several times. "There was a group of us, Azucena, Maria Adela, Dora, Juanita, but it didn't do any good. You go and they just give you another appointment for the next week and you don't find out anything. Sometimes it seems like they're just taking you for a ride. So a few weeks ago, Azucena suggested that we plan to meet in the Plaza, that way no one will get lost and we may even be able to gather a few more women on their way out of the Department of the Interior. And we wrote a letter to Videla, saying, all right, what's going on with our children? And so here we are."

The women continued passing around the paper and explained to two new faces who had just arrived. The city kept up its usual droning rhythm behind us, not seeming to notice us. Men rushed by to get to the banks before they closed, a few indifferent retirees relaxed in the sun. We could just as well have been a school alumni group planning our next meeting.

Most of them were about my age or around fifty. Azucena, although a bit older, jumped around from one spot to another like a young girl. The familiar first-name basis between us all affirmed our solidarity. I signed, a large firm signature so that the president would read my last name and it would make an impression on him. So that he would know, too, that I was not ashamed of my son's name.

We decided to gather again the following Thursday in the plaza. At that time, none of us ever imagined that waiting for our children would take more than a couple of months, nor that the initial search would one day be transformed into this painful story. None of these women kept track of the little details. Who would be the one to tell the story?

.   .   .

I had no idea that in those exact moments my life was preparing for a decisive turn. In those months, I didn't see that this activity—which I considered circumstantial as long as Jorge was missing—was going to change little by little into what the tired woman on the train had called it: work. For now, though, it was a question of how to manage two jobs.

My life, meanwhile, had turned into hell. My quiet routine without surprises was overshadowed by worries, bad news, trips to Buenos Aires that took four hours a day. I could no longer take care of the household or guarantee the continuity of our usual existence, as I had for the past twenty-five years. Everything was done hurriedly, at a feverish pace, which had the advantage of not leaving any time to go crazy with despair. Each of us, however, had to make certain adjustments.

Humberto, with his promotion at YPF,[1] and now with people working under him, had no time for these procedures. But there were new things that he had to learn: to be more fastidious in the house, to be Alejandrita's mother when I wasn't around, to assure our family's stability. At the same time, what I demanded of myself multiplied; I just could not let things go. I was divided into a hundred pieces. I waxed the floors at midnight, got up at 6:00 a.m. to leave dinner ready. I had to keep going at all costs, against all of my limits and fatigue.

Like this history, my own biography registered a profound break, in spite of my bold efforts to make sure that the woman who cooked and sewed wouldn't have to take a backseat. It was, however, unavoidable. For someone who only knew about the world through television and radio, this reality was a traumatic shock. I was full of fears. Although Humberto understood and supported me, how long would my wanderings last? Up to what point would he be able to adjust to the new situation? I also feared the lack of control inevitable in this change in routine. I ended up seeing less of Raúl; Alejandra demanded more time. Somehow I was being more of a mother for the child that wasn't there than for the ones still with me. I

felt nauseous, afraid, a little bit guilty. My head was spinning from the intense pain and such interminable demands.

.   .   .

I continued to be the pursuing activist, rushing back to take care of the household routine. But it was the search that kept me going; it truly was my life. However, my two worlds inevitably clashed. Public, external life was not in the end just an amplified replica of the neighborhood. The courts and offices never matched the efficient diligence of the united neighbors, pushing ahead full force. It was hard to accept that there were people so different from us, with an infinite capacity for evil. There was a totally different manner of speaking, way of dressing, that I would never have adopted personally except that it was necessary for our fight, for our sense of community. These new rules we learned in the plaza.

With our naivete, we opposed this living hell with more and more organized meetings. On Thursdays we would go to the plaza at 3:30, adding up our collective action. Azucena offered ideas. From Sarandi, her last name, Villaflor, evoked a long-standing militant tradition; she had a son and daughter-in-law missing. Little by little I began to take charge of organizing the mothers in La Plata; few of them dared to approach the pyramid of Buenos Aires, they preferred the quiet cathedral of their own town.

Toward the end of September 1977, we numbered more than fifty, and our confidence was growing with our numbers. Every day more joined us; every day we felt stronger and less fearful, more sure of ourselves, but every day more children were missing. We felt defiant in the plaza, sometimes, for a moment or two, almost invincible. The truth is that they did not know what to do with us. If any ounce of compassion was left in their hearts, it came from the stanza in some macho tango about "the poor old woman." That was protecting us for the time being. They thought we were emotionally shattered, that we would go away as soon as our varicose veins prevented our tired legs from standing up. It was horrible that our numbers kept growing, but now more eyes could see the balcony of the Casa Rosada. We had plenty of questions we would have liked to ask the assassin. We started to worry him, too; fifty women together week after week could not be any high school reunion. The bankers looked at our Thursday faces, and we always hoped someone would question us.

They sent a police officer over to us. "This is a demonstration, and we are under martial law, ladies. Circulate, circulate." We started walking in pairs, arm in arm, and they made us walk in a broad circle. The plaza was large, the couples spread out, and no one could distinguish us from other women just out to take a walk. We recognized that the best thing was to start gradually closing the circle, but so imperceptibly, slowly closing in on

the pyramid, so that they would have no time to notice. But every day there were more of them, and each Thursday we were also a bigger group, and they brought in reinforcements. They would stop in front of the pyramid and keep watch to keep us from getting too close. We kept conversing in pairs, looking at the backs of the women in front of us. Now, as if I were there, I can see the French woman, Alice Domon, walking with Mary Ponce. I can see her smiling, youthful face, and I know that she and I have the same God. Further along I see Azucena's strong face in the cold spring air, with her light sweater and her tough arms.

The police's ridiculously stupid treatment toward the Madres would reach its climax a couple of years later, after our first mandatory circle had turned into our weekly ritual. When they could no longer beat us or imprison us, they decided to form an enclosure with railings in the middle of the plaza, and we were supposed to show our white handkerchiefs in order to be let in. They isolated us from the plaza, to avoid the "contagion" that they so feared. We did our round, and then they let us leave.

·   ·   ·

In September 1977, in the space of a week, the Catholic community of Buenos Aires planned a march in Lujan. The small group of mothers working most intensely together was having tea in the Violetas Pastry Shop, a perfect place for us because it is always full of women having tea and pastries. Azucena said that it would be a good idea to all go together to Lujan, "because people talk a lot while they march, and we'll stand out." We all agreed. In any case, we had a lot of mothers in the group who wanted to say the rosary. The only problem was that not everyone wanted to walk, some would be meeting in Haedo, others in Moreno or Castelar. We had to make clear, definite plans about where to meet.

"I know," said Eva, a mother who almost never spoke up, "We should all wear something that is visible from far away. A handkerchief on our heads, for example."

"Or a mantilla;[2] but not everyone has one. A handkerchief is a better idea."

"Yes," said another, "Or even better, one of our children's muslin dia-pers; it looks like a handkerchief, but it will make us feel closer to the kids."

A baby's cloth diaper on our heads—we all had one packed away, so each of us retrieved one and carefully ironed it before the march. We were waiting for our moment. We would have to spread out a lot in La Plata, everywhere, so that we would seem like an enormous group. We had to drag out the mothers that wait around, sitting comfortably at home in their living rooms, and convince them to join us.

The husbands came along that day, but as usual they kept to themselves, a bit apart. All they needed was to be seen in that crowd and be identified

with their wives' white heads. The diapers shone in the sun in this sea of walking people. We started gathering, first two, then three; up ahead another small group united. The diapers multiplied, their whiteness calling out to the other marchers who could no longer consider us a coincidence. By the time we reached the plaza in front of the church we were a strong group. The first fifteen or twenty gathered in a circle to recite the rosary, for the children who weren't there. The missing children: the word *desaparecidos*, "those who disappeared," emerged naturally. What does that mean, someone asked. The term is explicit, anyone who wanted to would understand it. The word was repeated, multiplied by our voices, included in the "Hail Mary's" of the rosary, laden with horror. First one woman, then another, came over to talk to us. Their children were also *desaparecidos;* they wanted to meet with us again, to pray with us even without a diaper. The people watched, listening to the rosary in the plaza while in the church before the altar people were receiving communion or praying for world peace. In one way or another, we were the horrible worm that had crawled out of a brilliant Argentina "that advances." Where is it advancing? Toward those unmarked graves that have no crosses, toward the bottom of the sea.

My father and Toto used to sell wine in Lujan after work around that time, another of their little projects. A week after the march, all of Lujan was still talking about the women "with the white handkerchiefs." I told Azucena and the others. We decided that it identified us, made us visible, was a powerful sign. But we would have to stop using the muslin diapers; they were wearing out with use, and we wanted to save them. Using only half a yard of batiste, we could make handkerchiefs.

•   •   •

For those women who had never before gone out to fight in the streets, those of us who took care of our family's needs but only heard about the functioning of the world's rules and regulations secondhand, it was tough to get used to the official formalities, and sometimes it even seemed perverse. From Dique to Rome there was not only a distance of miles; it seemed truly as far removed from our reality as the moon. Changing money, managing the language, speaking with important people, these were all things that we faced naturally, and it was only after doing them that we became aware of our own learning process. And I was learning, but without the mystifying gaze of a student. Underneath this learning's lasting discovery lay our motive: our missing children.

However, our ease often clashed with much less naive, established rules. For example, our first petition, Mother's Day of 1977, was handwritten on a piece of notebook paper. Each of us brought along our name written on another piece of paper as a signature. The newspaper people looked at us, confused. "Well, then, lend us a typewriter," we said, but it seemed

unimaginable. Where were we going to get a typewriter? We had to alphabetize the signatures, list them all together on a page, and bring them back. *La Nación, La Prensa, El Día de La Plata* were not innocent. While all we wanted was our children's names to come out in the petition, they demanded, for mysterious motives, the mothers' identities. "We're not going to list the names of people who aren't here . . ." Their collaboration with the powers above was clear and obvious. Each one of our names, in order to come out in print in the petition, had to be accompanied by a residence certificate; thus, we each had to go to the police station where we would have to reveal, "I need a certificate for a petition." The information cycle worked without a hitch from one office to the next. They never needed to look for us, they already had us perfectly individualized and located. The worst of it is that at first we hardly even noticed.

The only way to counteract this pervasive system was to become a detective oneself. We followed the kidnappers, found the spies, identified police stations and concentration camps, listened in on conversations, even located our children's stolen cars and stereos. Like hounds, each one of us traced our children's lost steps until the very last.

Notes

1. Yacimientos Petrolíferos Fiscales (YPF) is the state-owned oil company.—Ed.
2. A mantilla is a piece of cloth (usually embroidered) that women wear as a head covering in church.—Ed.

María Borjas
& Rosario Antúnez

THE FOLLOWING INTERVIEWS, recorded and edited by Adriana Angel and Fiona Macintosh, form part of the growing corpus of texts that register personal experiences where social and collective meaning can be read. The different records of a mother and a daughter from Ocotal, province of Segovia, in Nicaragua near the border with Honduras, provide the basis for a comparison between lives that belong to two distinct periods of Nicaraguan history. Peasants and landowners, along with the "gringos," were the actors of a social drama of poverty and exploitation under the Somoza regime. When, in 1961, the Sandinista Front began organizing in the country, women like Rosario Antúnez (who directs the school of adult education in Estelí and is the Ocotal FSLN representative to the National Assembly) made their first and dangerous experiences in clandestine political work and guerrilla organization. By 1979 the Sandinistas overthrew the Somoza regime and began the reconstruction of the nation on new economic and social bases.

As militants, public officers, and organizers, women form part of a difficult process of change that affects not only the political sphere but longstanding habits and cultural configurations. The armed struggle against Somoza and the political training that many women underwent could not entirely eliminate traditional ideological patterns that have to be dealt with today through open debate, practical policies, education, and self-empowerment.

Mother . . .

I was born on March 31, 1901. I'm every bit as old as the century. I didn't have any schooling. It's the wear and tear of life and the passing of the years that have given me a few notions.

"Mother and Daughter in Nicaragua," from The Tiger's Milk: Women of Nicaragua. *Eds. Adriana Angel and Fiona Macintosh (New York: Henry Holt, 1987).*

I've always felt a lot for Sandino. My family was Liberal, so I grew up with that way of thinking. In those days in Ocotal there were few like us; people were mostly Conservatives. The Liberals eventually won their revolution when Moncada became president in 1929. It was around that time, in 1927, that Sandino appeared. He occupied the San Albino mine this side of El Jícaro and rose up in arms. By that time the old Somoza, father of Somoza Garcías, was already serving in the army.

Sandino passed by twice going from Jícaro, recruiting men. I used to see him because I lived off the turning from Ocotal in a place called El Deslizadero. The people were always saying, "Sandino is coming," but you never believed them until he finally did. He launched an attack from there on 16 July, and that morning I saw him from very nearby. Sandino's troops swarmed in on horseback. He killed a lot of marines; it was a massacre, but Sandino's men suffered more than the gringos. At dawn they were still fighting, and continued until three in the afternoon. Then Sandino withdrew. This was when I saw planes for the first time. They came in the morning and then later in the afternoon. And since then the planes have stayed.

But in the end the gringos defeated the poor fellow. He did a foolhardy thing: He went to make friends with the president, and that's how they murdered him. Somoza, the father of those crooks, was the one who killed him.

It was horrible when the Americans came to Nicaragua. It was then that prostitution sprang up. The gringos would kick the doors down and break in to rape us, so we had to arm ourselves with machetes and knives, whatever we could lay our hands on. That frightened them. But the gringos remained here, controlling the government. The government and the National Guard, along with the Somoza family, owned all the bars and the flesh of these poor prostitutes. When the gringos left, Somoza's men took over, licensing the bars and the women. It was a terrible time with drunken naked women getting up to all sorts of things in the streets. The Guard used to kill people, kill anyone at the drop of a hat. And it never made the news. They weren't punished because they were the Guard. Nobody talked about Sandino, nobody was allowed to, because they'd blow your head off if you did. You know, it was the Americans who cut up their victims.

The landowners took the peasants' lands and even their huts, everything. At that time my mother fled to Honduras, but I stayed behind in Ocotal. I was living alone, looking after my daughters. My father had died when I was sixteen. I worked for a rich family, grinding coffee, ironing, and making maize drinks and tortillas. Clothes and hardware were brought up to Ocotal on mules from León every six months. We used to save our pennies to buy the closely woven cloth. Everything was transported, including sugar and rice.

I was forty-seven when I met my second husband. I had nine children, but three of them died. The last two, María and Rosario, were by my late husband. When we had María, he said we should marry. Then when we had Rosario we separated, but we ended up together again, working to get by. We married until God separated us. He was a good father and husband. I breastfed all the children until they were fifteen to eighteen months old and by the time I had my last one I was ashamed to show my breasts.

. . . and Daughter in Nicaragua

I believe there are two main reasons why people become revolutionaries. One is when all your needs are answered for so you can look beyond yourself toward those less fortunate. The other reason is when you're brought up in abject poverty. To change this you realize that society has to change too, since your suffering is tied to that of your class. And that's what happened to me.

I was brought up on farms. Father came from Honduras. He'd fought the Honduran dictatorship, and was forced into exile. Although I think of him as an intellectual, when he came to Nicaragua he took what work he could find—as a farm laborer. But it wasn't long before he became the farm manager because he was honest, hardworking. But we were always very poor. I never had more than one dress, I remember; when mother washed it I had to wait naked at home until it was dry. Mother cooked for the seasonal coffee workers. The living conditions were terrible, and when a worker fell ill he'd be fired immediately. It would come as no surprise to hear that the poor man had dropped dead on the road. Of course the boss kept no medicines.

Then, after six years, the landlord sent Father to work on his other farm near Ocotal, where I began my primary education. I was older than my classmates but quickly caught up and did well in all my exams—so much so that I won a scholarship to go on to secondary school and study teaching in Estelí. I was there when the National Guard attacked the León students and a national students' strike was called in protest. Students from our college joined in. Father disapproved of me joining in and insisted that I return to my classes, but I refused to do as he said. Oddly enough, he was the one who'd put these rebellious ideas into my head with all his talk about justice.

Father did have conservative views where I was concerned and didn't want me to go and teach in Jalapa, up in the mountains. There was a great need for teachers in the rural areas, so I was determined to go, even though I dearly loved and respected my parents.

I had been teaching there a year when I met my husband. I had a girl by him but it wasn't long before we separated. He was so bossy. When I look back on why he was like that, I think it was really a product of the repressive atmosphere we lived in.

When I came back to Ocotal I found another teaching job and became involved in the teachers' union. I remember the textbooks I had to use at school were full of dreadful rubbish, like: "Thanks to our mother country, Spain, Nicaraguans now have a language." I was very concerned that my pupils see beyond this type of distortion so used to ask questions like, "What did the conquistadors bring but domination, syphilis, and poverty?" Well, you can imagine it wasn't long before the headmaster had me in his bad books! At times I used to feel isolated. But everything changed in 1974 after a friend visited me. He told me for the first time that he was a Sandinista. I can still remember our conversation: He asked me if I would join as well, and I said, "Well, why on earth didn't you ask me before? I'd have joined years ago."

With that sorted out, he said, "Then I will arrange for someone else to come and talk with you. It'd be best to make contact tonight after Mass when we leave the church. He'll pick us up there, by the shrine of the Virgin Guadalupe. But don't look for him, just keep talking to me." And all this happened to plan. The contact followed us back to my home.

When the man entered the living room I was overcome by such emotion. It was as though all my hopes and fears were coming true: the chance to join the struggle. I felt sad happiness—I imagined it was like falling deeply in love. The man introduced himself as Roberto. He said, "You must be María." This was to be my new name, and Roberto was really Carlos Manuel Morales. Today he runs the local government in Estelí.

That night he explained a lot, and he said, "There are two golden rules that you must never forget: Never ask for information, and never repeat to anyone what you've been told." My contact was to be a man called Octavio, pseudonym Alejandro, who would soon come with a woman called Ruth who'd work with me. Then he asked if I had friends who might collaborate. Before leaving he looked over the house in great detail.

I felt then that I'd found what I'd been looking for. As a teacher I'd felt so angry and sad to see my pupils arrive at school faint with hunger. Here now was the chance to end the people's humiliation. My fears dissolved; I felt totally committed. I didn't think I might die. Or worse, that my daughter would give her life for the revolution.

Ruth soon arrived. She's a wonderful person. Today we know her as Commander Monica Baltadano,[1] and we became very close friends. Our work was to build up the Sandinista Front in the region. My time was divided between the children and work. I became a messenger, recruited trusted friends, and found safe houses and transport. We would wake at

five to study philosophy and every night did physical training and checked out the movements of the Guard and State Security spies. I'd talk with market sellers, to people I knew, to tell them they too could collaborate if only they'd forget their fear. Gradually people did begin to join. My nephew and niece's husband and three neighbours joined our cell, and slowly the Ocotal network grew.

My home became like a crossroads. I'd be going to the bathroom in the middle of the night and suddenly bump into someone. It'd scare me rigid, but still I couldn't ask who they were.

Then someone very special to me came and stayed. He was very ill after an operation, but I was told that once he'd recovered he would be giving me political and military classes. We began by healing his wounds. I made sure he ate well. I knew the hardships endured by these *compas* in the mountains. He was so thin and tense; when he smoked you could see a nervous twitch in his throat. When the political classes finally started he had quite a time as I was always cracking jokes. There was no telling me off because that's the way I am. This man is Commander Omar Cabezas.[2] I adored these *compañeros;* they were my brothers. Another man came to stay who'd never sit still and was always laughing. He too was nervous, yet brave and intelligent, with an air of authority. We called him Oscar, but today he's Commander Bayardo Arce.[3]

At this time my nephew was doing military training at the Macuelizo Sandinista training camp when an informer revealed its whereabouts to the Guard. They attacked on August 27, but my nephew escaped. The Guard also discovered another center in El Sauce. Throughout the region there was dreadful repression, in Matagalpa, Jinotega, and Zelaya. It was at its worst in the Nueva Segovia and Chinandega regions. Peasants were captured and badly tortured, and this caused serious damage to the Sandinista structures. Although the cells were fundamentally strong, the weaker people cracked under this torture, and those informed on were caught.

Monica, Bayardo, and my nephew and I met with Pedro Arauz[4] to discuss what to do. It was decided that Bayardo and my nephew should leave the area immediately. I felt bad that I couldn't tell my sister where her son had gone; but in the end I didn't know either, because they didn't manage their getaway. Eight Guards stopped them on the Ocotal Bridge. The driver, a collaborator, was ordered out of the truck. He was so nervous that he took out the ignition keys, so the others couldn't speed off. When they searched them and found Bayardo's pistol, Bayardo grabbed the Guard's rifle and fired, and my nephew and Bayardo escaped. The two of them against eight Guards. But the collaborator was caught, and you can imagine what they did to that poor man.

Soon the Guard picked me up and took me in for interrogation. They really fired the questions about my nephew and the Sandinistas, and my

mind worked at a rate of knots. For hours, blindfolded, I somehow kept control of the situation. Eventually they let me go. But a month later I was arrested again and this time interrogated in the Guards' barracks. The colonel gave me hell, but I managed to stay calm and decided to try my luck. "I've committed no crime, so if you'll excuse me," I said, and got up and walked out. The man could easily have put a bullet through me, but he let me go. Well, he did get me in the end because I was sentenced to five years' imprisonment. This time my spirit felt broken. I couldn't face leaving my children and parents.

I was put into solitary confinement and after three days the attorney tried bribing me by saying my mother was critically ill in hospital: "If you tell us where your nephew is you can visit your mother." But it didn't work. They accused me of believing in communism, and I'd say, "I just believe in a world Christ dreamed of."

At four o'clock the next morning I was sent to prison. I was taken in an open truck with some other badly beaten-up prisoners through the pouring rain. At prison I met the Sandinista *compañeros* who'd carried out the assault on the Chema Castillo house, the residence of Somoza's minister of agriculture. This was in 1974 when they'd held a sixty-hour siege on a party there attended by the U.S. ambassador, Somoza's brother-in-law, and others associated with the dictator. Through this one act the Sandinistas had reestablished their political and military presence, and obtained the release of eighteen leading Sandinista prisoners.

Luckily I hadn't been long in prison when a lawyer was asked to investigate my case and discovered that in fact my release papers had been signed by the attorney, but because the colonel had chosen to ignore it I'd been sent down. But the nightmare wasn't over, for one month after my release I was held again for thirteen days. After this I felt I had to leave the country. But I worried for the children. I spent hours pacing the floor before finally making the decision that no, I'd stay on fighting for them and the struggle. But it wasn't easy; I had to move from one house to another: I was being watched and anyone who made contact would have been under suspicion. I kept my head low for a year. Still I managed to find work in a foreign-owned timber company and eventually gained the boss's respect enough to take over the running of the company store, and that way I managed to send tinned food out to the guerrillas.

After a year, in 1977, Monica reappeared and carefully we renewed the work of recruiting people, finding money, houses, ammunition, and medical supplies. In spite of all the constraints, we continued with the political education of our collaborators to explain the Sandinistas' objectives, the FSLN plan of action for a health service, education, agrarian reform, and much more.

It was a tense time. The timber company pulled out, leaving me penniless, and my youngest child was ill and needed an operation. My eldest daughter had joined the guerrillas, Front members were being captured, and it looked as if I would have to go. I felt at the breaking point, but kept going on my last reserves of energy.

By April 1979 the second insurrection was building up, and I received orders to leave the country immediately. I packed the girls' clothes and bottles, tins of powdered milk, and a few medicines, and with my baby still poorly after four operations, we left by bus for Honduras.

It was here that the final blow came. The last I'd heard of my eldest daughter was that she was fine in Costa Rica, and the Sandinista *compañeros* there were very proud of her. She had shown great maturity and, despite her age, had no time for admirers. But on arrival in Honduras I was told that she had entered Nicaragua and been killed in combat on May 17. I heard the news twenty-two days after her death; I was never to see her again. I didn't even have the consolation of having her body near me.

We stayed in Honduras looking after the Sandinistas' children. Every day we'd listen in to the radio to hear how the war was progressing in the south, southeast, and west. Victory was near. I so wished I'd been there in Nicaragua for July 19—I'd so longed for this moment. But I couldn't leave the children. I returned four days later to join my family, and my nephew had come home, too. It was a great day. There was a lot to be done, and at seven the next morning I started work, and to this day I haven't stopped!

When I think back—it's been such a terrible struggle . . . 50,000 killed in the battle to overthrow Somoza. But I suppose it was inevitable, as the most powerful nation, the United States, was supporting Somoza. Since the triumph, we've had to face not only natural disasters but also this aggression. Still, we are managing to reconstruct the country. Of course mistakes are made, but no one can deny our achievements. Only by closing your eyes can you shut out the sun.

After the triumph, my first job was to reassess education. I did this by bringing together teachers and parents in Ocotal to find out what they wanted for their children. Before, school authorities never consulted parents. It was nonstop work, helping in the Sandinista Workers' Union and talking with local women about childcare, trying to encourage them to send their children to school and not force them to work. I'd also go on walks with the teachers; we'd make a fire—it was great—and we'd talk about everything. All the plans we made! I knew this was a big break from their daily routine. I wanted them to see the teaching work as relevant to their own development.

But it was when I moved to Estelí to set up the adult education school that my greatest dream came true, to teach people who for centuries had been denied education. They're the people who strive to improve society

and have the moral authority to lead, whether in cooperatives, villages, Christian groups, or unions. But still the number of women who come are too few. I consider us women to be the backbone to society, and in time we'll cause the greatest changes to macho social values. We really are working toward a united and active society where both men and women will fight to overcome all forms of domination, from individual selfishness through to social systems of exploitation. It's a broad struggle, and not one aimed against men.

Through the revolution and this war you can see how women's strength has grown and gained recognition. We women do confront greater hardships, and this creates revolutionary consciousness faster than it does in men. After the triumph, women began by working to improve our communities and workplaces, but now as we organize as tractor drivers, union reps, as mothers and teachers, our militancy grows to demand our full equality. And this is taken up by the state in the form of the new constitution and legal changes to end discrimination. This is part of my work in the National Assembly. I am there to represent women, to defend our rights.

But we have a long way to go before reaching total freedom. The FSLN revolution created a space for women and all our oppressed people to strive for a full life, and that effort is bound to transform the home and society. In essence, the revolution is a love for life. Without this we lose the means to develop. This love is the energy of the revolution and demands from both men and women. Yet it is the men who must learn to give more if we are to reach this equality and free society.

For instance, there are couples who are very active in the revolution but who leave their children neglected, or, more commonly, men who will not allow their wives to participate. I remember the time when a friend asked me what she should do about her domineering husband. She couldn't attend the neighborhood meetings because her husband thought she would pick up another man there. So I related this story my religious godmother had told me. It's a fable.

She said that once a woman said, "My husband treated me so badly, he was so jealous he never took his eyes off me. No man could come near me. You see, I was very beautiful and all the men would fall in love with me. Of course I would take no notice of them, but still he'd give me no freedom. So what was I to do?"

One day after church she managed to talk to the priest and explain her predicament. He told her, "The only way your husband is going to change is if he drinks the milk of a tiger."

"But how on earth am I going to do this?" she exclaimed. "A wild fierce tiger with cubs would kill me for sure."

He said, "Well, I don't know, but this will change your husband's ways. You'll make a new man of him."

Wondering about this, she left the church. It wasn't long before the woman spied a tiger coming down to the river that ran through their farm. She came to drink the water with her cubs, and rested there every day. The woman began to leave meat on the river bank and as the days passed she crept nearer and nearer until she'd enough confidence to stroke the animal as it ate the meat. She then filled the bowl with the tiger's milk.

That evening, full of expectations, she gave her husband the milk and waited for the transformation. To her horror there was no change in him. He was just the same old beast.

She rushed back to the priest to ask him what could have gone wrong.

"But did you give your husband the milk?"

"Yes, I did."

"And he's just the same?"

"Yes, just the same," she said.

There was a pause. Then the old man said, "My dear, don't you see how blind you are? Just think, you have the resources and courage to get close to the wildest beast that easily could have killed you, even so close as to win its confidence and take its milk from the cubs, yet you cannot tame your husband to make him docile and good. If you have the strength and cunning and inner confidence to charm an irrational beast, then you certainly can get close enough to change any man you wish."

Notes

1. Monica Baltadano became involved in the FSLN in 1970 through her work in the Youth Christian Movement. In 1977 she was taken prisoner in Matagalpa but was released through popular pressure in 1978. As regional commander, she directed the final offensive in the taking of Granada. Following the triumph, she worked as the mass organizations' coordinator to become secretary of Nicaraguan regional affairs in 1982, and in 1984 she became undersecretary to the president's minister, René Nuñez.

2. Commander Omar Cabezas, in his book *Fire from Mountains,* gives a vivid portrayal of the harsh life he endured as a guerrilla leader from 1974 to 1979, when he helped to establish the Front's school in Estelí. Following the triumph, he was responsible for university education and today is political head in the Nicaraguan Home Office.

3. From 1969 Commander Bayardo Arce began work inside the FSLN to take responsibility for the Front's political/military structures in the Estelí and Matagalpa regions from 1974 to 1977, and to establish its school. He became part of the FSLN leadership in 1977, and during the final offensive commanded the Managua, Masaya, Carazo, and Granada regions. Today he is vice coordinator of the FSLN.

4. Pedro Arauz Federico was one of the Front's key organizers and teachers. He was killed by the National Guard in Masaya in 1977.

Carolina Maria de Jesus

In 1952 the Brazilian journalist Audalio Dantas visited the *favela* of Can-
indé, where he met Carolina Maria de Jesus, an extremely poor woman
who lived there with her children. A single mother, Carolina de Jesus made
a living, like thousands of marginals, through paid housework and the
collection of "useful" items from garbage dumps and industrial leftovers.
Carolina, born in 1913 in Minas Gerais, had learned to read and write.
After she moved to the *favela,* she began to compose short stories and
poems, which she kept to herself, and to register her daily experience,
mainly the painful strategies of survival by culling food from the dump.
Her diaries are the obsessive record of uninterrupted hunger, pain, and
humiliation, but Carolina de Jesus thought that her writing helped her to
preserve a rest of human dignity and personal independence in the world
of the *favela.* Audalio Dantas managed to gain her confidence, read the
diary, and convinced her to trust him with the twenty-six notebooks she
had filled. Published first in a national newspaper and then as a book,
Carolina de Jesus's diary, *Quarto de Despejo* (1960) was an immediate best-
seller. Ironically, the record of poverty she had kept became the bridge
through which she and her children moved from the *favela* to a brick house
and began a new life. This success story, no doubt an exception, produced
a second book, *Casa de Alvenaria* (1961), in which Carolina de Jesus
registered her experience in the new social world she had entered through
the unexpected path of a symbolic practice that was generally thought alien
to the urban and illiterate poor. She died in 1977. A third testimonial
volume, *Diário de Bitata,* was published posthumously in 1986.

Writing and Survival

July 17, 1955

A marvelous day. The sky was blue without one cloud. The sun was warm.
I got out of bed at 6:30 and went to get water. I only had one piece of

bread and three cruzeiros. I gave a small piece to each child and put the beans that I got yesterday from the Spiritist Center on the fire. Then I went to wash clothes. When I returned from the river the beans were cooked. The children asked for bread. I gave the three cruzeiros to João to go and buy some. Today it was Nair Mathias who started an argument with my children. Silvia and her husband have begun an open-air spectacle. He is hitting her and I'm disgusted because the children are present. They heard words of the lowest kind. Oh, if I could move from here to a more decent neighborhood!

I went to Dona Florela to ask for a piece of garlic. I went to Dona Analia and got exactly what I expected:

"I don't have any!"

I went to collect my clothes. Dona Aparecida asked me:

"Are you pregnant?"

"No, Senhora," I replied gently.

I cursed her under my breath. If I am pregnant it's not your business. I can't stand these *favela* women, they want to know everything. Their tongues are like chicken feet. Scratching at everything. The rumor is circulating that I am pregnant! If I am, I don't know about it!

I went out at night to look for paper. When I was passing the São Paulo football stadium many people were coming out. All of them were white and only one black. And the black started to insult me:

"Are you looking for paper, auntie? Watch your step, auntie dear!"

I was ill and wanted to lie down, but I went on. I met several friends and stopped to talk to them. When I was going up Tiradentes Avenue I met some women. One of them asked me:

"Are your legs healed?"

After I was operated on, I got better, thanks to God. I could even dance at Carnival in my feather costume. Dr. José Torres Netto was who operated on me. A good doctor. And we spoke of politics. When a woman asked me what I thought of Carlos Lacerda, I replied truthfully:

"He is very intelligent, but he doesn't have an education. He is a slum politician. He likes intrigues, to agitate."

One woman said it was a pity, that the bullet that got the major didn't get Carlos Lacerda.

"But his day . . . it's coming," commented another.

Many people had gathered and I was the center of attention. I was embarrassed because I was looking for paper and dressed in rags. I didn't want to talk to anyone, because I had to collect paper. I needed the money. There was none in the house to buy bread. I worked until 11:30. When I returned it was midnight. I warmed up some food, gave some to Vera

Eunice, ate, and lay down. When I awoke the rays of the sun were coming through the gaps of the shack.

July 18, 1955

I got up at 7. Happy and content. Weariness would be here soon enough. I went to the junk dealer and received 60 cruzeiros. I passed by Arnaldo, bought bread, milk, paid what I owed him, and still had enough to buy Vera some chocolate. I return to a Hell. I opened the door and threw the children outside. Dona Rosa, as soon as she saw my boy José Carlos, started to fight with him. She didn't want the boy to come near her shack. She ran out with a stick to hit him. A woman of 48 years fighting with a child! At times, after I leave, she comes to my windows and throws a filled chamber pot onto the children. When I return I find the pillows dirty and the children fetid. She hates me. She says that the handsome and distinguished men prefer me and that I make more money than she does.

Dona Cecilia appeared. She came to punish my children. I threw a right at her and she stepped back. I told her:

"There are women that say they know how to raise children, but some have children in jails listed as delinquents."

She went away. Then came that bitch Angel Mary. I said:

"I was fighting with the banknotes, now the small change is arriving. I don't go to anybody's door, and you people who come to my door only bore me. I never bother anyone's children or come to your shack shouting against your kids. And don't think that yours are saints; it's just that I tolerate them."

Dona Silvia came to complain about my children. That they were badly educated. I don't look for defects in children. Neither in mine nor in others. I know that a child is not born with sense. When I speak with a child I use pleasant words. What infuriates me is that the parents come to my door to disrupt my rare moments of inner tranquility. But when they upset me, I write. I know how to dominate my impulses. I only had two years of schooling, but I got enough to form my character. The only thing that does not exist in the *favela* is friendship.

·   ·   ·

May 10, 1958

I went to the police station and talked to the lieutenant. What a pleasant man! If I had known he was going to be so pleasant, I'd have gone on the first summons. The lieutenant was interested in my boys' education. He said the *favelas* have an unhealthy atmosphere where the people have more chance to go wrong than to become useful to state and country. I thought:

if he knows this why doesn't he make a report and send it to the politicians? To Janio Quadros, Kubitschek,[1] and Dr. Adhemar de Barros? Now he tells me this, I, a poor garbage collector. I can't even solve my own problems.

Brazil needs to be led by a person who has known hunger. Hunger is also a teacher.

Who has gone hungry learns to think of the future and of the children.

May 11

Today is Mother's Day. The sky is blue and white. It seems that even nature wants to pay homage to the mothers who feel unhappy because they can't realize the desires of their children.

The sun keeps climbing. Today it's not going to rain. Today is our day.

Dona Teresinha came to visit me. She gave me 15 cruzeiros and said it was for Vera to go to the circus. But I'm going to use the money to buy bread tomorrow because I only have four cruzeiros.

Yesterday I got half a pig's head at the slaughterhouse. We ate the meat and saved the bones. Today I put the bones on to boil and into the broth I put some potatoes. My children are always hungry. When they are starving they aren't so fussy about what they eat.

Night came. The stars are hidden. The shack is filled with mosquitoes. I lit a page from the newspaper and ran it over the walls. This is the way the *favela* dwellers kill mosquitoes.

May 13

At dawn it was raining. Today is a nice day for me, it's the anniversary of the Abolition. The day we celebrate the freeing of the slaves. In the jails the Negroes were the scapegoats. But now the whites are more educated and don't treat us any more with contempt. May God enlighten the whites so that the Negroes may have a happier life.

It continued to rain and I only have beans and salt. The rain is strong but even so I sent the boys to school. I'm writing until the rain goes away so I can go to Senhor Manuel and sell scrap. With that money I'm going to buy rice and sausage. The rain has stopped for a while. I'm going out.

I feel so sorry for my children. When they see the things to eat that I come home with they shout:

"*Viva Mama!*"

Their outbursts please me. But I've lost the habit of smiling. Ten minutes later they want more food. I sent João to ask Dona Ida for a little pork fat. She didn't have any. I sent her a note:

---

1. Juscelino Kubitschek: President of Brazil from 1956 to 1961.

"Dona Ida, I beg you to help me get a little pork fat, so I can make soup for the children. Today it's raining and I can't go looking for paper. Thank you, Carolina."

It rained and got colder. Winter had arrived and in winter people eat more. Vera asked for food, and I didn't have any. It was the same old show. I had two cruzeiros and wanted to buy a little flour to make a *virado*.[2] I went to ask Dona Alice for a little pork. She gave me pork and rice. It was 9 at night when we ate.

And that is the way on May 13, 1958, I fought against the real slavery—hunger!

May 15

On the nights they have a party they don't let anybody sleep. The neighbors in the brick houses near by have signed a petition to get rid of the *favelados*. But they won't get their way. The neighbors in the brick houses say:

"The politicians protect the *favelados*."

Who protects us are the public and the Order of St. Vincent Church. The politicians only show up here during election campaigns. Senhor Candido Sampaio, when he was city councilman in 1953, spent his Sundays here in the *favela*. He was so nice. He drank our coffee, drinking right out of our cups. He made us laugh with his jokes. He played with our children. He left a good impression here and when he was candidate for state deputy, he won. But the Chamber of Deputies didn't do one thing for the *favelados*. He doesn't visit us any more.

I classify São Paulo this way: The Governor's Palace is the living room. The mayor's office is the dining room and the city is the garden. And the *favela* is the backyard where they throw the garbage.

The night is warm. The sky is peppered with stars. I have the crazy desire to cut a piece of the sky to make a dress.

·   ·   ·

May 16

I awoke upset. I wanted to stay at home but didn't have anything to eat. I'm not going to eat because there is very little bread. I wonder if I'm the only one who leads this kind of life. What can I hope for the future? I wonder if the poor of other countries suffer like the poor of Brazil. I was so unhappy that I started to fight without reason with my boy José Carlos.

---

2. *Virado:* a dish of black beans, manioc flour, pork, and eggs.

A truck came to the *favela*. The driver and his helper threw away some cans. It was canned sausage. I thought: this is what these hardhearted businessmen do. They stay waiting for the prices to go up so they can earn more. And when it rots they throw it to the buzzards and the unhappy *favelados*.

There wasn't any fighting. Even I found it dull. I watched the children open the cans of sausages and exclaim:

"Ummm! Delicious!"

Dona Alice gave me one to try, but the can was bulging. It was rotten.

·   ·   ·

May 19

I left the bed at 5 a.m. The sparrows have just begun their morning symphony. The birds must be happier than we are. Perhaps happiness and equality reigns among them. The world of the birds must be better than that of the *favelados,* who lie down but don't sleep because they go to bed hungry.

What our President Senhor Juscelino has in his favor is his voice. He sings like a bird and his voice is pleasant to the ears. And now the bird is living in a golden cage called Catete Palace. Be careful, little bird, that you don't lose this cage, because cats when they are hungry think of birds in cages. The *favelados* are the cats, and they are hungry.

I broke my train of thought when I heard the voice of the baker:

"Here you go! Fresh bread, and right on time for breakfast!"

How little he knows that in the *favela* there are only a few who have breakfast. The *favelados* eat only when they have something to eat. All the families who live in the *favela* have children. A Spanish woman lives here named Dona Maria Puerta. She bought some land and started to economize so she could build a house. When she finished construction her children were weak with pneumonia. And there are eight children.

There have been people who visited us and said:

"Only pigs could live in a place like this. This is the pigsty of São Paulo."

I'm starting to lose my interest in life. It's beginning to revolt me and my revulsion is just.

I washed the floor because I'm expecting a visit from a future deputy and he wants me to make some speeches for him. He says he wants to know the *favelas* and if he is elected he's going to abolish them.

The sky was the color of indigo, and I understood that I adore my Brazil. My glance went over to the trees that are planted at the beginning of Pedro Vicente Street. The leaves moved by themselves. I thought: they are applauding my gesture of love to my country. I went on looking for paper. Vera was smiling and I thought of Casemiro de Abreu, the Brazilian poet,

who said: "Laugh, child. Life is beautiful." Life was good in that era. Because now in this era it's necessary to say: "Cry, child. Life is bitter."

•   •   •

May 21, 1958

I spent a horrible night. I dreamt I lived in a decent house that had a bathroom, kitchen, pantry, and even a maid's room. I was going to celebrate the birthday of my daughter, Vera Eunice. I went and bought some small pots that I had wanted for a long time. Because I was able to buy. I sat at the table to eat. The tablecloth was white as a lily. I ate a steak, bread and butter, fried potatoes, and a salad. When I reached for another steak I woke up. What bitter reality! I don't live in the city. I live in the *favela*. In the mud on the banks of the Tieté River. And with only nine cruzeiros. I don't even have sugar, because yesterday after I went out the children ate what little I had.

Who must be a leader is he who has the ability. He who has pity and friendship for the people. Those who govern our country are those who have money, who don't know what hunger is, or pain or poverty. If the majority revolt, what can the minority do? I am on the side of the poor, who are an arm. An undernourished arm. We must free the country of the profiteering politicians.

Yesterday I ate that macaroni from the garbage with fear of death, because in 1953 I sold scrap over there in Zinho. There was a pretty little black boy. He also went to sell scrap in Zinho. He was young and said that those who should look for paper were the old. One day I was collecting scrap when I stopped at Bom-Jardim Avenue. Someone had thrown meat into the garbage, and he was picking out the pieces. He told me:

"Take some, Carolina. It's still fit to eat."

He gave me some, and so as not to hurt his feelings, I accepted. I tried to convince him not to eat that meat, or the hard bread gnawed by the rats. He told me no, because it was two days since he had eaten. He made a fire and roasted the meat. His hunger was so great that he couldn't wait for the meat to cook. He heated it and ate. So as not to remember that scene, I left thinking: I'm going to pretend I wasn't there. This can't be real in a rich country like mine. I was disgusted with that Social Service that had been created to readjust the maladjusted but took no notice of marginal people. I sold the scrap at Zinho and returned to São Paulo's backyard, the *favela*.

The next day I found that little black boy dead. His toes were spread apart. The space must have been eight inches between them. He had blown up as if made out of rubber. His toes looked like a fan. He had no documents. He was buried like any other "Joe." Nobody tried to find out his name. The marginal people don't have names.

Once every four years the politicians change without solving the problem of hunger that has its headquarters in the *favela* and its branch offices in the workers' homes.

When I went to get water I saw a poor woman collapse near the pump because last night she slept without dinner. She was undernourished. The doctors that we have in politics know this.

Now I'm going to Dona Julita's house to work for her. I went looking for paper. Senhor Samuel weighed it. I got 12 cruzeiros. I went up Tiradentes Avenue looking for paper. I came to Brother Antonio Santana de Galvão Street, number 17, to work for Dona Julita. She told me not to fool with men because I might have another baby and that afterward men won't give anything to take care of the child. I smiled and thought: In relations with men, I've had some bitter experiences. Now I'm mature, reached a stage of life where my judgment has grown roots.

I found a sweet potato and a carrot in the garbage. When I got back to the *favela* my boys were gnawing on a piece of hard bread. I thought: for them to eat this bread, they need electric teeth.

I don't have any lard. I put meat on the fire with some tomatoes that I found at the Peixe canning factory. I put in the carrot and the sweet potato and water. As soon as it was boiling, I put in the macaroni that the boys found in the garbage. The *favelados* are the few who are convinced that in order to live, they must imitate the vultures. I don't see any help from the Social Service regarding the *favelados*. Tomorrow I'm not going to have bread. I'm going to cook a sweet potato.

June 16, 1958

José Carlos is feeling better. I gave him a garlic enema and some hortelá tea. I scoff at women's medicine but I had to give it to him because actually you've got to arrange things the best you can. Due to the cost of living we have to return to the primitive, wash in tubs, cook with wood.

I wrote plays and showed them to directors of circuses. They told me: "It's a shame you're black."

They were forgetting that I adore my black skin and my kinky hair. The Negro hair is more educated than the white man's hair. Because with Negro hair, where you put it, it stays. It's obedient. The hair of the white, just give one quick movement, and it's out of place. It won't obey. If reincarnation exists I want to come back black.

One day a white told me:

"If the blacks had arrived on earth after the whites, then the whites would have complained and rightly so. But neither the white nor the black knows its origin."

The white man says he is superior. But what superiority does he show? If the Negro drinks *pinga*,[3] the white drinks. The sickness that hits the black hits the white. If the white feels hunger, so does the Negro. Nature hasn't picked any favorites.

September 20, 1958

I went to the store and took 44 cruzeiros with me. I bought a kilo of sugar, one of beans, and two eggs. I had two cruzeiros left over. A woman who was shopping spent 43 cruzeiros. And Senhor Eduardo said: "As far as spending money goes, you two are equal."

I said:

"She's white. She's allowed to spend more."

And she said:

"Color is not important."

Then we started to talk about prejudice. She told me that in the United States they don't want Negroes in the schools.

I kept thinking: North Americans are considered the most civilized. And they have not yet realized that discriminating against the blacks is like trying to discriminate against the sun. Man cannot fight against the products of Nature. God made all the races at the same time. If he had created Negroes after the whites, the whites should have done something about it then.

December 8, 1958

In the morning the priest came to say Mass. Yesterday he came in the church car and told the *favelados* that they must have children. I thought: why is it that the poor have to have children—is it that the children of the poor have to be workers?

In my humble opinion those who should have children are the rich, who could give brick houses to their children. And they could eat what they wanted.

When the church car comes to the *favela*, then all sorts of arguments start about religion. The women said that the priest told them that they should have children and when they needed bread they could go to the church and get some.

For Senhor Priest, the children of the poor are raised only on bread. They don't wear clothes or need shoes.

---

3. *Pinga:* a white, fiery liquor made from sugar cane.

Elena Poniatowska

ELENA PONIATOWSKA is a well-known Mexican writer who has developed refined literary strategies in order to represent in her texts the unheard voices that echo in history and society. Her powerful novel *Hasta no verte, Jesús mío* recounts the life story of a *soldadera,* a woman who followed the army to provide food, drink, care, and other services to the soldiers during the Mexican Revolution. In *Querido Diego,* she imagines the feelings, hopes, and final despair of one of Diego de Rivera's mistresses, a painter who agonizes in poverty and loneliness in Paris after Diego has left for Mexico. In both books, Poniatowska has proven that new genres (that oscillate between fiction and nonfiction) can take up experiences originating in other social spaces or in the past. With *Massacre in Mexico,* she describes the political demonstrations and the cruel repression by both the army and the police that took place in 1968 in the Plaza de las Tres Culturas in Mexico City, resulting in hundreds of dead and wounded, among them students, women, and children. Poniatowska organizes the text according to the often fragmented discourse of people who were there, supplementing these with newspaper clippings and documents. Poniatowska pays keen attention to the voices of women, students, and political militants, giving them the verbal structure that allows them to persist. Through *Massacre in Mexico,* Poniatowska proves that a woman writing can cross the border of the private in order to place her practice in the public dimension, turning personal words and thoughts into a collective experience.

Voices from Tlatelolco

They are many. They come down Melchor Ocampo, the Reforma, Juárez, Cinco de Mayo, laughing, students walking arm in arm in the demonstration, in as festive a mood as if they were going to a street fair; carefree boys and girls who do not know that tomorrow, and the day after, their dead

For biographical information on Elena Poniatowska, see p. 80.

"*Voices from Tlatelolco,*" *from* Massacre in Mexico *by Elena Poniatowska. Translated by Helen R. Lane. Copyright © 1971 by Ediciones Era. English translation copyright © 1975 by The Viking Press. Used by permission of Viking Penguin, a division of Penguin Books USA, Inc.*

bodies will be lying swollen in the rain, after a fair where the guns in the shooting gallery are aimed at them, children-targets, wonder-struck children, children for whom every day is a holiday until the owner of the shooting gallery tells them to form a line, like the row of tin-plated mechanical ducks that move past exactly at eye level, click, click, click, "Ready, aim, fire!" and they tumble backward, touching the red satin backdrop.

The owner of the shooting gallery handed out rifles to the police, to the army, and ordered them to shoot, to hit the bull's-eye, and there the little tin-plated creatures were standing, open-mouthed with astonishment and wide-eyed with fear, staring into the rifle barrels. Fire! They fell, but this time there was no spring to set them up again for the next customer to shoot at; the mechanism was quite different at this fair; the little springs were not made of metal but of blood; thick, red blood that slowly formed little puddles, young blood trampled underfoot all over the Plaza de las Tres Culturas.

These youngsters are coming toward me now, hundreds of them; not one of them has his hands up, not one of them has his pants around his ankles as he is stripped naked to be searched; there are no sudden blows, no clubbings, no ill treatment, no vomiting after being tortured; they are breathing deeply, advancing slowly, surely, stubbornly; they come round the Plaza de las Tres Culturas and stop at the edge of the square, where there is a drop of eight or ten feet, with a view of the pre-Hispanic ruins below; they walk on toward me again, hundreds of them, advancing toward me with their hands holding up placards, little hands, because death makes them look like children's hands; they come toward me, row on row; though they are pale, they look happy; their faces are slightly blurred, but they are happy ones; there are no walls of bayonets driving them off now, no violence; I look at them through a curtain of raindrops, or perhaps a veil of tears, like the one at Tlatelolco; I cannot see their wounds, for fortunately there are no holes in their bodies, no bayonet gashes, no dum-dum bullets; they are blurred figures, but I can hear their voices, their footsteps, echoing as on the day of the Silent Demonstration; I will hear those advancing footsteps all the rest of my life; girls in miniskirts with their tanned young legs, teachers with no neckties, boys with sweaters knotted around their waists or their necks; they come toward me, laughing, there are hundreds of them, full of the crazy joy of walking together down this street, our street, to the Zócalo, our Zócalo; here they come; August 5, August 13, August 27, September 13; Father Jesús Pérez has set all the bells of the cathedral to ringing to welcome them, the entire Plaza de la Constitución is illuminated; there are bunches of *cempazúchitl* flowers everywhere, thousands of lighted candles; the youngsters are in the center of an orange, they are the brightest burst of fireworks of all. Was Mexico a sad country? I see

it as a happy one, a marvelously happy one; the youngsters are marching up Cinco de Mayo, Juárez, the Reforma, the applause is deafening, three hundred thousand people have come to join them, of their own free will, Melchor Ocampo, Las Lomas, they are climbing up through the forests to the mountaintops, *Mé-xi-co, Li-ber-tad, Mé-xi-co, Li-ber-tad, Mé-xi-co, Li-ber-tad, Mé-xi-co, Li-ber-tad, Mé-xi-co, Li-ber-tad.*

•   •   •

We shall probably never know what motive lay behind the Tlatelolco massacre. Terror? Insecurity? Anger? Fear of losing face? Ill-will toward youngsters who deliberately misbehave in front of visitors? Together with Abel Quezada,[1] we may ask ourselves, WHY? Despite all the voices that have been raised, despite all the eyewitness testimony, the tragic night of Tlatelolco is still incomprehensible. Why? The story of what happened at Tlatelolco is puzzling and full of contradictions. The one fact that is certain is that many died. Not one of the accounts provides an overall picture of what happened. All the people there in the plaza—casual bystanders and active participants alike—were forced to take shelter from the gunfire; many fell wounded. In an article entitled "A Meeting That Ended in Tragedy," which appeared in the Mexico City *Diario de la Tarde* on October 5, 1968, reporter José Luis Mejías wrote, "The men in white gloves drew their pistols and began indiscriminately firing at close range on women, children, students, and riot police. . . . And at that same moment a helicopter gave the signal for the army to close in by setting off a flare. . . . As the first shots were fired, General Hernández Toledo, the commander of the paratroopers, was wounded, and from that moment on, as the troops raked the crowd with a furious hail of bullets and pursued the sharpshooters as they fled inside the buildings, no one present was able to get an overall picture of what was happening. . . ." But the tragedy of Tlatelolco damaged Mexico much more seriously than is indicated in a news story entitled "Bloody Encounter in Tlatelolco," which appeared in *El Heraldo* on October 3, 1968, lamenting the harm done the country's reputation. "A few minutes after the fighting started in the Nonoalco area, the foreign correspondents and the journalists who had come to our country to cover the Olympic Games began sending out news bulletins informing the entire world of what was happening. Their reports—which in a number of cases were quite exaggerated—contained remarks that seriously endangered Mexico's prestige," the story read.

---

1. A famous Mexican editorial cartoonist, who after Tlatelolco drew a now celebrated cartoon in *Excélsior* with the caption WHY? (Translator's note.)

The wound is still fresh, and Mexicans, though stunned by this cruel blow, are beginning to ask themselves questions in open-mouthed amazement. The blood of hundreds of students, of men, women, children, soldiers, and oldsters tracked all over Tlatelolco has dried now. It has sunk once again into the quiet earth. Later, flowers will bloom among the ruins and the tombs.

I love love.
—Hippie button found in the Plaza de las Tres Culturas

I tugged at my brother's arm. "Julio, what's the matter?" I asked him. I tugged at his arm again; his eyes were half closed and there was a very sad look in them. And I heard him murmur the words "I think . . ."

My mind was a total blank. The tremendous crush of people screaming in panic made it hard for me to hear what he was saying. I thought later that if I'd known, if I'd realized that Julio was dying, I would have done something absolutely crazy right then and there.

Later some of the soldiers who had been shooting at the buildings around the plaza came over to us. The smell of gunpowder was unbearable. Little by little people made room for me so I could kneel down beside Julio.

"Julio, Julio, answer me, little brother," I said to him.

"He must be wounded," one woman said to me. "Loosen his belt."

When I loosened it, I could feel a great big wound. I found out later at the hospital that he had three bullet wounds: one in the stomach, one in the neck, and another in the leg. He was dying.

—Diana Salmerón de Contreras

That's enough of this! When are they going to stop all this?
—Pedro Díaz Juárez, student

"Hey, little brother, what's the matter? Answer me, little brother . . ."
—Diana Salmerón de Contreras

The hail of bullets being fired at the Chihuahua building became so intense that around seven p.m. a large section of the building caught on fire.

The fire burned for a long time. All the floors from the tenth to the thirteenth were enveloped in flames, and many families were forced to leave the unit, amid the heavy gunfire, carrying their children in their arms and risking their lives. We also saw many others struck by bullets fall to the ground.

—Jorge Avilés R., in a story entitled "Serious
Fighting for Hours Between Terrorists and Soldiers,"
El Universal, October 3, 1968

Little brother, speak to me. . . . Please, somebody get him a stretcher! I'm right here, Julio . . . a stretcher! . . . Soldier, a stretcher for somebody who's

been wounded. . . . What's the matter, little brother? . . . Answer me, little
brother. . . . A stretcher! . . .

—Diana Salmerón de Contreras

A number of dead bodies lying in the Plaza de las Tres Culturas. Dozens of
wounded. Hysterical women with their children in their arms. Shattered
windows. Burned-out apartments. The outer doors of the buildings destroyed. Water pipes in a number of buildings broken. Water leaking all over
many of them. Yet the shooting went on and on.

—News story entitled "Terrible Gun Battle in Tlatelolco.
The Number of Dead Not Yet Determined and
Dozens Wounded," *Excélsior,* October 3, 1968

Now that I'd managed to get to Julio and we were together again, I could
raise my head and look around. The very first thing I noticed was all the
people lying on the ground; the entire plaza was covered with the bodies of
the living and the dead, all lying side by side. The second thing I noticed was
that my kid brother had been riddled with bullets.

—Diana Salmerón de Contreras

This reporter was caught in the crowd near the Secretariat of Foreign Relations. A few steps away a woman fell to the ground—she had either been
wounded or had fainted dead away. A couple of youngsters tried to go to her
rescue, but the soldiers stopped them.

—Félix Fuentes, in a news story entitled "It All
Began at 6:30 p.m.," *La Prensa,* October 3, 1968

"Soldier, please have somebody bring a stretcher!"

"Shut up and stop pestering me or you'll be needing two of them!" was
the only reply I got from this "heroic Johnny," as our president calls the
soldiers in the ranks.

Just then a med student hurried over and said to this "heroic Johnny,"
"That boy there ought to be taken to the hospital right away!"

"Shut your trap, you son of a bitch," the soldier answered.

Everyone standing around watching began shouting in chorus, "A stretcher,
a stretcher, a stretcher!"

A couple of people made a makeshift stretcher out of some lengths of pipe
and an overcoat. But the med student who helped us was arrested.

—Diana Salmerón de Contreras

In a few minutes the whole thing became a scene straight out of hell. The
gunfire was deafening. The bullets were shattering the windows of the apartments and shards of glass were flying all over, and the terror-stricken families
inside were desperately trying to protect their youngest children.

—Jorge Avilés R., in "Serious Fighting for Hours
Between Terrorists and Soldiers"

"Please let me go with him—I'm his sister!" I begged.

They gave me permission to leave the Plaza with the stretcher-bearers. I climbed into the army ambulance with my brother.

—Diana Salmerón de Contreras

Why don't you answer me, *hermanito?*

—Diana Salmerón de Contreras

Everything was a blur—I don't know if it was because I was crying or because it had started to rain. I watched the massacre through this curtain of rain, but everything was fuzzy and blurred, like when I develop my negatives and the image begins to appear in the emulsion. . . . I couldn't see a thing. My nose was running, but I just snuffled and went on shooting pictures, though I couldn't see a thing: the lens of my camera was spattered with raindrops, spattered with tears. . . .

—Mary McCallen, press photographer

[As they started lining us up against the wall of the church] I saw two of my pals from *Excélsior* there, a reporter and a press photographer. They'd grabbed Jaime González's camera away from him. The reporter was saying, "I'm a journalist," but one of the soldiers answered, "Very pleased to meet you, but I couldn't care less whether you're a journalist or not; just stand over there against the wall." They'd slashed Jaime González's hand with a bayonet to get his camera away from him.

—Raúl Hernández, press photographer, in a story
entitled "The Gun Battle as the Press
Photographers Saw It," *La Prensa,* October 3, 1968

Before I climbed into the army ambulance, a "student" whom I'd seen at UNAM came up to me and said, "Your handbag, please. . . ."

"What do you want it for?" I asked.

The soldier who was with me was surprised too: "Who are *you?*" he asked him. But then he noticed a white handkerchief or something in the fake student's hand and said to him, "Oh, you're one of *them,* are you?"

The guy was an undercover agent posing as a student. I handed him my purse, and he searched through it and then gave it back to me. I have no idea to this day why he asked me for it.

They took my brother to the hospital then, and I waited there for hours to find out how the operation had gone. A male nurse kept coming in every so often, and one time he asked the women who were there waiting it out, just as I was, "Which one of you was with a boy in a blue suit?"

"He was with me . . . I came here with a boy in a blue suit," I said.

They took me to identify Julio's body and sign the necessary papers.

When we held the wake for Julio, I was deeply touched by his fellow students' loyalty to him and their concern for us. All the boys from Vocational 1 came to the house the minute they heard the tragic news of his death. They

had taken up a collection and offered us some five hundred pesos. My sister told them we didn't really need the money, and would prefer that they use it for the movement. "No," they all said. "The way we see it your brother *is* the movement. We'd like you to accept the money."

Julio was fifteen years old, a student at Vocational 1, the school out by the Tlatelolco housing unit. That was the second political meeting he'd ever gone to. He had asked me to go with him that day. The first meeting we went to together was the big Silent Demonstration. Julio was my only brother.

—Diana Salmerón de Contreras

Heleieth I. B. Saffiotti

First published in Brazil in 1969, Heleieth I. B. Saffiotti's *Women in Class and Society* is a theoretical and historical inquiry into capitalism and women's position in the economic structure. Saffiotti's hypothesis is that capitalism marginalizes large groups of people from the productive system and that women form part of these groups that enter the economic system only when it is structurally demanded. Ideology and the domestic functions for which they are held responsible prevent women's economic independence. After this general and ambitious description, Saffiotti develops the main chapters of a history of women in Brazil from slavery to the modern process of urbanization and the movements that, like suffragism, tried to organize women's action in favor of civil and political equality. Finally, she critically analyzes the ideological techniques that serve as powerful instruments of social control and the myths that steer human behavior along trends that, independently of voluntary conduct, reinforce social and economical dominance. Saffiotti's work combines highly abstract developments with the study of women's condition in Brazil through history and today. She has also done empirical research and published on women in the labor market and their subordinate position as a marginal force or labor reserve.

Feminism and the Feminist Mystique

Despite its limitations, there is no denying that petty-bourgeois feminism played an important role in competitive societies; it destroyed many prejudices, broke down long-standing taboos, and created a new perspective on women's social roles. Yet seen in terms of actual direct participation in the dominant system of production, women's situation in class society represents a step backwards compared with their participation in the economic life of other social formations, the productive forces of which were not so developed. In this respect, some of the rights won by women in

competitive society betokened merely a refinement of the techniques whereby large numbers of women are marginalized from the class structure. Hence, even though the goals of feminist movements generally may have been to institute real social equality between the sexes, the actual attainment of the goals feminist organizations set for themselves has been dependent on the structural possibilities offered by the major types of social formations, in which different degrees of equality obtain and in which the field of what has been historically possible is structured differently.

In most Western countries women now enjoy the same legal rights as men, with the exception of a few concerning marriage. Feminism was not brought to a standstill by the attainment of political rights. Tactics, of course, changed considerably when women began to gain seats in parliaments. Because of the aggressive features often assumed by the movement for women's suffrage, petty-bourgeois feminism historically has been identified with that movement and nothing else. It is true that a large number of women, even organized feminist militants, considered their job done when their political rights were won, but these were by no means all the women who had fought for the vote. Local and national organizations in many countries continued to struggle for the extension of all civil rights to women and for women's social advancement. They fought, for example, to obtain civil rights for married women, to establish equal educational opportunities for men and women, to broaden women's chances of employment, and to protect mothers and children. These national organizations formed themselves into a confederation, which took up the struggle for peace and for the destruction of Nazism and fascism, and strove to raise the level of women's personal aspirations across the world.

As formal rights became more and more universal in many industrial societies, and even in some countries less socially and economically developed, however, middle-class women found more and more ways to accommodate themselves to a situation of partial equality with men, while feminism itself, although still alive and active, was unable by word of mouth alone to maintain the spirit of struggle that had hallmarked the suffragist movement—and that spirit was essential if women were to advance further toward their goal of total liberation. Actually, these two developments were two sides of the same phenomenon. The acceptance by middle-class women in class society of a partial equality with men and petty-bourgeois feminist efforts to broaden women's rights and chances of employment without overstepping the structural limitations of capitalism represented differences of degree only in a social consciousness which basically accepted the class structure. The petty-bourgeois feminist movement may have been more progressive in the sense that it was not content merely with obtaining formal rights for women, but it was never able to see the question of sexual equality in terms of a social structure that negated

this equality. It strove and is today still striving to destroy the prejudices that have kept women subservient to men, and has always endeavored to get men to adopt a positive attitude toward female wage labor, altering women's own image in the process. But in so doing the petty-bourgeois feminists have neglected to promote those things which could really improve women's chances for integration into class society—in particular, they have failed to effect a change in the view taken of female labor. Work in capitalist society may be alienating and alienated, but a large number of jobs still give satisfaction. Nevertheless, the occasional chances that work offers for self-fulfillment are not so adequate as to produce a balanced human personality and achieve a harmonious union between the individual and society. The need to work must derive from more than a mere shortage of money; it must become an integral component of the female personality. If this were to happen, women would surely find incentives with men for the most diverse positions and the struggle against their discrimination on biological grounds.

This solution would certainly be a blow to the feminine mystique—for one thing it would increase the female employment rate—but it would not mean complete sexual equality, nor would it increase the irrationality of the capitalist system of production. In fact, it might even aggravate the social contradictions that are continually threatening the stability of competitive societies. Any reduction in the social differences between the sexes that would result from allowing more women into the class structure would accentuate the lines dividing the social classes; at least it would expose the class roots of a good number of personal successes and failures in the economic world. To abolish the various factors, such as the social discrimination of women, that camouflage the class structure of society would expose its basic contradictions to the view of groups that heretofore had been partially or wholly mystified by it, this also would help to sharpen the conflict between the social classes. Indeed, any reduction in the social differences between the sexes is a threat to the continued existence of capitalist society, just as a sharpening of the class struggle in general should ultimately bring about its destruction.

On the other hand, while making work into a genuinely felt need for women goes far beyond any other solution to the problem of female labor that has been proposed or applied so far, woman's unique position requires that she think about other aspects of her life as well. In competitive society a man's occupation is the focal point of his existence (as it would be for a woman if her labor were seen in a different light), but in woman's existence there are other factors which make her a worker of a fundamentally different kind (worker being defined here merely as a person doing work). A society based on the private appropriation of the economic surplus and subject to the recurrent crises of overproduction based on the private appropriation

of the economic surplus is not able to achieve full employment, nor can it afford to absorb the entire cost of maternity as a social expenditure; this being so, woman's special position must perforce detract from her ability to function as a worker in capitalist societies.

Of course, the social tensions generated by the market under capitalism are handled in different ways in different countries, and these therefore will tend to approach the woman problem in different ways, depending on their position within the overall international capitalist system. In the leading capitalist countries, which have become warfare or welfare states, it is easier to allay the social tensions caused by male and female employment. War places a temporary moratorium on crises of overproduction by employing more of the reserve work force; capital investment at the system's periphery siphons "international surplus value" to the system's center, where it is used to create a more just distribution of income, at the international level, and over the long term to build a social welfare state. This is perhaps the chief reason why women in the advanced countries enjoy more freedoms, although these freedoms also help to increase the alienation of all, both men and women, who receive the benefits of the social welfare state.

Yet, it is also true that while war and underdevelopment indirectly may bring a certain order and stability to the economics of the leading capitalist countries, they provide neither a total nor a sound and lasting solution to the woman problem. It is therefore only partially correct to say that economic growth and social development bring women more freedoms. The capitalist nations employed female labor rather flexibly as their economies grew, but its use may be restricted drastically as class society enters its maturity, since that flexibility was made possible by the social and economic underdevelopment of vast regions at the periphery of the international capitalist system.

Whatever revolutionary content there is in petty-bourgeois feminist praxis, it has been put there by the efforts of the middle strata, especially the less well-off, to move up socially. To do this, however, they sought merely to expand the existing social structures, and never went so far as to challenge the status quo. Thus, while petty-bourgeois feminism may always have aimed at establishing social equality between the sexes, the consciousness it represented has remained utopian in its desire for and struggle to bring about a partial transformation of society; this it believed could be done without disturbing the foundations on which it rested. As the productive forces of capitalist society developed, opportunities for women's employment grew, and large numbers of women workers entered the class structure. In this sense, petit-bourgeois feminism is not feminism at all; indeed it has helped to consolidate class society, by giving camouflage to its internal contradictions. There can be only one conclusion from this, namely, that the feminist position is wrong when conceived as autonomous.

In reality, there is no such thing as an autonomous feminism, independent of a class perspective. Petty-bourgeois feminism has always been, fundamentally and unconsciously, a feminism of the ruling class, since the "middle classes" could always be found tailing along behind one or the other of the classes opposing one another in the system of production. It thus played a broad role in sustaining the myth that competitive societies can exist without classes and without sexual biases. When analyzed carefully, petty-bourgeois feminism is found to be not a product of sharpened social tensions, but a means for allaying these tensions, both within particular nations and internationally.

As the tensions between the opposing classes abate in the countries at the center of the international capitalist system, the contradiction between underdevelopment and development becomes more acute. Here, classical feminism has had an additional role to play, in that it has helped ease the tensions between these two poles of the international capitalist system by making the woman problem contingent on an economic development that was supposed to be hindered by archaic structures rather than the interests of the highly developed countries in destroying precapitalist structures, in shaping new attitudes toward women taking work outside the home, or in restructuring patriarchal family patterns. The upshot of this has been of course that classical feminism believed that the key to the triumph of feminism lay in the recruitment of all women, independent of class. Hence, paradoxical as it may seem, feminism will only really find itself when it has become an integral part of a consciousness "capable of considering society from its central aspect, as a coherent whole, and hence capable of acting in a central way in changing reality."

Socialist feminism may not have been completely successful in its efforts to liberate women, but that petty-bourgeois feminism was even less so should be plain from the foregoing. The comparison, however, is not completely valid since socialism has yet to reach its maturity. The maturing process undergone by these two types of society must be viewed in the light of the premises on which the two systems are based: While capitalism moved toward its maturity (represented by its monopoly phase) along a path of privatization and concentration of property, the path of socialism was toward collectivized property. Complete sexual equality, therefore, may very well become a reality in socialist society as the socioeconomic relations in that society mature. On the other hand, in the case of capitalist society, as its productive forces mature, female employment stagnates at the same level or even declines. The degree of actual utilization of female labor during the periods of growth and the mature phase of economies based on private enterprise; the ebbs and flows in female employment and the arguments advanced in justification of these phenomena; and the level of consciousness which women and men attain of this process and the

forms which it takes in each particular society will vary, however, within the limits we have set down, as a function of specific historical conditions in which the capitalist system of production evolves in each case. Expressed in other terms, the analysis of each particular social formation will show how the general properties of that system are expressed in each set of concrete historical circumstances.

In its general contours, the analysis of feminism given in the first part of this work is equally valid for Brazilian feminism. A feminism which is concerned exclusively or primarily with the immediate problems of women, and outwardly independent of political ideologies, is essentially a product of a mode of thought with the following features:

1. It is *utopian,* if it really believes that a deep-going transformation of the roles and social standing of women in competitive societies is possible without at the same time altering the foundations on which social life is built. In this case, feminist consciousness is unconsciously compromised with the capitalist status quo.

2. It is *in conscious league with the competitive social order* if, in its struggle for total social equality between the sexes, it nevertheless accepts a partial equality which is fully compatible with the competitive order, even in countries where differences between men and women have been abolished in the formal legal sense, yet de facto inequalities still exist.

However forward-looking the progressive side of petty-bourgeois feminism may be, it is unable to exceed the bounds of its fealty, conscious or unconscious, to the social order of a capitalist economy. In this respect it is just as little a "pure" feminism—despite the claims of its apologists—as are "socialist feminist" movements or women's movements. Whatever the level of consciousness feminism is able to achieve with regard to women's problems, the problems themselves are always defined in the terms of a particular social order, and any possible solutions to these problems, short-term or long-term, must be gotten from within this order.

Socialist feminism, even in its most practical, nontheoretical forms, such as exist in Brazil, represents a much fuller consciousness than its petty-bourgeois counterpart. Since it proceeds from a vantage point that is highly critical of the capitalist status quo, it has been able to see the problems of women as merely one aspect of a social totality in which a number of other determining factors are also operative, and to discern among the latter those which call for immediate attention. In Brazil, however, left-wing feminism has almost always been clandestine, camouflaged under labels that would be palatable to the society at large. While this enabled it to penetrate into areas inaccessible by any other means, it made organizing extremely slow, even atomized, because it lacked continuity. The broader national political movement with which it linked up always lacked a solidly

based organization deriving its intellectual sustenance from a dynamic analysis of Brazilian realities, for which it substituted ready-made, and hence ineffective, formulae. Indeed, this was typical of all women's organizations, even those which, like the BFW [Brazilian Federation of Women], avoided decisionmaking from the top down at all costs.[1] The negative reaction of capitalist societies to socialist movements, whether or not they were engaged in the struggle for women's rights, set their main thrust against just that lack of a broad popular base, which had been so much a fact of life in Brazil. However, even if left-wing feminist organizations in Brazil had taken care to avoid this separation between leadership and base, they would still have suffered from the organizational vices of the broader political movement of which they were a part. This certainly does not explain all the flaws and failures in their efforts to politicize women, but it unquestionably has been a weighty factor in holding back women's awareness.

Another quite serious criticism of these women's organizations as both left-wing and feminist groupings is that they elected to give the defense of women's civil rights priority over the need for women to receive training for remunerative employment. This shows a negligence, if not ignorance, of the potential embodied in female labor outside the home for shaping an independent and critical awareness in women. Even more than this, it indicates an indifference to the possibilities offered women by a position in the class structure and by the social contacts and economic independence ensuing from that position.

Although feminism in general has never penetrated very deeply into Brazilian society, it has left an indelible mark on the personal histories of many urban women over two generations, and gained many civil rights. True, these rights have never, even today, had more than an abstract existence for a vast segment of the nation's female population; but they will always represent possibilities, to be used at any moment to facilitate women's entry into certain areas of social life by removing the legal obstacles in their way. The conquest of political and civil rights for women, however, also entailed certain reactions on the part of society, among which it is necessary to distinguish those which indicated a genuine acceptance of the new female roles, and those we might call "prejudicial refinements," that is, the intelligent use of social techniques to conceal prejudices against women.

The denial of civil and political rights to women belongs to a specific phase in the development of competitive society. In some countries, the concession of such rights may not fit the realities of everyday life, although they may be compatible with the general framework of ideas sustained by capitalist society as a whole. Thus, the hegemonic center determines the "level of reality" in the subsystems,[2] and sometimes actually widens the

gap between ideas and the actual state of things, or between the juridical structure of a nation and the social relations that have evolved within it. Aside from the serious discrepancies it creates between the various levels of social phenomena,[3] this determination of the "level of reality" from without in the dependent subsystems also gives rise to incongruities between different domains of human behavior. For example, it heightens the ambiguity of female roles, and thereby increases women's ambivalence toward them. This ambivalence, in turn, sustains and refuels mystificatory processes, and makes it difficult for women to distinguish functions they have qualifications to fulfill, and the limits beyond which their conduct would be considered deviant. Throughout their lives, then, women assume an attitude lacking the aggressiveness required by competitive society, which has come to see them as passive creatures. This characterization becomes a self-fulfilling prophecy that makes it difficult for women to develop a critical awareness.

As it is consciously or unconsciously compromised with the established order of class society, petty-bourgeois feminism is not capable of effecting a total mystification of feminine consciousness. It can find no other way to proceed than to attribute to the female gender a degree of autonomy it does not possess. Socialist feminism does provide a richer perspective from which to analyze the problems of women in competitive societies, but its simplifications leave much to be desired. Theoretically—and practically, perhaps for the same reason—it has failed to come up with any completely satisfactory solution to the feminine question. That is why the "feminine mystique" can still operate even in societies in which many women have become aware of their problems. Although this consciousness is precarious and simplistic in its petty-bourgeois form, in its socialist form it obliges class society to refine the processes by which it mystifies women. Once again, the workings of the hegemonic center of world capitalism are in evidence. The export of science, technology, and the techniques of social engineering to the periphery makes available ways and means to reelaborate and refine the prejudices and preconceptions that discriminate socially against the female population. Although science and technology did not originate at the periphery of world capitalism, they are used there with the object of establishing a level of reality compatible with the goals of the system at large. Thus although some of its features may vary from country to country depending on tradition, the feminine mystique has universal features, so that it may be ascribed to *class society* as a structural type. It is valid, therefore, to say that mystificatory processes, whatever their origin, have no nationality insofar as they are used rationally, with the end of maintaining and justifying a social order whose particular features may vary from one nation to another, yet have the same foundations in all cases—namely, class divisions. Considering the showcase effect exercised

by the dominant capitalist nations, social engineering techniques evolved in the economically and socially more advanced countries have all the basic requirements needed to make them universally applicable. In its universal aspect, the feminine mystique represents a refined social technique which in underdeveloped countries assumes exaggerated features as a result of the impressively effective ability of tradition to control the pace of change of women's social roles. On the other hand, however, the mystique serves as a foundation, and a substantively rational one, for traditional techniques of controlling the evolution of feminist consciousness.

Notes

1. The delegations who were to represent the BFW were not appointed by the executive board, but were chosen by elections held after a discussion of the problems by the municipal, state, and national organizations. For this reason, the Brazilian delegations which represented the BFW were always large in number, and the theses they presented reflected a variety of regional problems.

2. I am using this term as defined by Mannheim, as follows:

> By "reality level" we mean that every society develops a mental climate in which certain facts and their interrelations are considered basic and called "real," whereas other ideas fall below the level of "reasonably acceptable" statements and are called fantastic, utopian, or unrealistic. In every society there is a generally accepted interpretation of reality. In this sense every society establishes a set of respectable ideas through its conventions, and ostracizes any others as "diabolic," "subversive," or "unworthy." Being "real" or "less real" is always an *a priori* reason for ascribing more or less worth to certain facts. Whatever different schools of philosophy may think about this, and however much instrumentalism and logical positivism may consider such ontology fallacious, it is a sociological fact that public thinking unconsciously establishes such reality-levels, and a society is only integrated if its members roughly agree on a certain ontological order. (Karl Mannheim, *Freedom, Power and Democratic Planning*, pp. 138–139.)

3. See Georges Gurvitch, *Tratado de Sociologia*, ch. 5; and Marcel Mauss, *Sociologie et Anthropologie*, pp. 273–279.

Rigoberta Menchú

RIGOBERTA MENCHÚ WAS TWENTY-THREE YEARS OLD when, in 1982, she met Elisabeth Burgos-Debray, the Latin American anthropologist who edited some twenty hours of tape by Rigoberta. From the loyal and responsible work an autobiographical narrative was born. *I, Rigoberta Menchú, an Indian Woman in Guatemala,* first published in 1983, is in fact not an individual life story but rather a testimony on Quiché *campesinos* (peasants), the conditions of the life and their culture, seen through the eyes and conscience of a fighter and organizer. A peasant girl from an Indian family, Rigoberta inherited a treasure of cultural traditions, a language, the Quiché, and her family's commitment to their community. Her father, Vicente Menchú, died in 1980 in the massacre that followed the occupation of the Spanish embassy, launched by the *campesinos* in order to attract general attention to the struggle. Her mother met a horrible death in an Indian village, where she was tortured by the army. Rigoberta contributed from the very beginning to the organization of the CUC (Peasant Unity Committee) and her narrative gives detailed information about the making of a woman leader and the strategies of the peasant movement. Rigoberta's discourse, even when exploring remembrances of her childhood or the humiliations endured while working as a servant in the city, is a collective narrative, in the sense that her culture enables her to consider herself as part of a community and to build her subjectivity in close relation to its fate. Her voice resonates with the plurality of voices that come from the past (which she has learned not from books but from the lived experience and lore of the Indians) and echo in the present. She also knows how cultural boundaries can be crossed and different traditions interwoven. In the chapter included here, the reading of the Bible, as a collective activity in her village, runs parallel with her belief in the Indian's autonomy vis-à-vis the white and *ladino* (mestizo) world. At the same time, the stories of Judith and David are translated into the present needs of the *campesinos* who organize to face the blind repression of the army. Rigoberta's narrative offers a lesson on how a culture defends itself and how traditional values can acquire new political meanings.

The Bible and Self-Defense:
The Examples of Judith, Moses, and David

when the strangers who came from the East arrived, when they arrived; the ones who brought Christianity which ended the power in the East, and made the heavens cry and filled the maize bread of the Katun with sadness . . .

—Chilam Balam

Their chief was not defeated by young warriors, nor wounded by sons of Titans. It was Judith, the daughter of Maran, who disarmed him with the beauty of her face.

—*The Bible (Judith)*

We began to study the Bible as our main text. Many relationships in the Bible are like those we have with our ancestors, our ancestors whose lives were very much like our own. The important thing for us is that we started to identify that reality with our own. That's how we began studying the Bible. It's not something you memorize, it's not just to be talked about and prayed about, and nothing more. It also helped to change the image we had, as Catholics and Christians: that God is up there and that God has a great kingdom for us the poor, yet never thinking of our own reality as a reality that we were actually living. But by studying the Scriptures, we did. Take Exodus for example, that's one we studied and analyzed. It talks a lot about the life of Moses, who tried to lead his people from oppression and did all he could to free his people. We compare the Moses of those days with ourselves, the Moses of today. Exodus is about the life of a man, the life of Moses.

We began looking for texts which represented each one of us. We tried to relate them to our Indian culture. We took the example of Moses for the men, and we have the example of Judith, who was a very famous woman in her time and appears in the Bible. She fought very hard for her people and made many attacks against the king they had then, until she finally had his head. She held her victory in her hand, the head of the king. This gave us a vision, a stronger idea of how we Christians must defend ourselves. It made us think that a people could not be victorious without a just war. We Indians do not dream of great riches; we want only enough to live on. There is also the story of David, a little shepherd boy who appears in the Bible, who was able to defeat the king of those days, King Goliath. This story is the example for the children. This is how we look for stories and

---

"The Bible and Self-Defense," from I, Rigoberta Menchú. *Ed. Elisabeth Burgos-Debray (London: Verso, 1984). Translated by Ann Wright. Reprinted with permission.*

psalms which teach us how to defend ourselves from our enemies. I remember taking examples from all the texts which helped the community to understand their situation better. It's not only now that there are great kings, powerful men, people who hold power in their hands. Our ancestors suffered under them too. This is how we identify with the lives of our ancestors who were conquered by a great desire for power—our ancestors were murdered and tortured because they were Indians. We began studying more deeply and, well, we came to a conclusion. That being a Christian means thinking of our brothers around us, and that every one of our Indian race has the right to eat. This reflects what God himself said, that on this earth we have a right to what we need. The Bible was our principal text for study as Christians and it showed us what the role of a Christian is. I became a catechist as a little girl and I studied the Bible, hymns, the Scriptures, but only very superficially. One of the things Catholic Action put in our heads is that everything is sinful. But we came round to asking ourselves: "If everything is sinful, why is it that the landowner kills humble peasants who don't even harm the natural world? Why do they take our lives?" When I first became a catechist, I thought that there was a God and that we had to serve him. I thought God was up there and that he had a kingdom for the poor. But we realized that it is not God's will that we should live in suffering, that God did not give us that destiny, but that men on earth have imposed this suffering, poverty, misery, and discrimination on us. We even got the idea of using our own everyday weapons, as the only solution left to us.

I am a Christian and I participate in this struggle as a Christian. For me, as a Christian, there is one important thing. That is the life of Christ. Throughout his life Christ was humble. History tells us he was born in a little hut. He was persecuted and had to form a band of men so that his seed would not disappear. They were his disciples, his apostles. In those days, there was no other way of defending himself or Christ would have used it against his oppressors, against his enemies. He even gave his life. But Christ did not die, because generations and generations have followed him. And that's exactly what we understood when our first catechists fell. They're dead but our people keep their memory alive through our struggle against the government, against an enemy who oppresses us. We don't need very much advice or theories or documents: life has been our teacher. For my part, the horrors I have suffered are enough for me. And I've also felt in the deepest part of me what discrimination is, what exploitation is. It is the story of my life. In my work I've often gone hungry. If I tried to recount the number of times I'd gone hungry in my life, it would take a very long time. When you understand this, when you see your own reality, a hatred grows inside you for those oppressors that make the people suffer so. As I said, and I say it again, it is not fate which makes us poor. It's not

because we don't work, as the rich say. They say: "Indians are poor because they don't work, because they're always asleep." But I know from experience that we're outside ready for work at three in the morning. It was this that made us decide to fight. This is what motivated me, and also motivated many others. Above all the mothers and fathers. They remember their children. They remember the ones they would like to have with them now but who died of malnutrition, or intoxication in the *fincas* [farms], or had to be given away because they had no way of looking after them. It has a long history. And it's precisely when we look at the lives of Christians in the past that we see what our role as Christians should be today. I must say, however, that I think even religions are manipulated by the system, by those same governments you find everywhere. They use them through their ideas or through their methods. I mean, it's clear that a priest never works in the *fincas,* picking cotton or coffee. He wouldn't know what picking cotton was. Many priests don't even know what cotton is. But our reality teaches us that, as Christians, we must create a church of the poor, that we don't need a church imposed from outside which knows nothing of hunger. We recognize that the system has wanted to impose on us: to divide us and keep the poor dormant. So we take some things and not others. As far as sins go, it seems to me that the concept of the Catholic religion, or any other more conservative religion than Catholicism, is that God loves the poor and has a wonderful paradise in Heaven for the poor, so the poor must accept the life they have on earth. But as Christians, we have understood that being a Christian means refusing to accept all the injustices which are committed against our people, refusing to accept the discrimination committed against a humble people who barely know what eating meat is but who are treated worse than horses. We've learned all this by watching what has happened in our lives. This awakening of the Indians didn't come, of course, from one day to the next, because Catholic Action and other religions and the system itself have all tried to keep us where we were. But I think that unless a religion springs from within the people themselves, it is a weapon of the system. So, naturally, it wasn't at all difficult for our community to understand all this and the reasons for us to defend ourselves, because this is the reality we live.

As I was saying, for us the Bible is our main weapon. It has shown us the way. Perhaps those who call themselves Christians, but who are really only Christians in theory, won't understand why we give the Bible the meaning we do. But that's because they haven't lived as we have. And also perhaps because they can't analyze it. I can assure you that any one of my community, even though he's illiterate and has to have it read to him and translated into his language, can learn many lessons from it, because he has no difficulty understanding what reality is and what the difference is between the paradise up above, in Heaven, and the reality of our people here

on earth. We do this because we feel it is the duty of Christians to create the kingdom of God on earth among our brothers. This kingdom will exist only when we all have enough to eat, when our children, brothers, parents don't have to die from hunger and malnutrition. That will be the "Glory," a kingdom for we who have never known it. I'm only talking about the Catholic church in general terms because, in fact, many priests came to our region and were anticommunists, but nevertheless understood that the people weren't communists but hungry; not communists, but exploited by the system. And they joined our people's struggle too, they opted for the life we Indians live. Of course many priests call themselves Christians when they're only defending their own petty interests and they keep themselves apart from the people so as to endanger these interests. All the better for us, because we know very well that we don't need a king in a palace but a brother who lives with us. We don't need a leader to show us where God is, to say whether he exists or not, because, through our own conception of God, we know there is a God and that, as the father of us all, he does not wish even one of his children to die or be unhappy or have no joy in life. We believe that, when we started using the Bible, when we began studying it in terms of our reality, it was because we found in it a document to guide us. It's not that the document itself brings about the change, it's more that each one of us learns to understand his reality and wants to devote himself to others. More than anything else, it was a form of learning for us. Perhaps if we'd had other means to learn, things would have been different. But we understood that any element in nature can change man when he is ready for change. We believe the Bible is a necessary weapon for our people. Today I can say that it is a struggle which cannot be stopped. Neither the governments nor imperialism can stop it because it is a struggle of hunger and poverty. Neither the government nor imperialism can say: "Don't be hungry," when we are all dying of hunger.

To learn about self-defense, as I was saying, we studied the Bible. We began fashioning our own weapons. We knew very well that the government, those cowardly soldiers . . . perhaps I shouldn't talk of them so harshly, but I can't find another word for them. Our weapons were very simple. And at the same time, they weren't so simple when we all started using them, when the whole village was armed. As I said before, the soldiers arrived one night. Our people were not in their homes. They'd left the village and gone to the camp. They made sure that we hadn't abandoned the village altogether but thought it would be better to occupy it in the daytime. So sometime later, when we weren't expecting them, about fifteen days later, our lookouts saw the army approaching. We were in the middle of building houses for our neighbours. We needed some more huts there. We had two lookouts. One was supposed to warn the community and the other had to delay or stop the soldiers entering. They were aware that they

might have to give their lives for the community. At a time like this, if someone can't escape, he must be ready to accept death. The army arrived, and the first two to enter wore civilian clothes. But our children can easily recognize soldiers, by the way they walk, and dress, and everything about them, so the lookouts knew they were soldiers in disguise. They asked the names of certain *compañeros* in the community so they could take them away, kidnap them. One of the lookouts got away and came to warn the village that the enemy was nearby. We asked him if he was sure and he said: "Yes, they are soldiers, two of them. But as I was coming up here I saw others coming, further off, with olive green uniforms." The whole community left the village straight away and gathered in one place. We were very worried because the other lookout didn't appear. They were capable of having kidnapped him. But he did turn up in the end and told us how many soldiers there were, what each one was like, what sort of weapons they had, how many in the vanguard and the rearguard. This information helped us decide what to do, because it was daytime and we hadn't set our traps. We said: "What are we going to do with this army?" They came into the village and began beating our dogs and killing our animals. They went into the houses and looted them. They went crazy looking for us all over the place. Then we asked: "Who is willing to risk their lives now?" I, my brothers, and some other neighbors immediately put up our hands. We planned to give the army a shock and to show them we were organized and weren't just waiting passsively for them. We had less than half an hour to plan how were were going to capture some weapons. We chose some people—the ones who'd go first, second, third, fourth, to surprise the enemy. How would we do it? We couldn't capture all ninety soldiers who'd come to the village, but we could get the rearguard. My village is a long way from the town, up in the mountains. You have to go over the mountains to get to another village. We have a little path to the village just wide enough for horses . . . and there are big rivers nearby so that the path isn't straight. It bends a lot. So we said, "Let's wait for the army on one of those bends and when the soldiers pass, we'll ambush the last one." We knew we were risking our lives but we knew that this example would benefit the village very much because the army would stop coming and searching the village all the time. And that's what we did.

We chose a *compañera,* a very young girl, the prettiest in the village. She was risking her life, and she was risking being raped as well. Nevertheless, she said: "I know very well that if this is my part in the struggle, I have to do it. If this is how I contribute to the community, I'll do it." So this *compañera* goes ahead on another path to a place that the army has to pass on their way to the village. That's where we prepared the ambush. We didn't have firearms, we had only our people's weapons. We'd invented a sort of Molotov cocktail by putting petrol in a lemonade bottle with a few iron

filings, mixed with oil, and a wick to light it. So if the army got one of us, or if we couldn't do anything else, we'd set fire to them. This cocktail could burn two or three soldiers because it could land on them and burn their clothes. We had catapults, too, or rather, they were the ones we'd always used to protect the maize fields from the birds which would come into the fields and eat the cobs when they were growing. The catapults could shoot stones a long way and if your aim is good it lands where you want it to. We had machetes, stones, sticks, chile, and salt—all the different people's weapons. We had none of the weapons the army had. The community decided that the young girl who went on ahead would try to flirt with the last soldier and try to make him stop and talk to her. We all had numbers: who would be the first to jump, who would get him off balance, who would frighten him, and who would disarm him. Each of us had a special task in capturing the soldier.

First came the ones without weapons—they were members of the secret police, soldiers in disguise. Then came the others. The whole troop. They were about two meters away when the last one came. Our *compañera* came along the path. She paid no attention to the others. It was a miracle they didn't rape her, because when soldiers come to our area they usually catch girls and rape them—they don't care who they are or where they're from. The *compañera* was ready to endure anything. When she came to the last soldier, she asked him where they'd been. And the soldier began telling her: "We went to that village. Do you know what's happened to the people?" The *compañera* said: "No, I don't know." And he said: "We've been twice and there's no one, but we know they live there." Then one of our neighbors jumped onto the path; another came up behind the soldier. My job was to jump onto the path as well. Between us we got the soldier off balance. One of us said: "Don't move, hands up." And the soldier thought there was a gun pointing at his head or his back. Whatever he thought, he did nothing. Another *compañero* said: "Drop your weapon." And he dropped it. We took his belt off and checked his bag. We took his grenades away, his rifle, everything. I thought it was really funny, it's something I'll never forget, because we didn't know how to use it. We took his rifle, his big rifle, and a pistol, and we didn't even know how to use the pistol. I remember that I took the soldier's pistol away and stood in front as if I knew how to use it but I didn't know anything. He could have taken if off me because I couldn't use it. But, anyway, we led him away at gunpoint. We made him go up through the mountains so that if the others came back they wouldn't find the path. If they had it would have been a massacre. Two *compañeras* of about forty-five and a fifty-year-old *compañero* had taken part in the ambush. The little *compañera* who'd attracted the soldier was about fourteen.

We took the disarmed soldier to my house, taking all the necessary precautions. We blindfolded him so that he wouldn't recognize the house he was going to. We got him lost. We took him a roundabout way so that he'd lose his sense of direction. We finally arrived back. I found it really funny, I couldn't stop laughing because we didn't know how to use the gun. We were very happy, the whole community was happy. When we got near the camp, the whole community was waiting for us. We arrived with our captured soldier. We reached my house. He stayed there for a long time. We took his uniform off and gave him an old pair of trousers and an old shirt so that if his fellow soldiers came back—we tried to keep him tied up—they wouldn't know he was a soldier. We also thought that those clothes could help us confuse the other soldiers later on. Then came a very beautiful part when all the mothers in the village begged the soldier to take a message back to the army, telling all the soldiers there to think of our ancestors. The soldier was an Indian from another ethnic group. The women asked him how he could possibly have become a soldier, an enemy of his own race, his own people, the Indian race. Our ancestors never set bad examples like that. They begged him to be the light within his camp. They explained to him that bearing a son and bringing him up was a big effort, and to see him turn into a criminal as he was, was unbearable. All the mothers in the village came to see the soldier. Then the men came too and begged him to recount his experience when he got back to the army and to take on the role, as a soldier, of convincing the others not to be so evil, not to rape the women of our race's finest sons, the finest examples of our ancestors. They suggested many things to him. We told the soldier that our people were organized, and were prepared to give their last drop of blood to counter everything the army did to us. We made him see that it wasn't the soldiers who were guilty but the rich who don't risk their lives. They live in nice houses and sign papers. It's the soldier who goes around the villages, up and down the mountains, mistreating and murdering his own people.

The soldier went away very impressed; he took this important message with him. When we first caught him, we'd had a lot of ideas, because we wanted to use the gun but didn't know how. It wasn't that we wanted to kill the soldier because we knew very well that one life is worth as much as many lives. But we also knew that the soldier would tell what he'd seen, what he'd felt, and what we'd done to him, and that for us it could mean a massacre—the deaths of children, women, and old people in the village. The whole community would die. So we said: "What we'll do with this man is execute him, kill him. Not here in the village but outside." But people kept coming up with other ideas of what to do, knowing full well the risk we were running. In the end, we decided that, even though it might cost us our lives, this soldier should go and do what we'd asked him,

and really carry through the role he had to play. After about three hours we let him go, in his new disguise. His comrades, the troop of ninety soldiers, hadn't come back for him because they thought he'd been ambushed by guerrillas and they were cowards. They ran off as fast as they could back to town and didn't try to save the soldier left behind. We didn't kill the soldier. The army itself took care of that when he got back to camp. They said he must be an informer, otherwise how could he possibly have stayed and then returned. They said the law says that a soldier who abandons his rifle must be shot. So they killed him.

This was the village's first action and we were happy. We now had two guns, we had a grenade, and we had a cartridge belt, but we didn't know how to use them, nobody knew. We all wanted to find someone who could show us but we didn't know where or who, because whoever we went to, we'd be accused of being guerrillas using weapons. It made us sad to open the rifle and see what was inside, because we knew it killed others. We couldn't use it but it was the custom always to keep anything important. A machete that's not being used for instance, is always smeared with oil and wrapped in a plastic bag so it doesn't rust with the damp or the rain. That's what we had to do with the weapons because we didn't know how to use them. From then on the army was afraid to come up to our villages. They never came back to our village because to get there they would have to go through the mountains. Even if they came by plane they had to fly over the mountains. They were terrified of the mountains and of us. We were happy. It was the most wonderful thing that had happened to us. We were all united. Nobody went down to the *finca*, nobody went to the market, nobody went down to any other place, because they would be kidnapped. . . . The landowners were frightened to come near our village because they thought they would be kidnapped now that our village was organized. So they didn't come near us. The landowners went away and didn't threaten us like before. The soldiers didn't come any more. So we stayed there, the owners of our little bit of land. We began cultivating things so we wouldn't have to go down to town. It was a discipline we applied to ourselves in the village to save lives and only to put ourselves at risk when we had to. My village was organized from this moment on.

I couldn't stay in my village any longer because, now that it could carry on its struggle, organize itself, and take decisions, my role was not important. There was no room for a leader, someone telling others what to do any more. So I decided to leave my village and go and teach another community the traps which we had invented and which our own neighbors had used so successfully. It's now that I move on to teach the people in another village.

Lourdes Arizpe

IN THIS ESSAY the social scientist Lourdes Arizpe examines an ongoing issue in anthropology and women's studies: the modes and patterns of communication that women of very different social origins adopt in order to overcome silence imposed on them by culture and social or economic domination. Peasant women especially have suffered for centuries that their voices were not heard, and prejudice has elaborated the myth of their essential silence. Arizpe points out that peasant women were repressed in their identity since the colonial period, but that they have begun a process of self-consciousness and collective organization because agrarian development in Latin America has brought on a new type of crisis that deeply affects women in the traditional activities of agriculture. This essay belongs to an important trend in Latin American social sciences: Women researchers have learned that the objective status of their work can be successfully combined with a participating perspective that seeks to understand the deep patterns that underlie attitudes and reactions in peasant and Indian cultures. Arizpe's book, *Indígenas en la ciudad de México; el caso de las Marías* (1974), shows how this moral and epistemological standpoint can produce important results in social research.

Peasant Women and Silence

History has imposed a greater silence on peasant women than on any other social group. Perhaps it is the solitude of the plains or the obligatory circumspection of their gender or merely political repression, but circumstances combine to force them to live in a secret world. Doubtless there are those who would assert that their tie to nature leads them to express themselves with actions rather than words. But the male peasant lives in the natural world without being silenced.

Silence, when not deliberate (although, how can we be sure it isn't?) could be anger or wisdom or, simply, a gesture of dignity. When there is

"Peasant Women and Silence," from Fem *29 (December 1983). Translated by Laura Beard Milroy: copyright © 1991 by Sara Castro-Klarén.*

no one worth talking to, I stay silent. If someone doesn't want to recognize my existence, I stay silent. In the spectrum of invisibility that history has imposed on women, perhaps the most invisible of the invisibles have been the peasants.

When direct expression is not permitted, the possibility of knowledge is lost and we fill that disturbing vacuum with phantoms. It is therefore not surprising that the Mexican mentality is filled with myths and stereotypes about peasant women. There is the submissive Indian woman who is a product of condescending maternalism; the wild woman both fantasized about and feared by men; the brazen hussy of melodramatic soap operas; the faint-hearted but treacherous small-town woman invented by an urban mind. Silence is also created by everyone's desire to hear what they want to hear rather than listen to what women are trying to say.

Today it seems that everyone mouths concerns about peasant women without any sincere interest. It is a small difference that applies to all societal relationships, which can be divided two ways. One can speak in the name of a certain group or one can speak for the group. In the first case, a representation is made in an atmosphere in which the speaker is at ease. In the second case, the speaker, by taking over the role of the protagonist, employs one of the most effective forms of social control: the suppression of true expression by the illusion of representation.

When peasant women speak it is because the fabric of their lives has been torn: Rigoberta Menchú explains that in her village the women lived their lives quietly until history interrupted with the brutal repression of Guatemala. Dona Natalia Teniza, in Tlaxcala, has not been silenced. She has been speaking for centuries and she will continue repeating that if there is land to be divided up among the peasants so that they don't die of hunger, we must fight for it. Dona Esperanza also spoke to us in that vein. What is important today is to create opportunities for the peasant women to speak.

It is not that they have never spoken, only that their words have never been recognized. Because their words are discomforting when they denounce exploitation; disturbing when they display a deep understanding of the natural world not shared by their city sisters; strange when they describe an integrating vision of the universe; and because, being women's words, they are not important to androcentric history. Of all the marginalized peoples, peasant women are the most marginalized.

Who Are the Peasant Women?

Whenever we hear of peasants in Mexico, the image that comes to mind is that of a short, smiling woman with braids, bundled up in a black shawl. This image bears no resemblance to the haughty Tehuana or the fastidious

Yucateca, much less to the women of the north. The *ejido*[1] women of Monterrey are called "the millionaires" because of the value of their land, which has become part of the urban zone of Monterrey. They carry pistols to protect both their land and themselves.

We ought to characterize the distinct types of Mexican peasant women by region, as we do in other countries. So what does it mean that instead of talking about *norteñas* [from northern Mexico], *jarochas* [from Veracruz], *tapatías,* or *yucatecas* [from Yucatán], we speak of *peasants?*

We have already shown that referring to the Mazahuas, the Otomies, the Zapotecas, and the rest as Indians is a relic of the original colonial power. We erase their linguistic and cultural identities by lumping them all in the same category. It is an intellectual display of the power of the conqueror who only recognizes the other as the conquered. Thus a Nahuatl woman doesn't speak to me in a subtle, multihued language, from a complex culture with deep historical roots; she speaks to me as an "Indian" and as such, her language and her culture don't matter to me. Only her submission matters. For those of the old school, the greatest sin of the "Indian" has always been insolence—the simple act of rebelling against the silence imposed on her.

And when we speak of peasant women? There is an unpleasant colonial aftertaste when we use *peasant* in a derogatory sense: Peasant is equivalent to backward, lazy, inefficient. This manner of thinking traps us in circular reasoning: The current agricultural crisis in Mexico has been caused by the indolence, the backwardness, and the inefficiency of the peasant. In a nutshell, cause and the result are used to define each other. How convenient.

But there is a deeper significance to this talk about peasants. We can define as peasants those groups that have created a way of life based on the cultivation of the land. It is, certainly, a way of life older than written history itself. It is said that women discovered agriculture while men were out hunting.

But in modern times there is something incongruous about saying that peasants are people dedicated to farming. Because defining an activity as specialized, as separate from other social, political, and cultural activities, stems from the technological mentality that separates the subject from the object, alienating it. This is tantamount to thinking about the peasant from the fragmentary perspective of the urban industrial world. In fact, the peasant way of life became distinct only at the time when a dominant industrial society chose to ignore the historical and social complexity of the peasant society in order to recognize it only as the producer of foodstuffs.

---

1. Ejido = a parcel of land worked by communal labor.

If we step out of this urbocentric mind-set, we will see that the peasant woman is not only a farmer or a housewife. She is the central axle of a way of life that rejects the partialization characteristic of modern industrial society. From this distinct perspective, the agricultural crisis of Mexico would have to be explained from the viewpoint of the deliberate resistance (the rural exodus, the abandonment of seasonal crops, the low productivity) of the peasants to a market system that destroys their integrity and constantly takes the larger part of their earnings. The picture is not so simple.

Peasant Women and the Agricultural Crisis in Mexico and Latin America

The symptoms of the agricultural crisis in our subcontinent are very clear and have been repeated endlessly: the impoverishment of the peasants, ever smaller plots of land, immigration to the cities, and decreased agricultural production. Its causes, on the other hand, are subject to heated debate. There is no doubt that these different aspects are tied, in one way or another, to a great process of agrarian transformation linked to the development of capitalism in the countryside. Those that cling to the positive aspects of such a transformation—for example, the increase of agricultural productivity by means of technology and the accumulation of capital—insist that the great economic and social inequalities it creates are a bitter step, but one that is both temporary and necessary in the process of development. Others see these inequalities as inherent to capitalism and content themselves with predicting its imminent downfall. But the most important thing to know, within this process, is how the peasant economy helps the capitalist structure, what the agrarian reforms mean, and what alternatives for rural development exist for the future. Unfortunately, no one has bothered to study the role that peasant women play in these processes.

Nevertheless, evidence accumulates. The first symptom of pressure on the peasant economy is seen in the number of daughters leaving to work as servants in the city. Contrary to what is happening in African and Asian countries, but just like what happened in Europe during its period of industrialization, in Latin America the migration to the cities has been predominantly female. Since 1950 almost five million rural women have moved to the cities to work mostly in service positions (about 70 percent) and in the nonprofessional urban sector. Rural women have been key to three of the basic processes of the dependent development of Latin America: the rural exodus, the Third World status, and the marginality.

As members of peasant families whose incomes steadily decrease as agricultural prices drop, women have had to compensate for the deficit by increasing their workload. For women in the Henequenera region, their salaried work is vital to the survival of the family. In many cases, women

must work a full day on the farm in addition to their unpaid domestic work. Some add a salaried position in the neighborhood to their double load of domestic and farm work. Sometimes the woman must take over the supervision of the agricultural business when the husband is off on a seasonal job. And she cannot always depend on the technical and financial help offered by some institutions.

As salaried workers in the fields and in the mills, women gain a certain economic independence, but they are confronted with bosses who employ females because they are more easily exploited than males: for their lack of legal protection and union organization, for their inferior status in the workplace, and for the docility imposed upon them by society's rules of conduct. The women also face an undependable job market that is often controlled by intermediaries who expect sexual favors in return for job placement. In addition, the positive changes that could be expected in the relationship between the sexes as a result of women's greater economic independence are not always evident, as Marta Roldan shows in the case of the workers in Sinaloa.

What job alternatives are there then for the peasant women? Ideally, we should find them agroindustrial jobs in companies they could own. That is what the program of Women's Agricultural-Industrial Units (UAIM) has tried to do in the last twelve years. But Teresita De Barbieri has shown how difficult it is to put a theory into practice in a complicated web of economic, bureaucratic, political, and personal interests. Actually, the secretary of agrarian reform's assistant secretary for agrarian organization has proposed to try UAIM again within a broader program for the participation of women in rural development.

Certainly a protest must be lodged against needless wastes of resources in cutting off the National Program of Integration for Development (PRONAM) tied to the National Population Council (CONAPO). Several successful cooperative projects with peasant women will be abandoned, incurring, yet again, the distrust and alienation of the rural participants.

Meanings and Fears

It is often assumed that peasant women have no sense of themselves as women. More specifically, is it that they don't have that consciousness or, what is completely different, that they don't show an individualized consciousness? But it would be absurd to try to explain the consciousness of other people. There is no way to avoid self-awareness from the day a person questions her/his presence on earth. Religion, magic, rituals, and mythologies are all the result of that questioning and rural culture is woven from those beliefs. But the answer given by peasant women, in their very different

cultures, is very different from the one given by rational, individualized Western culture.

Within the wide range of integrating philosophies of the peasants, "being" is not restricted to humans. Each element derives its existence from its relation to the whole. The being is the whole. Thus is it difficult for them to understand a feminism that pleads for one part of the whole, for women. Nevertheless, they *are* women and we can understand their feelings and their fears about menstruation, virginity, and the experiences of marital relationships, births, and sexual violence. Violence can be named in a thousand different languages, but it is still violence.

As always, seemingly unavoidably, we have arrived at the same point: That there is a lot we must learn in order to understand and, even more, to empathize. As we have said so many times, breaking the silence is the first step.

Further Readings

Delmira Agustini

El libro blanco. Montevideo: O. M. Bertani, 1907.
Cantos de la mañana. Montevideo: O. M. Bertani, 1910.
Los cálices vacíos. Montevideo: O. M. Bertani, 1913
Obras completas de Delmira Agustini. Montevideo: Máximo García, 1924. 2 vols.
(*El rosario de Eros* and *Los astros del abismo*).

Secondary Sources

Burt, John R. "The Personalization of Classical Myth in Delmira Agustini." *Crítica Hispánica* 9 (1987): 115–127.
_____ . "Agustini's Muse." *Chasqui* 17 (1988): 61–65.
Fox-Lockert, Lucía. "Amorous Fantasies." *Americas* 39 (1987): 38–41.
Molloy, Sylvia. "Dos figuras del cisne: Rubén Darío y Delmira Agustini." *La sartén por el mango.* Eds. Patrícia González and Eliana Ortega, 57–70. San Juan: Ediciones Huracán, 1984.
Silva, Clara. *Genio y figura de Delmira Agustini.* Buenos Aires: EUDEBA, 1968.
Stephens, Doris T. *Delmira Agustini and the Quest for Transcendence.* Montevideo: Ediciones Géminis, 1975.
Valenti, Jeanette Y. "Delmira Agustini: A Reinterpretation of Her Poetry." Ph.D. diss., Cornell University, 1971.

Lourdes Arizpe

"Las campesinas y el silencio." *Fem* 29 (December 1983).
Indígenas en la ciudad de México; el caso de las Marías. Mexico, 1974.

Domitila Barrios de Chungara

"Si me permiten hablar" . . . *Testimonio de Domitila, una mujer de las minas de Bolivia.* Ed. Moema Viezzer. Mexico: Siglo XXI Editores, 1977.
_____ , and Moema Viezzer. *Let Me Speak! Testimony of Domitila, a Woman of the Bolivian Mines.* Trans. Victoria Ortiz. London: Monthly Review Press, 1978.

Secondary Sources

Rice-Sayre, Laura P. "Witnessing History: Diplomacy Versus Testimony." *Testimonio y literatura.* Ed. and introd. Rene Jara, 48–72. Minneapolis: Institute for Study of Ideologies and Literature, 1986.

Hebe de Bonafini

Historias de vida. Ed. Matilde Sánchez. Buenos Aires: Fraterna/ Del Nuevo Extremo, 1985.

Secondary Sources

Jelin, Elizabeth, ed. *Los nuevos movimientos sociales.* Buenos Aires: CEAL, 1985.

María Borjas and Rosario Antúnez

Interviews in *The Tiger's Milk: Women of Nicaragua.* Eds. Adriana Angel and Fiona Macintosh. New York: Henry Holt, 1987.

Julieta Campos

La imagen en el espejo. Mexico: Joaquín Mortiz, 1965.
Muerte por agua. Mexico: Fondo de Cultura Economica, 1965.
Celina o los gatos. Mexico: Siglo XXI, 1968.
Oficio de leer. Mexico: Joaquín Mortiz, 1971.
Tiene los cabellos rojizos y se llama Sabina. Mexico: Joaquín Mortiz, 1974.
El miedo de perder a Euridice. Mexico: Joaquín Mortiz, 1979.
"A Redhead Named Sabina." *Women's Fiction from Latin America.* Ed. and trans. Evelyn Picon Garfield, 243–251. Detroit: Wayne State University Press, 1988.

Secondary Sources

Agüera, Victorio G. "El discurso de lo imaginario en *Tiene los cabellos rojizos y se llama Sabina.*" *Revista Iberoamericana* 51, 132–133 (July–December 1985): 531–537.
Agustín, José. "Entrevista grabada con Julieta Campos." *Diagrama de la Cultura* 27 (March 1966): 3, 6.
Batis, Huberto. "*Muerte por agua,* la novela del tiempo detenido." *El Heraldo Cultural* 13, 6 (February 1966): 15.
Glantz, Margo. "Entre lutos y gatos: José Agustín y Julieta Campos." *Repeticiones: Ensayos sobre literatura mexicana,* 72–74. Veracruz: Universidad Veracruzana, 1979.
Martínez, Martha. "Julieta Campos o la interiorización de lo cubano." *Revista Iberoamericana* 51, 132–133 (July–December 1985): 793–797.
Pacheco, José Emilio. "Novela versus lenguaje poético." *Revista de la Universidad de México* 20, 10 (June 1966): 35.
Paley Francescato, Martha. "Un desafío a la crítica literaria: *Tiene los cabellos rojizos y se llama Sabina* de Julieta Campos." *Revista de Crítica Literaria Latinoamericana* 7, 13 (1981): 121–125.
Rivero Potter, Alicia. "La creación literaria en Julieta Campos: *Tiene los cabellos rojizos y se llama Sabina.*" *Revista Iberoamericana* 51, 132–133 (July–December 1985): 899–907.
Verani, Hugo. "Julieta Campos y la novela del lenguaje." *Texto Crítico* 2, 4 (1976): 132–149.

Rosario Castellanos

Apuntes para una declaración de fe. Mexico, 1948.
Sobre cultura femenina. Mexico: Ediciones de América, Revista Antológica, 1950.
Balún Canán. Mexico: Fondo de Cultura Económica, 1957.
Ciudad real. Xalapa, Mexico: Novaro, 1960.
Oficio de tinieblas. Mexico: Fondo de Cultural Económica, 1962.
Los convidados de agosto. Mexico: Era, 1964.
"La novela mexicana contemporánea y su valor testimonial." *Hispania* 47 (1964).
"Autobiografía." *Los narradores ante el público,* 89–98. Mexico: Joaquín Mortiz, 1966.
Juicios sumarios. Xalapa, Mexico: Universidad Veracruzana, 1966.
Album de familia. Mexico: Joaquín Mortiz, 1971.
Poesía no eres tú; obra poética, 1948–1971. Mexico: Fondo de Cultura Económica, 1972.
Mujer que sabe latín. Mexico: Secretaria de Educación Pública, 1973.
El uso de la palabra. Mexico: Excelsior, 1974.
El eterno femenino. Mexico: Fondo de Cultura Económica, 1975.
The Nine Guardians. Trans. Irene Nicholson. London: Faber and Faber, 1959. Reprint. New York: Vanguard Press, 1960.

Secondary Sources

A Rosario Castellanos, sus amigos. Mexico: Cía. Impresora y Litográfica Juventud, 1975.
Ahern, Maureen. "A Critical Bibliography of and About the Works of Rosario Castellanos." *Homenaje a Castellanos.* Valencia: Albatroz-Hispanófila, 1980.
Anderson, Helene M. "Rosario Castellanos and the Structures of Power." *Contemporary Women Authors of Latin America: Introductory Essays.* Eds. Doris Meyer and Margarite Fernández Olmos, 22–32. Brooklyn, N.Y.: Brooklyn College Press, 1983.
Beer, Gabriella de. "Feminismo en la obra poética de Rosario Castellanos." *Revista de Crítica Literaria Latinoamericana* 7, 13 (1981): 105–112.
Benedetti, Mario. "Rosario Castellanos y la incomunicación social." *Letras del continente mestizo,* 130–135. Montevideo: Arca, 1967. Reprint, 165–170. Montevideo: Arca, 1969.
Brushwood, John. *Mexico in Its Novel: A Nation's Search for Identity.* Notre Dame, Ind.: University of Notre Dame Press, 1971.
Campos, Julieta. "La novela mexicana después de 1940." *La imagen en el espejo,* 141–157. Mexico: Universidad Nacional Autónoma de México, 1965.
Palley, Julian, ed. and trans. *Meditation on the Threshold: A Bilingual Anthology of Poetry.* Tempe, Ariz.: Bilingual Press, 1988.

Ana Cristina Cesar

Cenas de Abril. 1979.
Correspondencia Completa. 1979.
Literatura Não é Documento. Rio de Janeiro: Edição FUNARTE, 1980.

A Teus Pés. São Paulo: Brasiliense, 1982.
Inéditos e Dispersos. São Paulo: Brasiliense, 1985.

Vanessa Droz

La cicatriz a medias. Río Piedras, Puerto Rico: Editorial Cultural, 1982.

Diamela Eltit

Lúmperica. Santiago de Chile: Ornitorrinco, 1983.
Por la patria. Santiago de Chile: Ornitorrinco, 1986.
El cuarto mundo. Santiago de Chile: Planeta, 1988.
El padre mio. Santiago de Chile: Editorial Francisco Zegert, 1990.

Secondary Sources

Castro-Klarén, Sara. "Del recuerdo y el olvido: el sujeto en *En breve cárcel y Lumpérica.*" *Escritura, transgresión y sujeto en la literatura latinoamericana,* 196–207. Mexico: Premiá, 1989.

Lygia Fagundes Telles

O Cacto Vermelho. Rio de Janeiro: Editôra Mérito, 1949.
Ciranda de Pedra. Lisbon: Editorial Minerva, 1956.
Histórias do Descontro. Rio de Janeiro: José Olympio, 1958.
Histórias Escolhidas. São Paulo: Boa Leitura, 1961.
Os Mortos. Lisbon: Casa Portuguesa, 1963.
Verão no Aquário. São Paulo: Martins, 1963.
O Jardim Selvagem. São Paulo: Martins, 1965.
Antes do Baile Verde. Rio de Janeiro: Edições Block, 1970.
As Meninas. Rio de Janeiro: José Olympio, 1973.
Seminário dos Ratos. Rio de Janeiro: José Olympio, 1977.
Filhos Pródigos. São Paulo: Livraria Cultura Editora, 1978.
A Disciplina do Amor; Fragmentos. 3rd ed. Rio de Janeiro, 1980.
Tigrela and Other Stories. Trans. Margaret Neves. New York: Avon, 1977.
The Girl in the Photograph. Trans. Margaret Neves. New York: Avon, 1982.
The Marble Dance. Trans. Margaret Neves. New York: Avon, 1986.

Secondary Sources

Angelini, Paulo, and Roberto Escudero. "A intertextualidade em Dois Contos Femeninos." *Cadernos de Lingüística e Teoría de Literatura* 8 (December 1982): 107–115.
Lucas, Fábio. "Misterio e magia: Contos de Lygia Fagundes Telles." *J. Letras* 23 (April 1971): 248.
Montfeiro, Leonardo. *Lygia Fagundes Telles: Seleção de Textos, Notas Estudos Biográfico, Histórico e Crítico e Exercícios.* São Paulo: Abril Educação, 1981.
Silva, Vera, and María Tietzmann. *A Metamorfose nos Contos de Lygia Fagundes Telles.* Rio de Janeiro, 1985.

Soledad Fariña

El primer libro. Santiago de Chile: Ediciones Amaranto, 1985.
Albricia. Santiago de Chile: Ediciones Archivo, 1988.

Rosario Ferré

El medio pollito. Río Piedras, Puerto Rico: Ediciones Huracán, 1976.
Papeles de Pandora. Mexico: Joaquín Mortiz, 1976.
La caja de cristal. Mexico: La máquina de escribir, 1978.
La muñeca menor. Río Piedras, Puerto Rico: Ediciones Huracán, 1980.
Sitio a Eros: trece ensayos literarios. Mexico: Joaquín Mortiz, 1980.
Los cuentos de Juan Bobo. Río Piedras, Puerto Rico: Ediciones Huracán, 1981.
La mona que le pisaron la cola. Río Piedras, Puerto Rico: Ediciones Huracán, 1981.
"La cocina de la escritura." *Literatures in Transition: The Many Voices of The Caribbean Area: A Symposium.* Ed. Rose S. Minc, 17–19. Gaithersburg, Md.: Hispamérica, 1982.
Fábulas de la garza desangrada. Mexico: Joaquín Mortiz, 1982.
"The Youngest Doll." Trans. Gregory Rabassa. *Kenyon Review* 2, 1 (1980).
"When Women Love Men." Trans. Cynthia Ventura. *Contemporary Women Authors of Latin America: Introductory Essays.* Eds. Doris Meyer and Margarite Fernández Olmos, 176–185. Brooklyn, N.Y.: Brooklyn College Press, 1983.
"The Glass Fox." Trans. by the author and Kathy Taylor. *Massachusetts Review* 3 and 4 (Fall–Winter 1986): 699–711.
Maldito amor. Mexico: Joaquín Mortiz, 1986.
"The Writer's Kitchen." Trans. by the author and Diana Vélez. *Feminist Studies* 12, 2 (Summer 1986).
Sweet Diamond Dust. Trans. by the author. New York: Ballantine, 1988.

Secondary Sources

Fernández Olmos, Margarite. "Sex, Color and Class in Contemporary Puerto Rican Women Authors." *Heresis* 4 (1982): 42–47.
———. "From a Woman's Perspective: The Short Stories of Rosario Ferré and Ana Lydia Vega." *Contemporary Women Authors of Latin America: Introductory Essays.* Eds. Doris Meyer and Margarite Fernández Olmos, 78–90. Brooklyn, N.Y.: Brooklyn College Press, 1983.
———. "Constructing Heroines: Rosario Ferré's *Cuentos infantiles* and Feminine Instruments of Change." *The Lion and the Unicorn: A Critical Journal of Children's Literature* 10 (1986): 83–94.
———. "Luis Rafael Sánchez and Rosario Ferré: Sexual Politics and Contemporary Puerto Rican Narrative." *Hispania* 70 (1987): 40–46.
Gelpí-Pérez, Juan. "Apuntes al margen de un texto de Rosario Ferré." *La sartén por el mango.* Eds. Patricia González and Eliana Ortega, 133–135. San Juan: Ediciones Huracán, 1984.
Lagos-Pope, María Inés. "Sumisión y rebeldía: El doble o la representación de la alienación femenina en narraciones de Marta Brunet y Rosario Ferré." *Revista Iberoamericana* 51, 132–133 (July–December 1985): 731–749.

López, Ivette. "La muñeca menor: Ceremonias y transformaciones en un cuento de Rosario Ferré." *Explicación de Textos Literarios* 11, 1 (1982–1983): 49–58.

Patricia Galvão (Mara Lobo, Pagú)

Parque Industrial. São Paulo, 1933.
"O álbum de Pagú." *Pagú, Patricia Galvão: Vida Obra.* Ed. Augusto de Campos. São Paulo, 1982.
"O Romance da Epoca Anarquista." *Pagú, Patricia Galvão: Vida Obra.* Ed. Augusto de Campos. São Paulo, 1982.
"A verdade e liberdade." *Pagú, Patricia Galvão: Vida Obra.* Ed. Augusto de Campos. São Paulo, 1982.
"A mulher do Povo." *O Homen do Povo.* São Paulo, 1931.
Galvão, Patricia, and Oswald Andrade. *O Homen do Povo.* São Paulo, 1931.

Secondary Sources

Besse, K. Susan, "Pagú: Patricia Galvão—Rebel." *The Human Tradition in Latin America.* Eds. William H. Beezley and Judith Ewell, 103–117. Wilmington, Del.: Scholarly Resources, 1984.
Jackson, David. "Patricia Galvão e o Realismo Social Brasileiro dos Anos 30." *A Famosa Revista.* Ed. Geraldo Ferraz. 1945.

Elena Garro

Un hogar sólido y otras piezas en un acto. Xalapa, Mexico: Universidad Veracruzana, 1958.
Los recuerdos del porvenir. Mexico: Joaquín Mortiz, 1963.
La semana de colores. Xalapa, Mexico: Universidad Veracruzana, 1964.
El árbol. Mexico: Editorial Peregrina, 1967.
Andamos huyendo Lola. Mexico: Joaquín Mortiz, 1980.
Testimonios sobre Mariana. Mexico: Grijalbo, 1981.
Reencuentro de personajes. Mexico: Grijalbo, 1982.
La casa junto al río. Mexico: Grijalbo, 1983.
Recollections of Things to Come. Trans. Ruth L. C. Simms. Austin: University of Texas Press, 1969.
"A Solid Home." Trans. Francesca Colecchia and Julio Matas. *Selected Latin American One-Act Plays.* Pittsburgh: Pittsburgh University Press, 1973.
The Lady on Her Balcony. Trans. Beth Miller. *Shantih* 3, 3 (Fall–Winter 1976): 36–44.
"The Day We Were Dogs." *Contemporary Women Authors of Latin America: Introductory Essays.* Eds. Doris Meyer and Margarite Fernández Olmos. Brooklyn, N.Y.: Brooklyn College Press, 1983.

Secondary Sources

Anderson, Robert K. "Myth and Archetype in *Recollections of Things to Come.*" *Studies in Twentieth Century Literature* 9, 2 (Spring 1985): 213–227.

Balderston, Daniel. "The New Historical Novel: History and Fantasy in *Los recuerdos del porvenir.*" *Bulletin of Hispanic Studies* 66 (1989): 41–46.

Boschetto, Sandra. "Romancing the Stone in Elena Garro's *Los recuerdos del porvenir.*" *Journal of Midwest Modern Language Association* 22 (1989): 1–11.

Callan, Richard J. "Analytical Psychology and Garro's *Los pilares de doña Blanca.*" *Latin American Theater Review* 16, 2 (Spring 1983): 31–35.

Carballo, Emmanuel. "La vida y la obra de Elena Garro." *Sábado* (supplement to *Unomásuno*) January 24, 1981: 2–5.

Duncan, Cynthia. " 'La culpa es de los Tlaxcaltecas': A Reevaluation of Mexico's Past Through Myth." *Crítica Hispánica* 7 (1985): 105–120.

Echeverría, Miriam B. "Texto y represión en 'Los Perros' de Elena Garro." *De la crónica a la nueva narrativa mexicana.* Eds. Merlin H. Foster and Julio Ortega, 423–429. Mexico: Oasis, 1986.

Franco, Jean. *Plotting Women: Gender and Representation in Mexico.* New York: Columbia University Press, 1989.

Fox-Lockert, Lucía. "The Meaning of Freedom in the Mexican Feminist Novel." *Sixth Conference of Ethnic Studies.* University of Wisconsin, La Crosse, 1978.

Johnson, Harvey L. "Elena Garro's Attitudes Toward Mexican Society." *South Central Bulletin* 40, 4 (Winter 1980): 150–152.

Larson, Ross. *Fantasy and Imagination in the Mexican Narrative.* Tempe: Arizona State University, Center for Latin American Studies, 1977.

Méndez Rodenas, Adriana. "Tiempo femenino, tiempo ficticio: *Los recuerdos del porvenir* de Elena Garro." *Revista Iberoamericana* 51, 132–133 (July–December 1985): 843–851.

Meyer, Doris. "Alienation and Escape in Elena Garro's *La semana de colores.*" *Hispanic Review* 55 (1987): 153–164.

Mora, Gabriela. "A Thematic Exploration of the Works of Elena Garro." *Latin American Women Writers Yesterday and Today,* 91–97. Pittsburgh: Carnegie-Mellon, 1975.

Orenstein, Gloria. "Surrealism and Women: The Occultation of the Goddess: Elena Garro, Joyce Mansour, Leonora Carrington." *The Theater of the Marvelous,* 99–147. New York: New York University Press, 1975.

Rosser, Harry L. "Form and Content in Elena Garro's *Los recuerdos del porvenir.*" *Revista Canadiense de Estudios Hispánicos* 2 (1978): 281–295.

―――. "Regional Reactions to Incursions from the Center: Yáñez, Garro, Mojarro." *Conflict and Transition in Rural Mexico,* 23–36. Boston: Crossroads Press, 1980.

Stoll, Anita K., ed. *A Different Reality: Studies on the Works of Elena Garro.* Lewisburg, Pa.: Bucknell University, 1990.

Margo Glantz

Las mil y una calorías (novela dietética). Mexico: Premiá, 1978.

Doscientas ballenas azules. Mexico: La máquina de escribir, 1979.

Intervención y pretexto. Mexico: Universidad Nacional Autónoma de México, 1980.

No pronunciarás. Mexico: Premiá, 1980.

Las genealogías. Mexico: Martin Casillas, 1981.

El día de tu boda. Mexico: Martin Casillas, 1982.
La lengua en la mano. Mexico: Premiá, 1983.
De la amorosa inclinación a enredarse en los cabellos. Mexico: Océano, 1984.
Erosiones. Mexico: Universidad Nacional Autónoma de México, 1984.
Síndrome de naufragios. Mexico: Joaquín Mortiz, 1984.
"Mi escritura tiene . . ." *Revista Iberoamericana* 51, 132–133 (July–December 1985): 475–478.

Secondary Sources

Otero-Krauthammer, Elizabeth. "Integración de la identidad judía en *Las geneaologías* de Margo Glantz." *Revista Iberoamericana* 51, 132–133 (July–December 1985): 867–873.
Valenzuela, Luisa. "Mis brujas favoritas." *Theory and Practice of Feminist Literary Criticism.* Eds. Gabriela Mora and Karen S. Van Hooft, 88–95. Ypsilanti, Mich.: Bilingual Press, 1982.

Lucila Godoy Alcayaga, see Gabriela Mistral

Carolina Maria de Jesus

Quarto de Despejo. Diário de uma Favelada. Rio de Janeiro: Francisco Alves Editora, 1960.
Diário de Bitita. Rio de Janeiro: Nova Fronteira, 1986.
Child of the Dark: The Diary of Carolina Maria de Jesus. Trans. David St. Clair. New York: E. P. Dutton, 1962.

Tamara Kamenszain

De este lado del Mediterráneo. Buenos Aires: Ediciones Noé, 1973.
Los no. Buenos Aires: Sudamericana, 1977.
El texto silencioso. Mexico: Universidad Nacional Autónoma de México, 1983.
La casa grande. Buenos Aires: Sudamericana, 1986.

Norah Lange

La calle de la tarde. Buenos Aires: Samet, 1925.
Los días y las noches. Buenos Aires: Sociedad de Publicaciones El Inca, 1926.
El rumbo de la rosa. Buenos Aires: Proa, 1930.
Cuadernos de infancia. Buenos Aires: Losada, 1937.
Antes que mueran. Buenos Aires: Losada, 1944.
Personas en la sala. Buenos Aires: Sudamericana, 1950.
Los dos retratos. Buenos Aires: Losada, 1956.
Estimados congéneres. Buenos Aires: Losada, 1968.

Secondary Sources

Lindstrom, Naomi. "Norah Lange, presencia desmonumentalizadora y femenina en la vanguardia argentina." *Crítica Hispánica* 5, 2 (1983): 131–148.

Molloy, Sylvia. "Dos proyectos de vida: *Cuadernos de infancia* de Norah Lange y *El archipiélago* de Victoria Ocampo." *Filología* (Buenos Aires) 20, 2 (1985): 279–293.

Sarlo, Beatriz. "Decir y no decir: erotismo y represión." *Una modernidad periférica: Buenos Aires, 1920–1930,* 69–93. Buenos Aires: Nueva Visión, 1988.

Henriqueta Lisbõa

Enternecimento. Rio de Janeiro: Pongetti, 1929.

Velário. Belo Horizonte: Impresa Oficial, 1936.

O Menino Poeta. Rio de Janeiro: Bedeschi, 1943. Reprint, Belo Horizonte: Secretaría da Educação e Cultura, 1975.

Flor da Morte. Belo Horizonte: Edições Calazans, 1949.

Convívio Poético. Belo Horizonte: Secretaría da Educação, Coleção Cultural, 1955.

O Alvo Humano. São Paulo: Editora do Escritor, 1973.

Vivência Poética, Ensalos. Belo Horizonte: Impresa Oficial, 1979.

Secondary Sources

Lucas, Fabio. "A poesía de Henriqueta Lisbõa." *Revista Hispánica Moderna* 41, 1 (1988): 69–74.

Guimarães Filho, Alphonsus de. "Lembrança de Henriqueta Lisbõa." *Coloquio/Letras* 94 (1986): 83–84.

Virgilio, Carmelo. "The Image of Woman in Henriqueta Lisbõa's 'Frutescencia.'" *Luso-Brazilian Review* 23, 1 (1986): 89–106.

Clarice Lispector

Perto do Coração Selvagem. Rio de Janeiro: A Noite, 1944.

O Lustre. Rio de Janeiro: Agir, 1946.

A Cidade Sitiada. Rio de Janeiro: A Noite, 1949.

Alguns Contos. Rio de Janeiro: Ministerio de Educação et Saúde, 1952.

Laços de Família. Rio de Janeiro: Francisco Alves, 1960.

A Maça no Escuro. Rio de Janeiro: Francisco Alves, 1961.

A Legião Estrangeira. Rio de Janeiro: Edição do Autor, 1964.

A Paixão Segundo G. H. Rio de Janeiro: Edição do Autor, 1964.

O Misterio do Coelho Pensante. Rio de Janeiro: José Alvaro, 1967.

A Mulher que Matou os Peixes. Rio de Janeiro: Sabiá, 1968.

Uma Aprendizagem ou o Livro dos Prazeres. Rio de Janeiro: Sabiá, 1969.

Felicidade Clandestina: Contos. Rio de Janeiro: Sabiá, 1971.

Agua Viva. Rio de Janeiro: Artenova, 1973.

A Via Crucis do Corpo. Rio de Janeiro: Artenova, 1974.

A Vida Intima de Laura. Rio de Janeiro: José Olympio, 1974.

Onde Estivestes De Noie. Rio de Janeiro: Artenova, 1974.

De Corpo Inteiro. Rio de Janeiro: Artenova, 1975.

A Hora da Estrela. Rio de Janeiro: José Olympio, 1977.

Um Sopro de Vida. Rio de Janeiro: Nova Fronteira, 1978.

Para não Esquecer. Rio de Janeiro: Atica, 1978.

Quase de Verdade. Rio de Janeiro: Rocco, 1978.

A Bela e a Fera. Rio de Janeiro: Nova Fronteira, 1979.

A Descoberta do Mundo. Rio de Janeiro: Nova Fronteira, 1984.

Family Ties. Trans. Giovanni Pontiero. Austin: University of Texas Press, 1972.

The Apple in the Dark. Trans. Gregory Rabassa. Austin: University of Texas Press, 1986.

An Apprenticeship, or the Book of Delights. Trans. R. Mazzara and L. Parris. Austin: University of Texas Press, 1986.

The Foreign Legion. Trans. G. Pontiero. Manchester: Carcanet, 1986.

The Hour of the Star. Trans. G. Pontiero. Manchester: Carcanet, 1986.

The Passion According to G. H. Trans. Ronald W. Sousa. Minneapolis: University of Minnesota Press, 1988.

Soulstorm. Trans. Alexis Levitin. Introd. G. Paley. New York: New Directions, 1989.

The Stream of Life. Trans. E. Lowe and E. Fitz. Foreword H. Cixous. Minneapolis: University of Minnesota Press, 1989.

Secondary Sources

Bryan, C.D.B. "Afraid to Be Afraid." *New York Times Book Review.* September 3, 1967: 22–23.

Cixous, Hélène. "Reaching the Point of Wheat, or a Portrait of the Artist as a Maturing Woman." *New Literary History* 19 (1987): 1–21.

Douglass, Ellen H. "Myth and Gender in Clarice Lispector: Quest as a Feminist Statement in *A Imitação da Rosa.*" *Luso-Brazilian Review* 25 (1988): 15–31.

Fitz, Earl. "Freedom and Self-Realization: Feminist Characterization in the Fiction of Clarice Lispector." *Modern Language Studies* 10, 3 (1983): 51–56.

———. *Clarice Lispector.* Boston: G. K. Hall, 1985.

———. "The Passion of Logo(centrism), or, the Deconstructionist Universe of Clarice Lispector." *Luso-Brazilian Review* 25 (1988): 33–44.

Froula, Christine. "Rewriting Genesis: Gender and Culture in Twentieth Century Texts." *Tulsa Studies in Womens' Literature* 7 (1988): 197–220.

Lastinger, Valerie C. "Humor in a New Reading of Clarice Lispector." *Hispania* 72 (1989): 130–137.

Lindstrom, Naomi. "Clarice Lispector: Articulating Women's Experience." *Chasqui* 8, 1 (1978): 43–52.

———. "A Feminist Discourse Analysis of Clarice Lispector's 'Daydreams of a Drunken Housewife.'" *Latin American Literary Review* 9 (1981): 7–16.

Nunes, Benedito. *O Mundo de Clarice Lispector.* Manaus, Brazil: Edições Governo do Stado do Amazonas, 1966.

Nunes, María Luisa. "Narrative Modes in Clarice's Lispector's *Laços e família.*" *Luso-Brazilian Review* 14, 2 (1977).

Palls, Terry L. "The Miracle of the Ordinary: Literary Epiphany in Virginia Woolf and Clarice Lispector." *Luso-Brazilian Review* 21 (1984): 63–78.

Peixoto, Marta. "*Family Ties:* Female Development in Clarice Lispector." *The Voyage In: Fictions of Female Development.* Eds. Elizabeth Abel, Marianne Hirsch, and Elizabeth Langland, 287–303. Hanover, N.H.: University Press of New England, 1983.

_____ . "Writing the Victim in the Fiction of Clarice Lispector." *Transformations of Literary Language in Latin American Literature from Machado de Assis to the Vanguards.* Ed. K. David Jackson, 84–97. Austin: Abaporu Press, 1987.

_____ . "Rape and Textual Violence in Clarice Lispector." *Representing Rape.* Eds. Lynn Higgins and Brenda Silver. New York: Columbia University Press, 1991.

Wheeler, A. M. "Animal Imagery as Reflection of Gender Roles in Clarice Lispector's *Family Ties." Critique: Studies in Modern Fiction* 28 (1987): 125–134.

Paulina Luisi

"Feminismo." *La mujer: encuesta feminista argentina.* Ed. Miguel A. Font. Buenos Aires, 1921.

Cecília Meireles

Nunca Mais. Rio de Janeiro: Editora Leite Ribeiro, 1923.

Viagem. Lisbon: Edições Occidente, 1939.

Vaga Música. Rio de Janeiro: Irmaos Pongetti Editora, 1942.

Mar Absoluto e Outros Poemas. Porto Alegre: Livraria do Globo, 1945.

Retrato Natural. Rio de Janeiro: Livros de Portugal, 1949.

Doze Noturnos da Holanda e O Aeronauta. Rio de Janeiro: Livros de Portugal, 1952.

Poemas Escritos na India. Rio de Janeiro: Livraria São José, [1953].

Romanceiro da Inconfidencia. Rio de Janeiro: Livros de Portugal, 1953.

Canções. Rio de Janeiro: Livros de Portugal, 1956.

Obra Poética. Rio de Janeiro: José Aguilar Editora, 1958; revised edition, 1967.

Metal Rosicler. Rio de Janeiro: Livros de Portugal, 1960.

Solombra. Rio de Janeiro: Livros de Portugal, 1963.

Ou Isto ou Aquilo. São Paulo: Editora Girofle, 1964.

Antología Poética. Lisbon: Guimarães Editores, 1968.

Poemas Italianos. Sao Paulo: Instituto Cultural Italo-Brasileiro, 1968.

Poesias Completas. Ed. Darcy Damasceno. Rio de Janeiro: Civilizaçao Brasileira, 1976.

Poems in Translation. Eds. Henry Keith and Raymond Sayers. Washington, D.C.: Brazilian American Cultural Institute, 1977.

Secondary Sources

Peixoto, Marta. "Poetic Identity and Female Gender in Cecília Meireles." *Identity in Portuguese and Brazilian Literature: Proceedings of a Mini Conference,* 50–57. Pittsburgh: University of Pittsburgh Press, 1982.

_____ . " 'Caminante do vazio': Voyages and Disconnection in Cecília Meireles." *Discurso Literario* 1, 2 (1984): 267–276.

_____ . "The Absent Body: Female Signature and Poetic Convention in Cecília Meireles." *Bulletin of Hispanic Studies* (January 1988): 87–99.

Sadlier, Darlene J. "Looking into the Mirror in Cecília Meireles' Poetry." *Romance Notes* 23 (1982): 119–122.

———— . *Imagery and Theme in the Poetry of Cecília Meireles: A Study of* Mar Absoluto. Madrid: José Porrúa Turanzas, Studia Humanitatis Series, 1983.

Sayers, Raymond S. "The Poetic Universe of Cecília Meireles." *From Linguistics to Literature: Romance Studies Offered to Francis M. Rogers.* Ed. Bernard Bichakjain, 243–258. Amsterdam: Benjamins, 1981.

Rigoberta Menchú

Me llamo Rigoberta Menchú y así me nació la conciencia. Barcelona: Editorial A. Vergara, 1983.

I, Rigoberta Menchú, an Indian Woman in Guatemala. Trans. Ann Wright. Ed. and introd. Elisabeth Burgos-Debray. London: Verso, 1984.

Secondary Sources

Beverly, John. "Anatomía del testimonio." *Revista de Crítica Literaria Latinoamericana* (Lima) 25 (1987): 7–16.

Feal, Rosemary. "Spanish American Ethnobiography and the Slave Narrative Tradition: *Biografía de un cimarrón* and *Me llamo Rigoberta Menchú. Modern Language Studies,* forthcoming.

Patai, Daphne. *Brazilian Women Speak: Contemporary Life Stories.* New Brunswick and London: Rutgers University Press, 1988.

Rice-Sayre, Laura P. "Witnessing History: Diplomacy versus Testimony." *Testimonio y literatura.* Ed. and introd. Rene Jara, 48–72. Minneapolis: Institute for Study of Ideologies and Literature, 1986.

Saffiotti, Heleieth I. B. *Women in Class and Society.* Trans. Michael Vale. New York and London: Monthly Review Press, 1978.

Sommer, Doris. "Not Just a Personal Story: Women's Testimonios and the Plural Self." *Life/Lines: Theoretical Essays on Women's Autobiography.* Eds. Celeste Schenk and Bella Brodzki, 107–130. Ithaca: Cornell University Press, 1988.

Gabriela Mistral

Desolación. New York: Instituto de las Españas, 1922.

Lecturas para mujeres. Mexico: Secretaría de Educación, 1923.

Ternura: canciones de niños. Madrid: Saturnino Calleja, 1924.

Tala. Buenos Aires: Sur, 1938.

Antología: selección de la autora. Prologue Ismael Edwards Matte. Santiago de Chile: Zig Zag, 1941.

Poemas de las madres. Santiago de Chile: Pacífico, 1950.

Todas íbamos a ser reinas. Santiago de Chile: Nacional Quimantu, 1971.

Antología general de Gabriela Mistral. Special issue of *Orfeo,* 23–27.

Cartas de amor de Gabriela Mistral. Ed. Sergio Fernández Larraín. Santiago de Chile: Andrés Bello, 1978.

Gabriela piensa en . . . Selección de prosas. Ed. Esteban Scarpa. Santiago de Chile: Andrés Bello, 1978.

Selected Poems. Trans. Langston Hughes. Bloomington: Indiana University Press, 1957.

Selected Poems of Gabriela Mistral. Trans. and ed. Doris Dana. Baltimore: Published for the Library of Congress by the Johns Hopkins University Press, 1971.
Crickets and Frogs: A Fable. Trans. and adapt. Doris Dana. New York: Atheneum, 1972.

Secondary Sources

Chase, Cida S. "Perfil ético de Gabriela Mistral." *Discurso Literario: Revista de Temas Hispánicos* 1, 2 (1984): 159–167.
Daydi-Tolson, Santiago. "La locura en Gabriela Mistral." *Revista Chilena de Literatura* 21 (1983): 47–62.
_____. "Las patrias de Gabriela Mistral." *Revista Chilena de Literatura* 27–28 (1986): 197–202.
_____. "Mater Dolorosa: A Feminine Archetype in Gabriela Mistral's Poetry." *Memory of Willis Knapp Jones,* 55–67. Reprint. *Retrospect: Essays on Latin American Literature.* Eds. Elizabeth Rogers and Timothy Rogers. York, S.C.: Sp. Lit. Pubs. Co., 1987.
Mandlove, Nancy B. "Gabriela Mistral: The Narrative Sonnet." *Revista/Review Interamericana* 12, 1 (1982): 110–114.
Vázquez, Margot Arce de. *Gabriela Mistral: The Poet and Her Work.* New York: New York University Press, 1964.
Virgilio, Carmelo. "Woman as Metaphorical System: An Analysis of Gabriela Mistral's Poem 'Fruta.'" *Woman as Myth and Metaphor in Latin American Literature.* Eds. Carmelo Virgilio and Naomi Lindstrom, 137–150. Columbia: University of Missouri Press, 1985.

Alicia Moreau de Justo

Evolución y educación. 1915.
La mujer en la democracia. Buenos Aires: El Ateneo, 1945.
Juan B. Justo y el socialismo. 1984.

Secondary Sources

Henault, Mirta. *Alicia Moreau de Justo.* Buenos Aires: CEAL, 1983.

Nancy Morejón

Mutismos. Havana, 1962.
Amor, ciudad atribuída. Havana: El Puente, 1964.
Richard trajo su flauta. Havana: Unión de Escritores y Artistas de Cuba, 1967.
Lengua de pájaro. Havana: Instituto del Libro Cubano, 1971.
Parajes de una época. Havana: Editorial Letras Cubanas, 1979.
Elogio de la danza. Mexico: Universidad Nacional Autónoma de México, 1982.
Octubre imprescindible. Havana: Unión de Escritores y Artistas de Cuba, 1982.
Cuaderno de Granada. Havana: Casa de las Américas, 1984.
Piedra pulida. Havana: Editorial Letras Cubanas, 1986.
Grenada Notebook (Cuaderno de Granada). Bilingual ed. Trans. Lisa Davis. New York: Círculo de Cultura Cubana, 1984.

Where the Island Sleeps Like a Wing (Selected poetry of Nancy Morejón). Trans. Kathleen Weaver. San Francisco: Black Scholar Press, 1985.
Fundación de la imagen. Havana: Editorial Letras Cubanas, 1988.

Secondary Sources

Captain-Hidalgo, Yvonne. "The Poetics of the Quotidian in the Works of Nancy Morejón." *Callaloo* 10 (1987): 506–604.

Davis-Lett, Stephanie. "The Image of the Black Woman as a Revolutionary Figure: Three Views." *Studies in Afro-Hispanic Literature* 2–3 (1978–1979): 118–133.

Marting, Diane. "The Representation of Female Sexuality in Nancy Morejón's *Amor, cuidad atribuída*." *Afro-Hispanic Review* 7 (1988): 36–38.

Willis, Susan. "Nancy Morejón: Wresting History from Myth." *Literature and Contemporary Revolutionary Culture* 1 (1984–1985): 247–255.

Silvina Ocampo

Viaje olvidado. Buenos Aires: Sur, 1937.
Espacios métricos. Buenos Aires: Sur, 1945.
Autobiografía de Irene. Buenos Aires: Sur, 1948.
Poemas de amor desesperado. Buenos Aires: Sudamericana, 1949.
Los nombres. Buenos Aires: Emecé, 1953.
La furia. Buenos Aires: Sur, 1959.
Las invitadas. Buenos Aires: Losada, 1961.
Lo amargo por dulce. Buenos Aires: Emecé, 1962.
Los días de la noche. Buenos Aires: Sudamericana, 1970.
Amarillo celeste. Buenos Aires: Losada, 1972.
El caballo alado. Buenos Aires: Ediciones de la Flor, 1972.
El tobogán. Buenos Aires: Angel Estrada, 1975.
La naranja maravillosa. Buenos Aires: Orión, 1977.
Y así sucesivamente. Barcelona: Tusquets, 1987.
Traidores. Buenos Aires: Ediciones Ada Korn, 1988.
Leopoldina's Dream. Trans. Daniel Balderston. Markham, Ontario: Penguin Books, 1987.

Secondary Sources

Araújo, Helena. "Ejemplos de la 'niña impura' en Silvina Ocampo y Alba Lucía Angel." *Hispamérica* 13, 38 (1984): 27–35.

Fox-Lockert, Lucía. "Silvina Ocampo's Fantastic Short Stories." *Monographic Review/Revista Monográfica* 4 (1988): 221–229.

Klingenberg, Patricia N. "Portrait of the Writer as an Artist: Silvina Ocampo." *Perspectives on Contemporary Literature* 13 (1987): 58–64.

———. "The Twisted Mirror: The Fantastic Stories of Silvina Ocampo." *Letras Femeninas* 13 (1987): 67–78.

———. "The Mad Double in the Stories of Silvina Ocampo." *Latin American Literary Review* 16, 32 (1988): 29–40.

Molloy, Sylvia. "Silvina Ocampo: la exageración como lenguaje." *Sur* 320 (1969): 15–24.

Ulla, Noemí. *Encuentro con Silvina Ocampo.* Buenos Aires: Editorial de Belgrano, 1982.

Victoria Ocampo

De Francesca à Béatrice à travers la Divine Comédie. Paris: Bossard, 1926. Translated as *De Francesca a Beatrice.* Madrid: Revista de Occidente, 1924.

Supremacía del alma y de la sangre. Buenos Aires: Sur, 1935.

Testimonios 1. Madrid: Revista de Occidente, 1935.

Testimonios 2. Buenos Aires: Sur, 1941.

Testimonios 3. Buenos Aires: Sudamericana, 1946.

Soledad sonora (Testimonios 4). Buenos Aires: Sudamericana, 1950.

Testimonios 5. Buenos Aires: Sur, 1957.

Testimonios 6. Buenos Aires: Sur, 1963.

Testimonios 7. Buenos Aires: Sur, 1967.

Testimonios 8. Buenos Aires: Sur, 1971.

Testimonios 9. Buenos Aires: Sur, 1975.

Testimonios 10. Buenos Aires: Sur, 1977.

Domingos en Hyde Park. Buenos Aires: Sur, 1936.

La mujer y su expresión. Buenos Aires: Sur, 1936.

Emily Brontë (terra incógnita). Buenos Aires: Sur, 1938.

Virginia Woolf, Orlando y Cía. Buenos Aires: Sur, 1938.

San Isidro. Buenos Aires: Sur, 1941.

338171 T.E. Buenos Aires: Sur, 1942.

Le vert paradis. Buenos Aires: Sur, 1944.

Soledad sonora. Buenos Aires: Sudamericana, 1950.

El viajero y una de sus sombras: Keyserling en mis memorias. Buenos Aires: Sudamericana, 1951.

Virginia Woolf en su diario. Buenos Aires: Sur, 1954.

La belle y sus enamorados. Buenos Aires: Sur, 1964.

Diálogo con Borges. Buenos Aires: Sur, 1969.

Diálogo con Mallea. Buenos Aires: Sur, 1969.

Autobiografía. 6 vols. Buenos Aires: Ediciones Revista Sur, 1979–1984.

338171 T.E. Trans. David Garnett. New York: Dutton, 1963. Reprint. London, Gollancz, 1973.

Secondary Sources

Bastos, María Luisa. "Dos líneas testimoniales: *Sur* y los escritos de Victoria Ocampo." *Sur* 348 (1981): 9–23.

Christ, Ronald. "Figuring Literarily: An Interview with Victoria Ocampo." *Review* 7 (1972): 5–13.

Cortázar, Julio. "Victoria Ocampo: *Soledad sonora.*" *Sur* 192–194 (1950): 194–197.

Estrella Gutiérrez, Fermín. "Victoria Ocampo y su aporte a la cultura argentina de nuestro tiempo." *Estudios literarios,* 323–330. Buenos Aires: Academia Argentina de Letras, 1969.

Falcoff, Mark. "Victoria Ocampo's *Sur.*" *The New Criterion* 7 (1988): 27–37.

Gallo, Marta. "Las crónicas de Victoria Ocampo: versatilidad y fidelidad de un género," *Revista Iberoamericana* 132–133 (1985): 679–686.

Greenberg, Janet Beth. "The Divided Self: Forms of Autobiography in the Writing of Victoria Ocampo." Ph.D. diss., University of California, Berkeley, 1986.

Meyer, Doris. *Victoria Ocampo: Against the Wind and the Tide.* New York: George Braziller, 1979.

———. "The Multiple Myths of Victoria Ocampo." *Revista/Review Interamericana* 12 (1982): 385–392.

———. " 'Feminine' Testimony in the Works of Teresa de la Parra, María Luisa Bombal and Victoria Ocampo." *Contemporary Women Authors of Latin America: Introductory Essays.* Eds. Doris Meyer and Margarite Fernández Olmos, 22–32. Brooklyn N.Y.: Brooklyn College Press, 1983.

Sarlo, Beatriz. "Decir y no decir: erotismo y represión." *Una modernidad periférica: Buenos Aires, 1920-1930,* 69–93. Buenos Aires: Nueva Visión, 1988.

Schultz Cazenueve de Mantovani, Fryda. *Victoria Ocampo.* Buenos Aires: Ediciones Culturales Argentinas, 1963.

Sergio, Lisa. "A Word with Victoria Ocampo." *Américas* 28, 5 (1976): 2–4.

Tuninetti, Beatriz. *Contribución a la bibliografía de Victoria Ocampo.* Buenos Aires: Universidad de Buenos Aires, 1962.

Victoria, Marcos. *Un coloquio con Victoria Ocampo.* Buenos Aires: Futura, 1934.

Viñas, David. "Desplazamiento y reinvidicación: entre Galvéz y Victoria Ocampo." *Literatura argentina y realidad política: de Sarmiento a Cortázar,* 323–327. Buenos Aires: Siglo Veinte, 1971.

Olga Orozco

Desde lejos. Buenos Aires: Losada, 1946.

Las muertes. Buenos Aires: Losada, 1952.

Los juegos peligrosos. Buenos Aires: Losada, 1962.

La oscuridad es otro sol. Buenos Aires: Losada, 1967.

Museo salvaje. Buenos Aires: Losada, 1974.

Veintinueve poemas. Caracas: Monte Avila, 1975.

Cantos a Berenice. Buenos Aires: Sudamericana, 1977.

Mutaciones de la realidad. Buenos Aires: Sudamericana, 1979.

Obra poética. Buenos Aires: Corregidor, 1979.

Poesía: Antología. Buenos Aires: CEDAL, 1982.

La noche a la deriva. Mexico: Fondo de Cultura Económica, 1983.

Secondary Sources

Liscano, Juan. "Olga Orozco y su trascendente juego poético." Introduction to *Veintinueve poemas,* by Olga Orozco. Caracas: Monte Avila, 1975.

Lindstrom, Naomi. "Olga Orozco: la voz poética que llama entre mundos." *Revista Iberoamericana* 51, 132–133 (July–December 1985): 765–776.

Running, Thorpe. "Imagen y creación en la poesía de Olga Orozco." *Letras Femeninas* 13 (1987): 12–20.

Elvira Orphée

Dos veranos. Buenos Aires: Sudamericana, 1956.
Uno. Buenos Aires: Fabril, 1961.
Aire tan dulce. Buenos Aires: Sudamericana, 1966.
En el fondo. Buenos Aires: Galerna, 1969.
Su demonio preferido. Buenos Aires: Emecé, 1973.
La última conquista de El Angel. Caracas: Monte Avila, 1977.
Las viejas fantasiosas. Buenos Aires: Emecé, 1981.
El Angel's Last Conquest. New York: Ballantine, 1986.

Secondary Sources

Díaz, Gwendolyn. "Escritura y palabra: 'Aire tan dulce' de Elvira Orphée." *Revista Iberoamericana* 51, 132–133 (July–December 1985): 641–648.
Garfield, Evelyn Picon. *Women's Voices from Latin America: Interviews with Six Contemporary Authors*. Detroit: Wayne State University Press, 1985.
Moctezuma, Edgardo. "Para mirar lejos antes de entrar: Los usos del poder en *Aire tan dulce* de Elvira Orphée." *Revista Iberoamericana* 49, 125 (1983): 929–942.

Cristina Peri-Rossi

Viviendo. Montevideo: Alfa, 1963.
Los museos abandonados. Montevideo: Arca, 1968.
El libro de mis primos. Montevideo: Marcha, 1969.
Indicios pánicos. Montevideo: Nuestra América, 1970.
Evohé. Montevideo: Girón, 1971.
"Alejandra Pizarnik o la tentación de la muerte." *Cuadernos Hispanoamericanos* 273 (1973): 584–588.
Descripción de un naufragio. Barcelona: Lumen, 1975.
Diáspora. Barcelona: Lumen, 1976.
La tarde XX del dinosaurio. Barcelona: Planeta, 1976.
Lingüística general. Valencia: Prometeo, 1979.
La rebelión de los niños. Caracas: Monte Avila, 1982.
El museo de los esfuerzos inútiles Barcelona: Seix Barral, 1983.
La mañana después del diluvio. Gijón, Spain: Noguera, 1984.
La nave de los locos. Barcelona: Seix Barral, 1984.
Una pasión prohibida. Barcelona: Seix Barral, 1986.
Solitario de amor. Madrid: Grijalbo, 1988.
Ship of Fools. Trans. P. Hugues. London: W. H. Allen, 1989.

Secondary Sources

Benedetti, Mario. *Literatura uruguaya del siglo XX*. Montevideo: Alfa, 1969.
_____ . "Peri Rossi: Vino nuevo en odres nuevos." *Marcha* (Montevideo), July 25, 1969: 31.
Chanady, Amaryll B. "Cristina Peri Rossi and the Other Side of Reality." *The Antigonish Review* 54 (1983): 44–48.

Deredita, John F. "El tiempo de los jóvenes: Cristina Peri Rossi." *Marcha* (Montevideo), December 27, 1968: 29.

————. "Desde la diáspora: Entrevista con Cristina Peri Rossi." *Texto Crítico* (Mexico) 9 (1978): 131–142.

Mora, Gabriela. "El mito degradado de la familia en *El libro de mis primos* de Cristina Peri Rossi." *The Analysis of Literary Texts.* Ed. Randolph D. Pope, 66–77. Ypsilanti, Mich.: Bilingual Press, 1980.

Rama, Angel. *La generación crítica, 1939–1969.* Montevideo: Arca, 1972.

Verani, Hugo. "Una experiencia de límites: La narrativa de Cristina Peri Rossi." *Revista Iberoamericana* 48, 118–119 (January–June 1982): 303–316.

Eva Perón

La razón de mi vida. Buenos Aires: Ediciones Peuser, 1951.

My Mission in Life. Trans. Ethel Cherry. New York: Vantage Press, 1953.

Secondary Sources

Taylor, J. M. *Eva Perón: The Myths of a Woman.* Chicago: University of Chicago Press, 1979.

Nélida Piñón

Guia-Mapa de Gabriel Arcanjo. Rio de Janeiro: Edições G.R.D., 1961.

Madeira Feita Cruz. Rio de Janeiro: Edições G.R.D., 1963.

Tempo das Frutas. Rio de Janeiro: José Alvaro, 1966.

Fundador. Rio de Janeiro, 1969.

A Casa da Paixão. Rio de Janeiro: Francisco Alves, 1972.

Sala de Armas. Rio de Janeiro: Sabiá, 1973.

Tebas do Meu Coração. Rio de Janeiro: José Olympio, 1974.

A Força do Destino. Rio de Janeiro: Francisco Alves, 1978.

O Calor das Coisas. Rio de Janeiro: Nova Fronteira, 1980.

A República dos Sonhos. Rio de Janeiro: Francisco Alves, 1984.

A Doce Canção de Caetana. Rio de Janeiro: Editora Guanabara, 1987.

"Brief Flower." *Contemporary Latin American Literature.* Trans. Gregory Rabassa. Eds. José Donoso and William Henkin, 309–316. New York: E. P. Dutton, 1969.

"House of Passion." *The Borzoi Anthology of Latin American Literature.* Vol. 2. Trans. Gregory Rabassa. Ed. Emir Rodriguez Monegal, 393–398. N.Y.: Knopf, 1977.

The Republic of Dreams. Trans. Helen Lane. New York: Knopf, 1989.

Secondary Sources

Guimarães, Denise A. D. "Uma Poética de Autor: Leitura de um Texto de Nélida Piñón." *Estudos Brasileiros* 5, 9 (1980): 39–55.

McNab, Gregory. "Abordando a Historia em *A República dos Sonhos.*" *Brasil/Brazil: Revista de Literatura Brasileira.* 1, 1 (1988): 41–53.

Further Readings

Moniz, Naomi Hoki. "*A Casa da Paixão:* Etica, Estética e a Condição Fer
  Revista Iberoamericana 50, 126 (1984): 129–140.
Paulino, María das Graças Rodrigues. "*Fundador:* A Subversão do Mito." *Cadern..
  de Lingüística e Teoría da Literatura* 2 (1979): 71–75.
Pontiero, Giovanni. "Notes on the Fiction of Nélida Piñón." *Review* (New York:
  Center for Inter-American Relations) 17 (1976): 67–71.

Alejandra Pizarnik

La tierra más ajena. Buenos Aires: Botella al Mar, 1955.
La última inocencia and *Las aventuras perdidas.* Buenos Aires: Botella al Mar, 1956.
Arbol de Diana. Buenos Aires: Sur, 1962.
Los trabajos y las noches. Buenos Aires: Sudamericana, 1965.
Extracción de la piedra de la locura. Buenos Aires: Sudamericana, 1968.
Nombres y figuras. Barcelona: Colección La Esquina, 1969.
La condesa sangrienta. Buenos Aires: López Crespo Editorial, 1971.
El infierno musical. Buenos Aires: Siglo XXI Argentina, 1971.
Los pequeños cantos. Caracas: Arbol de Fuego, 1971.
El deseo de la palabra. Barcelona: Ocnos, 1975.
Textos de sombra y últimos poemas. Buenos Aires: Sudamericana, 1982.
Alejandra Pizarnik: A Profile. Ed. and introd. Frank Graziano. Trans. María Rosa
  Fort, Frank Graziano, and Suzanne Jill Levine. Durango, Colo.: Logbridge-
  Rhodes, 1987.

Secondary Sources

Cámara, Isabel. "Literatura o la política del juego en Alejandra Pizarnik." *Revista
  Iberoamericana* 51, 132–133 (July–December 1985): 581–590.
DiAntonio, Robert E. "On Seeing Things Darkly in the Poetry of Alejandra
  Pizarnik: Confessional Poetics or Aesthetic Metaphor?" *Confluencia: Revista
  Hispánica de Cultura y Literatura* 2 (1987): 47–52.
Lasarte, Francisco. "Más allá del surrealismo: La poesía de Alejandra Pizarnik."
  Revista Iberoamericana 49, 125 (1983): 867–877.
Moia, Marta, and Susan Pensak. "Some Keys to Alejandra Pizarnik." *Sulfur* 8
  (1983): 97–101.
Running, Thorpe. "The Poetry of Alejandra Pizarnik." *Chasqui* 14, 2–3 (1985):
  45–55.

Elena Poniatowska

Palabras cruzadas. Mexico: Era, 1961.
Hasta no verte, Jesús mío. Mexico: Era, 1969. 23rd ed. Mexico: Era, 1984.
"Alusiones críticas al libro *Hasta no verte, Jesús mío.*" *Vida Literaria* (Mexico) 3
  (1970); 22–24.
"Un libro que me fue dado." *Vida Literaria* (Mexico) 3 (1970): 3–4.
La noche de Tlatelolco: Testimonios de historia oral. Mexico: Era, 1971.
"Las escritoras mejicanas calzan zapatos que les aprietan." *Los Universitarios* (Oc-
  tober 15–31, 1975): 4.

Querido Diego, te abraza Quiela. Mexico: Era, 1978.

Gaby Grimmer. Mexico: Grijalbo, 1979.

Fuerte es el silencio. Mexico: Era, 1980.

"La literatura de las mujeres en América Latina." *Revista de la Educación Superior Asociación Nacional de Universidades e Institutos de Enseñanza Superior* (Mexico) 19 (April–June 1981): 23–25.

El último guajalote. Mexico: Cultura, 1982.

"La literatura y la mujer en Latinoamérica." *Eco* (Bogotá) 3 (1983): 462–472.

De noche vienes. Mexico: Grijalbo, 1983.

¡Ay vida no me mereces! Mexico: Joaquín Mortiz, 1985.

Pablo O'Higgins. Mexico: Fondo de Cultura Económica, 1985.

Flor de lis. Mexico: Era, 1988.

"Love Story." *Latin American Literary Review* 3, 26 (July–December 1985): 63–73.

Massacre in Mexico. Trans. Helen R. Lane. Prologue Octavio Paz. New York: Viking Press, 1975.

Dear Diego. Trans. Katherine Silver. New York: Pantheon, 1986.

Until We Meet Again. Trans. Magda Bogin. New York: Pantheon, 1987.

Secondary Sources

Chevigny, Bell Gale. "The Transformation of Privilege in The Work of Elena Poniatowska." *Latin American Literary Review* 13, 26 (July–December 1985): 49–62.

Christ, Ronald. "The Author as Editor." *Review* 15 (1975): 78–79.

García Flores, Margarita. "Entrevista con Elena Poniatowska." *Revista de la Universidad de México* 30, 7 (1976): 25–30.

Hancock, Joel. "Elena Poniatowska's *Hasta no verte, Jesús mío:* The Remaking of the Image of Woman." *Hispania* 66, 3 (September 1983): 353–359.

Hinds, Harold. "Massacre in Mexico." *Latin American Review* 4, 9 (Fall 1976): 28–33.

Miller, Beth. "Interview with Elena Poniatowska." *Latin American Literary Review* (Carnegie-Mellon University) 4, 7 (1975): 73–78.

Ocampo de Gomes, Aurora M., and Ernesto Prado Velázquez. *Diccionario de escritores mexicanos.* Mexico: Universidad Nacional Autónoma de México, 1967.

Starcevic, Elizabeth O. "Breaking the Silence: Elena Poniatowska, A Writer in Transition." *Literatures in Transition: The Many Voices of the Caribbean Area: A Symposium.* Ed. Rose S. Minc, 63–68. Gaithersburg, Md.: Hispamérica, 1982.

Steele, Cynthia. "La creatividad y el deseo en *Querida Diego, te abraza Quiela* de Elena Poniatowska." *Hispamérica* 14, 41 (1985): 17–28.

——— . "La mediación en las obras documentales de Elena Poniatowska." *Mujer y literatura mexicana y chicana: Culturas en contacto.* Eds. Elena Urrutia and Aralia López. Mexico: Colegio de México, 1988.

Tatum, Charles M. "Elena Poniatowska's *Hasta no verte, Jesús mío.*" *Latin American Women Writers Yesterday and Today,* 45–58. Pittsburgh: Carnegie-Mellon, 1975.

Yampolsky, Marianne. *La casa en la tierra.* Mexico: Fonapás, 1980.

Magda Portal

El derecho de matar. Lima, 1926.
Una esperanza y el mar. Lima: Minerva, 1927.
El aprismo y la mujer. Lima, 1933.
Hacia la mujer nueva. Lima: Cooperative Aprista Atahualpa, 1933.
Costa sur. Santiago de Chile: Nueva, 1945.
Flora Tristán, precursora. Lima: Páginas Libres, 1945.
La trampa. Lima: Adimar Peruana, 1956.
Constancia del ser. Lima, 1965.
"Yo soy Magda Portal." *Ser mujer en el Perú.* Ed. Esther Andradi and Ana María
    Portal, 212–230. Lima: Tokapu Editores, 1979.

Adelia Prado

Bagagem. Rio de Janeiro: Nova Fronteira, 1976.
O Coração Disparado. Rio de Janeiro: Nova Fronteira, 1978.
Mulheres e Mulheres. Rio de Janeiro: Nova Fronteira, 1978.
Solte os Cachorros. Rio de Janeiro: Nova Fronteira, 1979.
Cacos para um Vitral. Rio de Janeiro: Nova Fronteira, 1980.
Terra de Santa Cruz. Rio de Janeiro: Nova Fronteira, 1981.
Os Componentes da Banda. Rio de Janeiro: Nova Fronteira, 1984.
Contos Mineiros. São Paulo: Atica & Minas, 1984.
O Pelição. Rio de Janeiro: Guanabara, 1987.
A Faco no Peito. Rio de Janeiro: Rocco, 1988.
The Headlong Heart. Trans. Ellen Watson. Livingston, Ala.: Livingston University
    Press, 1988.
The Alphabet in the Park. Trans. Ellen Watson. Middletown, Conn.: Wesleyan
    University Press, 1990.

Rachel de Queiroz

O Quinze. Fortaleza, 1930.
João Miguel. Rio de Janeiro: José Olympio, 1932.
Caminho de Pedras. Rio de Janeiro: José Olympio, 1937.
As Três Marias. Rio de Janeiro: José Olympio, 1939.
Lampião: Drama em Cinco Quadros. Rio de Janeiro: José Olympio, 1953.
A Beata Maria do Egipto: Peça em Três Atos e Quatro Quadros. Rio de Janeiro: José
    Olympio, 1958.
Quatro Romances: O Quinze, João Miguel, Caminho de Pedras, As Três Marias. Rio
    de Janeiro: José Olympio, 1960.
Seleta de Rachel de Queiroz. Rio de Janeiro: José Olympio, 1973.
As Menininhas e outras crónicas. Rio de Janeiro: José Olympio, 1976.
Obra Reunida. Rio de Janeiro: José Olympio, 1989.
"Metonymy or, The Husband's Revenge." *Modern Brazilian Short Stories.* Trans.
    and introd. William L. Grossman. Berkeley: University of California Press,
    1967.
Dora, Doralina. Trans. Dorothy Scott. New York: Dutton, 1984.

The Three Marias. Trans. and introd. Fred P. Ellison. Austin: University of Texas Press, 1985.

Secondary Sources

Bruno, Haroldo. *Rachel de Queiroz.* Rio de Janeiro: Catédra, 1977.

――――. *Rachel de Queiroz: Crítica, Biografia, Bibliografia, Depoimento, Seleção de Textos, Iconografia.* São Paulo, 1977.

Courteau, Joanna. "A Beata María do Egipto: An Anatomy of Tyranny." *Chasqui* 13, 2–3 (1984): 3–12.

――――. "The Problematic Heroine in the Novels of Rachel de Queiroz." *Luso-Brazilian Review* 22, 2 (1985): 123–144.

Ellison, Fred P. "Rachel de Queiroz." *Brazil's New Novel: Four Northeastern Masters.* Berkeley and Los Angeles: University of California Press, 1954.

Shade, George D. "Three Contemporary Brazilian Novels." *Hispania* 39, 4 (1956): 391–396.

Woodbridge, Benjamin M., Jr. "The Art of Rachel de Queiroz." *Hispania* 40, 2 (1957): 144–148.

Heleieth I. B. Saffiotti

Women in Class and Society. Trans. Michael Vale. New York and London: Monthly Review Press, 1978.

"Relationships of Sex and Social Class in Brazil." *Sex and Class in Latin America: Women's Perspectives on Politics, Economics and the Family in the Third World.* Eds. June Nash and Helen Icken Safa, 147–159. New York: J. F. Bergin, 1980.

"La modernización de la industria textil y la estructura de empleo femenino; un caso en Brasil." *Sociedad, subordinación y feminismo; Debate sobre la mujer en América Latina y el Caribe.* Ed. Magdalena León, 183–200. Bogotá: Asociación Colombiana para el Estudio de la Población, 1982.

Alfonsina Storni

La inquietud del rosal. Buenos Aires: Librería La Facultad, 1916.

El dulce daño. Buenos Aires: Sociedad Cooperativa Editorial Limitada, 1918.

Irremediablemente. Buenos Aires: Sociedad Cooperativa Editorial Limitada, 1919.

Languidez. Buenos Aires: Sociedad Cooperativa Editorial Limitada, 1920.

Ocre. Buenos Aires: Editorial Babel, 1925.

Poemas de amor. Buenos Aires: Editorial Nosotros, 1926.

Mundo de siete pozos. Buenos Aires: Editorial Tor, 1934.

Mascarilla y trébol. Círculos imantados. Buenos Aires: Librería El Ateneo, 1938.

Obra completa. Buenos Aires: Sociedad Editora Latinoamericana, 1976.

Selected Poems. Trans. M. Crow and M. Freeman. Clackamas, Oreg.: White Pine Press, 1988.

Secondary Sources

Cypess, Sandra Messinger. "Visual and Verbal Distances: The Woman Poet in a Patriarchal Culture." *Revista/Review Interamericana* 12, 1 (1982): 150–157.

Jehenson, Myriam Yvonne. "Four Women in Search of Freedom." *Revista/Review Interamericana* 12, 1 (1982): 87–99.

Jones, Sonia. *Alfonsina Storni.* Boston: Twayne, 1979.

———. "Alfonsina Storni's 'El amo del mundo.'" *Revista/Review Interamericana* 12, 1 (1982): 100–103.

Kirkpatrick, Gwen. "Alfonsina Storni: 'Aquel micromundo poético.'" *Modern Language Notes* 99 (1984): 386–392.

———. "Alfonsina Storni: New Visions of the City and the Body of Poetry." *The Dissonant Legacy of Modernismo: Lugones, Herrera y Reisig and the Voices of the Modern Spanish American Poetry,* 230–239. Berkeley and Los Angeles: University of California Press, 1989.

Phillips, Rachel. *Alfonsina Storni: From Poetess to Poet.* London: Támesis, 1975.

Sarlo, Beatriz. "Decir y no decir: erotismo y represión." *Una modernidad periférica: Buenos Aires, 1920–1930,* 69–93. Buenos Aires: Nueva Visión, 1988.

Blanca Varela

Ese puerto existe. Xalapa: Universidad Veracruzana, 1959.

Luz de día. Lima: Ediciones de la Rama Florida, 1963.

Valses y otras falsas confesiones. Lima: Instituto Nacional de Cultura, 1972.

Canto villano. Lima: Ediciones Arybalo, 1978.

Idea Vilariño

La suplicante. Montevideo: El Siglo Ilustrado, 1945.

Cielo cielo. Montevideo, 1947.

Paraíso perdido. Montevideo: Número, 1949.

Nocturnos. Montevideo, 1951. Reprint, Montevideo: Arca, 1986.

Por aire sucio. Montevideo: Número, 1951.

Poemas de amor. Montevideo, 1957. Reprint, Buenos Aires: Schapire, 1972.

Pobre mundo. Montevideo: Ediciones de la Banda Oriental, 1966.

Treinta poemas. Montevideo: Editorial Tauro, 1966.

No. Montevideo: Calicanto, 1980.

Segunda antología. Montevideo: Calicanto, 1980.

Poemas de amor. Nocturnos. Barcelona: Lumen, 1984.

Nocturnos del pobre amor. Havana: Casa de las Américas, 1989.

About the Book and Editors

IN THE LAST TWO DECADES Latin American literature has received great critical acclaim in the English-speaking world, although attention has been focused primarily on the classic works of male literary figures such as Borges, Paz, and Cortázar. More recently, studies have begun to evaluate the works of established women writers such as Sor Juana Inés de la Cruz, Delmira Agustini, Clarice Lispector, and Rosario Castellanos. A few experimental writers such as Nélida Piñón and Luisa Valenzuela have also had their novels translated. Nevertheless, a panoramic view of Latin American women's writing in the twentieth century has been sorely missing.

The selections included in this anthology center on three major aspects of women's writing: reflections on writing and its relation to the public self, the figuration of a female textual identity, and women as agents of history and ideology. The editors draw from a vast array of sources—in addition to fiction and poetry, they have included excerpts from the testimonial writing of peasant women, labor and social activists, and women writing about literature and the craft of writing.

This anthology makes available for the first time in English texts by Gabriela Mistral, Julieta Campos, Cristina Peri-Rossi, Magda Portal, Alicia Moreau de Justo, Paulina Luisi, Mara Lobo, Henriqueta Lisbõa, and Tamara Kamenszain, among others. The contiguity in the selections—one finds, for example, the testimony of Rigoberta Menchú placed next to a short story by Clarice Lispector—provides a reading bridge between the writing of women creating a social order in which they may function as full human beings and the writing of women attempting to fashion a self-identity within this imaginary.

A unique volume of readings, this book will be an invaluable resource for the student, scholar, and general reader interested in the contribution of women writers in Latin America.

Sara Castro-Klarén is professor of Latin American Literature at the Johns Hopkins University. **Sylvia Molloy** is Albert Schweitzer Professor in the Humanities at New York University. **Beatriz Sarlo** is professor of Latin American literature and culture at the University of Buenos Aires.